AF364666

Environmental Impact Assessment Methodologies

Third Edition

Environmental Impact Assessment Methodologies
Third Edition

Prof. Anjaneyulu. Yerramilli

Ph.D; FRSC(London)

Former Professor & Director Jackson State University, Jackson, MS, USA
Former Professor & Director, IST, JNT University Hyderabad, India.
Professor & Director, Center for Advanced Energy Studies.
K L University, Green Fields, Vaddeswaram, Guntur Dist., A.P., India; 522 502.

and

Valli Manickam, *Ph.D*

Centre for Atmospheric Sciences and,
Weather Modification Technologies,
Jawaharlal Nehru Technological University,
Hyderabad, India.

BSPBS Publications

A unit of BSP Books Pvt., Ltd.
4-4-309/316, Giriraj Lane,
Sultan Bazar, Hyderabad - 500 095.

Environmental Impact Assessment Methodologies *Third Edition*
by **Y. Anjaneyulu, Valli Manickam**
© 2021 by publisher, All rights reserved.

No part of this book or parts thereof may be reproduced, stored in a retrieval system or transmitted in any language or by any means, electronic, mechanical, photocopying, recording or otherwise without the prior written permission of the authors.

Published by:

 **BS Publications**

A unit of **BSP Books Pvt., Ltd.**
4-4-309/316, Giriraj Lane,
Sultan Bazar,Hyderabad - 500 095.
Phone :040 - 23445688, 23445600
e-mail :info@bspbooks.net
website: www.bspbooks.net

ISBN: 978-93-91910-49-5

Foreword

Satyanarayana Koneru, B.E.

PHONES   : 2575727, 2577715, 2577884
Fax   : 0866-2577717
e-mail   : chairman@klce.ac.in
29-36-38, Museum Road,
Governorpet, VIJAYAWADA-520 002.
A.P. INDIA

Foreword

Globally increasing pace of developmental activities are also associated with consequential deterioration of environmental components which need to be predicted by an EIA process for implementation of various development projects /policies so that proposed activities are economically viable, socially equitable and environmentally sustainable. EIA systematically examines both beneficial and adverse consequences of the project and ensures that these effects are taken care into account during project design.

As multi specialty EIA tools are now in extensive use nationally and internationally ,it is highly essential for all the stake holders like project engineers, decision makers regulators/environmental managers /environmental consultants to understand the intricacies of EIA analysis and technical aspects of EIA Tools for application to diverse projects/programs . To meet the fast growing demands for highly skilled man power/scientists/researchers/engineers in this emerging area , a number of institutes in Science and Technology are now offering specialized courses on environmental management adopting the new technologies.

I am happy to write Foreword for this book on "Environmental Impact Assessment Methodologies, authored by Prof. Anjaneyulu. Yerramilli,; Distinguished Professor and Director for our Center for Advanced Energy Studies along with his student Dr. T. Valli Manikkam

I understand this book in addition to basic concepts of EIA processes it covers the advanced information on Risk Assessment, application of Remote sensing (RS) & Geographic information Systems for EIA ,Strategic Environmental Assessment (SEA) , Biodiversity Impact Assessment (BIA) as a Planning Tool and some illustrative case studies which caters to the needs of Graduate/Post Graduate /Research students in science and Engineering tacking Environmental Management/EIA as Major and EIA practicing professionals, as a Text book/Reference material .

I congratulate Prof Anjaneyulu. Yerramilli and Dr Valli Manikkam for their great efforts in writing this very useful book on "Environmental Impact Assessment Methodologies " a topic of great relevance for sustainable development and wish them to bring some more books on various emerging technologies.

Koneru Satyanarayana

President,

Koneru Laksmaiah Education Foundation

(K.L Deemed to be University)

Chairman	*Managing Director*	*Director*	*CEO*
K.L.COLLEGE OF ENGINEERING	HAVISH TRANSPORTS (P) LTD.	HARNIKS PARK PVT. LTD.	HAVISH INFOSYSTEMS

Preface to Third Edition

EIA is a process with main objective of providing decision-makers precise information on future environmental consequences arising due to the project and necessary mitigation measures to be taken for restoration of environmental quality by proper planning and design. EIA is an important decision making tool in respect of various developmental projects and the methodologies for objective assessment of environmental implications for different categories of projects are constantly evolving and improving

A number of EIA tools are now in extensive use internationally for implementation various development projects so that proposed projects are economically viable, socially equitable and environmentally sustainable. Though many EIA methods have been developed and used in EIA process for projects, plans, programs and policies no single type of method can be used to satisfy the diversity of project activities and the key issue involves selecting appropriate methodology for specific needs in an impact study. So we felt that there is a great necessity to provide the details of all these EIA approaches systematically in the form of a book on "Environmental Impact Assessment Methodologies" to cater to the needs of Post Graduate students tacking Environmental Management/EIA as Major/ EIA practicing professionals. Our first edition of this book mainly covered basic information and systematic approach for different EIA methods, mitigation measures and regulatory requirements. To make it more comprehensive the second edition is restructured by adding new Chapters on Risk Assessment, Application of RS & GIS for EIA and some EIA case studies. As these two editions have received good response from students/professional users we are coming out with the third edition updating information on EIA methodologies adopting new technologies/approaches which have crept-in recently.

While all the chapters are updated with latest information in the third edition, we have added new topics like Strategic Environmental Assessment (SEA) in Chapter 1. Biodiversity Impact Assessment (BIA) as a Planning Tool in Chapter5., New Methodologies for Air Pollution Dispersion Modeling Approaches in chapter6.In addition some new case studies are also discussed to demonstrate the application ofEIA for various/industrial/development sectors.

We gratefully acknowledge the permission given by several publishers and authors whose names are given as sources at appropriate places in the text for according permission for using some generic figures, flow charts and data.

Our special thanks are due to Dr, Anil Kumar Dikshit (Center for Environmental Science and Engineering Indian Institute of Technology Mumbai), Mohan Lal Agrawal (Dept. of Civil Engineering, IIT Mumbai) and MrinalKanti Ghose, IIT Kharagpur Dr. Shibani Maitra) Geosoft Topographics (I) Pvt. Ltd. Mumbai and National Council of Applied Economic Research (NCAER), New Delhi, for their permission to include information from their articles in the case studies.

The administrative support and encouragement given by our Honble. President Koneru Satyanarayana garu and Vice Chancellor. Prof. L. S. S. Reddy Garu, K. L. University; Greenfields, Vaddeswaram, Guntur Dt, Andhra Pradesh, India is greatly appreciated.

-Authors

Preface to Second Edition

The First edition on EIA methodologies is mainly aimed at providing basic information and systematic approach on EIA to Post Graduate students pursuing Environmental Management/EIA as Major. However as EIA being an important decision making tool in respect of various developmental projects and the methodologies for objective assessment of environmental implications for different categories of projects are constantly evolving and improving there is a necessity to update with new approaches. The good response to the first Edition made us to revise it with updated information as very few books are available which deal with Indian perspective.

The Second Edition of EIA provides systematic approach on how basic data on each environmental component is collected, how impacts are assessed, predicted and how mitigation measures are implemented for minimizing the impacts. Further each Chapter on impact assessment prediction and mitigation of soil, water, air, socio-economic and health environments is presented with additional information on practical approaches.

To make it more comprehensive the second edition is restructured by adding new Chapters on Risk Assessment, Application of RS & GIS for EIA and some EIA case studies which provide information on the application of new technologies and concepts

This revised version is aimed to serve the needs of students who take courses on EIA, Environmental Management, and Environmental Planning, Ecology, Geography and Environmental Managers and EIA practitioners.

We gratefully acknowledge the permission given by several publishers and authors whose names are given as sources at appropriate places in the text for according permission for using some generic figures, flow charts and data.

Our special thanks are due to Dr, Anil Kumar Dikshit (Center for Environmental Science and Engineering Indian Institute of Technology Mumbai), Mohan Lal Agrawal (Dept. of Civil Engineering, IIT Mumbai) and Mrinal Kanti Ghose, IIT Kharagpur Dr. Shibani Maitra, Geosoft Topographics (I) Pvt. Ltd. Mumbai and National Council of Applied Economic Research (NCAER), New Delhi, for their permission to include information from their articles in the case studies.

The support and encouragement given by Dr K. Rajagopal, Vice Chancellor, JNT University, Hyderabad, India and Dr Mark. G. Hardy, Dean College of Science Engineering and Technology, Jackson State University, MS,USA, for writing the Foreword for this revised version are highly appreciated.

- Authors

Contents

CHAPTER 1

Fundamental Approach to Environmental Impact Assessment (EIA)

CHAPTER 2

EIA Methodologies

CHAPTER 3

Prediction and Assessment of Impacts on Soil and Groundwater Environment

CHAPTER 4

Prediction and Assessment of Impacts on Surface Water Environment

CHAPTER 5

Prediction and Assessment of Impacts on Biological Environment

CHAPTER **6**

Prediction and Assessment of Impacts on the Air Environment

CHAPTER 7

Prediction and Assessment of Impacts of Noise on the Environment

CHAPTER 8

Prediction and Assessment of Socio-Economic and Human Health Impacts

CHAPTER 11

EIA Case Studies

CHAPTER 1

Fundamental Approach to Environmental Impact Assessment (EIA)

1.1 Introduction

1.1.1 What is an EIA?

Many developmental activities degrade the bio-geochemical environment resulting in health problems and Environmental Impact Assessment (EIA) is an impact assessment tool to predict the nature and extent of impact which will be useful to formulate appropriate preventive and operational programs to reduce the impacts and make necessary legislative measures to control the same.

Thus, any developmental/project activity requires not only the analysis, but also an estimate of the monetary costs and benefits involved, but also most importantly, it requires a consideration and detailed assessment of the effect of a proposed project on the environment.

EIA will greatly help in the execution of development projects by incorporating ecofriendly designs, process, procedures throughout the life cycle of the projects (1, 2). Thus, EIA is an exercise to be carried out before any project or major activity is undertaken to ensure that it will not in any way harm the environment on a short-term or long-term basis.

For each development project a number of alternatives can be examined by carrying-out EIA a planning tool (3,4) to forecast and assess the impacts of each alternative to understand and decide which is environmentally benign. Further EIA will also be useful to enlighten the stakeholders about the likely impacts of each alternative and critical environmental issues need to be considered in arriving at final decisions. Sharing of information on the possible impacts of various alternatives will help in achieving in coordination among stakeholders and decision makers.

In a rational model of planning and decision making, planners, engineers and decision makers based on detailed EIA and cost benefit analysis, prepare different alternatives and their possible impacts on the environment. In these efforts EIA will be of great value as a planning tool to select best environmentally sound alternative.

1.1.2 Basic Principles of EIA (5)

(i) Proactive Planning and Decision Tool

EIA is a proactive planning tool need to be integrated with the final decision process as it will be useful to plan properly to reduce negative impacts of any proposed project.

(ii) Avoidance, Pre-emption and Prevention of Adverse Environmental Consequences

To inform the stakeholders about the disastrous environmental consequences of any development project proposed so that appropriate corrective action can be planned for their avoidance or prevention will be the main aim of any EIA process. EIA process should also provide alternatives in circumstances where the impacts cannot be completely avoided. Alternatively, it can help to reduce and control the extent and strength of impacts within certain limits of acceptable criteria.

(iii) Making Positive Influence on Decision Making at the Earliest Possible Opportunity and Thinking Proactively about Options and Alternatives

It is always beneficial to carry out EIA at the beginning of the project planning stage rather than after starting the project as this will help to increase environmental performance of the project. Prevention of adverse impacts/consequences at the start of the project will save time and costs or one may have to spend more due to expensive or time-consuming remedial measures at a later stage. Thus, by critically analyzing various alternatives, options, costs and their adverse environmental impacts the project can be termed as environmentally benign or not.

(iv) Living Process throughout the Project Cycle

It is necessary to design, EIA process with continuous and dynamic protocols/process for the entire project cycle which can identify/predict the environmentally degrading impacts so that proper planning can be carried-out to implement measures for prevention/reduction of scale and scope of the disastrous impacts within acceptable limits/criteria.

(v) Making EIA Recommendations Enforceable: The final recommendations of any EIA should be very simple, sensible, effective and practically implementable control process. They should clearly specify the responsibility of each of the 5Ws (viz. What mitigation measures need to be implemented, Who, When, Where, and What requirements).The mitigation measures proposed in the final recommendations of EIA should mainly aim in the prevention of adverse impacts than remedial measures after environmental degradation occurs.

(vi) Flexibility Amidst Robustness and Transparency, with Public Participation and with the Ability to Adapt to Change: All the process with the EIA need to be flexible so that they can be adopted to changing circumstances without compromising on basic environmental concerns/issues and should be transparent so that all the stakeholders by their participation in community meetings are properly informed about various environmental consequences.

(vii) Seeking Practical Environmental Outcomes for the Environment and Community: The EIA process should be designed to provide and inform realistic adverse /disastrous consequences for the environment and the community.

(ix) Efficiency amidst Effectiveness: An EIA will have high productivity if it is designed and developed with efficient and effective approach, well defined aims, objectives and methodical protocols.

(x) Transparent Agreement among Relevant Parties, Clear Expectations of what needs to be done and what the Performance will be, and Explicit Resolution of any Conflicts: For sound management of Environment after project execution, it is necessary to clearly communicate to all the stakeholders all the agreements, expectations, performance requirements and any conflict resolutions etc in an open and frank manner. This will greatly help to avoid misunderstanding between stakeholders and get their active participation and help in executing remedial measures recommended in an EIA.

1.1.3 When is an EIA Required? (5)

It is very important to define the guidelines on whether a project needs EIA or not before starting the project as communities, developers and local authorities think from different perspectives. Two types of schedules are provided by EIA regulations. While industries categorized in schedule I need compulsory EIA, industries in category II need to be examined based on factors like size, location, environmental sensitivity to the local environment.

All Projects categorized under schedule I need compulsory EIA

Schedule I projects

- Major power plants
- Chemical works
- Waste disposal incineration
- Major Roads Schemes
- Major Irrigation projects

All projects categorized in Schedule II, require EIA needs to be carried out if they are expected to impact the environment due to their sensitivity, size or location.

Examples of Schedule II Projects include:

- Infrastructure projects
- Waste treatment plants
- Rubber industries
- Foundries and forges
- Diary product production units
- Coke ovens
- Brewing

- Livestock
- Storage of fossil fuel
- Textile operations
- Tourism projects/Golf projects

In assessing which projects in schedule II need EIA it is necessary to examine

1. Whether the specific project falls under schedule II
2. Does the project activities exceed threshold limits of scheduled II
3. If the project is located in environmentally sensitive areas like the National Park, Hospital zone, School zone etc.
4. Whether the project is expected to have significant effect due to it's size or location

If the answer to all three of those questions is 'yes' then an EIA is required.

If the answer to any of those questions is 'no' then an EIA is not required.

1.1.4 Environmental Audit

In the context of EIA, audit refers to (a) the organization of monitoring data to record change associated with a project and (b) the comparison of actual and predicted impacts. An audit can be applied to both pre-project and post-project approval stages. EIA necessarily does not reject a project but does in rare cases. By conducting an early EIA, a timely and suitable modification in the project can be incorporated which ultimately may help the project itself.

1.1.5 Importance of Scientific Knowledge on the Impacts of Various Activities

The development of natural resources for economic benefit is desirable. Whether resource development programs prove to be beneficial or destructive depends largely on how far scientific knowledge is obtained in their formulation and the ability of the government agencies to control their implementation.

Development projects go hand in hand with environmental impact and hence before any project is undertaken, the damages in relation to its benefits should be assessed. EIA is not negatively oriented towards development of a project. EIA has found wide utility both in developed and developing countries in achieving development in an environmentally sound manner, either at national or regional scale or at the level of individual development project. Environmental impact is any change to the environment, whether adverse or beneficial, wholly or partially resulting from an organization's activities, products or services.

For example, as a simple case, let us examine impacts of any land clearing activity for any proposed development project. Usually these projects need changing/removing the native land cover like trees, bushes, boulders, etc., and one needs examining if this is conducive to the environment.

Land clearing is largely in the form of agriculture or urban/industrial use often may need changing /destruction of native cover like trees, bushes and boulders /rocks etc. to convert the land into usable form. These activities may result into water pollution, erosion, habitat loss, among others (6-10).

Major impacts of typical Land Clearing Activities (L.C.A) project on the environment are shown in Fig. 1.1.

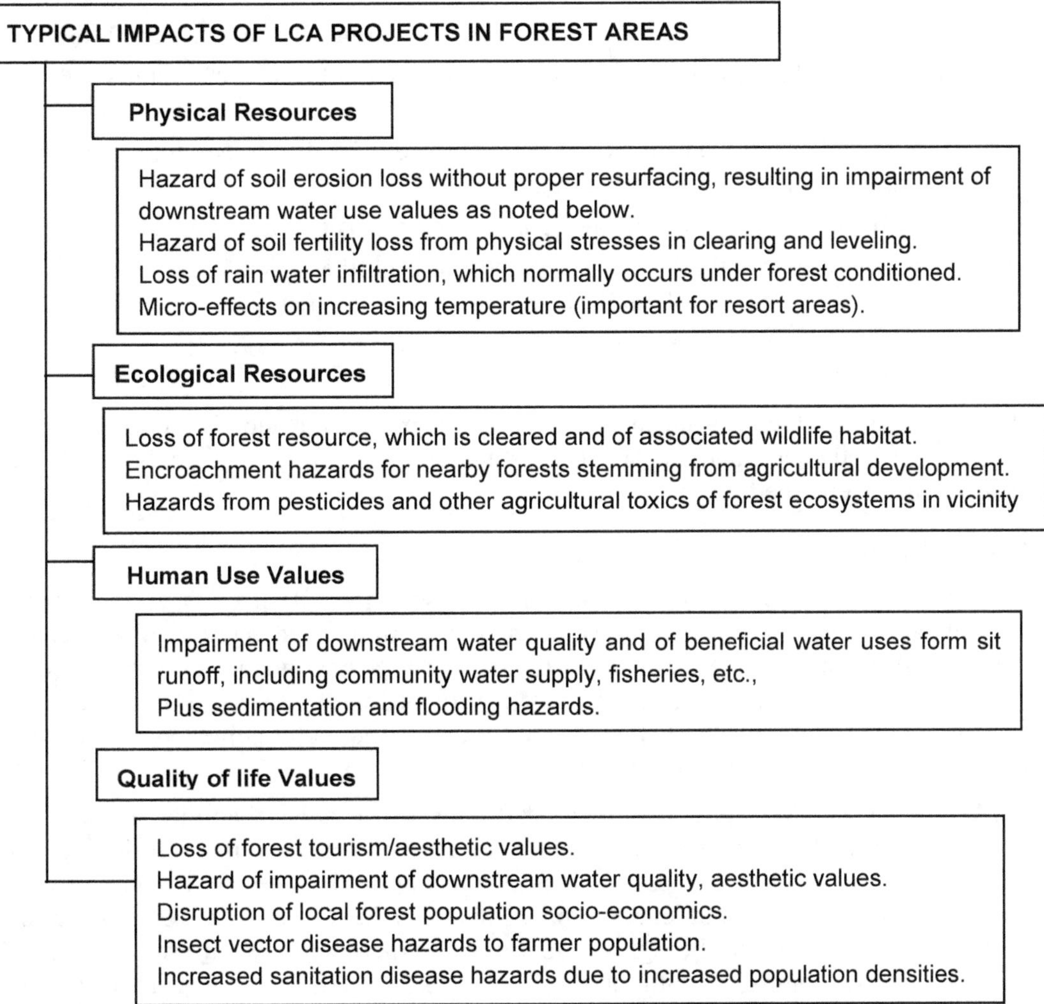

Fig. 1.1 Some major impacts of typical LCA project on environment.

1.1.5.1 Impact of land clearing on climate (11-16)

As an example, land clearing has been shown to affect temperature and rainfall patterns in regional Australia, and land clearing contributes significantly to greenhouse gas emissions. In fact native vegetation absorbs heat and releases moisture into the air, and when removed leads to more hot days, less rainfall and more severe droughts. Australia's greenhouse gas emissions from the land are currently skyrocketing, with land-sector emissions almost doubling in the past three years, and projected to further increase.

Many research reports (Pitman et al. 2004 (15), McAlpine et al. 2007 (14), Deo et al. 2009, (11)) indicate that land clearing and loss of vegetation cover have triggered strong regional climatic effects like

(i)   Surface temperature increased in eastern Australia and southwestern Western Australia, particularly in summer;

(ii) Southwestern, Western Australia experienced warming and decreased rainfall

(iii) Eastern Australia experienced decreased summer rainfall (4-12%) and southwestern Western Australia (4-8%);

(iv) In the period 2002/2003 an average of 2 degree rise in temperature was observed resulting drought in eastern Australia;

(v)   In the Murray-Darling basin decrease in rainfall intensity was observed

(vi) Southern (New Southern Wales) NSW and northern Victoria have observed decrease in wet days and rainfall around 10-30 mm per year while coastal NSW experienced an increase number of dry days

(vii) Amplification of El Niño events.

Considerable research has been carried out on procedural and methodological issues related to EIA in the past, and an acceptable standard of practice, against which EIA can be reviewed has not emerged.

1.1.6   Criticism on EIA

However, despite ample evidence to support the usefulness of EIA, its effectiveness and efficiency are being increasingly questioned. Criticism leveled against EIA include (a) Tokenism (b) unrealistic time constraints (c) Failure to accommodate uncertainty (d) Poor coordination and poorly stated objectives (e) Inadequate research (f) limited use of protective techniques and limited study of indirect and cumulative consequences and (g) being too descriptive and voluminous.

EIA is being criticized for becoming an end in itself and rather than the means to a more balanced process of decision-making. More specifically, the accuracy and precision of impact prediction are being questioned as is the appropriateness of mitigation and the effectiveness of its implementation. A number of studies have, therefore, been undertaken to review EIA methodology in the light of operational experience. Actual effects caused by a project are being compared with predicted effects. Models are being reevaluated and appropriate methodologies and models are being used. Follow-up or post operational studies are being conducted.

1.2   EIA Procedure (17-20)

The entire EIA procedure can be divided into two complementary tasks or sub-reports, (i) the Initial Environmental Examination (IEE) and (ii) the Full-Scale Environmental Impact Assessment (EIA).

1.2.1   Initial Environmental Examination (IEE)

IEE is a means of reviewing the environmental integrity of projects to help determine whether or not EIA level studies can be undertaken. In this sense IEE can be used for project screening to determine which projects require a full-scale EIA. IEE will have several other uses for ensuring project-oriented environmental management as well as minimizing the effort, expense, and delay in carrying out such planning. IEE involves assessing the potential environmental effects of a proposed project that can be carried out within a very limited budget and will be based on the available recorded information or on the professional judgment of an expert. If the IEE results indicate that a full-scale EIA is not required, then, any environmental management parameters, such as, environmental protection measures or a monitoring program can be adapted to complete the EIA for such a project.

If on the other hand, full-scale EIA is required, IEE can be of great help as a mechanism to determine and identify key issues that merit full analysis in EIA and to designate the issues that deserve only a cursory discussion. It may also identify other environmental review and consultation requirements so that necessary analyses or studies can be made concurrently with EIA. This would reduce delay and eliminate redundant or extraneous discussion of EIA reports. IEE is a means of providing the most efficient and feasible preparation of adequate environmental management plans with or without the requirement of a full scale EIA. Therefore, for most Industrial Development Projects, IEE is desirable simply from the economic point of view.

1.2.2 Full Scale Environmental Impact Assessment (EIA)

A multidisciplinary approach to environmental impact analysis is crucial to the decision-making process and to an equal consideration of all areas of potential impact, when the tradeoffs of particular alternatives are evaluated. Therefore, the professional assessing impacts within a particular area of impact, such as, natural resources, air quality, and neighborhood effects, must be educated and quantified within the disciplinary area.

Impact assessment methods are classified into following analytical functions: **Scope identification, prediction, and evaluation.**

Methods of identification of environmental impacts can assist in specifying the range of impacts that may occur, including their special dimensions and time frame. This usually involves the components of the environment affected by the activities of the project. The natural environment of man consists of air, water, land, noise, flora and fauna, etc., while the man-made environment consists of socioeconomic aspects, aesthetics, transportation etc.

Predictive methods will define the quantity or special dimensions of impact on an environmental resource. It can differentiate between various project alternatives in terms of questions covering "how much?" or "where?" the impact may occur.

Methods of evaluation determine the groups (facility users or populations) that may be directly affected by the project or action. They will communicate to the decision maker what the deficiencies (trade-offs) are between possible alternatives or courses of action and the impacts associated with each alternative. But the number of available tools and techniques for E.I.A, only a few look simple and suitable for developing countries.

1.2.2.1 Analytical Functions Associated with the Environmental Impact Assessment

Analytical functions associated with the environmental impact assessment are

(a) *Defining scope of an EIA*

  1. Important issues and concern,

  2. Areas of less concern for the present acts, and

  3. Regulations requirement.

(b) *Identification*

  1. To exhaustively describe the present project environment

  2. To examine and fix the principle project components

  3. To critically analyze and describe various environmental components likely to be impacted/modified by various project activities

(c) *Prediction*

  1. From critical studies fix the significant environmental modification likely to occur due to various project activities

  2. On the expected changes in the environment likely to occur, predicting and presenting the intensity and spatial dimensions of impacts

  3. Providing estimates of the extent of impacts likely to occur in a given time period

(d) *Impact Evaluation and Analysis*

  1. Examining and assessing the relative impacts of various project alternatives and fixing the best alternative with low impacts

  2. By exhaustive and critical studies the significance of various impacts

  3. Preparation of draft and final impact statement

1.2.2.2 Defining the Scope of EIA

It is necessary to define the scope of EIA at the early stages of environmental impact assessment so as to reinforce a commitment to an organized, and systematic program of agency and public participation in the environmental process. The public must be made aware in order to be able to make informed choices. Scoping refers to early coordination with interested and affected agencies and the public.

Scoping identifies important issues and concerns, areas of no concern for a particular project or action, and other legislative or regulatory requirements.

Purpose of Scoping

Scoping is used to:

- Define the proposed action,
- Enlist the cooperation of agencies,
- Identify what's important,
- Identify what's not important,
- Set time limits on studies,

- Determine requirements of the study team,
- Collect background information,
- Identify required permits,
- Identify other regulatory requirements, and
- Determine the range of alternatives.

The scoping process should be specifically designed to suit the needs of the individual project or action being proposed. It can be a formal, extensive process or an informal, simple process. There are many options for the extent and format of meetings, mailings, and agency and local group contacts.

Scoping is also used to identify the cost and time of the assessments and therefore is a very important step both in identifying the impacts and controlling the limits of the EIA process. As a systematic approach, the commonly used techniques for scoping include Checklist, Matrix, Networks and Overlay methods.

1.2.2.3 Baseline Data Collection

Following scoping, the next step is the baseline data collection. All relevant environmental data of the project area need to be collected which is termed as baseline studies. Baseline studies are based on primary data, secondary data, relevant reports and research papers and cover everything important that is expected to have a potential impact on the environment in and around the project.

1.2.2.4 Full Scale Impact Assessment

This requires interpretation of the significance of the impacts to provide a conclusion, which can ultimately be used by decision-makers in determining the fate of the project. This step is generally considered the most crucial in the entire EIA process. Its complexity increases as not every impact, especially natural and social impacts, can be quantified. A well balanced approach is essential to help take a final decision regarding the impact of the project on the economic, social and environmental conditions in the region.

1.2.2.5 Identification of Impacts on the Environment by Preliminary Overview Assessment

Often the first step in an environmental impact assessment has been a preliminary overview of the proposed project alternatives and locations. Several steps are included in the overview. First, the project alternatives and characteristics must be reviewed with reference to the following pertinent questions. Is the project a building, a highway, a park, or a land-use plan? What are the characteristics of the setting? Is the potentially affected area urban or rural, natural or made by human beings?

This step is also used to ensure that the proponent has considered other feasible approaches, including alternative project locations, scales, processes, layouts, operating condition including the no-action option.

The purpose of the preliminary assessment is to identify the potential for significant environmental impacts of the initial set of alternatives. The results then function to refine the alternatives and to determine the appropriate subsequent environmental documentation. A few examples of the types of questions included in an initial assessment overview, in areas of potential physical, biological, social and economic impacts, are as follows: Will the proposal either directly or indirectly:

- Modify a channel or a river or a stream?
- Reduce the habitat of any unique, threatened, or endangered species?
- Divide or disrupt an established community?
- Require the displacement of businesses or farms?

In the identification of impacts one should establish the already existing state and clearly identify:

1. What will happen if the project does not come into existence?

2. What will happen if the project comes up? The impacts of a project can be depicted only through certain parameters.

Some typical expected changes in the environment and human aspects of various project activities are presented in Tables 1.1 and 1.2.

Table 1.1 Possible impacts of various project activities on the various components of environment.

Component	Important Considerations
Air	Degradation, type of emissions released and the extent to which they affect air quality, creation of excess noise and the effect on man.
Water	Availability, use and quality of water, effects on the aesthetics and aquaculture potential of the ecosystems, effect on the canal system, depletion of ground water, pollution of waters by hazardous and toxic substances, effect on temperature and siltation capacity.
Solid waste facilities	Excess generation of solid waste. Stress on the existing facilities..
Vegetation	Destruction of forest cover, depletion of cultivable land, changes in biological productivity, changes in the species diversity and hastening the disappearance of important species.
Energy and natural resources	Effects on physio–chemical characteristics of soils, effect on stability or instability of soils.
Soils and local geology	Impact on availability of energy sources in the area. Thermal power generation, natural gas consumption, and effect on local natural resources.
Processes	Floods, erosion, earth quake, depositions, stability, and air movements.
Man-made facilities and activities	Structures, utility networks, transportation, and waste disposal.
Cultural status	Employment situation, life style of people, and health services.
Ecological relationship	Food chain, diseases/vectors.

Table 1.2 Impacts of various project activities on certain human aspects.

Economic and occupational	Displacement of population, reaction of population in response to employment opportunities, services and distribution patterns,: property values.
Social pattern or life style	Resettlement; rural depopulation: population density : food; housing, material goods, nomadic; settled: pastoral clubs; recreation ; rural; urban.
Social amenities and relationships	Family life styles, schools; transport, community feelings; disruptions, language, hospital clubs, neighbours.
Psychological features	Involvement, expectations, stress, work satisfaction challenges, national or community pride, freedom from chores, company or solution; mobility.
Physical amenities (intellectual, cultural, aesthetic and seasonal)	National parks; wild life, art galleries, museums, historic and archaeological monuments, beauty, Land scape; wilderness, quiet; clean air and water.
Health	Freedom from molestation; freedom from natural disasters.
Personal security	Changes in health, medical services, medical standards.
Regional and traditional belief	Symbols; taboos; values
Technology	Security hazards, safety measures: decommissioning of wastes; congestion, density.
Cultural	Leisure; fashion and clocking changes; new values.
Political	Authority, level and degree of involvement priorities, structure of decision-making responsibility and responsiveness, resources allocation: local and minority interest: defence need contributing or limiting factors;
Legal	Restructuring of administrative management: changes in taxes; public policy.
Aesthetic	Visual physical changes, moral conduct, sentimental values.
Statutory laws	Air and water quality standards; nation building acts; noise abatement byelaws.

Some of the selected relevant environmental parameters are:

1. Crop productivity,

2. Air quality,

3. Water quality of aquatic resources,

4. Nutrient status of water,

5. Drinking water quality and

6. Availability of agricultural land.

1.2.2.6 Classification and Prediction of Impacts (21-24)

Impact Types

Environment impacts arising from any development projects fall into three categories:

(i) Direct impacts,

(ii) Indirect impacts; and

(iii) Cumulative impacts.

These three groups can be further broken down according to their nature, into

- Positive and negative impacts;
- Random and predictable impacts;
- Local and widespread impacts; and
- Short – and long term impacts.

An interdisciplinary approach helps in assessing environmental impacts. The analysis considers potential consequences which may be long-term and short-term; direct and indirect, secondary, individual and cumulative; beneficial and adverse. Environmental issues are interdisciplinary, interactive, biological and probabilistic.

Indirect, or secondary effects are those that may occur remote as they are far in distance or time from the actual proposed project. An example is the construction of a major employment center, which may have direct effects related to aesthetics in the area, traffic at nearby intersections, removal of natural vegetation, or interference with natural waterways. Additional employment opportunities in the location, however, may prompt additional housing or commercial uses to support employees. Potential impacts of this housing or additional business activity would then be a secondary, or indirect effect of the construction of the employment center and should be evaluated to the best extent possible in the environmental analysis.

Cumulative impacts occur in those situations where individual projects or actions may not have a significant effect, but when combined with other projects or actions, the individual project's incremental contribution of adversity may cause an overall adverse cumulative effect.

Impacts of some typical projects are discussed below for clear understanding.

1.3   Examples of Various Types of Impacts that Occur in a Typical Road Development Project

1.3.1   Direct Impacts

Road formation itself involves different land use accompanied by destruction of vegetation and farm land. Further use of gravel in road construction which is removed from a nearby pit will make a direct impact on the pit area. However, direct impacts can be easily identified, assessed, and controlled.

1.3.2   Indirect Impacts

Secondary or tertiary or chain impacts are known as indirect impacts which normally arise due to various project activities and many times are associated with significant impacts on the environment. Further indirect impacts over time can affect larger areas than predicted and are difficult to measure.

Examples include degradation of surface water quality by the erosion of land cleared as a result of a new road Fig. 1.2 and urban growth near a new road. Another common indirect impact associated with new roads is increased deforestation of an area, stemming from easier (more profitable) transportation of logs to market, or the influx of settlers. In areas where wild game is plentiful, such as Africa, new roads often lead to the rapid depletion of animals due to poaching.

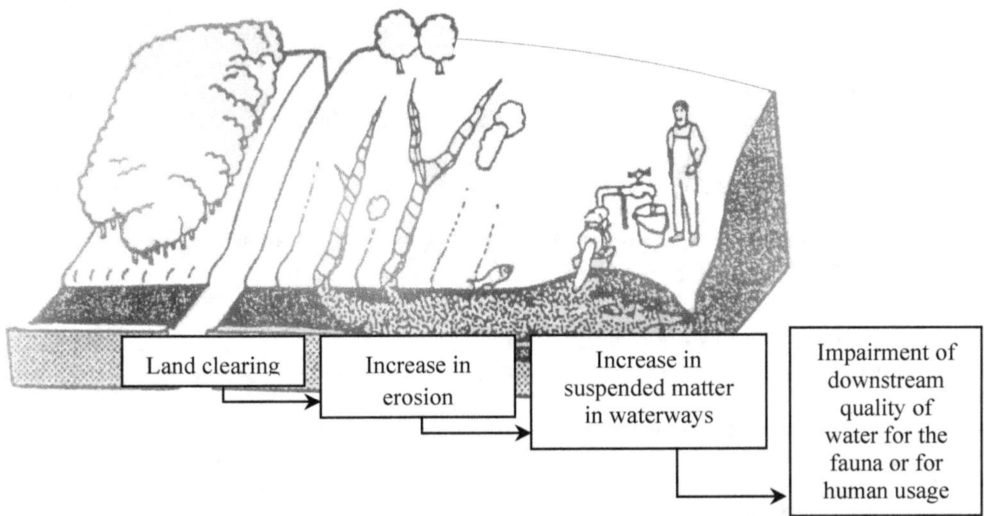

Fig. 1.2 Indirect impacts : the example of land clearing.

Environmental impacts should be considered not only as they pertain to road rights-of-way, but also to sites associated with the road project, which include deposit and borrow sites, materials treatment areas, quarries, access roads and facilities provided for project work.

Some potential Direct and Indirect impacts are summarized in Table 1.3.

Table 1.3 Potential direct and indirect environmental impacts of a typical road construction project in mangroove swamp and rice growing area.

Direct Impact (D); indirect impact (1)

Soils
   Compaction of alluvial soils by earth moving equipment (D)
   Erosion and modification of surface relief of borrow zones (275,000 square meters) (D)
   Loss of topsoil (165 hectares) in the borrow areas(D)
   Over-exploitation of agricultural soils due to future development in a zone sensitive to erosion (I)
   Irreversible salinization and acidification of mangrove swamp soils (I)

Water
   Modification of flowing surface water in borrow areas, causing erosion and silation (I)
   Modification of water flows during construction (stream diversion, modification of water table recharging) (D)
   Sedimentation near crossings of presently cultivated flood plain (D)
   Modification of surface and subterranean water flows and resulting drying or flooding (I)
   Pollution of water tables by equipment lubricants, fuels, and detergents (D)
   Displacement of salinity threshold into the mangrove swamp zone: effect on fauna and flora, impregnation of soils with tannin, erosion of coastline (I)

Flora

260 hectares of deforestation and undergrowth clearance (D)

Destruction of plantings (28, 00 oil palms, 1,600 various trees)(D)

Reduction of cornice forests around swamps, from modified water flow and increased agricultural use (I)

Disappearance of reproduction and food zones for species of fish, aquatic and migratory birds (I)

Reduction of mangrove plant population (habitat for fauna, purifying micro fauna, firewood (D)

Erosion of the coastline (I)

Increase in farming activity, reduction of fallow times, and impoverishment of the soils (I)

Fauna

Reduction in mangrove fauna (crabs, shrimps, egrets, herons, kingfishers, spoonbills, ibises, terns, and other species (I)

Increase in poaching during the works period, and subsequent hunting and fishing (I)

Increase in tourism (Tristan Island, the center for many migratory birds)(I)

People

Loss of farms and homes (1,300 square meters) (D)

Reduction in agricultural production per surface unit (over-exploitation, impregnation of soils with tannin)(I)

Increase in consumption of wood, particularly from the mangrove swamps: erosion (I)

Reduction in fishing potential (I)

Increase in land tenure conflicts, and conflicts between farmers and nomad cattle breeders (I)

Increase in speed of propagation of endemic disease (I)

Positive Impacts

Providing all weather road link for coastal population with major urban markets, institutions and goods (D).

Sale of dried fish products (90 percent of national production) increased through quicker transport and access (D).

More effective sale of rice from industrial growers (35, 00 hectares) and small-scale growers (D).

Creation of jobs, Improved access to medical help etc. (I)

Source: SETRA

1.4   Impact Prediction and Assessment

1.4.1   Main Objectives of any Environmental Impact Assessment are:

- To predict significant particular environmental impact
- To determine the significant residual environmental impact predicted
- To examine and select the best from the project options available
- To identify and incorporate into the project plan appropriate abatement and mitigating measures
- To identify the environmental costs and benefits of the project to the community

Thus, Impact prediction and assessment is the major step in any environmental assessment process. It involves projection of environmental setting into the future with the proposed action and predicting the impact and assessing the consequences.

Taking a holistic approach of impacts is very important as many times synergistic relationship between impacts occur which have to be closely examined, since indirect effects frequently lead to synergistic impacts.

It is with indirect effects that impact linkages between the natural and social environment often take place. For example, the appropriation of land to build a road may displace farmers, and may interfere with their cropping pattern and force them to use another water supply. This change could result in a depletion of a groundwater aquifer, intensification of new land clearing, erosion, water runoff contamination with added fertilizers and pesticides, etc.

1.4.2 Cumulative Impacts (21)

The process of cumulative environmental change can arise from any of the four following types of events:

(i)   Single large events, i.e., a large project;

(ii) Multiple interrelated events, i.e., road project with a region;

(iii) Catastrophic sudden events, i.e., a major landslide into a river system; and

(iv) Incremental, widespread, slow change, such as a poorly designed culvert or drainage system along a long road extending through a watershed.

These can generate additive, multiplicative or synergistic effects, Fig. 1.3 which can then result in damage to the function of one or several ecosystems (such as the impairment of the water regulation and filtering capacity of a wetland system by construction of a road across it), or the structure of an ecosystem (such as placement of a new road through a forest, leading to in-migration or land clearing which results in severe structural loss to the forest).

A cumulative impact, in the context of road development, might be the de-vegetation and eventual erosion of a roadside pullout. Roadside vegetation is damaged by vehicle and foot traffic, and the soil is left unprotected. Subsequent rainfall causes erosion and siltation of nearby watercourses. The vegetation never has enough time to recover (because of high traffic volume on the road), and the problem is exacerbated over time.

As this example illustrates, cumulative effects assessment is a complex process which requires extensive knowledge of ecological principles and ecosystem response mechanisms on the following aspects.

➢ It is necessary to define the spatial and temporal boundaries for the assessment of the impacts

➢ Of the various variables of the project activities, it is necessary to identify which are measurable

➢ Establish the relationship between choosing variables

Evaluate the cumulative effects of the proposed road project by

The cumulative effects of the proposed road project on the local environment can then be evaluated by:

➢ Identifying the list of important project activities

➢ Assessing the impacts of various project activities on the measured variables

➢ Assessing the spatial and temporal boundaries of the impacts on the measurable impacts

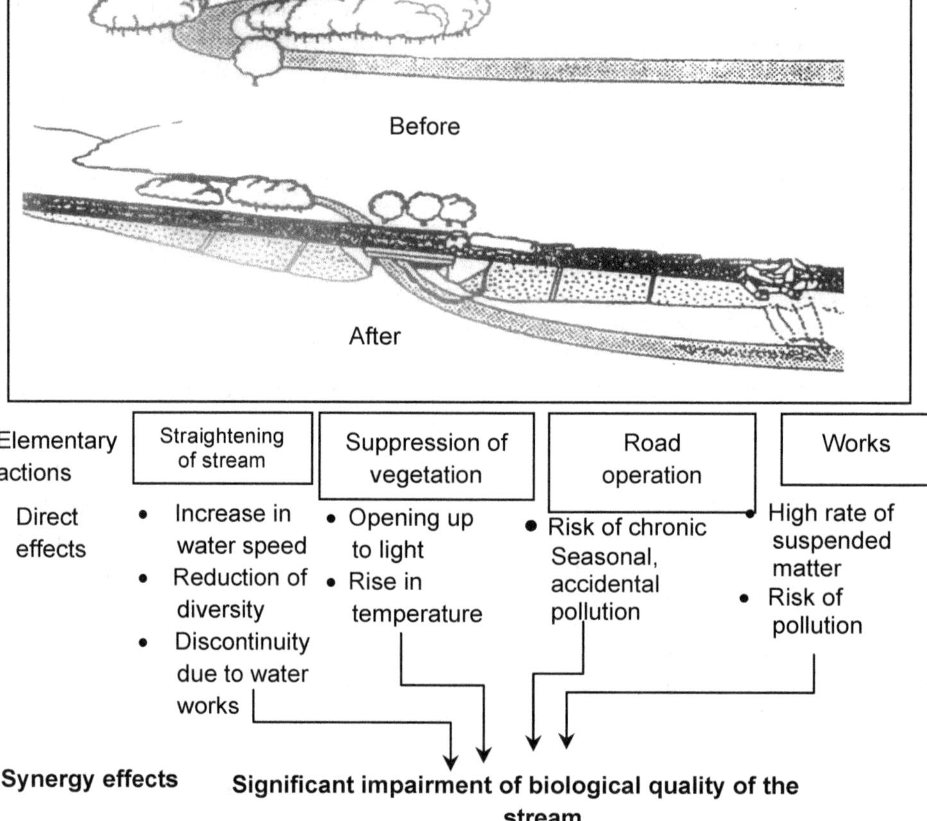

Each-elementary action produces a certain effect or a risk that can be limited, but the combination of such actions and therefore their consequences may be the source of significant effects. In this example, steps can be envisaged with reference to each elementary action, in order to avoid the synergy effect.

Fig. 1.3 Cumulative impacts : the example of a stream.

Cumulative effects assessment is an effective impact assessment tool, but it must be carried out properly in order to produce reliable results.

1.4.3   Ecosystem Function Impacts

Technically a subset or variant of cumulative impacts, ecosystem function impacts, which disable or destabilize whole ecosystems are the most dangerous and often the least likely to manifest themselves over a short period of time. Many road-related examples deal with roads which need to traverse watersheds in which surface and subsurface water movement is complex. One striking example is the highway constructed across a mangrove forest (100 ha in size) along the Caribbean coast. It was not fully understood at the planning stage to what extent the fresh and sea water needed to mix in order for the healthy forest to survive on both sides of the road. As a result, most of the forest has died. On one side the waters were not saline enough, and on the other there was not enough mixing with fresh water. The effect on the ecosystem was devastating and the impact on the local population, which used the mangrove forest area was severe. Almost certainly, no sign of this impact appeared until two to three years after the road was built. A second example, could develop in situations where roads bisect wildlife migration routes, which can inflict stress on the migratory population for many generations, or even permanently, and cause instability, increased mortality, and possibly catastrophic decline.

1.4.4 Assessment of Significance of an Impact

The determination of *significance* is defined in terms of context and intensity. *Context* refers to the geographical setting of a proposed project or action. When a proposed shopping center is evaluated, the context for the determination of significance in the immediate setting and the general community or area of influence, but not any country as a whole.

Intensity refers to the severity of impact

- The degree to which the proposed action affects public health or safety
- The presence of unique characteristics in the geographic setting or area, such as, cultural resources, parklands, wetlands, ecologically critical area, or wild and scenic rivers
- The degree at which the effects are likely to be highly controversial
- The degree at which the action would establish a precedent for further actions with significant effects.
- The degree at which the possible effects will be highly uncertain or involve risks
- The degree of effect on the sites listed in the Central Court Register of Historic Places
- The degree of effect on the threatened or endangered species or their habitats
- Whether the action conflicts with other Central, State, or local laws or requirements.

Initial the environmental settings of the project area needed to be described to predict and assess the impacts the impacts associated with various project activities. This provides base line environmental information as input to comprehensive EIA and prediction and assessment of impacts of project activities will be made with the reference baseline environment.

1.4.5 Impact Evaluation and Analysis (22, 24)

Environmental impact evaluation and analysis is meant for relative evaluation and assessment of different alternatives in project activities which also includes what will happen to the environment if no action/no- build alternative exists. In this exercise prediction, analysis and judging the environmental impacts are the major steps which involve the following aspects

1. Formulation of major activities of the project
2. Identification of important environmental components
3. Identifying various types of impacts likely to occur
4. Assessing the possibilities and or probabilities of occurrences

5. Assessing the intensity and the time frame of the impacts

6. Identifying positive, negative and neutral impacts

7. Assessing and evaluating tradeoffs among activities and impacts

For understanding various deleterious effects of the proposed project activities likely to impact the environment, EIA provides in the form of critical Environmental information to the decision makers. However, as EIA involves predictive techniques and evaluation methods which have uncertainty in error margins, the EIA results need to be checked /corrected by a feedback mechanism from decision makers and other stakeholders. This should involve actual transfer of knowledge about the actual environmental effects of a project action instead of predictions. Environmental Audit (EA) is an exercise which involves post project monitoring environmental quality through which feedback mechanism is incorporated in the EIA process.

Mitigation Measures

Mitigation measures to be implemented by project proponent are normally included in any EIA in order to reduce the magnitude/intensity of the impacts affecting the environment. Appropriate mitigation measures based on their cost effectiveness and environmental protection need to be implemented to make the project environmentally sustainable and economically viable.

1.4.6   Evaluation of Least Environmentally Damaging Alternatives

One of the most important contributions of an initial overview assessment is the early input of environmental considerations for the design or development of the project, action, or plan. If coordination is efficiently among the various members of the team for the project or action, the information provided by an initial overview can lead to better projects with fewer environmental impacts. These "least environmentally damaging" alternatives are then the ones evaluated in the subsequent detailed environmental studies and public and agency review process.

The development and analysis of alternatives form the very core of environmental impact assessment which is nothing but a comparative analysis of alternatives. Environmental Impact Statements are often titled Draft (or Final) Environmental Impact Assessment Alternatives Analysis. The driving impetus for conducting environmental impact studies is to make a comparative study of the effects of the proposed alternatives so as to be able to arrive at a better decision-making.

Due to its importance in the impact analysis, the study of alternatives should be a thorough and systematic process. It should include input from Central and State governments, local agencies and the general public. Decisions made in every phase of analysis should be logical and documented on the bases of a solid platform of evaluation criteria. The alternatives section of the Environmental Assessment/Finding of no significant Impact or the Draft and Final Environmental Impact Statements is the most noteworthy portion of the environmental document.

1.4.6.1 Examination of Project Alternatives

The necessity to develop alternatives is warranted by the deficiencies, if any, in the existing position. Similarly, the need for transportation projects is based on the deficiencies of the existing transportation system, such as, lack of safety, and inability to handle existing or projected traffic volumes, and meet air quality standards for a region. A National Forest Management Plan may need updating because of a regulatory requirement for periodical reevaluation, a change in use, demand or objectives, or because the present management techniques may not be producing the desired results. For instance, a more spacious jail may be proposed since the present jail is congested. Similarly a new low-income housing project may become imperative on account of the shortage of houses as against the demand.

Thus a need–based project should take into account the following:

1. The deficiencies in the existing circumstances
2. The present projected and specific needs
3. The goals and objectives of these needs.

The first section of any Environmental Assessment (EA) or Draft Environmental Impact Statement should thus be a consideration of the purpose and need. It should logically lead to the adopted list of goals and objectives for a proposed project or action plan. Depending on the type and size of this project or action plan, review of and concurrence with the purpose-and-need summary should be obtained from Central or State Govt., or local agencies.

1.4.6.2   Developing a Preliminary Range of Alternatives

The development of an initial range of alternatives will logically follow an analysis of purpose-need activity. For this purpose, all possible alternatives that satisfy the goals and objectives, as well as action plans even if they are outside the jurisdiction of the project sponsoring agency must be considered.

For example, these alternatives identified to correct transportation deficiency may include the following:

- Constructing a new highway at the location of the problem
- Constructing a new highway or widening an existing route at another location that may divert traffic away from the problem area
- Widening existing highways
- Providing HOV (high occupancy vehicle) lanes
- Increasing bus services
- Constructing or extending commuter rail systems
- Revising traffic signal timing, adding left-turn lanes or other such measures to improve traffic flow
- Implementing inspection and maintenance programs to check vehicles for emissions
- Switching to natural gas vehicles to limit air pollutants
- Encouraging major employers to offer incentives for corporate employees
- Encouraging major employers to implement staggered work hours
- Recommending that major traffic generators such as shopping centers or housing developments be located at alternative areas or sites.
- Coordinating with local planning officials in tackling potential future traffic problems through reasoning or limiting permits.

1.4.7   Factors to be considered for taking decisions based on Assessment of Significance of an Impact

There are six factors that should be taken into account when assessing the significance of an environmental impact arising from a project activity. The factors are interrelated and should not be considered in isolation. For a particular impact some factors may carry more weight than others, but it is the combination of all the factors that determines the significance.

1. *Magnitude*: Will the impact be irreversible? If irreversible, what will be the rate of recovery or adaptability of an impact area? Will the activity preclude the use of the impact area for other purposes.

2. *Prevalence*: Each action taken separately, might represent a localized impact of small importance and magnitude, but a number of actions could result in a widespread effect,

3. *Duration and Frequency*: The significance of duration and frequency is reflected in the following questions: Will the activity be long-term or short-term? If the activity is intermittent, will it allow for recovery during inactive periods?

4. *Risk*: To accurately assess the risk, both the project activity and the area of the environment impacted must be well known and understood.

5. *Importance*: This is defined as the value that is attached to an environmental component.

6. *Mitigations*: Are solutions to problems available? Existing technology may provide a solution to a silting problem expected during construction of an access road, or to bank erosion resulting from a new stream configuration.

The possible assessment decisions, using the above criteria are:

1. No impact

2. Unknown and potential adverse impact

3. Significant impact

1.4.8   Critical Assessment Criteria

The EIA methodology constitutes the use of assessment criteria concerned with the utilization of precious irreplaceable resources. The methodology includes questions such as (a) how is the project justified if it results in the loss of precious/irreplaceable natural resources? (b) whether the project will sacrifice important long-term environmental resources and values (ERVs) for the sake of immediate gains, (c) if it creates environmental issues which are likely to be highly controversial, (d) if the project endangers survival of species, then how is it justified, (e) whether the project will establish a precedent for future actions involving sensitive environmental issues (f) whether the project, while in itself, not causing serious impacts, will be related to other actions where the accumulated total effects could be serious, (g) whether the project is consistent with national energy policies, (h) whether the project is consistent with national foreign exchange policies, and (i) whether due consideration has been given in the project feasibility study, to alternative projects which could realize the desired development objective, and whether any of these alternatives might offer a better overall solution when all applicable project constants including environmental effects have been considered.

The steps involved in the assessment are: (i) description of the study area, that island/water areas affected by the project, including all significant environmental resources and values (ERVs) in the area, (ii) description (at the feasibility study level) of the project (proposed or existing) including the project properly and operations involved in transporting materials to and from the project vicinity, (iii) description and quantification of the impacts or effects of the project on ERVs, including legal implications, field investigations and sampling/analyses for obtaining other additional information required, and (iv) development of conclusions and recommendations on the environmental integrity of the project and on feasible measures which should be considered by the project planners to modify the project plan in order to offset or minimize adverse effects on ERVs.

1.4.9 Public Participation

Public participation and awareness about the social and environmental impacts of a project is extremely crucial. At numerous points throughout the process it is essential to engage with various stakeholders for successful grounding of the project. It should be a two-way exchange of information and views. Public participation may consist of informational meetings, public hearings, and opportunities to provide written comments about a proposed project.

1.5 Systematic Approach for using EIA as a Planning Tool for Major Project Activities

1.5.1 Introduction

The concept of EIA as a planning tool requires that it be concerned with, all phases of project development, including (i) planning, (ii) final design/construction start-up, and (iii) project operations. Fig. 1.4 illustrates the relationship between the various stages of a project development and the timing for the tasks to be included in the EIA process.

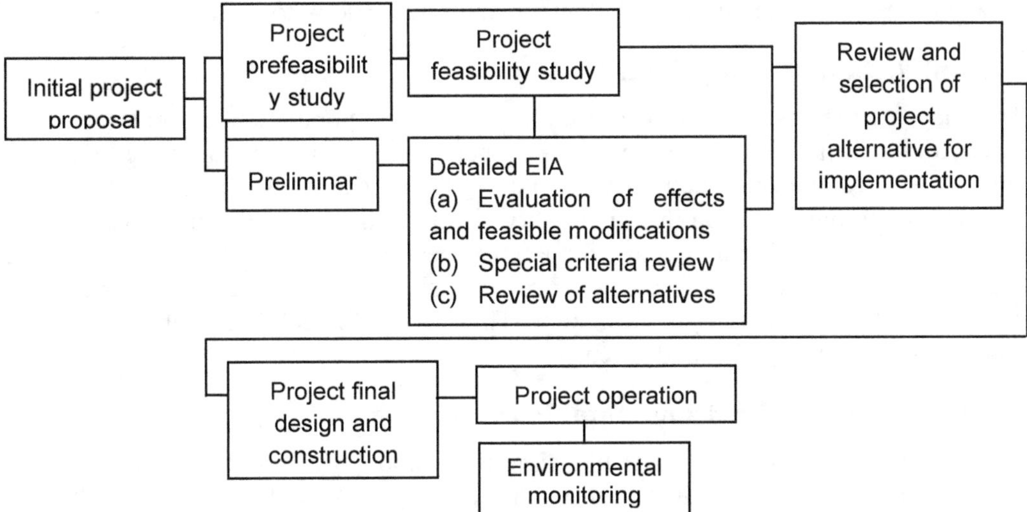

Fig. 1.4 Relationship of EIA process to project planning and implementation.
Source: Environmental Impact Assessment.

**Guidelines for Planning and Decision Makers,
UN Publication ST/ESCAP/351, ESCAP, 1985 (1)**

For the EIA being of optimal value in influencing the overall project's impact on the environment, the EIA itself should be a part of step (i) of the planning activity.

In respect of step (iii) project operations, the EIA will be mostly concerned with the provision of continuing to monitor the project's impacts, with feedback, so that this information can be used for making improvements in the project as shown by the monitoring data. However, for assessing the impact of the project on environmental values, such as water quality, some initial monitoring may be needed in the pre construction period for establishing a "baseline" picture of the pre-project situation and preparing environmental baseline impacts. Environmental inventory is a complete description of the environment as it exists in an area where a particular action is being considered. It is included in impact statement and serves as the basis for evaluating the potential impacts on the environment, both beneficial and adverse of a proposed action.

1.5.2   Preparation of Environmental Base Map (EBM)

An important requirement is the preparation of an Environmental Base Map (EBM) or maps showing the salient information as in (i) and (ii). This includes the essential background information on the environmental situation so that the reviewer, by referring to this, can readily interpret the report text and especially the conclusions and recommendations. For an Industrial Development Project EIA thus usually includes demography, land use, infrastructure, receiving water, ground water and soil conditions, other industries and their waste streams, institutions, ecological resources, areas of cultural, archaeological and tourist interest, and meteorological conditions.

The EBM should be portrayed as simply as possible (it should not include extraneous information which may obscure the presentation) and for this purpose a schematic-type drawing will usually be more appropriate than a map drawn strictly to scale.

1.5.3   Identification of Study Area

The EIA study area should include water bodies, land, and population centers where the project activities will have significant effects : General environmental parameters likely to be affected by developmental activities include: ground water hydrology and quality; surface water hydrology and quality; air quality; land quality and land uses; vegetation; forests; fisheries; aesthetics; public and occupational health and socio-economics. The size of the study area will vary according to the type and size of the project activities and the characteristics of the surrounding environment. The meteorological conditions would also be considered in determining the study area.

1.5.4   Classification of Environmental Parameters

Most EIA guidelines follow the relatively simple methodology in which environmental resources or values are classified into four general categories, namely, (a) natural physical resources, (b) natural ecological resources, (c) human/economic development resources, and (d) quality-of-life values including aesthetic and cultural values which are difficult to assess in conventional terms.

1.5.5   Formation of EIA Study Team

Since most EIAs involve consideration of environmental parameters covering many disciplines, to produce a meaningful EIA will require inputs of expertise from all the disciplines involved in a particular project. This does not mean that a large team must be organized to include inputs from each discipline. The key point is that the individual in charge of the EIA must have certain skills so that findings from the environmental studies can be used appropriately for modifying the project plan to obtain a more optimal economic-cum-environmental development project. The composition of the team should depend on the nature of the activity. This can be determined only after the key users have been identified. In any use the team should include persons familiar with the particular type of operations. The number of persons required will depend on the size and complexity of the activity to be investigated.

1.5.6   Preparation of Terms of Reference

The first step in undertaking any EIA is to carry out a preliminary evaluation of the situation. If done by a skilled environmental analyst within a short period, say two weeks, it is possible to size up the situation, identify the beneficial uses which are likely to be significantly affected, make preliminary estimates of the magnitudes of these effects and preliminary delineation of the feasible measures which will be needed to minimize/offset degradation, and draw conclusions on (a) whether a detailed EIA follow-up study is needed, and if so, to prepare the Terms of Reference (TORs) and recommended budget, and (b) if not, to prepare a report on the initial work which in itself becomes the final EIA for the project.

1.5.7   Preparation of an EIA Report

Numerous techniques are available for the assessment of environmental impacts and preparation of EIA reports. Alternative assessment techniques are continuously developed and utilized. The project proponent is free to select the method most appropriate for the specific situation.

The manual presents a recommended standard format for the organization of EIA reports. Essential steps to complete an environmental impact assessment include:

1. Describe the proposed project as well as the options
2. Describe the existing environment
3. Select the impact indicators to be used
4. Predict the nature and extent of the environmental effects
5. Identify the relevant human concerns
6. Assess the significance of the impact
7. Incorporate appropriate mitigating and abatement measures into the project plan
8. Identify the environmental costs and benefits of the project to the community
9. Report on the assessment.

The sequence may be repeated for a number of project options and for a selected project concept with mitigating or abatement measures incorporated.

However, the following is a standard format for EIA reports as per Central Pollution Control Board of India.

(a) *Introduction*: This constitutes the purpose of the report, the extent of the EIA study, and a brief outline of the contents and techniques.

(b) Description of the project.

(c) *Description of the existing environment*: This first requires identification of the project "area of influence". The environmental resources within the "area of influence" are then identified as physical resources, ecological resources, human and economic values, development, and quality-of-life values.

(d) Anticipated environmental impacts and plans for protection as follows:

   (i) *Item-by-item review*: impacts resulting from project implementation are evaluated and quantified wherever possible;

   (ii) *Mitigating and offsetting adverse effects*: a plan is presented for offsetting or compensating for significant adverse impacts and for enhancement of positive impacts;

   (iii) Identification of irreversible impacts and irretrievable commitments of resources;

   (iv) Identification of impacts during construction and appropriate protection measures.

(e) *Consideration of alternatives*: for each alternative considered the probable adverse impacts are identified and related to the proposed project and other alternatives.

(f) *Monitoring program*: this is so designed that the environmental agency receives monitoring reports which will ensure that all necessary environmental protection measures are being carried out as listed in the approved project plan.

(g) *Summary and conclusions*: the summary and conclusions section is prepared in such a way that it is a complete and comprehensive document in itself. This section includes;

   (i) A review of gains versus losses in environmental resources and values, and of the overall net gains which presumably justify the project

   (ii) An explanation of how unavoidable adverse impacts have been minimized, offset and compensated for

   (iii) An explanation of use of any replaceable resources

   (iv) Provision for follow-up surveillance and monitoring.

1.5.8 Environmental Monitoring and Management Plan

An appropriate plan should be developed and described for constant monitoring to ascertain the impact of the project on those applicable environmental parameters, which are especially sensitive for the project under consideration. These will usually include environmental resources within the industrial plant (for example, occupational health) and those in the region affected by plant establishment and operations.

It is recognized that most developing countries generally have expressed little interest in funding and implementing monitoring programs of this type probably because of the lack of appreciation by decision-makers of their vital role in ensuring optimal overall economic and environmental project benefits.

1.5.9   Draft and Final Environmental Impact Statements

The most detailed procedure for analyzing potential environmental impact of the alternatives of a proposed project or action is the Environmental Impact Statement process. The DEIS contains the final results of environmental studies of proposed alternatives which are available for public and agency review. The DEIS is a "draft" because it compares all proposed alternatives and is the document upon which the decision to proceed with any particular alternative is made. The DEIS is also the tool through which public and agency input is incorporated into this decision-making process. The E.I.S represents a summary of environmental inventory and the findings of environmental assessments.

The alternatives section of the DEIS contains a detailed description of each proposed alternative, including physical characteristics, operating features, costs, schedule, description of the construction process, and all other relevant features of the proposed action. Certain basics, which are required to accomplish an environmental assessment, are related to description of the environmental setting, impact prediction and assessment and preparation of E.I.S.

The *Affected Environment* section of the DEIS contains information on the existing setting. Although the organization and format vary, the following areas may be included.

Land Use and Zoning

Social and neighborhood characteristics

- Demographic characteristics
- Housing
- Travel patterns
- Stability
- Pedestrian and bicycle travel
- Community activities and services (fire, police, hospitals, schools, churches, day care and so on)
- Recreational facilities

Economic factors

- Taxes
- Existing business community
- Proposed developments

Traffic and Transportation Energy

Historic and archaeological resources, Visual resources, Air quality, Noise levels, Geology and soils, including farmland Environmental health and public safety (hazardous wastes)

Water Resources

- Groundwater
- Surface water
- Water supply and wastewater systems

- Wild and scenic rivers
- Wetlands Flood plains and coastal zones
- Vegetation and wildlife.

The *Environmental Consequences* section of the DEIS contains the results of the assessment of impacts. The assessment can be organized by impact category or by alternative; the usual format is by impact category.

This section focusses on relevant environment issues and impacts. Some areas of potential effect must be included regardless of expected impact. Resources protected by statute, regulation, or executive order must be addressed in all the environmental documents. When such protected resources do not exist within the area or will not be affected, the EIS must document that the resource was considered in compliance with the applicable regulation, and statements must be made as to why the resource will not be affected the regulation does not apply.

1.5.10 Impact Analysis

Analysis of environmental impacts begins with a description of the existing environment, the assembly of relevant information and data and finally the evaluation and analysis of the degree of impact. Considered impacts must include direct and indirect effects, cumulative effects, and long-term and short-term effects. In the analysis process, potential mitigation measures are developed and explored.

The preparation of separate methodologies and technical reports supporting the DEIS has to be in accomplice with the area of discipline and contain the detailed information on existing conditions, methodologies, analysis, and results. The technical reports are then summarized in the DEIS.

Technical reports supporting a DEIS can be prepared for:

- Socioeconomic impacts, which include community impacts, land use, economic impacts, visual effects, relocations, traffic and pedestrian and bicycle travel
- Natural resources, which include water quality, vegetation, wildlife, scenic rivers, floodplains, wetlands, and coastal zones, and
- Air quality

1.5.11 Format and Content of a
Draft Environmental Impact Statement (DEIS)

After completing the analysis DEIS should have at least the following components:

Cover sheet, Summary, Table of Contents

(i)   Purpose of and Need for Proposed Action

(ii)   Alternatives

(iii) Affected environment

(iv) Environnemental Conséquences

List of Agencies, Organizations, and Persons to whom copies of the DEIS are sent should be given as an Index in Appendices

The language of DEIS must be concise and clear, and the data and the information must be relevant.

1.5.12 DEIS Processing

When the DEIS is completed, it is circulated among the Central, State, and Local agencies concerned. In some cases the summary of the DEIS can be circulated instead of the entire document. Notices have to be published in newspapers to notify the public of the availability of the DEIS and the locations in the community where it will be reviewed.

After the public hearing and the review period, the comments received are evaluated, and a required additional analysis is conducted. Alternatives and mitigation measures may be revised based on the comments received and the responses are prepared to each substantive comment.

Based on the review of the comments and the results of additional studies, the sponsoring agency selects the preferred alternative. This selection process should be a systematic evaluation procedure. The process then continues for the preparation of the Final EIS.

1.5.13 Final Environmental Impact Statement (FEIS)

The FEIS document is the preferred alternative consisting of the DEIS with modifications. In some cases, where minor changes are required, the abbreviated form of the FEIS can be used which merely attaches the required changes or findings to the DEIS.

A new section is added at the end of the document. It can be titled *Comments Received on the DEIS and Responses*. It documents the public hearing and summarizes the major comments. It also contains copies of all written comments received from agencies or the public, with written responses to all the substantive comments.

Upon completion, the FEIS is circulated among all interested agencies and persons. A notice indicating the availability of the FEIS should be published or advertised in local newspapers.

1.6 Comparative Evaluation of Alternatives from EIA Studies

1.6.1 Selecting a Preferred Alternative

The Environmental Assessment or Draft Environmental Impact Statement should be made available to the public and other interested agencies for comments and the comments thus received should be summarized. Subsequently, any additional environmental analysis required should be conducted, and then the alternatives considered should be reevaluated for possible changes so as to further minimize the impacts, or respond to comments received.

The revised summaries of the impacts of each alternative should be compared, using the evaluation criteria and measurement parameters. The next task is the selection of the preferred alternative. In some cases, the preferred alternative may be obvious, and the selection process brief. Other proposed projects or actions, a more thorough analysis and process will be required.

Documentation should be prepared of the decisions made and the reasons that prompted each decision. The following is an effective system to use for fairly involved projects or actions.

Each member of the team should prepare a brief summary of the impacts and comments received within his or her discipline, such as, air quality, noise, social effects, and wildlife. These summaries should be circulated among all the members of the team for review. A meeting of all team members can then be held to discuss the pros and cons of each alternative in each area of potential impact.

A good approach is to compare the build, or action, alternatives first. The least environmentally damaging alternative, with mitigation in place, should be identified. If any build alternatives are less responsive to the identified project purpose and need, they should be eliminated first. There is little sense in proceeding with a proposed project or action if it cannot accomplish the basic goals and objectives to meet the established needs.

The next step is to compare the remaining build or action alternatives for legislative or regulatory restrictions. Numerous types of potential impacts are regulated by specific guidelines to prohibit selection of a particular alternative under certain conditions, such as the existence of a feasible and prudent alternative, or a less-environmental-impact alternative, in the remaining set of alternatives. There may also be circumstances where a jurisdictional agency has indicated a future denial of a necessary permit for a particular alternative. Any alternative not meeting the regulatory requirements must be eliminated from further consideration.

The remaining build or action alternatives are then compared in detail, including such criteria as an opportunity for mitigation of adverse effects, project costs, severity of impact in any particular area, public and political opinions, and other established evaluation standards. Through the interaction of the interdisciplinary team, an alternative is selected as the preferred action alternative.

The next step after the preferred build, or an action alternative is selected is to directly compare it with the no-build alternative. The team is now at the final stage of build versus no build. This is the phase where trade-offs should be clearly presented and evaluated. The analysis of benefits versus costs, with the incorporation of any agency specific feasibility criteria, will finally decide whether the identified preferred alternative is the selected build alternative or the no-action alternative.

With the selection of a preferred alternative and completion of the Final Environmental Impact Statement and Record of Decision, the environmental impact study process gets completed. Committed mitigation monitoring programs will continue with the project or action through construction. Other considerations may, however, still prevent the proposed project or action from proceeding with the construction or implementation. A summary of the major factors, which enter the decision-making process for selection of a preferred alternative and for ultimate project completion is illustrated in Fig.(1.6).

Following completion of all appropriate environmental impact assessment studies, the major task is to make the completed analyses productive in the decision-making process. The evaluation of alternatives must result in a clear and concise comparison that easily illustrate the tradeoffs involved between the build and no-build alternatives and the distinguishing degree of impact among the various build or action alternatives.

1.6.2 Conceptual Basis for Trade-Off Analysis

As a systematic approach for deciding upon right alternatives, it is desirable to use trade-off analysis. Trade-off analysis involves the comparison of a set of alternatives relative to a series of decision making factors. The following approaches can be used to complete the trade-off matrix.

1. A qualitative approach, in which descriptive, synthesized and integrated information on each alternative relative to each decision factor is presented in the matrix.

2. A quantitative approach, in which quantitative, synthesized and integrated information on each alternative relative to each decision factor is displayed in the matrix; or a combination of qualitative-quantitative approach.

3. A ranking, rating or scaling approach, in which the qualitative or quantitative information on each alternative is summarized by using the assignment of a rank rating or scale value relative to each decision factor is presented in the matrix

4. A weighted approach, in which the importance of weight of each decision factor is considered, and the resultant decision of the information on each alternative (qualitative, quantitative, or ranking, rating, or scaling,) is presented in Fig 1.5 in terms of the relative importance of the decision factors.

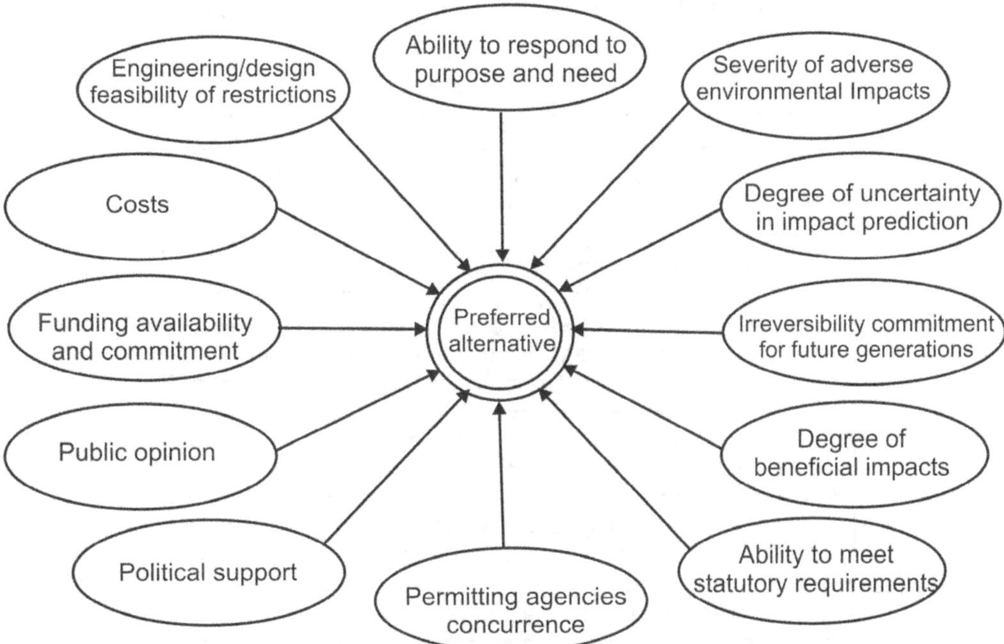

Fig. 1.5 Factors affecting selection of a preferred alternative and ultimate project of action implementation. (Some factors often will be more important than others in the decision-making process).

5. A weight-ranking, rating, or scaling approach, in which the importance of weight for each decision factor is multiplied by the ranking, rating or scale of each alternative, and the resulting products for each alternative are then totaled to develop an overall composite index or score for each alternative; the index may take the form of

$$Index = \sum_{i=j}^{n} IW_i R_{ij}$$

where

$Index_j$ = composite index for j_{th} alternative

n = number of decision factors

IW_I = importance weight of i_{th} decision factor

R_{ij} = ranking, rating or scale of j_{th} an alternative for i_{th} decision factor

Decision-making in relation to selecting the proposed action from alternatives, which have been analyzed and compared should take place in relation to an overall planning model, which is also called the "rational planning model," as shown in Fig.1.6. An illustration of the application of this model to the selection of a "best practicable environmental option" (BPEO) (in this case, for pollution control) is shown in Fig.1.7. Decision-focused checklists can be used in the "Analysis of alternatives" step in Fig. 1.5, and the "Select preferred option" step in Fig.1.6. Finally, McAllister (2), Fig 1.8., has suggested that evaluation of an alternative can be divided into two phases: analysis, in which the whole is divided into parts, and synthesis, in which the parts are reformed into a whole.

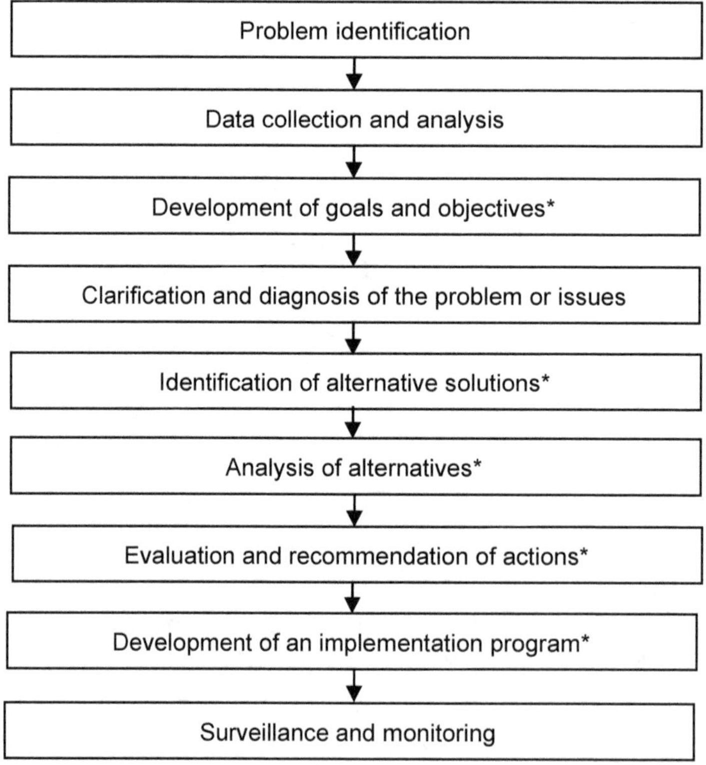

*Denotes components of what is frequently called the rational planning model

This information could be used to prepare a trade-off analysis and select the proposed action. If the qualitative and/or quantitative approach is used for completion of the matrix, information for this approach relative to the environmental impacts should be based on impact prediction. This information would also be needed for impact ranking, rating or scaling.

1.6.3   Importance-Weighting of Decision Factors

If the importance-weighting approach is used in decision-making, the critical issue is the use of an effective method to assign importance weights to the individual decision factors or, at least, to arrange the factors in a rank ordering of importance. Table 1.4 lists some structured importance-weighting or ranking techniques that could be used in numerous EIS decision-making efforts.

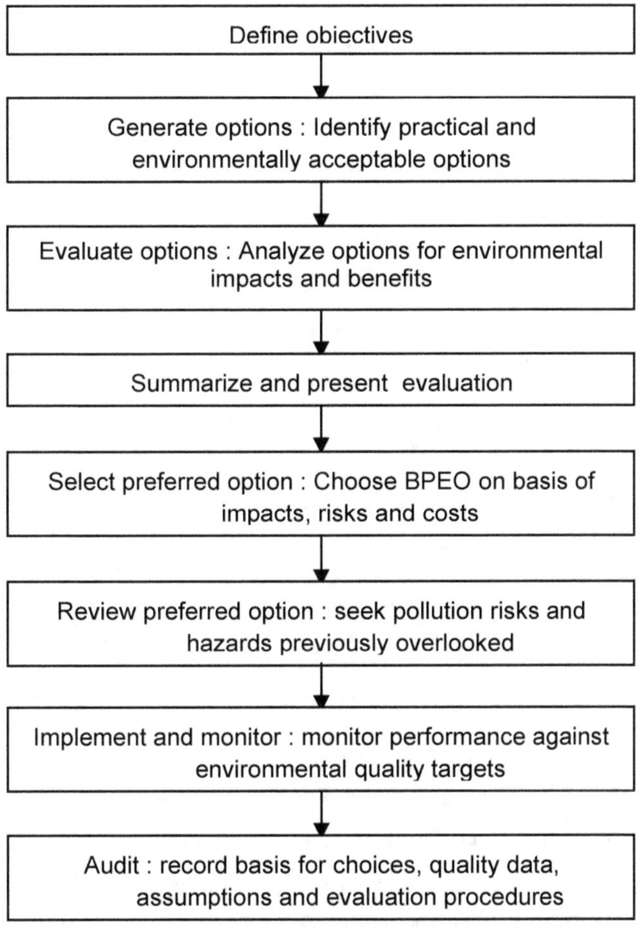

Fig. 1.7 Steps in selecting a best practicable environmental option (BPEO) using the rational planning model (Selman, (3)).

Table 1.4 Examples of types of importance-weighting techniques used in environmental impact studies.

Ranking
Nominal-group process
Rating Predefined importance scale
Multiattribute (or multicriterion) utility measurement
Unranked pairwise comparison
Ranked pairwise comparison
Delphi study

These ranking methods assist the environmental analyst in developing project-specific evaluation methodologies for the particular projects or actions being considered. The actual method, however, should include local factors and opinions of local and state agencies.

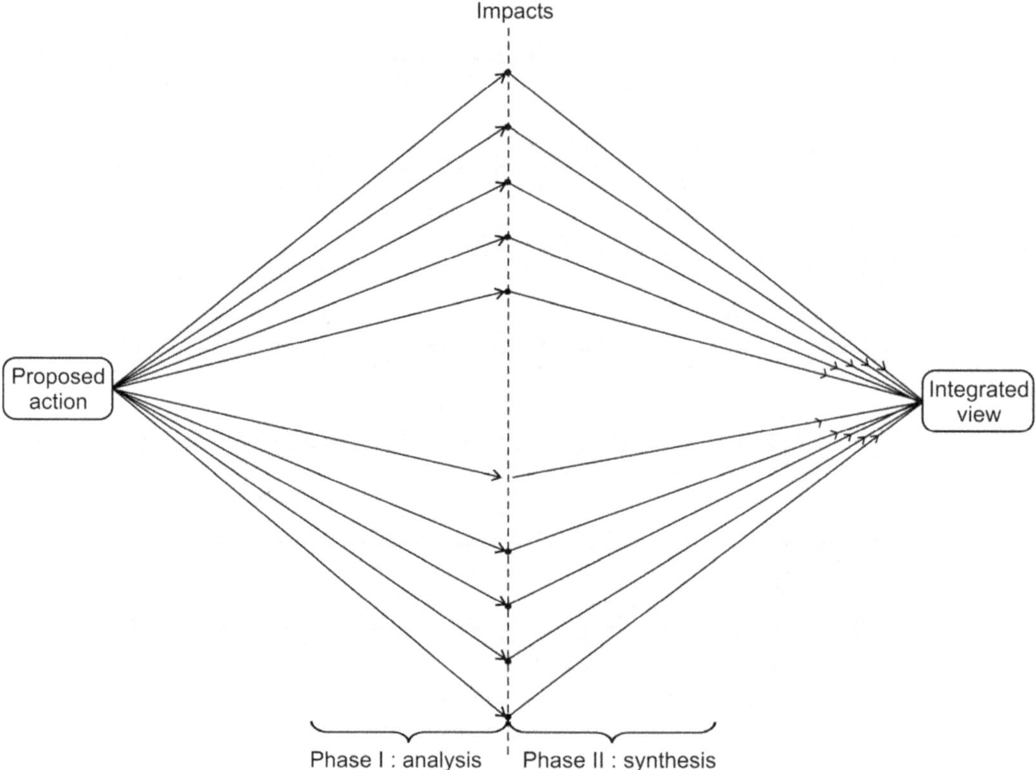

Fig. 1.8 The two phases of the alternative evaluation process (McAllister).

An overview of the total EIA process to assist applicants, stakeholders and public is presented in Table 1.5 giving various systematic steps to be followed for determining whether a new developmental project needs EIA or not (20-25).

1.7 Strategic Environmental Assessment (SEA)

1.7.1 What is Strategic Environmental Assessment (SEA)?

Systematic assessment/evaluation of significant impacts likely to occur whenever a new strategic plan/program/policy is proposed/decided to be implemented is called Strategic environmental assessment (SEA). A number of new welfare/policy decisions are being implemented by various Governmental/Statutory agencies as a part of governing/developmental process. In this context SEA provides a proactive and comprehensive assessment which identifies/evaluates and predicts significant environmental impacts/sustainability risks most likely to occur if the policy or program is implemented. SEA, thus greatly helps policy/program implementing authorities to take into cognizance various environmental issues/sustainability risks most likely to occur and address in the early stages of project/policy implementation (26, 27).

Therivel *et al.*, 1992, (28) defined" SEA "as the formalized, systematic and comprehensive process of evaluating the environmental impacts of a policy, plan or program and its alternatives, including the preparation of a written report on the findings of that evaluation, and using the findings in publicly accountable decision-making.

Sadler and Verheem, 1996 (29) defined SEA as "a systematic process for evaluating the environmental consequences of a proposed policy, plan or program initiatives in order to ensure they are fully included and appropriately addressed at the earliest appropriate stage of decision-making on par with economic and social considerations.

SEA as a continuous, proactive, integrated decision support process (rather than production of a report) attempts to highlight with broad scope, sustainability policy issues and focus on visions and initiatives rather than on concrete actions and outcomes (30).

Strategic Environmental Assessment (SEA) and Environmental Impact Assessment (EIA) effectively promote sustainable development by mainstreaming the environment into economics. The main differences between EIA and SEA are that in the former the likely environmental impacts of various physical activities in a project are examined/ assessed/predicted while in SEA the environmental consequences/risks will be assessed for a new plan/policy/project proposed to be implemented which are need to be visualized. For planners/decision makers, SEA will become an important tool as it provides important information and analysis on environmental issues likely to arise in short, medium, and at long range if certain policies/programs /plans are implemented and what are regional environmental changes likely to occur for different land uses.

Table 1.5

DETERMINING WHETHER A PROJECT NEEDS ENVIRONMENTAL IMPACT ASSESSMENT (EIA)

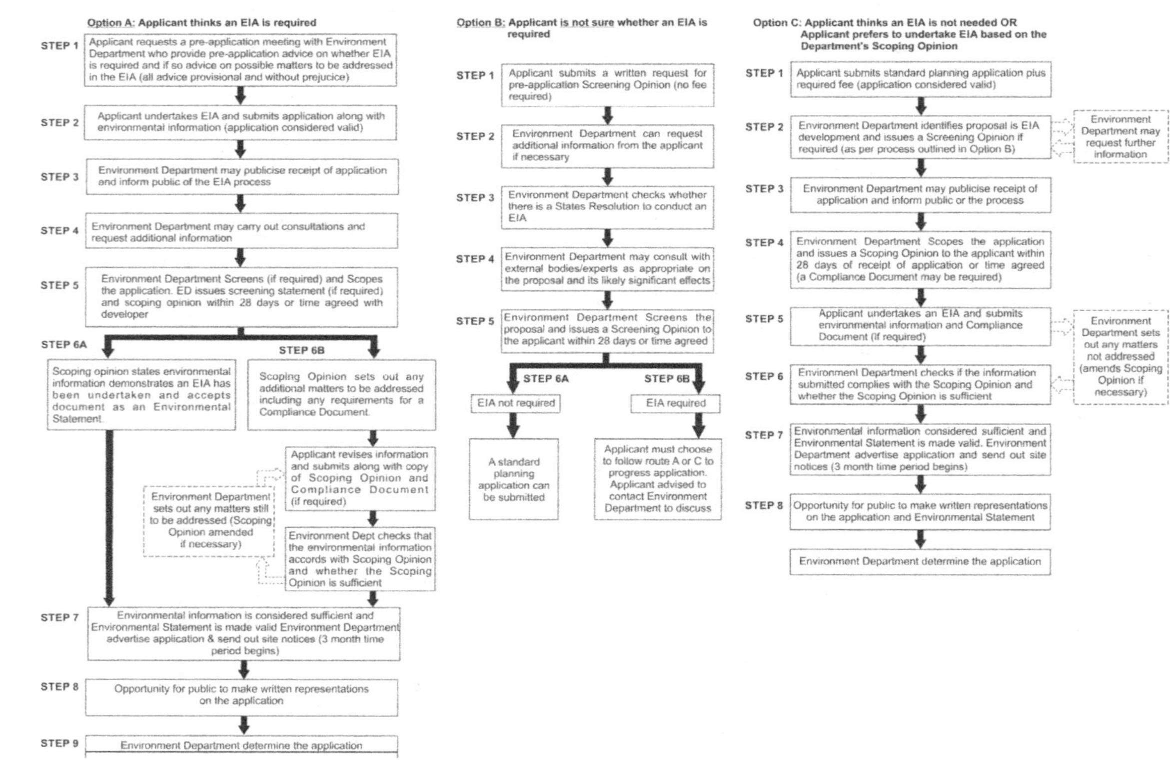

34

1.7.2 Strategic Environmental Assessment (SEA) : A Decision Support System

1.7.2.1 SEA as structured a Decision support System

For ensuring that all environmental risks sustainability issues likely to occur are properly addressed in the policy/plan/program planning level, SEA serves as a systematic and structured decision support tool for planners/decision makers/private bodies with the following features

- SEA offers a structured, rigorous, participative, open and transparent Environmental Impact Assessment (EIA) Tool.
- SEA provides a participative, open and transparent, possibly non-EIA-based tool, that can be applied in a more flexible manner to policies
- SEA will be a versatile and simple tool that can be applied in a flexible way to legislative proposals and other policies, plans and programs in political/cabinet decision-making.
  For achieving sustainable development with good governance, SEA serves as a structured and tiered decision support tool which will be more effective, efficient, simple and flexible. This approach will provide substantive focus regarding questions, issues and alternatives to be considered in policy, plan and program (PPP) making.

1.7.2.2 Good Practice SEA

For achieving good output from SEA it is necessary to adopt the following guidelines

- Instead of justifying the proposed policy it is necessary discussed it critically
- All feasible policies and planning options (alternatives) need to discuss critically and compared in terms of their environmental risks and benefits
- In the policy preparation process itself the environmental risks need to be described exhaustively and addressed effectively
- Strategic sustainability assessment need be simple with a clear assessment of environmental impacts
- By involving public at all levels SEA should reflect the view of all stakeholders on possible environmental risks.
- Information on environmental consequences of a new policy should be informed with good communication systems

1.7.3 Background and Origin of SEA Process

1.7.3.1 Chronological Order of Development of SEA

For integrating environmental, economic and social issues in various developmental activities in public/governmental/private decision making, many organizations have started working on the improvement of environmental performance and sustainability issues in these aspects, resulting in significant changes in environmental policy changes. In this context Strategic Environmental Assessment (SEA) has been emerged. The term SEA was first referred in The National Environmental Policy Act (NEPA) 1969. To bring about substantive environmental reform through the US federal bureaucracy, they were directed to prepare \ environmental impact statement for "legislation and other major federal actions significantly affecting the quality of the human environment" as SEA, which has become a part of (Section 102(2) (c), National Environmental Policy Act of 1969).

The Principal International Instrument on SEA process implementation was promulgated in 2003 the Protocol on Strategic Environmental Assessment in a Transboundary Context (Kiev Protocol) and the European Union (EU) Directive on SEA being a key regional example of its implementation in policy.

The objective of the EU Directive on SEA is stated as being "to provide for a high level of protection of the environment and to contribute to the integration of environmental considerations into the preparation and adoption of plans and programs with a view to promoting sustainable development, by ensuring that an environmental assessment is carried out of certain plans and programs which are likely to have significant effects on the environment" (EU Directive, Article 31).

The Kiev Protocol describes some of the key elements in an SEA process in its definition of SEA. These include the evaluation of the likely environmental and health effects of a policy, plan or program through the determination of the scope of an environmental report and its preparation, the carrying out of public participation and consultations, and the taking into account of the environmental report and the results of the public participation and consultations in the policy plan or program (Kiev Protocol, Articles 6-12).

1.7.3.2 Relevant Provisions of the Planning and Development Act 2000

Although the SEA Directive was not formally adopted until 2001, its imminent arrival was anticipated by certain provisions of the 2000 Act. The Act required that when Regional Planning. Guidelines, Development Plans, Local Area Plans or Strategic Development Zone (SDZ) planning schemes are being made by the relevant authority, they must be accompanied by information about the likely significant effects on the environment of implementing such plans.

1.7.3.3 Transposition of the SEA Directive into Planning Law

The European Communities (Environmental Assessment of Certain Plans and Programs) Regulations 2004 (S.I. 435 of 2004) amended certain provisions of the Planning and Development Act 2000 to provide the statutory basis for the transposition of the Directive in respect of land-use planning. These amendments facilitated the making of the Planning and Development (Strategic Environmental Assessment) Regulations 2004 (S.I. No. 436 of 2004) which give effect to the SEA Directive in the land-use planning sector.

The latter Regulations, which integrate SEA into current plan-making procedures as far as possible are:

(i)   Require SEA in the case of all Regional Planning Guidelines;

(ii)   Require SEA in the case of Development Plans, Variations of Development Plans and Local Area Plans likely to give rise to significant environmental effects;

(iii)   Require SEA in the case of Planning Schemes in respect of a Strategic Development Zone (SDZ);

(iv) Set out the procedural requirements for the preparation and consideration of the Environmental Report, including scoping and public consultation, and the integration of these new requirements with existing plan preparation/review processes; and designate the Environmental Authorities to be consulted at various stages during the SEA process.

These guidelines relate to the application of the SEA Directive to certain plans prepared under the Planning and Development Act 2000.

The guidelines are addressed to:

(a) Regional Authorities: in relation to the preparation or review of Regional Planning Guidelines;

(b) Planning Authorities: in relation to the preparation or review of Development Plans and Local Area Plans;

(c) Relevant Development Agencies: in relation to the preparation of a Planning Scheme in respect of a Strategic Development Zone (SDZ).

Where the word "plan" is used throughout these guidelines, it will refer to Regional Planning Guidelines, Development Plans, Local Area Plans and Planning Schemes in respect of SDZs for which SEA will be required. Similarly, "plan-making authorities" includes regional and planning authorities, and in the case of a Planning Scheme in respect of an SDZ, the relevant development agency.

A further element in any SEA process is ongoing monitoring of the implementation of particular policies, plans and programs to determine any adverse unforeseen environmental or health effects so as to undertake remediation (32).

However, from a global perspective, SEA for all components of the environment, including the marine environment, is still at an early stage of implementation, Table 1.6 lists a series of key events that have contributed to the evolution and consolidation of SEA.

Table 1.6 Chronological development of SEA Concepts

1969	The National Environmental Policy Act (NEPA) passed by the U.S. Congress mandating all federal agencies and departments to consider and assess the environmental effects of proposals for legislation and other major projects.
1978	US Council for Environmental Quality (USCEQ) issues regulations for NEPA which apply to USAID and specific requirements for programmatic assessments.
1989	The World Bank adopted an internal directive (O.D. 4.00) on EIA which allows for the preparation of sectoral and regional assessments.
1991	The UNECE Convention on EIA in a Transboundary Context promotes the application of EA for policies, Plans and programmes.
1990	The European Economic Community issues the first proposal for a Directive on the Environmental Assessment of Policies, Plans and Programmes.
1991	The OECD Development Assistance Committee adopted principles calling for specific arrangements for analyzing and Monitoring Environmental Impacts of Programme Assistance.

Table 1.6 *contd...*

1995	The UNDP introduces the environmental overview as a planning tool.
1997	The Council of the European Union adopts a proposal for a Council Directive on the assessment of the effects of certain plans and programmes on the environments.
2001	The UNECF issues a draft protocol on strategic Environmental Assessment applying to Policies, Plans and Programmes.
2001	Council of the European Union adopts the Council Directive 2001/42/CE on 27 June on the assessment of the effects of certain plans and programmes on the environment.

1.7.4   SEA's Contribution to Sustainability and Success Factors

1.7.4.1 SEA's Contribution Towards Sustainability

1. SEA helps to get a broader environmental perspective of any policy

2. SEA provides critical information on various environmental risks and sustainability issues likely to arise due to implementation of any policy/planning activity/govt decision which help in incorporating mitigation measures at the policy/project formation stage itself.

3. SEA predicts serious/significant environmental risks/sustainability issues likely to arise due to project/policy implementation

4. SEA helps in formulating a number of environmentally sound chain actions for environmentally sustainable development

5. SEA helps in integrating policy making/planning with various environmental sound mitigation measures for sustainable development.

SEA is an evidence-based instrument, aiming to add scientific rigor to PPP making by using suitable assessment methods and techniques.

1.7.4.2 Success Factors in SEA

- Legal basis, administrative order, policy or recommended requirements are basic success factors

- Achievable and clear environmental policy objectives

- Non biased Environment reporting

- Robust planning process with well structured operations

- Responsibility for compliance

- Proponent commitment and accountability

- Multiple organizations that work together

- Objectives, criteria and quality standards framework
    - to assess proposal need and justification
    - to assess environmental effects (losses/changes)

- Guidelines for good practice

- Resources availability

- Access to information

- Public interest and involvement of non-governmental organizations
- Independent oversight and review of the implementation and performance (quality control)
- Inputs for decision making: are SEA results timely, relevant and influential? (Use versus non-use of SEA in policy design/approvals).

1.7.5 Aims and Objectives of SEA (33)

The following should be the aims and objectives of any SEA for achieving sustainable development:

i. To help achieve environmental protection and sustainable development by:

- Assessment and evaluation of environmental risks /sustainability issues arising out of implementation of the new policy/projects/plans
- Consideration and evaluation of various project options to identify environmentally sound and practicable option
- Develop inbuilt early warning systems for cumulative effects and large-scale changes in the environment due various strategic activities

ii. To strengthen and streamline project EIA by:

- To assess the scale, spatial and temporal variations of the impacts likely to occur due to project activities
- To obtain statutory clearances for various strategic issues and concerns related to the justification of proposals
- Reducing the time for expert reviews

iii. To integrate the environment into sector-specific decision-making by:

- Supporting environmentally sound and sustainable proposals
- Developing objective and rapid decision systems

1.7.6 Elements of SEA and Priority needs for Good Practice of SEA

1.7.6.1 Elements of SEA

In practice, SEA is generally understood as comprising of a flexible framework of key elements which support decisions on development by integrating environmental considerations into the decision making process (30).

SEA instruments and policy directives such as the Kiev Protocol and the EU directive on SEA contain provisions prescribing policies, plans and programs which are subject to SEA processes. These are known as screening provisions and refer to those policies, plans and programs likely to have significant effects on the environment. In some cases, they provide a more specific listing of those plans and programs for which a SEA process is mandatory.

SEA typically involves the setting of an overarching environmental vision and objectives for a particular geographic region and activities within that region (33). A broad range of alternative courses of action to achieve the specified objectives are then developed and each alternative is assessed against specific criteria within the context of the broader environmental vision and objectives.

Alternative courses of action are then assessed against criteria such as sustainability measures and acceptable levels of environmental change for particular species, habitats and ecosystems.

On the basis of this assessment, the most desirable courses of action are selected and implemented in policies, plans and programs in that area.

Elements in a SEA process include an array of "analytical and participatory approaches" designed to "integrate environmental considerations into policies, plans and programs and evaluate the interlinkages with economic and social considerations" (34).

Different tools can be employed at different stages as part of a SEA according to the context of the policy, plan or program being assessed.

These include tools to predict environmental and socioeconomic effects, tools to ensure full participation of stakeholders and tools for analyzing and comparing options.

An OECD Guide to Understanding SEA in the development context provides some useful examples of the different types of tools which can be employed:

1.7.6.2 Priority needs for Good Practice SEA

➢ Policy context (sustainability policy, objectives and strategies)

➢ Accountable decision-making systems

➢ Adaptivity nature of decision-making processes

➢ Be integral and well coordinated with policy-making

➢ Simple, interactive and flexible approaches

➢ Integrated approaches regarding scope and cross-interaction of relevant factors

➢ Guidance and perhaps minimum regulatory context

➢ Demonstration of benefits - examples of good and bad

➢ Practice

➢ Participated process, including multiple agents and consideration of public priorities and preferences changing attitudes, overcoming prejudices, new routines in decision-making.

1.7.7   Benefits of SEA

One of the main benefits of SEA is that it provides a means of anticipating and avoiding cumulative adverse impacts on the environment (32):

➢ SEA can play a significant role in enhancing the integration of environmental concerns in policy and planning processes (30). It helps to incorporate sustainability principles in the policymaking process.

➢ SEA is intended to provide the framework for influencing decision-making to address environmental issues/concerns at an earlier stage when many govt. plans and programs are mainly dominated by sole economic aim.

➢ SEA facilitates sustainable development through the systematic assessment of policy options.

➢ SEA specifically designed to address cumulative impacts of individual projects while EIA is not always best placed to address cumulative impacts.

➢ SEA makes the plan-making process effective by facilitating the identification and assessment of alternative plan strategies:

➢ SEA brings awareness of the environmental impacts of plans: while it will not always be possible to eliminate all potentially significant negative effects in balancing policy options, SEA at least helps to clarify the likely consequences of such choices, and makes specific provision for mitigation measures where some negative impacts cannot be avoided.

➢ SEA encourages the inclusion of measurable targets and indicators: which will facilitate effective monitoring of the implementation of the plan, and thus make a positive contribution to subsequent reviews.

➢ SEA design will facilitate the principles of sustainability to be carried down from policies to individual projects

➢ Credible and feasible strategic options allow evaluation of a decision based on comparable rather than in absolute value

1.7.8 Methods and Tools used in SEA for Predicting Environmental and Socio-Economic Effects

1.7.8.1 SEA Methods

SEA is carried-out in two phases: to identify and predict the potential impacts of any new policy/project/plant on the physical, biological, ecological, socioeconomic, cultural environment and on human health.

The first phase is called Initial Environmental and Sustainability Evaluation (IESE) and the second phase is Environmental and Sustainability Impact Studies (ESIS).

ESIS is used to identify and evaluate the environmental and sustainability consequences, both beneficial and adverse impacts to ensure that the environmental and sustainability impacts were taken into consideration in organization's planning and decision making process.

The following are a list of some of the SEA methods and techniques adopted for the sustainable project formulation:

1. Expert judgment and stakeholders' sentiments
2. Checklist and matrices
3. Multi criteria analysis
4. Case comparisons
5. Simulation models
6. Software and information system
7. Questionnaires

8. Group discussions
9. Delphi approach
10. Flow charts and decision trees
11. Contingency analysis
12. Overlays
13. Fuzzy logics.

1.7.8.2 Computation Tools for SEA

Modelling or forecasting of direct environmental effects

➢ Matrices and network analysis
➢ Participatory or consultative techniques
➢ Geographical information systems (GIS) as a tool to analyze, organize and present information.

The following are some important steps included in SEA to ensure stakeholder participation

➢ Stakeholder analysis to identify those affected and involved in the policy, plan or program decision
➢ Consultation surveys
➢ Consensus building processes
➢ Tools for analyzing and comparing options
➢ Scenario analysis and multi-criteria analysis
➢ Risk analysis or assessment
➢ Cost benefit analysis
➢ Opinion surveys to identify priorities (34).

As per the context of the SEA the above tools can be selected by each of the broad general categories being represented in the mix.

1.7.9   SEA and EIA

SEA proactively examines a wide range of alternatives for policies, plans and programs and selects the preferred course of action with a broader environmental and planning vision in mind. In contrast, EIA is more confined and concrete in focus determining the likely environmental impacts of a particular project or development. EIA examines some alternatives for strategic project activities and mitigation measures to address environmental concerns/risks likely to arise due to project /development.

SEA is a flexible concept than EIA allowing for a more comprehensive and forward looking assessment of environmental concerns at the policy, planning and program level (32-34).

Thus EIA analyses, environmental impacts on certain locations within a given time, whereas SEA covers the environmental impacts of a wide range of project, strategic activities over a large geographic areas as an institutionalized part of decision making on a long term basis. For better understanding of Environmental impacts, vertical integration of

SEA and EIA for various environmental issue/concerns are considered at the policy/planning level and flow down to project level (35).

In descriptions of the relationship between SEA and EIA, EIAs are often described as being nested within a particular SEA. In practice and in the past this has not always happened, with EIAs for specific projects often occurring in the absence of a broader environmental vision for the particular marine region and associated activities and industries.

1.7.9.1 SEA and EIA Relationship

Several authors have discussed comparisons in terms of advantages and disadvantages of SEA with respect to EIA (36-39) which justify the necessity of SEA before any policy/planning decision implementation. The main differences in EIA and SEA are tabulated in Table 1.7.

Table 1.7 Main differences between SEA and EIA (40)

Nature of Action	SEA	EIA
Focus	Critical decision moments (decision windows) along decision processes	Construction/operation actions
Level of decision	Policy, planning	Project
Relation to decision	Facilitator	Evaluator, often administrative requirement
Alternatives	Spatial balance of location, technologies, fiscal measures, economic, social or physical strategies	Specific alternative locations, design, construction, operation
Scale of impacts	Macroscopic, mainly global, national, regional	Microscopic, mainly local
Scope of impacts	Sustainability issues, economic and social issues may be more tangible than physical or ecological issues	Environmental with a sustainability focus, physical or ecological issues, and also social and economic
Time scale	Long to medium term	Medium to short-term
Key data sources	State of the Environment Reports, Local Agenda 21, statistical data, policy and planning instruments	Field work, sample analysis, statistical data
Data	Mainly descriptive but mixed with quantifiable	Mainly quantifiable
Rigor of analysis uncertainty	Less rigor/more uncertainty	More rigor/less uncertainty
Assessment benchmarks	Sustainability benchmarks (criteria and objectives)	Legal restrictions and best practice
Outputs	Broad brush	Detailed
Public perception	Vague/distant	More reactive (NIMBY)
Post-evaluation	Other strategic actions or project planning	Objective evidence / construction and operation

Summary

The general concepts of EIA and the salient features of EIA process are presented in this chapter. In the introduction, the concepts of EIA in different contexts are explained and ten basic principles of EIA are also presented. The criteria to decide which projects need EIA and how to determine is also discussed. Concept of Environmental audit is explained and the importance of scientific knowledge of environmental changes in assessing impacts is explained. As a simple example, various impacts occurring in any land clearing projects which is one of the main activities in any development project are presented. The impact of land clearing activities on climate is also discussed.The various steps involved in the entire EIA procedure are described in detail.

Initial Environmental Examination (IEE) and Full Scale Environmental Impact Assessment which is complimentary tasks of EIA are presented giving the scope and various operational functions.

The various analytical functions to be studied to carry out the full scale EIA of any major project activity like fixing the scope, identification of impacts on ecological sensitive resources, impact prediction, impact evaluation & analysis are discussed with suitable examples.

Systematic approach to be adopted for incorporation of EIA as a planning tool in different phases of major project activities and its advantages are discussed. The usefulness of various components of this approach like environmental base map preparation, delineation of the study area, identification of critical resources likely to have impacts, prediction of impacts, formation of the interdisciplinary study team, preparation of Terms of Reference (TOR), format for the presentation of the EIA report, environmental monitoring and management plan and preparation of draft and final environmental impact statement(EIS), for making EIA as a valuable tool for effectively assessing overall impacts of any major project activity is discussed with examples.

The methodology to be adopted for comparative evaluation of various project alternatives, which are very important in final decision-making, are discussed. A good approach is to compare the build, or action, alternatives first. The least environmentally damaging alternative, with mitigation in place, should be identified.

As a systematic approach for deciding upon right alternatives, it is desirable to use trade-off analysis. Trade-off analysis involves the comparison of a set of alternatives relative to a series of decision-making factors. The basic concepts of tradeoff analysis of various project alternatives and ranking and weighing factors are also discussed.

Strategic Environmental Assessment (SEA) proactively examines environmental impacts of a wide range of alternatives for policies, plans and programs and selects the preferred course of action with a broader environmental and planning vision in mind. This makes SEA as an important tool for planners and decision makers to understand what will happen to an area if there are different land uses. It will try to provide information and analysis of the consequences of different actions and their environmental impacts in the short, medium and long term.

The background and origin of SEA process- and chronological order of development of SEA is discussed. SEA contribution to sustainability and success factors, aims and objectives of SEA, elements of SEA and priority needs for good practice of SEA, benefits of SEA, methods and tools used in SEA for predicting environmental and socioeconomic effects are discussed in detail.

The technical details of Strategic Environmental Assessment (SEA) and in association with Environmental Impact Assessment (EIA) and how it effectively promotes sustainable development is discussed.

In the SEA process how to protect the environment in the planning, economic development and integrating green economy targets into strategic and project-related decision-making are explained.

References

1. EIA guidelines for planning and decision making. U. N. Publications. ST/ESCAP/351, ES.CAP, 1985.

2. Quotes Munn, Robert Edward (1979), Environmental impact assessment:

   (a) Principles and procedures Scope (Chichester), Wiley.

3. McAllister, D.M. (1986), Evaluation in Environmental planning. The MIT press. Cambridge Mass. p 6-7.

4. Selman. P. (1992), Environmental Planning. Paul Chapman. London, p.176.

5. EIAO Guidance Note No. 1/2010 EIA Ordinance Register Office of EPD

6. David P. Lawrence (2003), Environmental Impact Assessment-Practical Solutions to Recurrent Problems; Published by John Wiley & Sons, Inc., Hoboken, New Jersey.

7. Duffy, P. (1998), Environmental Impact Assessment Training for Sustainable Agriculture and Rural Development: A Case in Kenya. SD Dimensions, FAO, Rome (also available at http://www.fao.org/sd/epdirect/epan0012.htm).

8. C. Jones, Baker, M, Carter, J, Jay, S, Short, M and Wood, CM (eds) (2005), Strategic environmental assessment and land use planning: Sweden. 15 pp. Strategic Environmental Assessment and Land Use Planning: an International Evaluation.

9. Duffy, P. & DuBois, R. (1999), Environmental Impact Guidelines. FAO Investment Centre. 12 pp. Published

10. Duffy, P. & Tschirley, J. (2000), Use of Environmental Impact Assessment in Addressing Chronic Environmentally Damaging Agricultural and Rural Development Practices: Examples from Kenya and Cambodia. Impact assessment and project appraisal, Vol. 18, no. 2, pages 161–167.

11. Deo, R. C., J. I. Syktus, C. A. McAlpine, P. J. Lawrence, H. A. McGowan, and S. R. Phinn. (2009), Impact of historical land cover change on daily indices of climate extremes including droughts in eastern Australia. Geophysical Research Letters 36.

12. Maron, M., B. Laurance, R. L. Pressey, C. P. Catterall, J. Watson, and J. Rhodes (2015), Land clearing in Queensland tripes after policy ping pong, <https://theconversation.com/land-clearing-in-queensland-triples-after-policy-ping-pong-38279>.

13. McAlpine, C., J. Syktus, R. Deo, J. Ryan, G. McKeon, H. McGowan, and S. R. Phinn. (2009), An Australian continent under stress: A conceptual overview of processes, feedbacks and risks associated with interaction between increased land use pressures and a changing climate. Global Change Biology.

14. McAlpine, C. A., J. Syktus, R. C. Deo, P. J. Lawrence, H. A. McGowan, I. G. Watterson, and S. R. Phinn. (2007), Modeling the impact of historical land cover change on Australia's regional climate. Geophysical Research Letters 34.

15. Pitman, A. J., G. T. Narisma, R. A. Pielke, and N. J. Holbrook. (2004), Impact of land cover change on the climate of southwest Western Australia. Journal of Geophysical Research: Atmospheres 109.

16. Taylor, M., and C. R. Dickman. (2015), Bushland destruction rapidly increasing in Queensland, New South Wales Native Vegetation Act saves Australian wildlife,

17. UNEP. EIA Training resource manual, 2nd edition
(available at http://www.unep.ch/etb/publications/eiaman2edition

18. World Bank. (1991), Environmental assessment sourcebook, three volumes and updates. Environment department technical papers 139–140, Washington, DC, USA.

19. World Bank. (2006), Environmental impact assessment regulations and strategic environmental assessment requirements: Practices and lessons learned in east and southeast Asia. Environment and social development Safeguard dissemination note no. 2.

20. Leonard Ortolano & Anne Shepherd (1995), Environmental Impact assessment: Challenges and Opportunities, Impact Assessment, 13(1): 3-30 DOI:10.1080/07349165.1995.9726076

21. Burris, R. K. and L. W. Canter (1997), Cumulative impacts are not properly addressed in environmental assessments. Environmental Impact Assessment Review 17(1): 5-18.

22. Hegmann, G., C. Cocklin, R. Creasy, S. Dupuis, A. Kennedy, L. Kingsley, H. Ross and D.Stalker (1999), Introduction to environmental impact assessment. London, Spon.

23. Hilding-Rydevik, T. and M. Fundingsland (2005), Cumulative Effects Assessment, Practitioners Guide. C. E. A. Agency. Hull, Quebec. Earthscan, London.112

24. Kvale, S. (1997). Den kvalitativa forskningsintervjun An introduction to qualitative research interviewing). Lund, student litterateur.

25. MacDonald, L. H. (2000). "Evaluating and Managing Cumulative Effects: Process and Constraints." Environmental Management 26(3): 299-315.

26. Olausson, I., A. Oscarsson and I. Palm (2004). MKB för detaljplan - användning och kvalitet (EIA of Detailed Development Plans - Application and Quality). Swedish EIA Centre, Swedish University of Agricultural Sciences, Uppsala.

27. Verheem and Tonk (2000), "Strategic environmental assessment: one concept, multiple forms", Impact Assessment and Project Appraisal 18(3):177-182.

28. Bram F. Noble (2000), "Strategic environmental assessment: What is it? & what makes it strategic?", Journal of Environmental Assessment Policy and Management 2(2): 203–224.

29. Therivel, R.; Wilson, E.; Thompson, S.; Heaney, D. and D. Pritchard, (1992), Strategic Environmental Assessment. London, Earthscan

30. Sadler, B. and R. Verheem, (1996), Strategic Environmental Assessment - status, challenges and future directions. The Hague. Ministry of Housing, Spatial Planning and the Environment of the Netherland

31. Partidário, M. R., (1999), Strategic Environmental Assessment - principles and potential, ch 4, in Petts, Judith (Ed.), Handbook on Environmental Impact Assessment, Blackwell, London: 60-73.

32. Directive 2001/42/EC of the European Parliament and of the Council of 27 June 2001 on the assessment of the effects of certain plans and programs on the environment.

33. Mandy Elliott (2014), Environmental Impact Assessment in Australia. Theory and Practice 6th Edition (Federation Press, Sydney).

34. UK-DETR, International Seminar on SEA, Lincoln, May 1998

35. OECD (2006), Applying Strategic Environmental Assessment. Good Practice Guidance for Development Co-operation.

36. Clark, R. (2000), Making EIA Count in Decision-Making, in Partidário and Clark (eds). 2000: 15-27.

37. Wood, C. and Djeddour, M., (1992), Strategic Environmental Assessment: EA of Policies, Plans and Programs. Impact Assessment Bulletin 10 (1): 3-21.

38. Lee, N. and F. Walsh, 1992, "Strategic Environmental Assessment: an overview", Project Appraisal, 7(3): 126-136.

39. EC-DGVII, (1997), Common Methodology for Multi-modal Trans-European Transport Networks (COMMUTE) - Deliverable 1. COMMUTE-MEET Consortium. Brussels.

40. Partidário, M.R., (2000), Elements of an SEA framework – improving the added-value of SEA, Environmental Impact Assessment Review, 20(6): 647-663.

Further Reading

1. Meinhard Doelle (2009), "Role of Strategic Environmental Assessments in Energy Governance: A Case Study of Tidal Energy in Nova Scotia's Bay of Fundy", Journal of Energy and Natural Resources Law 27(2):112-144.

2. Meinhard Doelle, Nigel Bankes and Louie Porta (2013), "Using Strategic Environmental Assessments to Guide Oil and Gas Exploration Decisions: Applying Lessons Learned from Atlantic Canada to the Beaufort Sea" Review of European Community and International Environmental Law (RECIEL), 22(1): 103-116.

3. Fanny Douvere (2008), "The importance of marine spatial planning in advancing ecosystem based sea use management", Marine Policy, 32(5): 762-771.

4. Robert B. Gibson, Meinhard Doelle and John Sinclair (2016), "Fulfilling the Promise: Basic Components of Next Generation Environmental Assessment", Journal of Environmental Law and Practice, 29, 257–283.

5. Thomas Greiber and Marissa Knodel (with comments from Robin Warner), "An International Instrument on Conservation and Sustainable Use of Marine Biodiversity in Areas beyond National Jurisdiction. Exploring Different Elements to Consider. Paper VII - Relation between Environmental Impact Assessments, Strategic Environmental Assessments and Marine Spatial Planning" (IUCN, Commissioned by the German Federal Agency for Nature Conservation, 2015).

6. Simon Marsden (2012), "Coordinating Strategic Environmental Assessments of Marine and Terrestrial Plans: Australian Experience in the Sub-Antarctic" in Robin Warner and Simon Marsden (editors), Transboundary Environmental Governance: Inland, Coastal and Marine Perspectives (Ashgate Publishing Limited, Farnham, Surrey).

7. Maria do Rosario Partidario (2012), Strategic Environmental Assessment Better Practice Guide – methodological guidance for strategic thinking in SEA (Portuguese Environment Agency and Redes Energetic as Nacionais, Lisbon).

8. Protocol on Strategic Environmental Assessment to the Convention on Environmental Impact Assessment in a Transboundary Context, adopted 21 May 2003, entered into force 11 July 2010 (Kiev Protocol).

9. Ricardo Roura and Alan Hemmings (2011), "Realising Strategic Environmental Assessment in Antarctica" Journal of Environmental Assessment Policy and Management, 13(3): 483-514.

Questions

1. What is Environmental Impact Assessment (EIA)?

2. List out the ten basic principles that describe the characteristics of EIA.

3. Discuss the conditions that determine which project activities need EIA.

4. How EIA be useful as a planning tool for Environmental Protection in various developmental projects?

5. What is Initial Environmental Examination (IEE)? Why it is necessary before going for final EIA?

6. Explain the various analytical functions of an EIA.

7. Write short notes on (a) Direct impacts, (b) Indirect Impacts, (c) Cumulative Impacts and (d) short term and long term impacts.

8. Discuss various direct and indirect impacts likely to occur for typical (a) Land Clearing Activity and (b) Road Construction Activity.

9. Discuss the main features of Impact Evaluation and Analysis? What should be the important objectives of any effective EIA?

10. Explain what is meant by the terms significance and intensity of an impact. What are the various factors to be considered for assessing the significance of impact of any project activity?

11. What are the critical assessment criteria in any EIA methodology?

12. Explain various steps involved in adopting EIA as a planning tool for any major project activity.

13. Discuss the following terms in an EIA process

    (a) study area (b) base map (c) terms of reference and (d) study team.

14. Explain the criteria for formalizing various alternatives for any project. How do you make a comparative evolution of different alternatives? Explain trade off analysis?

15. Explain the different STEPS to be followed in a total EIA process to assist applicants, stake holders and public.

16. What is Strategic Environmental Assessment (SEA)? Explain how it is useful as a decision support system.

17. Discuss the evaluation of SEA process. Give the Chronological order of development of SEA

18. What are the ideal aims and objectives of any SEA process. Discuss various success factors.

19. Discuss the various elements of SEA and explain priority needs for good practice of SEA.

20. Explain the benefits of any SEA process.

21. Discuss various Methods and Tools used in SEA for predicting environmental and socio-economic effects.

22. Explain the main differences between SEA and EIA.

CHAPTER 2

EIA Methodologies

2.0 Introduction

In this chapter some simple and widely used EIA methods are described along with criteria to be followed for choosing the most appropriate method in a given situation.

2.1 Essential Broad Characteristics of EIA Methods

EIA methodologies are mainly to **identify**, **predict** and **value** environmental changes of an action. Reflected in the sequence of activities or steps, as well on the range of environmental issues considered are physical, chemical, biological, socioeconomic, cultural, landscape values and processes. EIA uses its methods and techniques to quantify or to qualify those changes. All aspects and variables can be measured, the problem is to value them.

Many times an EIA analyst or the person charged with the preparation of an EIA report, is faced with a vast quantity of raw and usually unorganized data. Hence, each technique and method for the evaluation of impacts should have the following qualities and characteristics (1, 2):

1. Systematic in approach;
2. Ability to organize a large mass of heterogeneous data;
3. Able to quantify the impacts;
4. Capable of summarizing the data;
5. The ability to aggregate the data into sets with the least loss of information because of the aggregations;
6. It should have a good predictive capability;
7. It should extract the salient features, and
8. It should finally be able to display the raw data and the derived information in a meaningful fashion.

Each of the different methodologies for the assessment of environmental impacts of development projects has their advantages and disadvantages and their utility for a particular application is largely a matter of choice and judgment of the analyst.

2.2 Task Specific EIA Methods

There is no single best methodology for EIA. Characteristics of a methodology such as types of impacts, projects covered and resources required may be important in some instances while some other in another instance. Only the user can determine which tools may be best fit a specific task. For selecting appropriate method, the following basics need to be considered.

1. *Application*: One has to consider whether the EIA is meant for primarily for a decision or for an information. A decision EIA needs to be the best course of action while an information EIA need to provide impacts of best course of single choice. A decision EIA should consider identification of key issues, quantification of impacts and direct comparison of alternatives while information EIA requires comprehensive analysis and interpretation of significance of the broad spectrum of impacts.

2. *Alternatives*: Two types of alternatives exist: fundamentally different and incrementally differently. For fundamentally different alternatives the impact significance need to be measured against some absolute standards than their direct comparison as the impacts will differ in size and in kind. Incrementally different alternatives require precise quantification as different alternatives require different levels of analysis to discriminate between them.

3. *Public Participation*: Public participation can be either a token review or very critical review of highly concerned public which need substantial preparation. This requires the use of more complex techniques like computer modeling, statistical analysis, with a greater degree of quantification of impact significance incorporating issues of public concern.

4. *Resources*: The extent of availability of skills, time, money, data and computational facilities and actual requirement for each method need to be examined.

5. ***The familiarity of EIA analyst with project area and techniques/tests/actions to be conducted on alternative methods***: Greater familiarity will improve the precision of impact significance assessment.

Some objective criteria also exist in making a final choice and these are stated below under the key areas that involve the assessment process.

2.3 Criteria for the Selection of EIA Methodology

2.3.1 General

(a) *Simplicity:* The methodology should be simple so that the available manpower with limited background knowledge can grasp and adopt it without much difficulty.

(b) *Manpower time and budget constraints:* The methodology should be applied by a small group with a limited budget and under time constraints.

(c) *Flexibility:* The methodology should be flexible enough to allow for necessary modifications and changes through the course of the study.

2.3.2   Impact Identification

(a) *Comprehensiveness*: The methodology should be sufficiently comprehensive to contain all possible options and alternatives and should give enough information facilitate proper decision-making.

(b) *Specificity:* The methodology should identify specific parameters on which there would be significant impacts.

(c) *Isolation of project impacts:* The methodology should suggest procedures for identifying project impacts as distinguished from future environmental changes produced by other causes.

(d) *Timing and duration:* The methodology should be able to identify accurately the location and extent of the impacts on a temporal scale.

2.3.3   Impact Measurement

(a) *Commensurate units*: The methodology should have a commensurate set of units so that comparison can be made between alternatives and criteria.

(b) *Explicit indicators*: The methodology should suggest specific and measurable indicators to be used to qualify impacts on the relevant environmental parameters.

(c) *Magnitude*: The methodology should provide for the measurement of impact magnitude, defined as the degree of the extensiveness of the scale of the impact, as distinct from impact importance, defined as the weighting of the degree of significance of the impact.

(d) *Objective criteria*: It should be based on explicitly stated objective criteria.

2.3.4   Impact Interpretation and Evaluation

(a) *Significance*: The methodology should be able to assess the significance of measured impacts on a local, regional and national scale.

(b) *Explicit criteria*: The criteria and assumptions employed to determine impact significance should be explicitly stated.

(c) *Portrayal of "with" and "without" situation*: The methodology should be able to aggregate the vast amounts of information and raw input data.

(d) *Uncertainty*: Uncertainty of possible impacts is a very real problem in environmental impact assessment. The methodology should be able to take this aspect into account.

(e) *Risk*: The methodology should identify impacts that have a low probability of occurrence, but a high potential for damage and loss.

(f) *Depth of analysis*: The conclusions derived from the methodology should be able to provide sufficient depth of analysis and instill confidence in the users, including the general public.

(g) *Alternative comparison*: It should provide a sufficiently detailed and complete comparison of the various alternatives readily available for the project under study.

(h) *Public involvement*: The methodology should suggest a mechanism for public involvement in the interpretation of the impacts and their significance.

2.3.5 Impact Communication

(a) *Affected parties*: The methodology should provide a mechanism for linking impacts to specific effected geographical or social groups.

(b) *Setting description*: It should provide a description of the project setting to aid the users in developing an adequate comprehensive overall perspective.

(c) *Summary format*: It should provide the results of the impact analysis summarized in a format that will give the users, who range from the common public to the decision makers, sufficient details to understand it and have confidence in its assessment.

(d) *Key issues*: It should provide a format for highlighting the key issues and impacts identified in the analysis.

(e) *Compliance*: One of the most important factors in choosing a methodology is whether it is able to comply with the terms of reference established by the controlling agency.

2.4 EIA Methods

Nearly 22 EIA methodologies (tools) are in use over the last 25 years to meet the various activities required in the conduct of an environmental impact study. Table 2.1 provides the summary of these 22 methods and the time period of Emphasis and specific activities they are used **(1)**

Table 2.1 Summary of 22 EIA methods and their study activities (1)

Types of Methods in EIA	Define Issues (Scoping)	Impact Identification	Describe Affected Environment	Impact Prediction	Impact Assessment	Decision Making	Communication of Results
Analogs (look-alikes) (case studies)	X	X		X	X		
Checklists (simple, descriptive, questionnaire)		X	X				X
Decision-focused checklists (MCDM, MAUM, DA, Scaling/rating/ranking, weighting)					X	X	X
Environmental cost-benefit analysis				X	X	X	
Expert opinion (professional Judgment, Delphi, adaptive environmental assessment, simulation modeling)		X		X	X		
Expert systems (Impact identification, prediction, assessment, decision making)	X	X	X	X	X	X	
Indices or indicators	X		X	X	X		X
Laboratory testing and scale models		X		X			
Landscape evaluation			X	X	X		
Literature review		X		X	X		
Mass balance calculations (inventories)				X	X		X
Matrices (simple, stepped, cross-impact, scoring)	X	X		X	X	X	X
Monitoring (baseline)			X		X		
Monitoring (field studies of receptors near analogs)				X	X		

Table 2.1 *Contd...*

Types of Methods in EIA	Define Issues (Scoping)	Impact Identification	Describe Affected Environment	Impact Prediction	Impact Assessment	Decision Making	Communication of Results
Networks (impact trees/chains, cause/effect or consequence diagrams)		X	X	X			
Overlay mapping via GIS			X	X	X		X
Photographs/Photomontages (Historical and current)			X	X			X
Qualitative modeling (conceptual)			X	X			
Quantitative modeling (media, ecosystem, visual, archaeological, socio-economic, and simulation)			X	X			
Risk assessment (relative or quantitative and probabilistic)	X	X	X	X	X		
Scenario building				X		X	
Trend extrapolation			X	X			

2.4.1 Details of some Important EIA methods of Wide Usage

The following are the important methodologies of recent utility for assessing the impacts of development activities on the environment:-

1. Adhoc methods
2. Checklists methods
3. Matrices methods
4. Networks methods
5. Overlays methods
6. Geographic information systems (GIS)
7. Task-specific computer modeling-Simulation Modeling Workshops
8. Expert systems
9. Cost/benefit analysis
10. Environmental Medium Quality Index factor analysis Method
11. Rapid Assessment of Pollution Sources Method
12. Predictive or Simulation methods

Impact assessment methodologies range from simple to complex and are also progressively changing from a static, piecemeal approach to the one that reflects the dynamism of nature and the environment (3). Consequently, the trend is away from the mere listing of potential impacts towards more complex modes whereby the methodology can identify feedback paths, higher order impacts than merely those apparent, first order ones, and uncertainties. In short, the methodological trend is approaching an overall management perspective requiring different kinds of data in different formats and varying levels of expertise and technological inputs for correct interpretation. It is important to understand their drawbacks in order to determine which of the methods are most appropriate. An evaluation of various methodologies (4) is presented in Table 2.2.

Table 2.2 Summary of current EIA methodology evaluation.

Criteria	Check lists	Over-lay	Net-work	Matrix	Environ-mental index	Cost/benefit analysis	Simulation modeling workshop
1. Comprehensiveness	S	N	L	S	S	S	L
2. Communicability	L	L	S	L	S	L	L
3. Flexibility	L	S	L	L	S	S	L
4. Objectivity	N	S	S	L	L	L	S
5. Aggregation	N	S	N	N	S	S	N
6. Replicability	S	L	S	S	S	S	S
7. Multi-function	N	S	S	S	S	S	S
8. Uncertainty	N	N	N	N	N	N	S
9. Space-dimension	N	L	N	N	S	N	S
10. Time-dimension	S	N	N	N	S	S	L

Table 2.2 Contd...

Criteria	Check lists	Over-lay	Net-work	Matrix	Environ-mental index	Cost/benefit analysis	Simulation modeling workshop
11. Data requirement	L	N	S	S	S	S	N
12. Summary format	L	S	S	L	S	L	L
13. Alternative comparison	S	L	L	L	L	L	L
14. Time requirement	L	N	S	S	S	S	N
15. Manpower requirement	L	S	S	S	S	S	N
16. Economy	L	L	L	L	L	L	N

Legend : L = Completely fulfilled, or low resource need.

S = Partially fulfilled, or moderate resource needs.

N = Negligibly fulfilled, or high resource need.

Source: *Environmental Impact Assessment: Guidelines for Planners and Decision Marker, UN Publication S1/1 SCAP/351/ESCAP, 1985 (4)*

2.4.2   Ad hoc Methods (3-10)

Ad hoc methods indicate broad areas of possible impacts by listing composite environmental parameters (Ex: flora and fauna) likely to be affected by the proposed project and each parameter is considered separately and the nature of impacts (long term or short term, reversible or irreversible) are considered.

Types of Ad hoc methods are:

1. Opinion poll

2. Expert opinion and

3. Delphi methods

Ad hoc methods involve assembling a team of specialists to identify impacts in their area of expertise. In this method, each environmental area, such as, air, and water, is taken separately and the nature of the impacts, such as, short-term or long term, reversible or irreversible are considered. Ad hoc methods are for rough assessment of total impact given the broad areas of possible impacts and the general nature of these possible impacts. For example, the impacts on animal and plant life may be stated as significant but beneficial.

In the ad hoc methods, the assessor relies on intuitive approach and makes a broad-based qualitative assessment. This method serves as a preliminary assessment which helps in identifying more important areas like

1. Wildlife
2. Endangered species
3. Natural vegetation
4. Exotic vegetation
5. Grazing
6. Social characteristics
7. Natural drainage
8. Groundwater
9. Noise
10. Air Quality
11. Visual description and services
12. Open space
13. Recreation
14. Health and safety
15. Economic values
16. Public facilities

The adhoc methods, while being very simple can be performed without any training, it is merely presenting the relevant information on a project's impact on the environment without any sort of relative weighting or any cause-effect relationship. It provides minimal guidance for impact analysis while suggesting broad areas of possible impacts. It does not give the actual impacts on specific parameters that will be affected.

The adhoc method has the following drawbacks:

(a) It gives no assurance that it encompasses a comprehensive set of all relevant impacts;

(b) It lacks consistency in analysis as it may select different criteria to evaluate different groups of factors; and,

(c) It is inherently inefficient, as it requires a considerable effort to identify and assemble an appropriate panel for each assessment.

As the expert judgement in assessing the primary impacts is done in an ad hoc manner, it cannot be replicated, making it to review or analyze the conclusions in EIA. As a considerable amount of information about the social, economic, biological and physical environment is to be collected and analyzed in EIA of any project activity ad hoc methods fail to do this in any meaningful way.

Due the above drawbacks, it is not recommended as a method for impact analysis. It is after all ad hoc method and has utility only when other methods cannot be used for lack of expertise, resources and other necessities.

2.4.3 Checklist Methodologies (4-7)

Introduction

Checklist methodologies range of listings of environmental factors in highly structured approaches involving importance weightings for factors and application of scaling techniques for the impacts of each alternative on each factor.

Checklists in general are strong in impact identification and are capable of bringing them to the attention and awareness of their audiences. Impact identification is the most fundamental function of an EIA and in this respect, all types of checklists, namely simple, descriptive, scaling and weighting checklists do equally well.

Checklists are of four broad categories and represent one of the basic methodologies used in EIA. They are:

(a) ***Simple Checklists***: that are a list of parameters without guidelines provided on how to interpret and measure an environmental parameter.

(b) *Descriptive Checklists*: that includes an identification of environmental parameters and guidelines on how parameter data are to be measured.

(c) *Scaling Checklists*: that are similar to descriptive checklist with the addition of information basis to subjective scaling or parameter values.

(d) *Scaling Weighting Check Lists*: are capable of quantifying impacts.

"Simple checklists" represent lists of environmental factors, which should be addressed; however, no information is provided on specific data needs, methods for measurement, or impact prediction and assessment. "Descriptive checklists" refer to methodologies that include lists of environmental factors, along with information on measurement and impact prediction and assessment.

Scaling and weighting inherent in the latter types of checklists facilitates decision-making. Such checklists, apart from being strong in impact identification, also incorporate the functions of impact measurement and to a certain degree of interpretation and evaluation, and it is those aspects that make them more amenable to decision-making analysis.

But the impact of scaling and weighting is, nevertheless, subjective and this poses the danger that society holds all diverse impacts to be equally important. Further, it implicitly assumes that numerical values assigned to impacts can be derived on the basis of expert knowledge and judgement alone.

Scaling and weighting checklists, while capable of quantifying impacts reasonably well, albeit using subjective estimates, make no provision for assessing dynamic probabilistic trends or for mitigation, enhancement and monitoring programs. Identification of higher order effects, impacts and interactions are outside their scope. But simple and descriptive checklists offer no more than this. They merely identify the possible potential impacts without any sort of rating as to their relative magnitudes.

Methods that involve scaling and weighting and the consequent aggregation remove decision making from the hands of decision makers. Further, they incorporate into one number various intrinsically different impacts and this deprives the decision maker of the possibility of tradeoffs.

In check lists methods, impacts will be tabulated in the form of cells with information either in the descriptive form which give information about the possibility or potential existence of an impact while in the scaling or weighing methods the magnitude or importance of the impact is as shown in Table 2.3.

2.4.3.1 Simple Checklists

Simple checklists represent a valid approach for providing systemization to an EIS and Table 2.3 presents a list of environmental factors to be considered in construction and operational phases. The checklist also includes information on mitigation.

Methods of Identifying Impacts

Table 2.3 Environmental factors to be considered in construction and operating phase.

Check List Method

			Construction Phase			Operating phase	
	Beneficial	Adverse effect	No. effect	Beneficial effect	effect	Adverse effect	No. effect
(A)	**Land Transportation and Construction**						
	(a) Compaction and settlement						
	(b) Erosion						
	(c) Ground cover						
	(d) Deposition						
	(e) Stability (slides)						
	(f) Stress-strain (earth peaks)						
	(g) Floods						
	(h) Waste control						
	(i) Drilling and blasting						
	(j) Operational failure						
(B)	**Land Use**						
	(a) Open space						
	(b) Recreational failure						
	(c) Agricultural						
	(d) Residential						
	(e) Commercial						
	(f) Industrial						
(C)	**Water Resources**						
	(a) Quality						
	(b) Irrigation						
	(c) Ground water						
(D)	**Air Quality**						
	(a) Oxides (Sulfur, carbon, nitrogen)						
	(b) Particulate matter						
	(c) Chemical						
	(d) Odors						
	(e) Gases						
(E)	**Service System**						
	(a) Schools						
	(b) Police						
	(c) Fire protection						
	(d) Water and power system						

Table 2.3 Contd…

	Construction Phase			Operating phase		
Beneficial	Adverse effect	No. effect	Beneficial effect	Adverse effect	No. effect	
(e) Sewerage system						
(f) Reuse disposal						
(F) Biological conditions						
(a) Wild life						
(b) Trees, shrubs						
(c) Grass						
(G) Transportation systems						
(a) Automobiles						
(b) Truckling						
(c) Safety						
(d) Movement						
(H) Noise and Vibration						
(a) On-site						
(b) Off-site						
(I) Aesthetics						
(a) Scenery						
(1) Structures						

2.4.3.2 Descriptive Checklists

Descriptive checklists are widely used in environmental impact studies. For example, Carstea (5)) developed a descriptive checklist approach for projects in coastal areas. The methodology addresses the following issues, actions, and projects: rip rap placement, bulkheads: groins and jetties, piers, dolphins, mooring piles, and ramp construction; dredging (new and maintenance); outfalls, submerged lines, and pipes; and aerial crossings. For each of the items, environmental impact information was provided on potential changes in erosion, sedimentation, and deposition; flood heights and drift; water quality; ecology; air quality; noise; safety and navigation; recreation; aesthetics; and socio-economics.

Several descriptive checklists have been developed for water resources projects. For example, Canter and Hill (7) suggested a list of about 65 environmental factors related to the environmental quality account used for project evaluation in the United States. For each factor, information is included in its definition and measurement, prediction of impacts, and functional curves for data interpretation (where one was available or easily developed).

A portion of a descriptive checklist containing several factors for housing and other land development projects is shown in Table 2.4. The basis for the estimate column presents a simplified, brief listing of key data models needed, if any, for the factor.

Table 2.4 Descriptive checklist for land development projects.

Factor	Bases for Estimates
I. Local economy	
Public fiscal balance Net change in government fiscal flow (revenue less expenditures)	Public revenues: expected household income, by residential housing type; added property values Public expenditures: analysis of new-service demand, current costs, available capacities by service
Employment Change in numbers and percent Employed, Unemployed, and Underemployed, by skill level	Direct from new business, or estimated from floor space, local residential patterns, expected immigration, current unemployment profiles
Wealth Change in land values	Supply and demand of similarly zoned land, environmental changes near property
II. Natural environment **Air quality** Health Change in air pollution concentrations by frequency of occurrence, and number of people at risk.	Current ambient concentrations, current and expected emissions, dispersion models, population maps
Nuisance Change in occurrence of visual (smoke, haze) or olfactory (odor) air quality nuisances, and number of people affected	Baseline citizen survey, expected industrial processes, traffic volumes
Water quality Changes in permissible or Tolerable water uses, and Number of people affected for each relevant body of water	Current and expected effluents, current ambient concentrations, water quality model
Noise Change in noise levels and infrequency of Occurrence, and Number of people bothered.	Changes in nearby traffic or other noise sources and in noise barriers; noise-propagation model or nomographs relating noise levels to traffic, barriers, etc.; baseline citizen survey or current satisfaction with noise levels

2.4.3.3 Important Characteristics of Simple and Descriptive Checklists

1. Simple and descriptive checklists consider environmental factors and/or impacts, which can be helpful in planning and conducting an EIS, particularly if one or more checklists for the specific project type can be utilized.

2. Published agency checklists and/or project specific checklists represent the collective professional knowledge and judgement of their developers; hence, they have professional credibility and usability.

3. The Checklists provide a structured approach for identifying key impacts and/or pertinent environmental factors for consideration in impact studies. More-extensive lists of factors of impacts do not necessarily represent better lists, since relevant factors or impacts will need to be selected. Checklists can be easily modified (items can be added or deleted) to make them more pertinent to particular project types in giving locations.

4. Checklists can be used to stimulate or facilitate interdisciplinary team discussions during the planning, conduction, and/or summarization of EISs.

5. In using a checklist, it is important to carefully define the utilized spatial boundaries and environmental factors. Any special impact codes or terminology used within the checklist should also be defined.

6. Documentation of the rationale basics for identifying key factors and/or impacts should be accomplished. In this regard, factor-impact quantification and comparison to pertinent standards can be helpful.

7. Factors and/or impacts from a simple or descriptive checklist can be grouped together to demonstrate secondary and tertiary impacts and/or environmental system interrelationships.

8. Important weights could be assigned to key environmental factors or impacts; the rationale and methodology for such importance weight assignments should be clearly delineated.

9. Key impacts, which should be mitigated, can be identified through the systematic usage of a simple or descriptive checklist.

2.4.3.4 Scaling Checklists

Simple and descriptive checklists in general are strong in impact identification and are capable of bringing them to the attention and awareness of their audiences. Impact identification is the most fundamental function of an EIA and in this respect, all types of checklists, simple, descriptive scaling and weighting checklists do well. But simple and descriptive checklists offer no more than this. They merely identify the possible potential impacts without any sort of rating as to their relative magnitudes. As a result, they are most applicable at the IEE stage of an assessment.

The Oregon Scaling Checklist methods go a step further and provide an idea of the nature of the impact by means of assigning a textual rating of the impact as long-term, direct, and so on. Nevertheless, this approach is not suitable for impact measurement and does not aid much in the decision - making process. Rather, it identifies the impacts and leaves the interpretation to the decision makers.

The element of scaling and weighting that is inherent in the latter types of checklists makes it easier for decision-making. Such checklists, apart from being strong in impact identification, also incorporate the functions of impact measurement and with a certain degree those of interpretation and evaluation and it is these aspects that make them more amenable to decision-making analysis.

Scaling and weighting checklists, while capable of quantifying impacts reasonably well, albeit using subjective estimates make no provision for assessing dynamic probabilistic trends or for mitigation, enhancement and monitoring programs. Identification of higher order effects, impacts and interactions are outside their scope.

Methods that involve scaling and weighting and the consequent aggregation remove decision-making from the hands of decision-makers. Further, they incorporate into one number various intrinsically different impacts and this deprives the decision-maker of the possibility of trade-offs.

2.4.3.4.1 Weighting and Scaling Checklist Methods

As descriptive checklists cannot rank various alternatives, various methods were developed for selecting alternatives based on the following criteria

1. An appropriate set of environmental factors which are likely to be significant for the activity for which EIA has to be carried out are to be fixed (for example, wildlife, habitat etc.):

2. The units of measurement for each factor (e.g., hectares conserved) have to be determined

3. Data on a fixed unit (100 or 1000 hectares) with reference to various sets of environmental factors have to be collected

4. The interval scale (0-0.1) for each environmental factor has to be fixed and the data is converted into an environmental factor index by normalizing the scale over the maximum and minimum values and determining the weight of each environmental factor.

5. Establish the method of aggregation across all the factors established.

The following example where two factors (Wildlife habitat in hectares and employment increase in jobs) for two alternatives are considered will explain how scaling, weighing method can be applied. In this example the environmental factor data have been scaled to an index (0 is worst and 1 is best). Scaling was done by dividing the factor data by maximum values for both alternatives. Two aggregation methods were followed:

(a) Assuming all factors are equally weighted, the following simple addition indicates alternative 2 should be preferred.

(b) In weighing scale weights of 0.8 for employment and 0.2 on wildlife make first alternative preferable (Table 2.5).

Table 2.5 Addition and weighting of factor indices for two alternatives.

Factors	Weights	Alternative one			Alternative two		
		Raw data	Scaled	Weighted	Raw data	Scaled	Weighted
Wildlife Habitat preserved (ha.)		5000			10000		
Employment increase (jobs)		5000			3000		
Wildlife Habitat index	1		0.5			1	
Employment increase index	1		1			0.6	
Wildlife habitat weighted index	0.2			0.1			0.2
Employment increase weighted index	0.8			0.8			0.48
Grand index		n/a	1.5	0.9	n/a	1.6	0.66

For preparing checklists, information, expertise at different levels are required. While simple checklists require information about impacts on general environmental factors, scaling weighing checklists require more detailed expert knowledge.

The assumptions made with respect to:

(a)   Environmental factors under consideration

(b)   The methodology followed for calculating the index

(c)   Weightage assigned to each factor

(d)   Aggregation methods adopted across all factors will make weighing scaling check lists methods to differ one from the other.

The various types of scales used in EIA methods are presented in Table 2.6

Table 2.6 Different Scales Used in EIA Methods.

Scale	Nature of scale	Examples	Permissible mathematical Transformation	Measure of location	Permissible statistical analysis
Nominal	Classifies objects	Species classification, coding soil types	One-to-one substitution	Mode	Information statistics
Ordinal	Ranks objects	Orderings: - minimum to maximum - worst to best - minor to major	Equivalence to non-monotonic functions	Median	Non parametric
Interval	Rates objects in units of equal difference	times (hours) temperature (degrees)	Linear transformation	Arithmetic mean	Parametric
Ratio	rates objects in Equal difference and equal ratio	height, weight	multiplication or geometric division by a constant or other ratio scale value	Arithmetic mean	Parametric mean

(**Source** : Westman 1985-5).

It is very important to understand which scale has to be used in dealing with different types of data. Nominal scales are used when dealing with descriptive information which is categorized while evaluative information is analyzed by ordinal, interval or ratio scales and interval and ratio scales are used to aggregate information into an overall grand index the scale used should be properly defined for clarity. To construct environmental quality Dee et. al., 1972 (8) suggested the following procedure:

(a) Data relating to the quality of the environment and various factors have to be collected and arranged the environmental factor scale (x axis) such that low or worst value corresponds to zero on the environmental quality scale (y axis).

(b) The Environmental quality scale has to be divided into equal intervals varying between 0 and 1 and appropriate value of the factor for each interval should be fixed and this process has to be continued until a reasonable curve is obtained.

(c) The above steps a and b have to be repeated by different experts independently such that average values produce group curves.

(d) A review has to be performed if there are large variations.

(e) Steps 'a' to 'd' have to be repeated by different groups of experts for testing reproducibility.

Using this technique, graphs can be constructed for understanding the relationship between factor index and environmental variable.

2.4.3.4.2 Battelle Environment Evaluation System" (EES), was developed by Battelle Laboratories of Columbus, for the US Bureau of Land Reclamation, is an early, weighting-scaling checklist methodology for water-resources projects, which deals with the environmental factors, as shown in Fig. 2.1. This method was specifically intended for use in the assessment of water resources projects, but is potentially applicable to other types of development. Each of the elements will be assigned an importance weight using the ranked pairwise-comparison technique; resultant importance-weight points (PIUs) are shown in Fig. 2.2 by the numbers adjacent to the four environment categories, in the right-hand corner of the boxes representing the intermediate components, and in the parentheses in front of each environmental factor. The higher the number, the greater the relative importance. Impact scaling in the Battelle EES is accomplished through the use of functional relationships for each of the 78 factors (9).

The basic concept of the Battelle EES is that an index expressed in environmental impact units (EIUs) can be developed for each alternative and baseline environmental conditions. The mathematical formulation of this index is as follows:

$$\mathrm{EIU}_i = \sum_{i=1}^{n} \mathrm{EQ}_{ij}\, \mathrm{PIU}_i$$

EIU_j = environmental impact units for j^{th} alternative

EQ_{ij} = environmental-quality-scale value for i^{th} factor and j^{th} alternative

PIU_I = parameter importance units for i^{th} factor

Usage of the Battelle EES consists of obtaining baseline data on the 78 environmental factors and, through the use of their functional relationships, converting the data into EQ scale values.

Checklist Battelle

4 Environmental components:

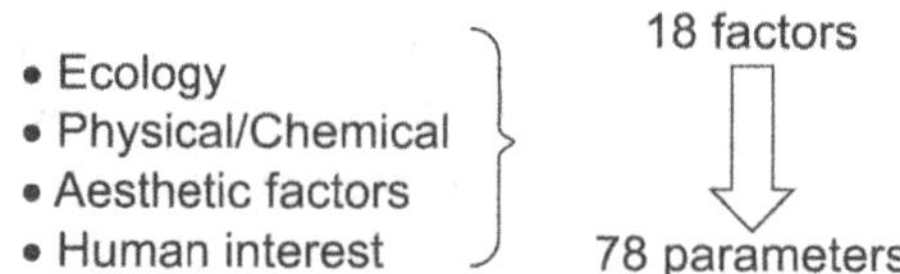

Fig. 2.1 Environmental components of Battelle Check list

Each Parameter has a Weightage or Environmental Importance

These scale values are then multiplied by the appropriate PIUs and aggregated to obtain a composite EIU score for the baseline setting. For each alternative being evaluated, it is necessary to predict the anticipated changes in the 78 factors. The predicted-factor measurements are then converted into EQ scale values using the appropriate functional relationships. Next, these values are multiplied by the PIUs and aggregated to arrive at a composite EIU score for each alternative. This numerical scaling system provides an opportunity for displaying system provides an opportunity for displaying trade-offs between the alternatives in terms of specific environmental factors, intermediate components, and categories. Professional judgement to be exercised in the focus should be on comparative analyses, rather than on specific numerical values. Battelle EES is thus based on a hierarchical checklist of 78 environmental parameters. To overcome the problem of comparing and summing up impacts, parameters were weighted so that at they would be related to each other in terms of relative importance predevelopment parameter estimates are transformed into measures of environmental quality. Providing a quantified representation of environmental quality, which can be used in comparison with the post-impact situation. Environmental quality is scaled from 0 (very bad) to 1 (very good) and can be defined in a number of ways. The transformation of a parameter estimate into environmental quality is achieved by using "value functions" devised by a group of experts. Changes that might occur if development were to proceed are projected using predictive techniques. Projected parameter values are used to determine the environmental quality score. Each parameter quality score is multiplied by the number of "parameter impact units" allotted to that parameter, giving a final score for each parameter in term of ''Environmental impact units''. This method is less cumbersome, but the definition of environmental quality is arbitrary and decided by a panel of experts.

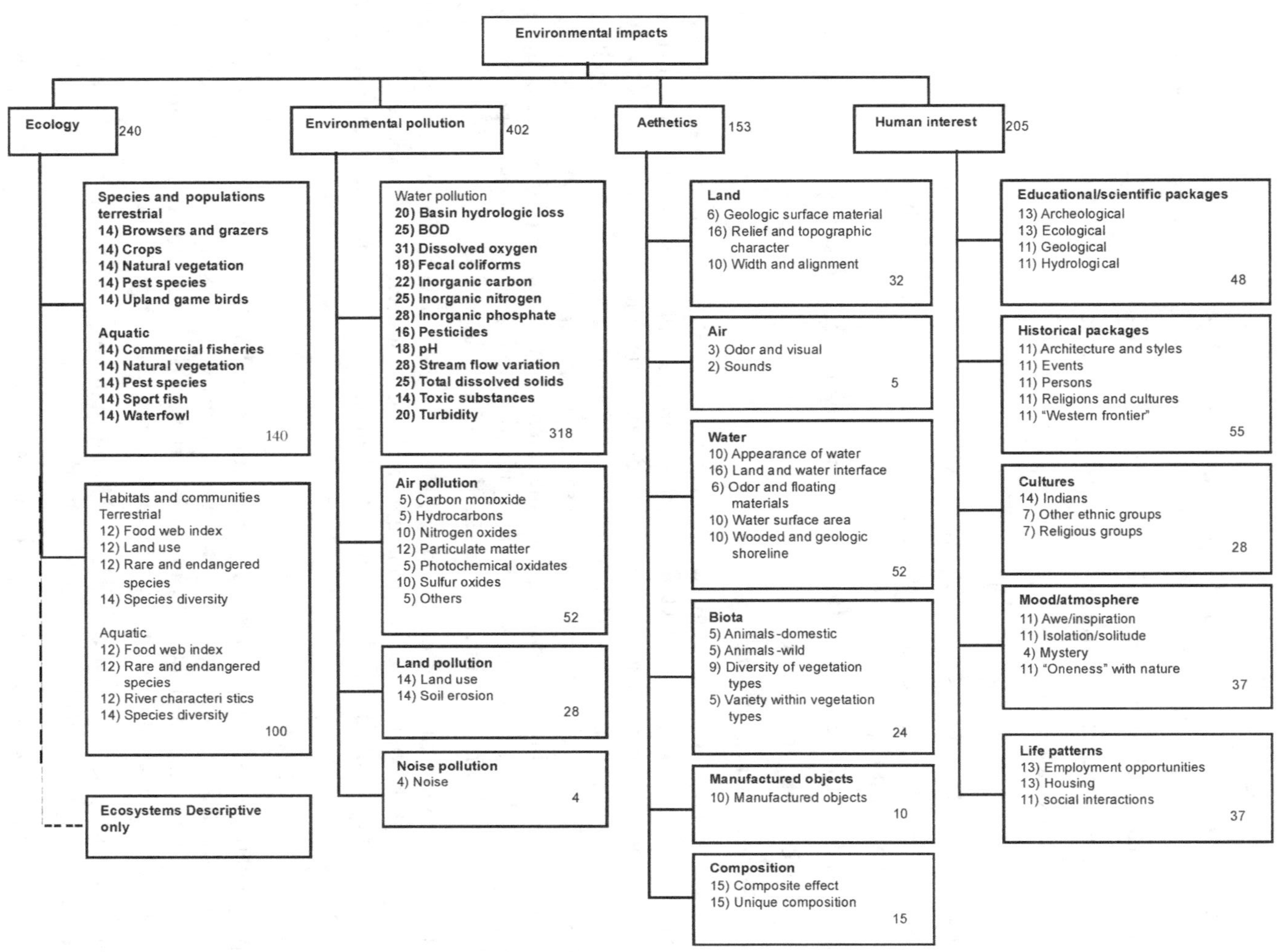

Fig. 2.2 Battelle Environment Evaluation System

Note : Numbers denote PIUs.

2.4.3.5 Advantages of Check List Methods

(a)　Checklists are mainly useful for summarizing information to make it accessible to experts in different fields or decision makers who have little technical knowledge.

(b)　Preliminary analysis will be available in scaling check lists.

(c)　Information on ecosystem functions can be clearly understood from weighing methods.

Some of the drawbacks of checklists are (Westman 1985) (10)

1. They are too general or incomplete

2. They do not illustrate interactions between effects

3. The number of categories to be reviewed can be immense, which will create confusion about significant impacts

4. Involves the identification of effects which are qualitative and subjective.

2.4.4　Matrix Methods (11-14)

2.4.4.1 General Characteristics

In matrix methods, interactions between various activities and environmental parameters will be identified and evaluated. Matrix methods are basically generalized checklists where one dimension of a matrix is a list of environmental, social and economic factors likely to be affected by a project activity. The other dimension is a list of actions associated with development. These relate to both the construction and operational phases. Making cells representing a likely impact resulting from the interaction of a facet of the development with an environmental feature identify impacts. With some matrices qualitative representation of impact importance and magnitude are inserted into individual cells.

Matrices provide cause-effect relationships between the various project activities and their impacts on the numerous environmentally important sectors or components. Matrices provide a graphic tool for display impacts to their audience in a manner that can be easily comprehended.

2.4.4.2 Matrices Types

Simple, interaction matrices largely overcome this limitation. But such matrices are generally useful for depicting ecological interactions only for the sake of documentation. While the scale of the interaction is identified, the individual actions of the project are not correlated with the resulting impact on the environmental components.

The most serious criticism of such weighting matrices, which can also be extended to scaling and weighting checklists, is that

(a)　They require a large amount of information about the environmental components and project activities.

(b)　Through the inherent aggregation process, decision-making is, in effect, removed from the hands of the decision-makers and the public concerned. A great deal of information that is valuable to decision-making is lost in the conversion to number.

(c)　Weights are assigned to environmental components and consequently to impacts without any guarantee that such weights and rating will represent the actual impacts that will be apparent once the project is implemented and operational;

What is generally called an objective procedure, the assignment of weights and the subsequent quantification is, in fact, an arbitrary assignment of scales of "environmental quality" based on the value judgment of "experts".

(d) Aggregation of numerical impacts through suitable transformation functions results in the combination of inherently different items into a single index or number and leads to loss of information about the various impacts from the numerous project actions, thereby precluding the possibility of tradeoffs by the decision makers.

Matrices are strong in identifying impacts and unlike checklists, can also represent higher order effects and interaction. Some of the dynamic nature of impacts can also be identified. They can also provide the functions of impact measurement, interpretation and evaluation, and can communicate the results in an easily understandable format to their audiences. But they cannot compare alternatives in a single format, and different alternatives need to be assessed and presented separately. The purpose of a matrix is to help the project planner to

1. Identify specific sources of potential environmental impact

2. Provide a means of comparing the predicted environmental impacts of the various project options available

3. Communicate in graphic form the (i) Potentially significant adverse environmental impact for which a design solution has been identified (ii) Adverse   environmental impact that is potentially significant, but about which insufficient information has been obtained to make a reliable prediction (iii) Residual and significant adverse environmental impact and (iv) Significant environmental impact

2.4.4.3 Salient Features of Matrices Methods

1. It is necessary to define the spatial boundaries of environmental factors, the temporal phases and specific actions associated with the proposed project; and the impact rating or summarization scales used in the matrix.

2. A matrix should be considered a tool for purposes of analysis, with the key need being to clearly state the rationale utilized for the impact ratings assigned to a given temporal phase and project action, and a given spatial boundary and environmental factor. If a particular activity is likely to cause an effect on any environmental factor, it will be noted at the intersection point in the matrix. Matrices are two dimensional table which will facilitate identification of impacts arising from interaction between certain project activities and specific environmental components (Table 2.7).

Table 2.7 Matrix Components

Environmental Components	Phase I				Phase II	
	Project Activities					
Water Quality						
Air						
Biodiversity						
Standard of Living						

The development of one or more preliminary matrices can be useful as a simple matrix technique in discussing a proposed action and its potential environmental impacts (Table 2.8). This can be helpful in the early stages of a study to assist each team member in understanding the implications of the project and developing detailed plans for more extensive studies on particular factors and impacts.

Table 2.8 Simple Matrices

Environmental Component	Project action				
	Construction				
	Unities	Residential and commercial buildings	Residential buildings	Commercial buildings	Parks and Open spaces
Soil and geology	x	x			x
Flora	x	x			X
Fauna	x	x			X
Air quality				x	
Water quality	x	x	x		
Water quality	x	x	x		
Population density			x	x	
Employment		x		x	
Traffic	x	x	x		
Housing			x		
Community structure		x	x		x

4. The interpretation of impact ratings should be carefully and critically considered, particularly when realizing that there may be large differences in spatial boundaries as well as temporal phases of a proposed project.

5. Interaction matrices can be useful for delineating the impacts of the first and second or multiple phases of a two-phase or multiphase project; the cumulative impacts of a project when considered relative to the other past, present, (Table 2.9) and reasonably foreseeable future actions in the area; and the potential positive effects of mitigation measures.

Table 2.9 Time Dependent Matrices

Environmental Component	Project action				
	Construction (3 Years)		Operation (25 Years, evens out after 4 Years)		
	Unities	Residential and commercial buildings	Residential buildings	Commercial buildings	Parks and Open spaces
Soil and geology	211	321	0000	0000	0001
Flora	221	422	1223	1111	1123
Fauna	221	311	1100	1100	1122
Air quality	000	000	0123	0034	0011
Water quality	010	022	1223	0111	0000
Population density	011	112	2244	0222	0011
Employment	120	342	1111	1334	1111
Traffic	220	332	2233	2333	1111
Housing	010	121	2344	0000	0000
Community structure	010	232	2344	1111	1233

6. If interaction matrices are used to display comparisons between different alternatives, it is necessary to use the same basic matrix in terms of spatial boundaries and environmental factors, and temporal phases and project actions for each alternative being analyzed. Completion of such matrices can provide a basis for tradeoff analysis.

7. Impact qualification and comparisons to relevant standards can provide a valuable basis for the assignment of impact ratings to different project actions and environmental factors.

8. Color codes can be used to display and communicate information on anticipated impacts. For example, beneficial impacts could be shown by using green or shades of green; whereas, adverse effects could be depicted with red or shades of red. Impact matrices can be used without the incorporation of number, letter, or color ratings. For example, circles of varying size could be used to denote ranges of impacts. The magnitude of separate or combined effects and their importance considerations will also be considered.

Table 2.10 Magnitude Matrices

No.	Environmental Items	Rehabilitation of Wadi Gaza into a protected natural National Park								Improvement of Agriculture Infrastructure in the Wadi Gaze Area		
		1	2	3	4	5	6	7	8	9	10	11
Historic and Cultural Heritage												
	Archeological and cultural heritage sites		•	•	☺	☺		•	☺	•		•
Human Health												
	Public and Local Community health risks	☺	☺					☺				☺
	Workers health and safety	•	•	•				☺ •		•		•
	Noise		•					•				•

9. One of the concerns relative to interaction matrices is that project actions and/ or environmental factors are artificially separated, when they should be considered together. It is possible to use footnotes in matrix to identify groups of actions, factors, and/or impacts which should be considered together. This would allow the delineation of primary and secondary effects of projects.

10. The development of a preliminary interaction matrix does not mean that it would have to be included in a subsequent EA or EIS. The preliminary matrix could be used as an internal working tool in the study, planning and development.

11. It is possible to utilize importance weighting for environmental factors and project actions in a simple interaction matrix. If this approach is chosen, it is necessary to carefully delineate the rationale upon which differential importance weights have been assigned. Composite indices could be developed for various alternatives by summing up the products of the importance weights and the impact ratings.

12. Usage of an interaction matrix forces the consideration of actions and impacts related to a proposed project within the context of other related actions and impacts. In other words, the matrix will prevent overriding attention being given to one particular action of environmental factors.

2.4.4.4 Interaction-Matrix Methodologies

In interaction matrix method project actions or activities will be displayed along one axis with appropriate environmental factors listed along the other axis of the matrix.

Simple matrices are:

2.4.4.4.1   Simple Interaction Matrix Method

For a simple interaction matrix method, the one developed by Leopold (15) will serve as an example. In this method approximately 100 specified actions and 90 environmental items can be examined. Table 2.11 presents the list of the actions and environmental items. In the uses of the Leopold matrix, each action and its potential for creating an impact on each environmental item will be considered. Where an impact is anticipated, the matrix is marked with a diagonal line in the appropriate interaction box.

Table 2.11 Action and Environmental Items in Leopold Interaction Matrix

Actions		Environmental items	
Category	Description	Category	Description
(a) Modification of regime	(a) Exotic fauna introduction (b) Biological controls (c) Modification of habitat (d) Alternation of ground (e) Alternation of groundwater hydrology (f) Alternation of drainage (g) River control and flow modification (h) Canalization (i) Irrigation (j) Weather modification (k) Burning (l Surfacing or paving (m) Noise and Vibration	(a) Physical and chemical characteristics 1. Earth 2. Water	(a) Mineral resources (b) Construction material (c) Soils (d) Landform (e) Force fields and background radiation (f) Unique physical (a) Surface (b) Ocean (c) Underground (d) Quality (e) Temperature (f) Recharge (g) Snow, ice, and permafrost
(b) Land transformation and construction	(a) Urbanization (b) Industrial sites and buildings (c) Airports (d) Highways and bridges (e) Roads and trails (f) Railroads (g) Cables and lifts (h) Transmission lines, pipe lines and corridors (i) Barriers, including fencing (j) Channel dredging and straightening (k) Channel revetments (i) Canals	3. Atmosphere 4. Processes	(a) Quality (gases, particulates (b) Climate (micro, macro) (c) Temperature (a) Floods (b) Erosion (c) Deposition (sedimentation, precipitation) (d) Solution (e) Sorption (ion exchange, complexing) (f) Compaction and settling (g) Stability (slides, slumps) (h) Stress-strain (earthquakes) (i) Air movements

Table 2.11 *Cotd…*

Actions		Environmental items	
Category	**Description**	**Category**	**Description**
	(m) Dams and impoundment's (n) Piers, seawalls, marinas, and sea terminals (o) Offshore structures (p) Recreational structures (q) Blasting and drilling (r) Cut and fill (s) Tunnels and underground structures	(B) Biological conditions 1. Flora	(a) Trees (b) Shrubs (c) Grass (d) Crops (e) Microflora (f) Aquatic plants (g) Endangered species (h) Barriers (i) Corridors
(C) Resource extraction	(a) Blasting and drilling (b) Surface excavation (c) Subsurface excavation and retorting (d) Well dredging and fluid removal (e) Dredging (f) Clear cutting and other lumbering (g) Commercial fishing and hunting	2. Fauna	(a) Birds (b) Land animals including reptiles (c) Fish and shellfish (d) Benthic organisms (e) Insects (f) Microfauna (g) Endangered species (h) Barriers (i) Corridors
(D) Processing	(a) Farming (b) Ranching and grazing (c) Feed lots (d) Dairying (e) Energy generation (f) Mineral processing (g) Metallurgical industry (h) Chemical industry		

Table 2.11 *Cotd…*

Actions		Environmental items	
Category	Description	Category	Description
	(i) Textile industry (j) Automobile and aircraft (k) Oil refining (l) Food (m) Lumbering (n) Pulp and paper (o) Product storage	(C) Cultural factors 1. Land	(a) Wilderness and open spaces (b) Wetlands (c) Forestry (d) Grazing (e) Agricultural (f) Residential (g) Commercial (h) Industry (i) Mining and quarrying
(E) Land alteration	(a) Erosion control and terracing (b) Mine sealing and waste control (c) Strip-mining rehabilitation (d) Landscaping (e) Harbor dredging (f) Marsh fill and drainage	2. Recreation	(a) Scenic views and vistas (b) Wildness qualities (c) Open-space qualities (d) Landscape design (e) Unique physical features (f) Parks and reserves (g) Monuments
(F) Resource renewal	(a) Reforestation (b) Wildlife stocking and management (c) Groundwater recharge (d) Fertilization application (e) Waste recycling	3. Aesthetics and human interest	(h) Rare and unique species or eco-systems (i) Historical or archaeological sites and objects (j) Presence of misfits

Table 2.11 *Cotd…*

Actions		Environmental items	
Category	Description	Category	Description
(g) Changes in traffic	(a) Railway (b) Automobile (c) Trucking (d) Shipping (e) Aircraft (f) River and canal traffic (g) Pleasure boating (h) Trails (i) Cables and lifts (j) Communication (k) Pipeline	5. Manufactured facilities and activities	(a) Cultural-patterns (life-style) (b) Health and safety © Employment (d) Population density
(h) Waste emplacement and treatment	(a) Ocean dumping (b) Landfill (c) Emplacement of tailings, spoils, and Overburden (d) Underground storage (e) Junk disposal (f) Oil well flooding (g) Deep well emplacement (h) Cooling water discharge		(a) Structures (b) Transportation network (movement, access) (c) Utility networks (d) Waste disposal (e) Barriers (f) Corridors

Table 2.11 Cotd...

Actions		Environmental items	
Category	Description	Category	Description
	(i) Municipal waste discharge including spray irrigation (j) Liquid effluent discharge (k) Stablization and oxidation ponds (l) Septic tanks, commercial and domestic (m) Stack, and exhaust emission (n) Spent lubricants	(D) Ecological relationships	(a) Salinization of water resources (b) Eutrophication (c) Disease and insect vectors (d) Food chains (e) Salinization of surficial materials (f) Brush encroachment
(I) Chemical treatment	(a) Fertilization (b) Chemical deicing of highways, etc. (c) Chemical stabilization of soil (d) Weed control (e) Insect control (pesticides)		
(J) Accidents	(a) Explosions (b) Spills and leaks (c) Operational failure		
(K) Others			

Source : Compiled using data from Leopold (15)

Actions causing impact

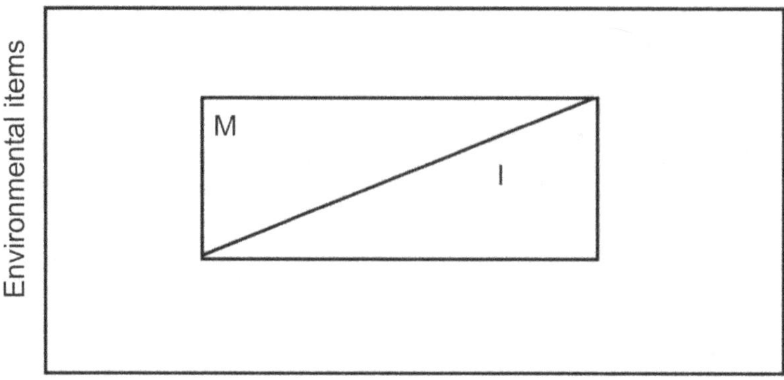

Fig. 2.3 Leopold interaction Matrix : M = magnitude; I = importance (Leopold (15)

Next the interaction in terms of its magnitude and importance will be considered in the method. The "magnitude" of an interaction is its extensity or scale and is described by the assignment of a numerical value from 1 to 10, with 10 representing a large magnitude and 1 a small magnitude. Values near 5 on the magnitude scale represent the impacts of intermediate extensity. Assignment of numerical values for the magnitude of an interaction should be based on an objective evaluation of the facts related to the anticipated impact. The "importance" of an interaction is related to its significance, or an assessment of the probable consequences of the anticipated impact. The scale of importance also ranges from 1 to 10, with 10 representing a very important interaction and 1 an interaction of relatively low importance. Assignment of a numerical importance value is based on the subjective judgement of the expert group, or interdisciplinary team working on the study.

Leopold (15) outlined a procedure for evaluating the environmental impact of development projects. This is an environmental matrix which is primarily checklist designed to show possible interactions between developmental activities and a set of environmental characteristics. One hundred different types of impacts and 88 environmental characteristics are identified in the system, giving a total of 8800 possible interactions, but in practice it can usually be quickly reduced to a fewer number of related items.

Leopold matrix can be expanded or contracted, that is, the number of actions can be increased or decreased from the total of about 100, and a number of environmental factors can be increased or decreased from about 90. The Leopold matrix method is very useful as a gross screening tool for impact identification purposes, and it can provide a valuable means for impact communication by providing a visual display of the impacted items and the major actions causing impacts.

Summation of the number of rows and columns designated as having interactions can offer insight into impact assessment.

Leopold matrix can be employed to identify impacts at various temporal phases of a project; for example, construction, operation, and post-operation phases, and to describe impacts associated with various spatial boundaries of the site and in the region. The Leopold matrix can also be utilized to identify beneficial as well as detrimental impacts through the use of appropriate designators, such as plus and minus signs.

Further, three levels of magnitude and importance can also be assigned to the extent of impact. Major interactions can be assigned a maximum numerical score, with minor interactions being assigned minimal scores. Intermediate-level interactions can be assigned values between the major and minor scores.

Apart from the environmental factors discussed in Table 2.6, other environmental factors oriented to the socioeconomic environment should also be changed in the Leopold matrix method.

Information, expressed by means of ranks other than numerical values for magnitude and importance, can be included in the impact scales associated with identification of an interaction.

Another approach for impact rating in a matrix involves the use of a pre-defined code denoting the characteristics of the impacts and the possibility of mitigating certain undesirable features.

Simple interaction matrices can be used for analyzing the impacts of some other types of projects; examples include flood-control and/or hydropower, highway, transmission line, offshore oil lease, coal mine, power plant, industrial plant, industrial park, pipeline, housing development, tourism, and coastal development projects.

2.4.4.4.2 Procedure for the Development of Specific and Simple Interaction Matrix

It is considered better to develop a specific interaction matrix for the project, plan, program, or policy being analyzed, rather than using a generic matrix. The following steps can be used by an individual or an interdisciplinary team in preparing a simple interaction matrix:

1. List all anticipated project actions and group them according to temporal phase, such as construction, operation, and post-operation.
2. List all pertinent environmental factors from the environmental setting, and group them (a) according to physical-chemical, biological, cultural, and socioeconomic categories, and (b) based on spatial considerations such as site and region, or upstream, and downstream.
3. Discuss the preliminary matrix with study team members and/or advisors to the team or study manager.
4. Decide on an impact-rating scheme (for example, numbers, letters, or colors) to be used.
5. Talk through the matrix as a team, and make ratings and notes in order to identify and summarize impacts (documentation).

2.4.4.4.3 Other Types of Matrix Method

The simple baseline method can also be used to propose other than impact identification. Table 2.12 shows a matrix framework which can be used to summarize baseline environmental conditions. In this example, relative factor's importance, present condition, and extent of management will be taken into consideration.

Table 2.12 Concept of an environmental baseline matrix.

Identification	Evaluation		
Environmental elements/units	**Scale of Importance**	**Scale of present condition**	**Scale of management**
	1 2 3 4 5 low high	**1 2 3 4 5** low high	**1 2 3 4 5** low high

Biological
  Flora
  Fauna
  Ecological relationships
Physical-chemical
  Atmosphere
  Water
  Earth
Cultural
  Households
  Communities
  Economy
  Communications
Bio-cultural linkages/units
  Resources
  Recreation
  Conservation			

Source: Fischer and Davies, (15).

2.4.4.4.4 Stepped Matrix Method

For enclosing secondary method and tertiary impacts of initiating actions, stepped matrix method can be used. In this method, one set of environmental factors is displayed against another set of environmental factors. The effects of initial changes of some factors on the other factors will be evaluated in water reservoir project activities. As many as 92 environmental attributes are listed in Table 2.13.

Table 2.13 Simple and stepped interaction matrix for water resources reservoir projects.

Clearing	**Air quality Microclimate**
Grubbing	(A) Air movement
Stripping	(B) Air temperature
Excavation	(C) Relative humidity
Stockpiling	(D) Incident radiation
Loading-hauling	**Soil conditions**
Placement of materials	(A) Temperature
Grading Compaction	(B) Soil moisture

Table 2.13 *contd….*

Removal of materials	(C) Soil Structure
Blasting Concrete placement	(D) Soil flora
Surfacing Building erection	(E) Soil fauna
Building movement	**Ecological relationships**
Building demolition	(A) Terrestrial ecosystems
Pavement demolition	1. Change in ecostructure
Batch and aggregate plants	2. Trophic structure
Temporary buildings	3. Pollution of land
Vehicle and equipment maintenance	4. Rare or unique ecotypes
Restoration	5. Diversity of ecotypes
Filing reservoir	6. Biogeochemical cycles
Flood-control operation	(B) Aquatic ecosystems
	Fauna
	(A) Terrestrial animals
	1. Mammals
	2. Birds
	3. Other vertebrates
	4. Mosquitoes
	5. Other invertebrates
	6. Rare and endangered species
	7. Species diversity, etc
	8. Nuisance species
	(B) Aquatic animals
	Flora
	(A) Terrestrial plants
	1. Natural vegetation
	2. Rare and endangered species
	3. Species diversity
	4. Primary productivity
	5. Weedy species
	6. Detritus
	B. Aquatic flora Groundwater hydrology
	(A) Depth
	(B) Movement
	(C) Recharge rates
	Surface-water hydrology
	(A) Elevation
	(B) Flow pattern
	(C) Stream discharge
	(D) Velocity
Construction and operation activities (x axis) Environmental attributes (y axis)	
	Land forms and processes
	(A) Compaction of soil
	(B) Topography

Table 2.13 *contd….*

	(c) Stability of land forms
	(D) Water erosion of soil
	(E) Silt deposition
	(F) Wave movement of soil
	(G) Wind movement of soil
	Outdoor recreation
	(A) Land-base
	(B) Water-based
	Preservation of natural resources
	(A) Fauna
	(B) Flora
	(C) Natural ecosystem types
	(D) Open and green space
	(E) Water supply
	(F) Agricultural land
	Special-interest areas Aesthetics
	(A) Air quality
	(B) Construction scars
	(C) Man-made features
	(D) Scenic views
	(E) Landscape diversity
	(F) Vegetation
	(G) Water quality
	(H) Noise
	Surface-water quality
	(A) Physical attributes
	1. Color
	2. Discharge
	3. Redox potential
	4. Turbidity
	5. Water temperature
	(B) Chemical attributes
	1. Carbondioxide
	2. COD
	3. Dissolved oxygen (DO)
	4. Nitrate
	5. Phosphorus
	6. Sulphur

Source : Adapted from Johnson and Bell, (16)

The Matrix methods are flexible and valuable tool for explaining impacts by presenting a visual display of the impacts and their causes and can be employed to identify impacts during various stages of the entire project.

2.4.5 Network Methods

Networks are capable of identifying direct and indirect impacts, higher order effects and interactions between impacts, and hence are able to identify and incorporate mitigation and management measures into the planning stages of a project. They are suitable for expressing ecological impacts but of lesser utility in considering social, human and aesthetic aspects. This is because weightings and ratings of impacts are not features of network analysis. Development of network diagrams (Fig. 2.4) present the potential impact pathways as casual chains will be very useful for displaying first, secondary, tertiary and higher order impacts.

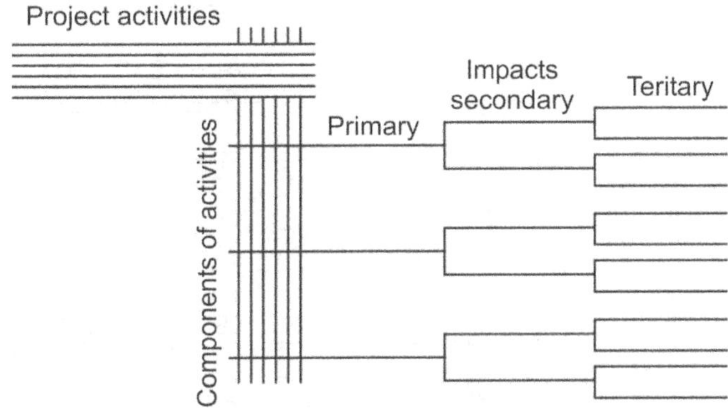

Fig. 2.4 Conceptual model of impact networks.

To develop a network, a series of questions related to each project activity (such as what are the primary impact areas, the primary impacts within these areas, the secondary impact areas, the secondary impacts within these areas and so on) must be answered. In developing network diagram the first step is to identify the first order changes in environmental components. The secondary changes in other environmental components that will result from first order changes will be then identified. In turn third order changes resulting from secondary changes will then be identified. This process will be continued until the network diagram is completed to the experts' satisfaction. Network analyses are particularly useful for understanding the relationship between environmental components that produce higher order impacts, which are often overlooked in some major projects. Networks can also aid in organizing the discussion of anticipated project impacts. Network displays are useful in communicating information about an environmental impact study to an interested public.

2.4.5.1 Stepped Matrix Technique For Networks

This technique developed by Sorenson (1971) (17) was applied to Nong Pla Reservoir (Fig. 2.5).

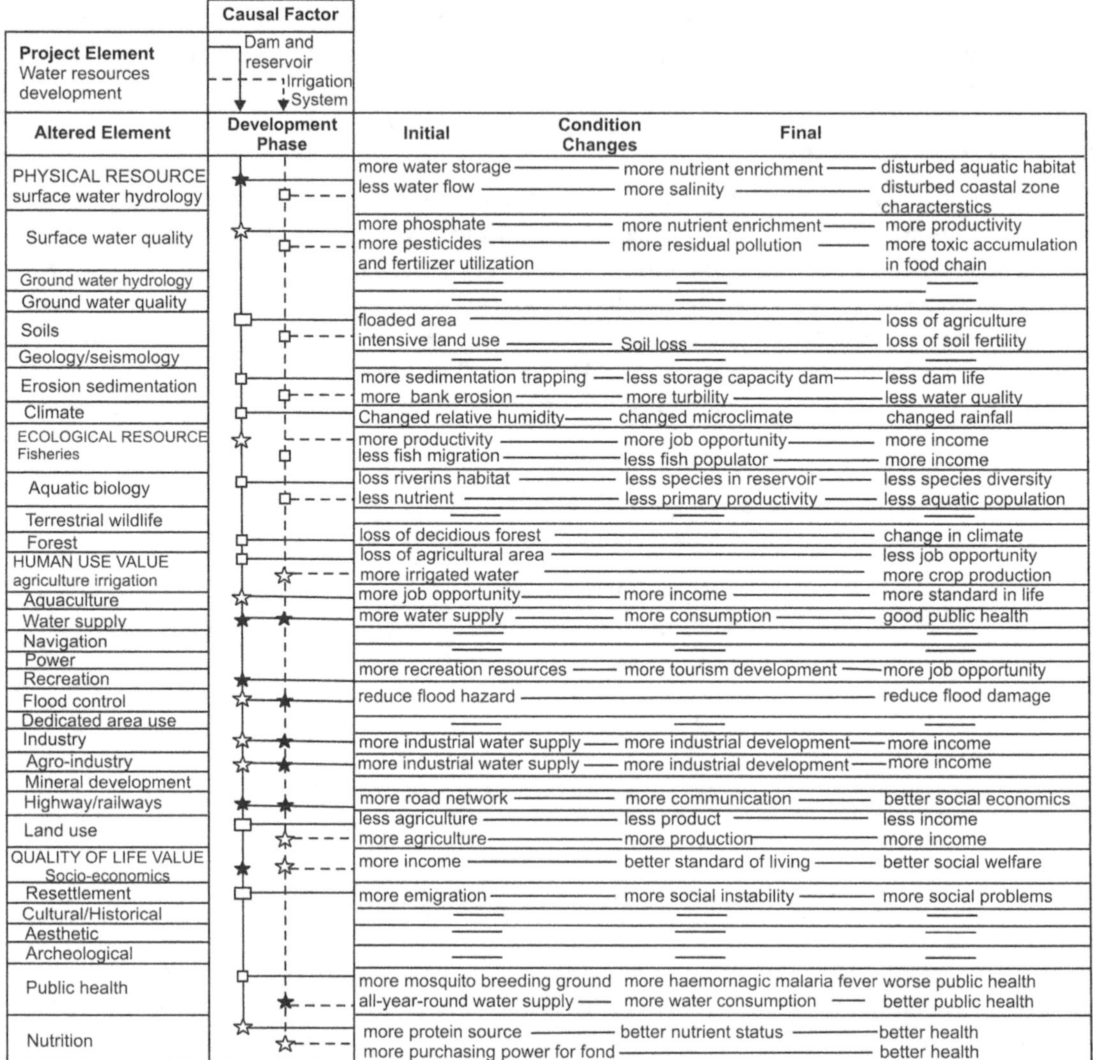

Fig. 2.5 Stepped matrix for Nong Pla reservoir.

The interpretation of results in the above figure will be as follows:

(a)   At the upper left hand corner the project elements are entered while the casual factor that may result is shown as an impact under the dam and reservoir.

(b)   Reading downwards the impacts are presented as **Major positive, *minor Positive, ##major negative, #minor negative.

Fig. 2.6 presents the network diagram for a dredging project (Sorenson 1971) (17) while Fig. 2.7 presents the network impacts of a Pulp Mill (Lohani and Halim 1983) (18).

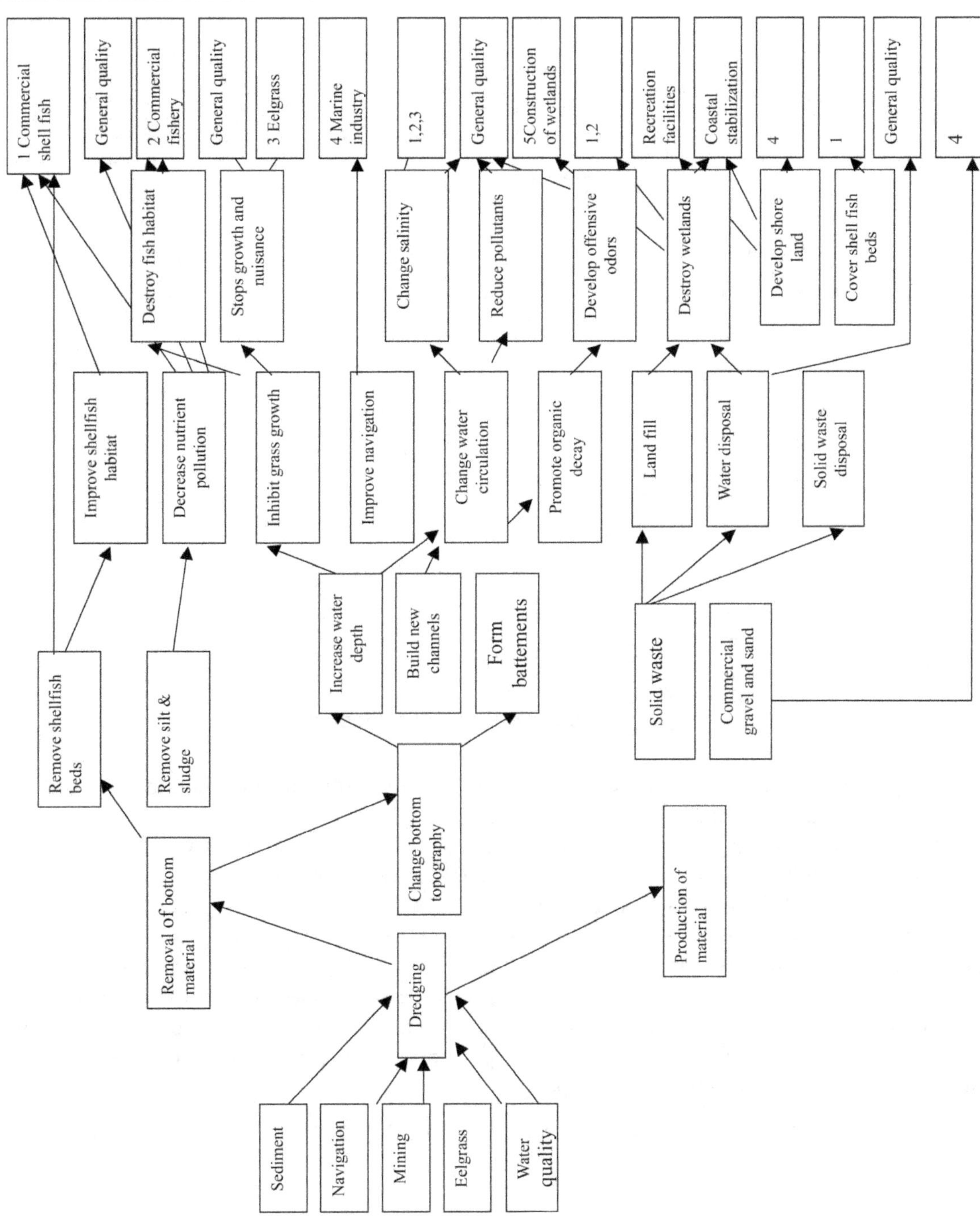

Fig. 2.6 Network analysis of impacts of a dredging project.

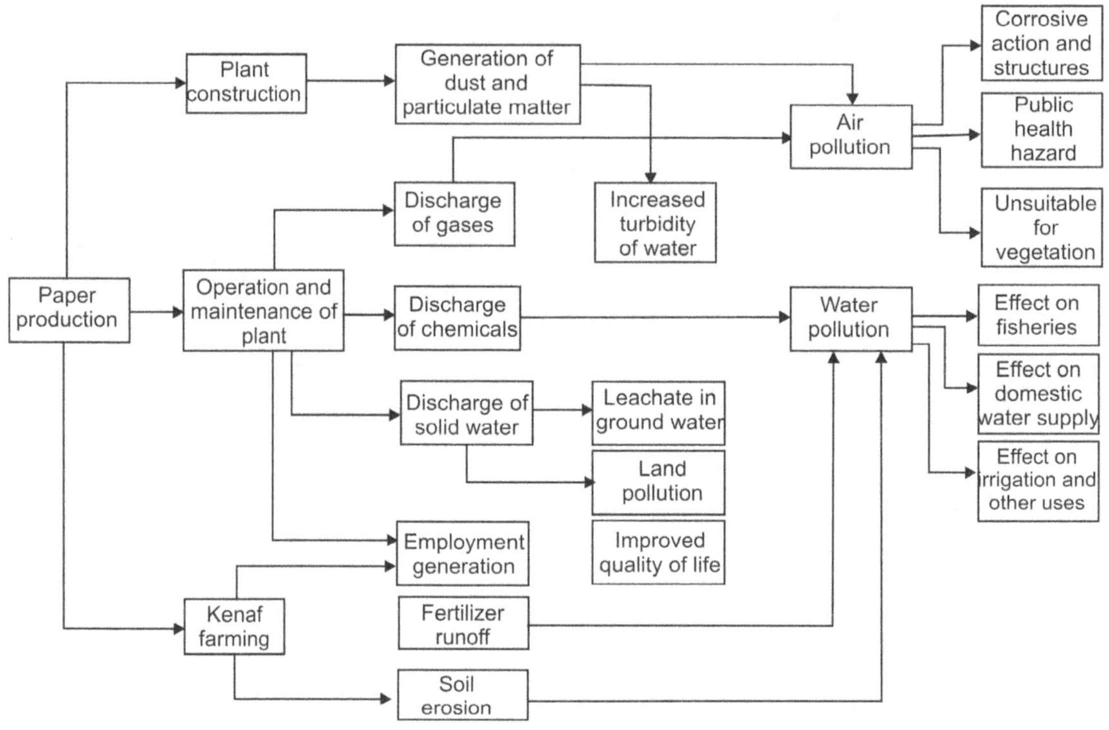

Fig. 2.7 Network of pulpmill impacts.

The primary limitation of the network approach is the minimal information provided on the technical aspects of impact prediction and the means for comparatively evaluating the impacts of alternatives. In addition, networks can become very visually complicated. Networks generally consider only adverse impacts on the environment and hence decision - making in terms of the cost and benefit of a development project in a region is not feasible by network analysis. Temporal considerations are not properly accounted for and short term and long term impacts are not differentiated to the extent required for an easy understanding. While networks can incorporate several alternatives into their format, the display becomes very large and hence unwieldy when large regional plans are being considered. Further, networks are capable of presenting scientific and factual information, but provide no avenue for public participation.

The typical networking of the impacts of an aerial application of Herbicide program is shown in Fig.2.8.

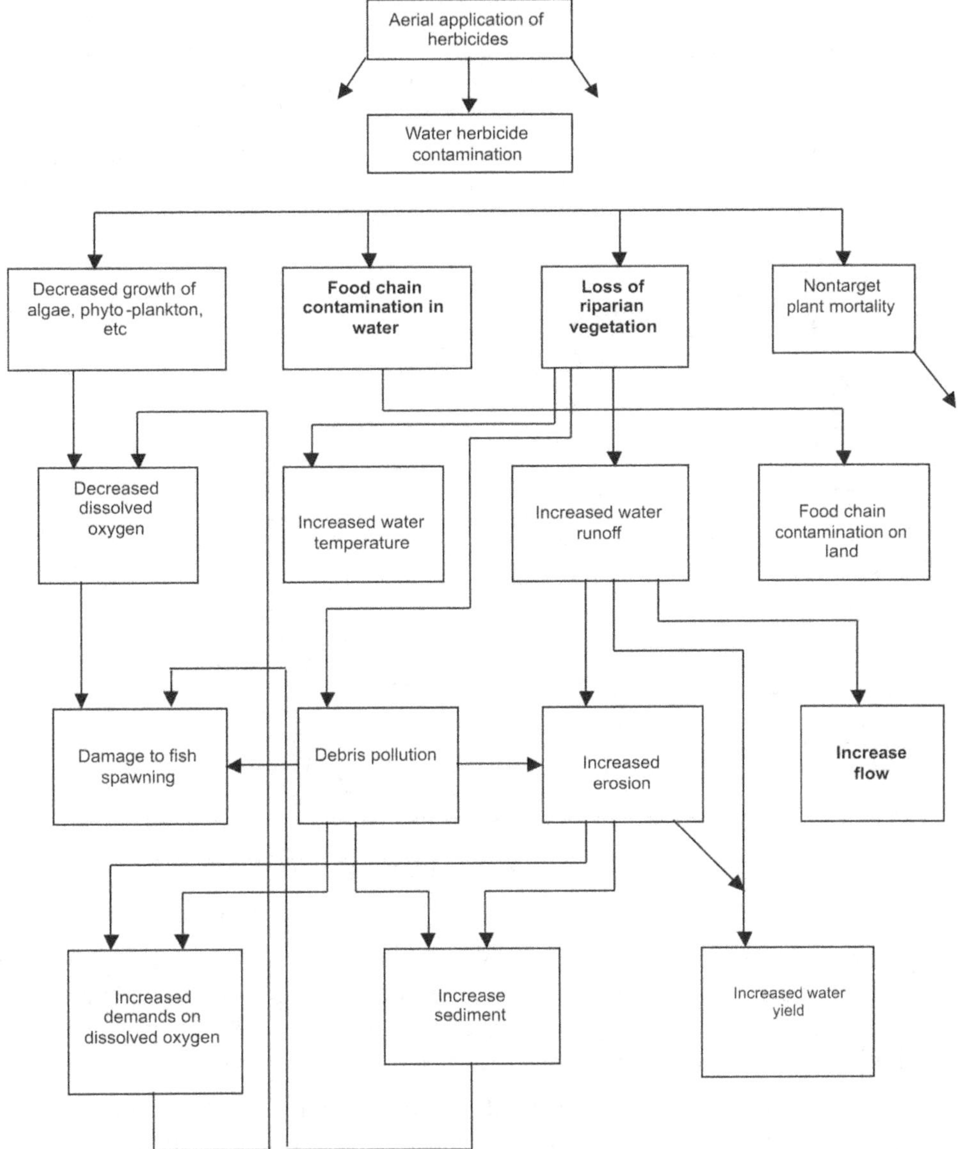

Fig. 2.8 Measure diagram for the aerieal applications of herbicide.

2.4.6 Overlay Methods

Overlay methods involve preparation of a set of transparent maps, which represent the spatial distribution of an environmental characteristic (e.g., Extent of dense forest area). Information on a wide range of variables will be collected for standard geographic units within the study area which will be recorded on a series of maps typically one for each variable. These maps will be overlaid to produce a composite (Fig 2.9). The resulting composite maps characterize the area's physical, social, ecological, land use and other relevant characteristics relative to the location of the proposed development. To evaluate the degree of associated impacts many project alternatives can be located on the final map and validity of the assessment will be related to the type and number of parameters chosen. Normally to have some clarity the number of parameters that can be overlayed in a transparency map is limited to 10. These methods are widely used for assessing visually the changes in the landscape before and after the activity. Secondly, it can be used for preparing combined mapping with an analysis of sensitive areas or ecological carrying capacity. As these methods are spatially oriented they can very clearly show the spatial aspects of cumulative impacts.

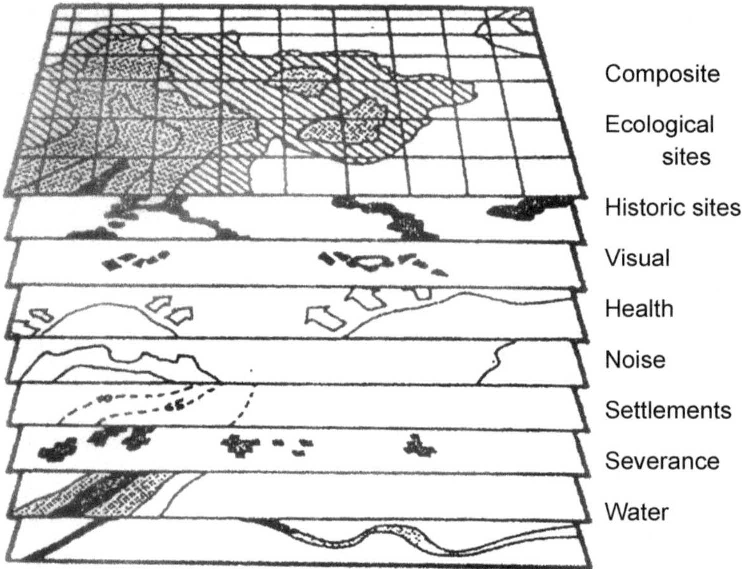

Fig. 2.9 Presentation of array of variables in overlay method.

Source: Wathern 1988 (19)

Overlays are very subjective in that they rely on the judgement of the analyst to evaluate and assess questions on compatibility relating to the existing land use patterns and the prospects of the development activity. In practice, overlays are self-limiting because there is a practical limit on the number of transparencies that can be overlaid.

Overlays are useful when addressing questions of site and route selection. They provide a suitable and effective mode of presentation and display to their audiences. But overlay analysis cannot be the sole criterion for environmental impact assessment.

There is no provision for quantification and measurement of the impacts nor is it assured that all impacts will be covered. The considerations in overlay analysis are purely spatial, temporal considerations being outside its scope. Social, human and economic aspects are not accorded any consideration. Further, higher order impacts cannot be identified. The methodologies rely on a set of maps of environmental characteristics (physical, social, ecological, aesthetic) for a project area. These maps are overlaid to produce a composite characterization of the regional environment. Impacts are identified by noting the impacted environmental characteristics lying within the project boundaries. The approach seems more useful as a method of screening alternative project sites or routes, before detailed impact analysis.

Overlays can be useful for industrial EIA of any project for comparing land capabilities, existing and projected land uses, road route alternatives and other under parameters, and alternative levels of air quality conditions along with pollution control.

The overlay approach is generally effective for selecting alternatives and identifying certain types of impacts; however, it cannot be used to quantify impacts to identify secondary and tertiary interrelationships.

2.4.7 Geographic Information System for EIA

EIA is a decision process, which aims to both identify and anticipate impacts on the natural environment. The interface between these two components produces several effects, which will generate specific impacts. Geographic Information Systems are a set of tools for collecting, storing, retrieving at will, transforming and displaying spatial data from the environment (20, 21). GIS databases include a wide variety of information, including geographic, social, political, environmental, and demographic. GIS is employed within the EIA process to improve different features, mainly related to data storage and access, to the analytical capabilities and to the communicability of the results.

Given the spatial nature of many environmental impacts, Geographical Information Systems can have a wide application in all EIA stages, acting as an integrative framework for the entire process, from the generation, storage, and display of the thematic information relative to the vulnerability and/or sensitivity of the affected resources, to impact prediction and finally their evaluation of decision support (22). GIS uses layers also known as "themes" to overlay different forms of information, with each theme representing a category (in some cases categories) of information, such as population, roads, vegetation, settlements, forest cover, land use etc.

Computer modules can be used to store the characteristics of the proposed developments and the surrounding area. This enables us to introduce impact weightings into the assessment. The computer can perform the complex mathematical operation required when a large number of variables are weighted. GIS brings to the EIA process in a new way of analyzing and manipulating spatial objects and an improved way of communicating the results of the analysis, which can be of great importance during the public participation process where the results from the public consultation and social surveys can be imported into a GIS for spatial and non-spatial analysis, and display in a format that is easily understood by both technical and nontechnical stakeholders.

A significant application of GIS is the construction of real world models based on digital data. Modeling can analyze trends, identify factors that are causing them, reveal alternate paths to solve the given problem and indicate the implications or consequences of decisions. GIS can show how a natural resource will be affected by a decision. Based on satellite data, areas that suffer most from deforestation may be identified and analyzed on the basis of overlaying data on soil types, the species required, the likely growth and yield and impact of regulatory measures applicable to that area. The impact of various development plans on the environment can be assessed by integrating data on land use with topographical and geological information. Similarly, satellite imagery can be periodically used to update maps of irrigated land. The spectral features of irrigated and non-irrigated fields can be combined with other data on the fields to derive estimates of demands for irrigation water and devise land management plans. GIS can be used to assess the risk of drought in choosing areas for rain fed crops.

With these capabilities, GIS is effectively used in. **Overlay, Checklist, Matrix** and **Network Methods** (23).

GIS becomes the ultimate tool for the **Overlay method** as in this EIA involves heavily on graphical display overlaying of various layers of environmental components like topological data, air dispersal patterns, land use data, wildlife, surface and ground water intakes etc. This method depends on data and as such in EIA. Also, the overlay of baseline information, maps with project layouts is frequently used for impact identification (24-26).

GIS provides a computer platform in **Checklist method** for organizing, storing and analyzing environmental components, attributes and processes, which are categorized under different categories, e.g., geology, vegetation and air in qualitative understanding the impacts.

In **Matrix method** also GIS provides a powerful tool for organizing, analyzing and storing matrices, particularly in the process of relating specific project activities to specific types of impacts considering both the magnitude and importance of impacts (25).

A network of possible impacts that may be triggered by project activities in the **Network method** involves the identification of project actions along with the direct and indirect impacts. From the network methodology, direct, secondary, tertiary and other higher order impacts of actions are to be identified. There is need for complex data analyses using this methodology and on a GIS platform large volumes of data can be better analyzed within a short period of time.

Thus, GIS is a powerful management tool and has the capability for site impact prediction (SIP), wider area prediction (WAP), cumulative effect analysis (CEA), and environmental audits and for generating trend analysis within an environment. Furthermore, Eedy (1995, 26) and Joao (1996, 27) stresses on some of the advantages of the use of GIS in EIA, namely for data management, overlay and analysis, trend analysis, as sources of data sets for mathematical impact models, habitat and aesthetic analysis, and public consultation. GIS have also been used for the presentation of environmental baseline information and project description, through the preparation of thematic maps for the several environmental impact assessments.

The evaluation of various geospatial methods with reference to various assessment processes is presented in Table 2.14.

Table 2.14 Evaluation of geospatial EIA method.

Key Area of the Assessment Process	Criteria	L denotes Criteria completely satisfied. P denotes Criteria Partially satisfied. N denotes criteria not satisfied
Cost/Time Effectiveness Criteria	1. Expertise requirements	L
	2. Data requirements	P
	3. Time requirement	L
	4. Flexibility	L
	5. Personnel level of effort	P
Impact identification	6. Comprehensiveness	N
	7. Indicator-based	P
	8. Discriminative	N
	9. Time dimension	P
	10. Spatial dimension	L
Impact Measurement	11. Commensurate	L
	12. Quantitative	L
	13. Measurement changes	L
	14. Objective	L
Impact Assessment	15. Credibility	L
	16. Replicability	L
	17. Significance-based	N
	18. Aggregation	P
	19. Uncertainity	N
	20. Alternative comparison	P
Communication	21. Communicability	L
	22. Summary format	L

2.4.8   Task-specific Computer Modeling

Simulation Modeling Workshops

System analysts have developed an approach to environmental impact assessment and management commonly referred to as Adaptive Environmental Assessment and Management (AEAM), which combines various simulation models to predict impacts. This approach broadens the potential of simulation models to evaluate the impacts of alternatives and is beneficial for project planning.

The AEAM approach uses small interdisciplinary teams interacting through modeling workshops over a relatively short time to predict impacts and evaluate alternatives including management measures. The adaptive assessment process can be divided into three types of workshops:

- The initial workshop
- The second phase workshop
- The transfer workshop

The AEAM technique largely overcomes the shortcomings of most other methods in that the other methods assume unchanging conditions or project impacts in a single time frame on statistically described environmental conditions (28).

It also overcomes a built-in bias towards compact mentalization, and fragmentation of the relationship between project actions, environmental characteristics and likely impacts, while the reality may be that the impacts alter the scale and direction of change within environmental and social systems.

The AEAM technique can handle higher-order impacts and interactions between impacts. But it depends on a small group of experts and has no avenue for public participation. This aspect is particularly significant for large-scale development where the opinions of interest groups are important.

The technique can be time-consuming and may impose a severe burden on the monetary resources available for the purpose of environmental assessment. Simulation models, especially of ecosystems, are still in an embryonic stage of development and their accuracy and predictive capacity is yet to be proved. The use of this technique requires the input of people trained in its use and functions. This may lead to the need for expatriate expertise in proportions greater than required for other techniques and this may be the limiting constraint in developing countries.

2.4.9 Expert Systems

Expert opinion, which can also be referred to as professional judgement represents a widely used type of method within the EIA process. This type of method is typically used for addressing the specific impacts of a proposed on different components of the environment. Specific tools within the category of expert opinion, which can be used to delineate information include Delphi studies and the use of the adaptive environmental assessment process. In this approach groups of experts identify appropriate information and build qualitative/quantitative models for impact prediction or to simulate Environmental impact process.

Expert systems are an emerging type of methods (29) which consists of drawing upon professional knowledge and judgement of experts in particular areas. Such knowledge is encoded via a series of rules into Expert System shells in computer software. Expert Systems are typically user friendly and require the user to answer a series of questions to conduct a particular type of analysis. Increasing attention is being given to the development of a more comprehensive expert system for the EIA process.

2.4.10 Cost/Benefit Analysis - Economic Techniques

Economic techniques have been developed to try to value the environment (30-39). The most commonly used methods of project appraisal are social cost-benefit and cost-effectiveness analysis as it's not easy to incorporate environmental impacts into traditional cost-benefit analysis, principally because of the difficulty in quantifying and valuing environmental effects. The information of baseline data collected in an EIA report may be used in the evaluation of environmental cost-benefit analysis. This approach has been adopted in a few countries across the world. However, the majority of the decisions on approval of major projects is even today done only on social/project cost benefit. Cost effectiveness analysis can also be used to determine the most efficient or least-cost method of meeting a given sustainable development goal.

Valuing the environment, resources is complex and over the years, many attempts have been made to develop standard procedures for pricing environmental resources. It is complex because it also involves an intrinsic value which are difficult to quantify such values. Developing countries are using "Effect on Production" (EOP) and "Preventive Expenditure and Replacement Costs" (PE/RC) methods for valuation. The EOP method attempts to represent the value of change in output that results of the environmental impact of the development.

The PE/RC method makes an assessment of the value that people place on preserving the resource by estimating their willingness to pay to prevent its degradation or to restore its original state after it has been damaged (replacement cost). Both methods have their strengths and weaknesses and must be used judiciously.

Environmental cost-effectiveness analysis is, however a useful tool in the selection of mitigating or control measures, for appraisal of the impact on human health and monetary values.

Cost/benefit analysis provides the nature of expense and benefit accrual from a project in monetary terms as a common practice in traditional feasibility studies and thus enables easy understanding and aids decision-making. The principal methods available for placing monetary values (costs and benefits) on environmental impacts, a taxonomy of valuation methods, and steps involved in economic evaluation of environmental impacts are discussed under this category. The role of environmental economics in an EIA can be divided into three categories, namely:

1. The use of economics for "benefit-cost analysis" as an integral part of project selection;

2. The use of economics in the assessment of activities suggested by the EIA; and

3. The economic assessment of the environmental impacts of the project.

Environmental economics can aid in the selection of projects in that benefit-cost analysis can be used in the prescreening stage of the project, and the environmental components can be brought into the process of presenting various options and selecting from among them (31-34). Doing so eventually leads to a project selection process, which takes the environment into consideration. In the second role, the economic assessment is focused on the cost assessment of environmental mitigation measures and management plans suggested in the EIA. The economic analysis in the EIA may include a summary of the project costs and how such cost estimates would change due to the activities proposed under the EIA. This component can be considered as an accounting of the environmental investment of a project. The third role, which is the economic assessment of the environmental impacts of a project, is geared towards seeking the economic values (of both costs and benefits) of the environmental impacts. These impacts are neither mitigated, nor taken into account in traditional economic analysis of projects. They should be identified by the EIA and sufficient quantitative and qualitative explanations should be given in EIA documents.

The difficulty encountered in the use of these techniques will be that, impacts have to be transformed and stated in explicit monetary terms, and this is not always possible, especially for intangibles like the monetary value of health-related impacts of industrial development.

Cost/benefit analysis of the type of assessment of natural systems is not merely concerned with the effects on environmental quality, but rather, it seeks the conditions for sustainable use of the natural resources in a region. This type of approach is not useful for small scale development projects, but is better suited for the analysis and evaluation of a regional development plan. Even thought it may not be possible to place an economic value on environmental losses or gains resulting from a developmental project, decision makers should take into account implied environmental values in their decision-making.

To facilitate the decision making process, therefore, assessors conducting environmental impacts, should not just identify environmental impacts, they should also provide information on the implied value of the environmental losses and gains.

The evaluation of site/sites and major design options should be taken together, within the economic and technical limitations imposed by the aim of the project, the combination of the project site and project design needed to produce no significant environmental impacts, should also incur the least economic cost to the community.

If, for the preferred site and project option, the assessor has predicted, potentially significant environmental impacts, he should consider the cost to the community of any mitigating or abatement measures and their alternatives before adopting them into the project plan. Whenever there is a choice of measures to mitigate or abate a significant potential impact he should select the solution that will incur the least economic cost to the community (35, 36).

2.4.10.1 Steps in Economic Valuation of Environmental Impacts

Economic analysis of environmental impacts is important in project preparation to determine whether the net benefits of undertaking the project are greater than the alternatives, including the non-project scenario. Project alternatives often vary in their economic contribution and environmental impacts. Economic assessment of different alternatives in the early stages of project planning should provide important inputs to improve the quality of decision-making. The economic analysis of the environmental impacts of the selected projects also allows for a more complete assessment of the project's costs and benefits. A general procedure that can be followed in economic analysis of environmental impacts is presented in ref (30-39).

In the case of internal or mitigated impacts, there is no need to look for an extensive monetization of environmental impacts. They are already treated as part of the project. However, one has to assure that they are properly costed or valued in economic terms and appropriately incorporated into the economic cost benefit streams of the project. As explained earlier, this is the first part of the economic analysis section of the EIA report. A good EIA report should include a section on all economic aspects of the project, including the results of a cost benefit analysis of the overall project after incorporating the internal or mitigated impacts when qualitative assessment and documentation is important. The second component of the economic analysis of the EIA which gives an assessment of environmental impacts that can be quantified. At a minimum, the following six tasks need to be completed in the economic analysis of environmental impacts:

1. Determine the spatial and conceptual boundaries of the analysis;

2. Identify environmental impacts and their relationships to the project;

3. Quantify environmental impacts and organize them according to importance — the impacts described qualitatively, if they cannot be expressed in quantitative terms;

4. Choose a technique for economic valuation;

5. Economic valuation (place monetary values) of environmental impacts identified; and

6. Set an appropriate time frame and perform the extended benefit cost analysis.

The boundary of the economic analysis refers to the conceptual and physical limits of the analysis. It may consider on-site and off-site environmental impacts that are consequences of project activities. Another consideration is the type of goods and services that should be included in the analysis. The complexities of a project's environmental impacts may cause some difficulty in establishing the spatial and conceptual boundary of the economic analysis. The rule is to start the analysis with directly observable and measurable impacts.

Thus the evaluation of the environmental and development benefits and costs is an essential aid to decision making. The information that an assessor should provide detailed assessment is an annotated list of economic costs and benefits to the community that arise from:

1. His selection of technically and financially feasible site options and major design, operations (process options);

2. His selection of any technically and financially feasible measures to mitigate or abate significant environmental impacts predicted;

3. The total project plan;

A successful EIA report should provide the required information for economic analysis of the environmental impacts. The necessary output of Tasks 1, 2, and 3 shows a list of all possible environmental impacts of the project. Thus, the EIA should identify and completely document all impacts, providing sufficient quantitative and qualitative descriptions. This list becomes the basis for the economic valuation carried out in Task 5. Valuing environmental impacts in monetary terms is the most difficult part of the economic analysis. This necessitates the use of valuation techniques appropriate to the environmental impacts being investigated. Choosing the appropriate valuation techniques is itself a difficult task, requiring an expert judgment from economists and environmental specialists.

2.4.11 Environmental Medium Quality Index factor analysis Method

2.4.11.1 Generic Steps

Several generic steps are associated with the development of numerical indices or classification of environmental quality, or the pollution potential of human activities. These include factor identification, assignment importance weights, establishment of scaling sections or other methods for factor evaluation, termination and implementation of the appropriate aggregation approach, and application of field verification.

Factor identification basically consists of delineating key factors that can be used as indicators of environmental quality, susceptibility to pollution, or the pollution potential of the source type.

Factor identification should be based on the collective professional judgement of knowledgeable individuals relative to the environmental media or pollution-source category. Organized procedures, such as, the Delphi approach, can be used to aid in the solicitation of this judgement and the aggregation of the results (40).

The second step in the development of an index is the assignment of relative-importance weights to the environment-media and/or source-transport factors, or at least the ranking of these factors in order of importance. Some techniques, which should be used to achieve this step include the Delphi approach, the unranked pairwise comparison. Several approaches have been used to scale and evaluate the data associated with factors in the methodologies. Examples of techniques of scaling or evaluation for this purpose include use of 1. linear scaling or categorization used on the range of data, 2. letter or number assignments designating data categories, 3. notional curves, or 4. the unranked pairwise comparison technique. The development of scaling or evaluation approaches should be used on the collective professional judgement of individuals knowledgeable in areas related to environmental-media or pollution-source.

2.4.11.2 Air Pollution Index

In most natural situations the effects of air pollution are a result of the combined effects of various different pollutants rather than only one single pollutant. The magnitude of air pollution in such cases is difficult to be assessed on the basis of the concentration of individual pollutants alone, but the goal can be achieved by applying certain indices. The kind and number of pollutants for calculating the indices can be selected depending upon their predominance in the ambient air.

Any pollution index should indicate the gross level of pollution with reference to the standard limits of the individual pollutants. It should be easy to understand and should include the major air pollutants. If the use of indices is to be made on a national/local level, care has to be taken to include always the same air pollutants to maintain parity.

2.4.11.2.1   Calculation of Air Pollution Indices

There are a number of methods, which are used to calculate the air pollution indices. Some of these are mentioned below.

1. In one of the methods the important individual pollutants are compared with their ambient air quality standards in terms of percentage and the air pollution index (API) is calculated by taking their average. If there are three pollutants, then

$$API = 1/3 \sum_{i=1}^{3} A_1$$

$$A_1 = C_1 / S_1 \times 100$$

where C_1 is the concentration of the pollutant I and S_1 is the standard value of the pollutant in the ambient air.

2. In this method, ratios of three important air pollutants to their ambient air quality standards are first obtained and then the average of their sum is multiplied by 100.

$$API = \left\{ \frac{I_{PM}}{S_{SPM}} + \frac{I_{SO}}{S_{SO}} + \frac{I_{CO}}{S_{CO}} \right\} \times 100 \qquad .....(2.1)$$

where I_{PM}, I_{SO}, I_{CO} represent individual values of particulate matter, sulphur dioxide and carbon monoxide and S_{sPM}, S_{SO}, S_{CO} their ambient air quality standards.

3. In another method, subindices (Al) are first obtained for five air pollutants by giving them arbitrary numbers according to their range of ambient concentrations as indicated in the table 2.15. The air pollution index is the then calculated as the sum of these subindices.

$$API = \sum_{I=1}^{5} A_I \qquad .....(2.2)$$

Table 2.15 Sub index value (Al) for five important air pollution parameters to calculate air pollution index.

Air pollution parameters	Sub index values (A_1)					
	2	4	8	12	16	20
CO(ppm)	0-1	1-2	2-4	4-6	6-8	8-35
NO$_2$	0-0.005	0.005-0.01	0.01-0.02	0.02-0.06	0.06-0.10	0.10-0.20
Oxidants (ppm) coefficients of haze	0-0.5	0.5-1.0	1-2	2-3	3-4	4-5
Visibility	12-24	8-12	6-8	4-6	2-4	0-2

4. The air pollution index can also be calculated on the basis of a single parameter taking into consideration only the dominant pollutant. For example, the index for sulphur dioxide can be represented as follows.

$$API = \frac{I_{SO}}{S_{SO}} \times 100 \qquad .....(2.3)$$

where I_{SO} and S_{SO} are ambient SO$_2$ concentrations, and the air quality standard respectively. Similarly, the ozone index can be calculated in the highly congested traffic areas. For an industrial areas index can also be calculated by taking two parameters into account, namely, sulphur dioxide and particulate matter.

As all above indices have been calculated on a percent basis, a common rating scale given in Table 2.16 can be used to indicate the air quality.

Table 2.16 Air quality rating scale.

Index range	Quality of air
0 – 25	Clean air
26 – 50	Light air pollution
51 – 75	Moderate air pollution
76 – 100	Heavy air pollution
Above 100	Severe air pollution

2.4.11.2.2 The Pollution Standards Index

In 1976, as a result of the coordinated efforts of a number of federal agencies, a Pollution Standards Index (PSI) was developed (Fig. 2.10). This index makes it possible to compare the air quality of different urban areas and its potential threat to human health. The EPA has established acceptable standards of air pollution, that is, National Ambient Air Quality Standards (NAAQS). If a pollutant in a city is at the standard, it is given an index rating of 100. Any index value below 100 suggests relatively clear air with minimal health effects. However, as the index value rises to above 100, the air is progressively more polluted and the health effects correspondingly severe. Any index value of 400 or above suggests air pollution. It may cause premature death of the sick and elderly. Under those conditions, all persons are advised to remain indoors. PSI values for various atmospheric contaminants are frequently reported in daily local newspapers in major urban areas (Fig 2.10).

The actual calculation of a days PSI value will be done based on the values reported in Table 2.17 which gives individual PSI numbers corresponding to various pollutant concentrations. Individual PSI sub-indices are computed for each of the pollutants in the table using linear interpretation between indicates break points. The highest PSI sub- index determines the overall PSI. It can be seen from Table 2.17 that in addition to T.S.P (Total Suspended Particulate), the product of T.S.P $\times$ SO_2 is also given on the two pollutant, when present together they act synergistically producing higher effects than the individual effect do.

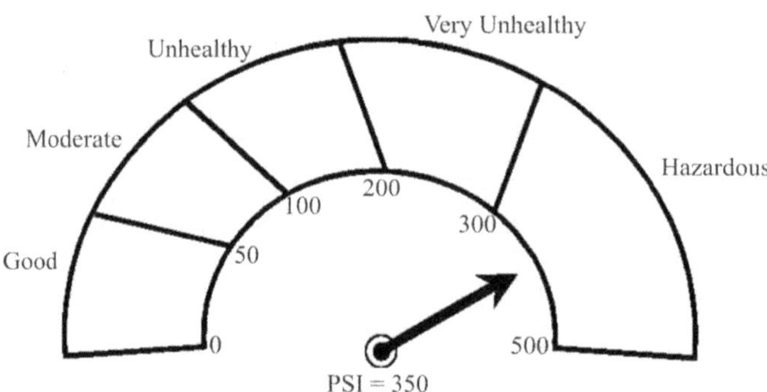

Fig. 2.10 Pollution standard index.

Table 2.17 Pollutant standard index (PSI) breakpoints.

Index	1 hr O_3 ug/m^3	8 hr CO ug/m^3	24 hr TSP ug/m^3	24hr SO_2 ug/m^3	TSP $\times$ SO_2 10^3 (ug/m^3)	1 hr NO_2 ug/m^3
0	0	0	0	0	0	—
5 0	118	5	7 5	8 0	—	—
100	235	1 0	260	366	—	—
2 0 0	4 0 0	1 7	3 7 5	8 0 0	6 5	1130
3 0 0	8 0 0	3 7	6 2 5	1600	2 6 1	2260

2.4.11.3 Water Qualities Index (WQI) Based on Expert Opinion

For calculating the water quality index the nine individual variables of greatest importance identified are dissolved oxygen (DO), fecal coliformed, which formed pH, 5-day bio-chemical oxygen demand (BOD), nitrates (NO_3), phosphates (PO_4), temperature deviation, turbidity (in JTU), and total solids (TS). The resultant importance weights based on the ratings for each variable are listed in Table 2.18 and 2.19. The weights have a public health focus based on using the water for human consumption.

The water rating curve for each of these parameters would number with 0-100 water rating scale on the y-axis and actual level of each parameter on the x-axis, Fig. 2.10. Solid lines indicate the arithmetic meanwhile the diameter line indicates 80% confidence limit.

To calculate the aggregate water quality index either a weighted leaner sum of the subindices (WQ Ia) or a weighted project aggregate function (WQ Im) can be used. These are expressed mathematically as

$$WQ\ Ia = \sum^{n} wi\ Ii \qquad\qquad .....(2.4)$$

$$C = 1$$

$$WQ\ Im - \Pi^{n}\ Ii\ Wi \qquad\qquad .....(2.5)$$

$$1 = 1$$

The typical WQIa and WQIm values for a set of water quality parameters are shown in Table 2.19, while the interpretation of the resultant water quality index is presented in Table. 2.20.

Table 2.18 Pollutant standard index (PSI) breakpoints.

	Pollutant levels (in micrograms per liter) index value	Air quality level	General health effects	Cautionary statements
Hazardous	500	Significant harm	Premature death of sick and elderly. Healthy people will experience adverse symptoms that effect their normal activity.	All persons should remain indoors. All persons should minimize physical exertion and avoid traffic
	400	Emergency	Premature onset of certain disease, in addition to signification of symptoms and decreased exercise tolerance in healthy persons	Elderly and persons with existing diseases should stay indoors and avoid physical exertion. General population should avoid outdoor activity

Table 2.18 *contd...*

	Pollutant levels (in micrograms per liter) index value	Air quality level	General health effects	Cautionary statements
	300	Warming	Significant aggravation of symptoms and decreased exercise tolerance in persons with heart or lung disease, with widespread symptoms in the healthy population	Elderly and persons with existing heart or lung disease should stay indoors and reduce physical activity
Unhealthy	200	Alert	Mild aggravation of symptoms in susceptible persons, with irritation symptoms in the healthy population	Persons with existing heart or respiratory ailments should reduce physical exertion and outdoor activity.

Table 2.19 Example calculations for water quality index.

Variable	Measurement	I_1	w_1	I_1w_1	I_1w_i
DO	60%	60	0.17	10.2	2.5
Fecal coliforms	103	20	0.15	3.0	1.5
pH	7	90	0.12	10.8	Subject:1.7
BOD5	10	30	0.10	3.0	1.4
NO_3	10	50	0.10	5.0	1.4
PO_4	5	10	0.10	1.0	1.2
Temperature deviation	5	40	0.10	4.0	1.4
Turbidity	40 JTU	44	0.08	3.5	1.3
Total solids (TS)	300	60	0.08	4.8	1.3
				WQIa = 45.3	WQIa = 38.8

$$\text{Sus index } I_1 = \frac{\text{Equality Qualities}}{\text{Environmental Quality Standard}} \times 100$$

Table 2.20 Description words and colors suggested for reporting the example WQI.

Descriptor words	Numerical range	Colour
Very bad excellent	0-25	Blue/red
Bad good	26-50	Green/orange
Medium	51-70	Yellow
Good bad	71-90	Orange/green
Excellent very bad	91-100	Red/blue

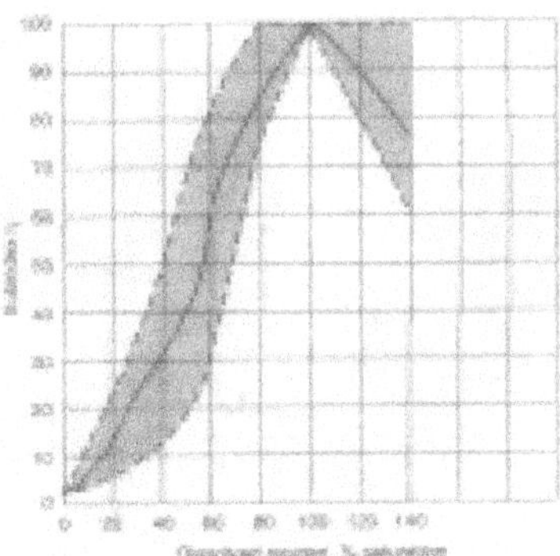

Fig. 2.11(a) Subindex Function for DO in the WQI

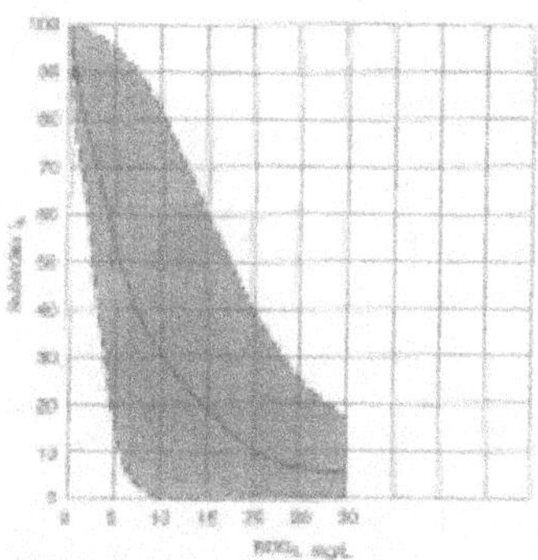

Fig. 2.11(b) Subindex Function for Fecal Coliforms (average for DO < 140% I_1 = 50) 10) number of organisms per 100 ml) in the WQI (for fecal coliforms > 10^5/100 ml, I_2 = 2) (Ott, 10).

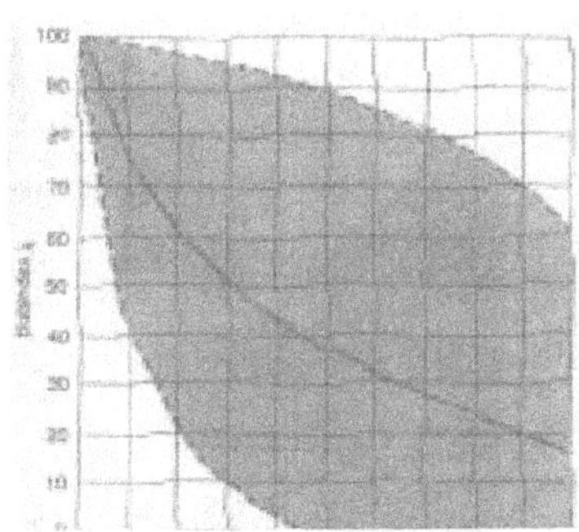

Fig. 2.11(c) Subindex function for pH in the WQI

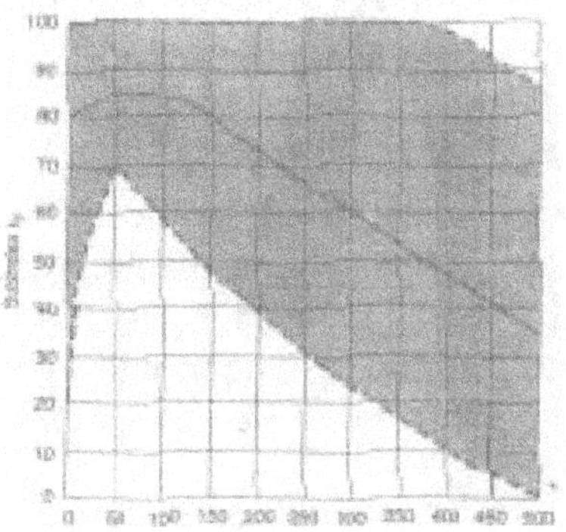

Fig. 2.11(d) Subindex function for BOD5 in the WQ

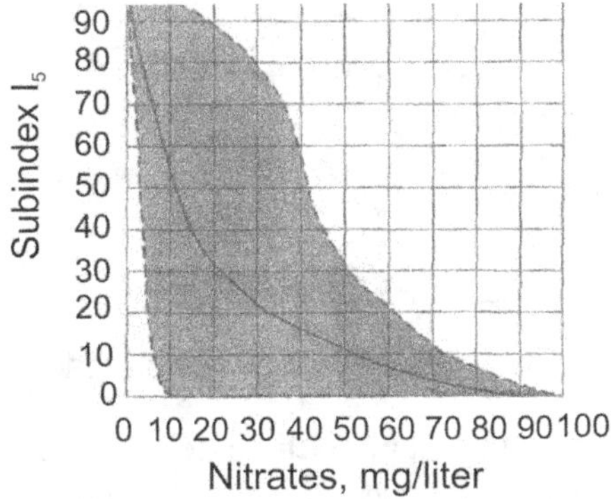

Fig. 2.11(e) Subindex function for nitrates in the WQI (for nitrates > 100 mg/L, I5 = 1) (Ott, 10).

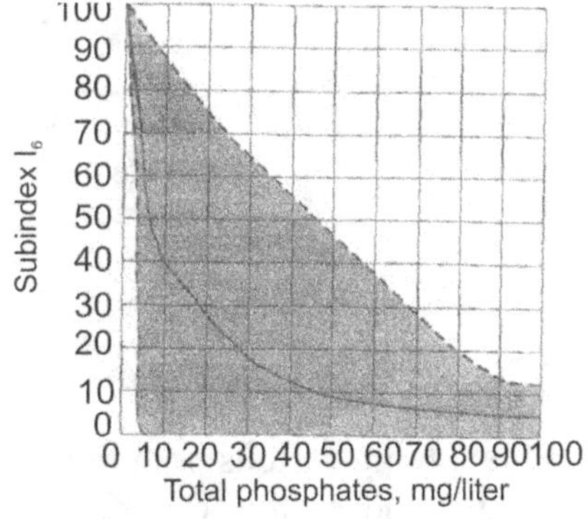

Fig. 2.11(f) Subindex function for total phosphates in the WQI (for total phosphates > 10 mg/L, I6 = 2) (Ott, 10).

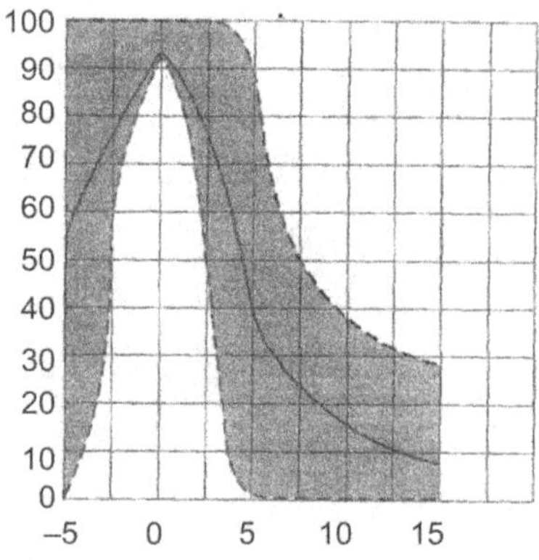

Fig. 2.11(g) Subindex function for temperature deviation from equilibrium (ΔT) in the WQI (for DT > 150C, 17 = 5) (Ott, 10).

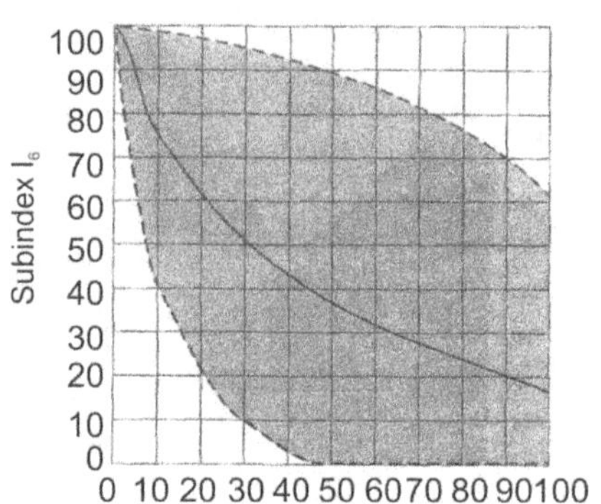

Fig. 2.11(h) Subindex function for turbility (Jackson turbility units) in the WQI (for turbility > 100 JTU, 18 = 5) Ott, 10).

2.4.12 Rapid Assessment of Pollution Sources Method

Rapid assessment of pollution procedure developed by WHO 1982 allows for quick estimation of releases of pollutants to the environment for information on existing pollution sources for a given study area (Fig. 2.12). Rapid estimates of different types pollution loads will be made based on the information of the quantities of consumption and outputs of various industrial and urban process industrial production figures, fuel usage, number of motor vehicles, number of houses connected to sewers etc., multiplied by predetermined waste load factors which will give a worst case estimate of the amount of pollutant that is being released to the environment.

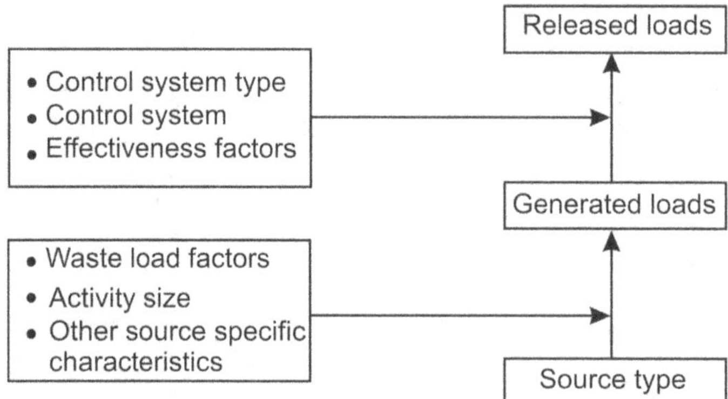

Fig. 2.12 Schematic of rapid assessment procedure for estimation pollution loads.

By applying various pollution control measures one can calculate the extent of reduction in pollution loads that can be achieved and manage the system with appropriate control measures such that only acceptable loads are released into the environment. In Table 2.21 the load factors of various activities as per UN SIC system which account for most of the industrial pollution sources (Economopoulos 1993) (40,41) are given which may be used as guide to identify major pollution sources.

Table 2.21 List of Activities which produce pollution (14).

	Emissions	effluents	solid wastes
0 Activities not Adequately defined			
Consumer solvent use	-		
Surface coating	-		
1 Agriculture, hunting, Forestry and Fishing			
11 Agriculture and hunting			
111 Agriculture and livestock production	-	-	-
12 Forestry and Logging	-		
121 Forestry			
2 Mining and Quarrying			
21 Coal mining	-	-	
22 Crude petroleum and natural gas production	-		-
23 Metal ore mining	-		-
24 Other mining	-		-
2 Manufacturing			
31 Manufacture of food, beverages and tobacco			
312 Food Manufacturing			
3111 Slaughtering preparing and preserving meat	-	-	-
3112 Manufacture of dairy products		-	-
3113 Canning and preserving fruits and vegetables		-	-
3114 Canning, preserving and processing of fish		-	-
3115 Manufacture of vegetables and animal oils and fats		-	-
3116 Grain mill products		-	-
3117 Bakery products		-	
3118 Sugar factories and refineries			-
3121 Food products not elsewhere classified		-	-
3122 Alfalfa dehydrating		-	
313 Beverage industries			
3131 Distilling, rectifying and blending spirits		-	-
3132 Wine industries		-	
3133 Malt liquids and malt		-	-
3134 Soft drinks		-	
32 Textile, wearing apparel and leather			
321 Manufacture of textiles			-
3210 Manufacture of textiles			
322 Manufacture of wearing apparel, except footwear			
3211 Spinning, wearing and finishing textiles	-		-
3214 Carpet and rug manufacture			-
323 Manufacture of leather and products of leather			
3231 Tanneries and leather finishing	-	-	

Table 2.21 Contd...

Code	Industry	Emissions	effluents	solid wastes	
34	*Paper and paper products, printing and publishing*				
341	Manufacture of paper and paper products	-	-	-	
342	Printing, publishing and allied industries	-		-	
35	*Manufacture of chemicals, and chemical, petroleum, coal, tuber and plastic products*				
351	Manufacture of industrial chemicals				
3511	Basic industrial chemicals except fertilizers	-	-	-	
3512	Manufacture of fertilizers and pesticides	-	-		
3513	Resins, plastics and fibers except glass		-	-	
352	Manufacture of other chemical products				
3521	Manufacture of paints, varnishes and lacquers	-		-	
3522	Manufacture of drugs and medicines		-	-	
3523	Manufacture of soap and cleaning preparations		-	-	
3529	Chemical products not elsewhere classified	-	-		
353	Petroleum refineries	-	-	-	
354	Manufacture of miscellaneous products of petroleum and coal		-	-	-
355	Manufacture of rubber products				
3551	Tire and tube industries		-	-	
36	*Non-mettalic mineral products, except products of petroleum and coal*				
361	Manufacture of pottery, china and earthenware			-	
362	Manufacture of glass and glass products		-	-	
369	Manufacture of other non-metallic mineral products				
3691	Manufacture of structural clay products		-		
3692	Cement, lime and plaster	-			
3699	Products not elsewhere classified	-			
37	*Basic metal industries*				
371	Iron and steel basic industries		-	-	-
372	Non-ferrous metal basic industries		-	-	-
38	*Fabricated metal products, machinery and equipment*				
381	Fabricated metal products, except machinery		-	-	
384	Manufacture of transport equipment				
3841	Ship building and repairing			-	
4	**Electricity, Gas and Water**				
41	Electricity, gas and steam				
4101	Electricity, light and power	-	-	-	
6	**Wholesale and Retail Trade**				
61	Wholesale trade		-		
62	Retail trade		-	-	
63	Restaurants and hotels				
631	Restaurants, cafes and other eating and drinking activities				-
632	Hotels, rooming houses, camps and other lodging			-	
7	**Transport, Storage and Communication**				
711	Transport and storage		-		
712	Water transport		-		
713	Air transport		-	-	
719	Services allied to transport				
7192	Storage and warehousing	-	-		

	Emissions	effluents	solid wastes
9 Community, Social and personal Services			
92 Sanitary and related community services	-	-	-
93 Social and related community services			-
931 Education services			-
932 Medical, dental and other health services			-

Waste Load Factors

The waste load factors for air, water and solid wastes per unit loading (Economopoulos 1993a) (40) are presented in Tables 2.22 (a), (b), and (c).

Table 2.22(a) Natural gas-model for air emissions inventories and control.

Process	Unit(U)	TSP kg/U	SO$_2$ kg/U		NO$_2$ kg/U		CO kg/U	VOC kg/U
Gaseous fuels								
Natural gas								
Utility boiler	1000 Nm^2	0.048	15.6	S	8.8	f	0.64	0.028
	T	0.061	20	S	11.3	f	0.82	0.036
Industrial boiler	1000 Nm^2	0.048	15.6	S	2.24		0.56	0.092
	T	0.061	20	S	2.87		0.72	0.18
Domestic furnaces	1000 Nm^2	0.048	15.6	S	1.6		0.35	0.127
	T	0.061	20	S	2.05		0.41	0.163
Stationary gas turbines	1000 Nm^2	0.224	15.6	S	6.62		1.84	0.673
	T	0.287	20	S	8.91		2.36	0.863

(**Source** : Economopoulos, (40,41,)).

Table 2.22(b) Petroleum refineries-model for liquid waste inventories and control.

Major Division 3. Manufacturing Division 35. Manufacture of Chemicals and of Chemical, Petroleum, Coal, Rubber and Plastics Products SIC# 353 Petroleum Refineries.

Process	Unit (U)	Waste Volume m^3/U	BOD$_5$ kg/U	TSS kg/U	Tot N kg/U	Tot P kg/U	Other pollutants	Load kg/U
Topping refinery	1000 m^3 of crude	484	304	11.7	1.2		Oil	8.3
							Phenol	0.034
							Sulfide	0.054
							Cr	0.007
Cracking refinery	1000 m^3 of crude	605	72.9	18.2	28.3		Oil	31.2
							Phenol	1.0
							Sulfide	0.94
							Cr	0.25
Petrochemical refinery	1000 m^3 of crude	726	172	48.6	34.2		Oil	52.9
							Phenol	7.7
							Sulfide	0.086
							Cr	0.234

Table 2.22(b) Contd...

Process	Unit (U)	Waste Volume m³/U	BOD₅ kg/U	TSS kg/U	Tot N kg/U	Tot P kg/U	Other pollutants	Load kg/U
Lube oil refinery	1000 m³ of crude	1090	217	71.5	24.1		Oil Phenol Sulphide Cr	120 8.3 0.014 0.046
Integrated refinery	1000 m³ of crude	1162	197	58.1	20.5		Oil Phenol SulPhide Cr	74.9 3.8 2.0 0.49

(**Source** : Economopoulos. 1993(a).(40)

Table 2 .22(c) Petroleum refineries-model for solid and hazardous waste inventories.

Major Division 3. Manufacturing Division 35. Manufacture of Chemicals and of Chemical, Petroleum, Coal, Rubber and Plastics Products SIC# 353 Petroleum Refineries.

Process	Unit (U)	Oily kg/U
Topping refinery	1000 m³ of crude	1311
Low cracking refinery	1000 m³ of crude	1675
High cracking refinery	1000 m³ of crude	3303
Lude oil refinery	1000 m³ of crude	6140

Note : The major problem is oily which are often contaminated by heavy metals.

2.4.12.1 Application of Rapid Assessment Procedure in EIA

For a number of new development projects the EIA can be carried out using Rapid assessment procedure. The use of waste load factors enables the prediction of the approximate pollutant loadings generated by various activities of the project, which with the information on existing pollution levels allows to make a preliminary estimate of the degree to which the project activities will adversely affect the prevailing conditions of the proposed site. Rapid assessment studies will provide the following information to regulatory agencies as per WHO report

(a) Identify high priority control actions

(b) To conduct detailed pollution source survey very effectively

(c) To conduct accurate environmental pollution monitoring programs

(d) To precisely estimate and evaluate the impacts of proposed control strategies

(e) To assess the impacts of new industrial development projects

(f) To develop a decision support system for site selection and for control measures.

The evaluation of Rapid assessment methods for EIA is presented in Table 2.23

Table 2.23 Evaluation of rapid assessment method.

Key Area of the Assessment Process	Criteria	L denotes Criteria completely satisfied. P denotes Criteria Partially satisfied. N denotes criteria not satisfied.
Cost/Time Effectiveness Criteria	1. Expertise requirements	L
	2. Data requirements	P
	3. Time requirement	L
	4. Flexibility	L
	5. Personnel level of effort	P
	6. Comprehensiveness	N
	7. Indicator-based	P
	8. Discriminative	N
	9. Time dimension	N
	10. Spatial dimension	N
Impact Measurement	11. Commensurate	L
	12. Quantitative	L
	13. Measurement changes	L
	14. Objective	L
Impact Assessment	15. Credibility	L
	16. Replicability	L
	17. Significance-based	N
	18. Aggregation	P
	19. Uncertainity	N
	20. Alternative comparison	P
Communication	21. Communicability	L
	22. Summary format	L

2.4.13 Predictive Models for Impact Assessment

2.4.13.1 Introduction

The EIA methods described earlier to give information in identifying impacts pathways based on underlying conceptual models linking project activities to changes in environmental components. In the application of these methods, predictions of the degree of change assessed may be qualitative which depend heavily on expert judgment or quantitative which rely on mathematical models developed by experts. In this section some of the technical and scientific methods of quantitative models for prediction of environmental changes are described. Predictive methods require collection of environmental information to set baseline values for the model variables and to determine the environmental values of a computer model parameter.

2.4.13.2 Models and Modeling

Modeling is a step-by-step process by which models are developed and/or applied. The three most common types of models used in EIA are physical models, experimental models, and mathematical models.

2.4.13.2.1   Physical Models

Physical models are small-scale models of the environmental system under investigation on which experiments can be carried out to predict future changes. Two types of physical models are discussed here:

(a)   Illustrative or visual models, and working physical models (ERL, 1984 1)(42).

(b)   Illustrative/visual models depict changes to an environmental system caused by a proposed development activity using pictorial images developed from sketches, photographs, films, photo montages, three-dimensional scale models, and from digital terrain models or digital image processing systems. Physical models simulate the processes occurring in the environment using reduced scale models so that resulting changes can be observed and measured in the model. Such models, however, cannot satisfactorily model all real-life situations; faults may occasionally arise as a result of the scaling process.

2.4.13.2.1.1   Steps in Physical Modeling

The basic steps in developing physical models are:

1. Define the environmental system to be modeled, the system's salient features, and the effect requiring prediction.

2. Select a suitable existing model facility or construct a special facility. Activities may include photographing the proposed site, and then sketching a new storage terminal on the photographs to determine the visual effect of the development, or using an existing wave chamber to predict water and sediment movements in an estuary after the construction of a new dock.

3. If no appropriate model or facility exists, one may be constructed — for example, one could construct a model of the mentioned estuary, simulating hydrological conditions in the estuary (for example flows, density, currents, waves, etc.), using an existing chamber facility. In such a case, data on morphology, hydrological conditions, and sediment movements in the estuary should be collected in order to construct a model with similar conditions.

4. Test the validity of the model by comparing its behavior with observations in the field. Adjust the model as necessary after observations.

5. Simulate the source and the conditions in the surrounding environment using appropriate methods and observe or measure the relevant changes in the model. Extrapolate the observations or measurements to predict the effects in the real environment.

6. Interpret the results, taking into account simplification of the real world made by the model.

2.4.13.2.1.2 Resource Requirements

In some cases, physical modeling exercises may be carried out in existing facilities of public and private organizations. If such facilities do not exist and if funding permits facilities may be constructed for prediction purposes. This is, however, rarely possible. Many illustrative models require less effort and expense than working physical models, although the more sophisticated computerized visual simulation models available are substantially more costly. Technical expertise and large quantities of data are required to construct working physical models that adequately simulate the behavior of the real environment. Validation and interpretation of the results of modeling may also require time and technical expertise.

2.4.13.2.2 Experimental Models

Scientific data from laboratory or field experiments provide basic information on the relationships between environmental components and human activities. Research results are used to construct empirical models that can infer the likely effects of an activity on an environmental component. Examples of experiments in which the environmental system is modeled and tested in the laboratory include toxicological tests on living organisms using polluted air, water, food, etc., micro-ecosystem experiments and pilot-scale plant tests.

Examples of experiments in which tests are carried out in the actual environment include *in situ* tracer experiments to monitor the movement of releases into the environment; controlled experiments in small parts of potentially affected ecosystems; noise tests to determine levels of disturbance; and pumping tests on groundwater.

2.4.13.2.2.1 Steps in Experimental Modeling

The basic steps in experimental modeling are:

1. Define the environmental system to be modeled, the system's salient features, and the effect requiring prediction.

2. Select a suitable experimental approach and define the specific method to be employed. Experimental activities may range from a simple laboratory determination of the level of a specific contaminant in a river and its consequent effects on fish behavior, to an *in situ* tracer experiment approach to predict the dispersion of a pollutant from a proposed sea outfall.

3. Collect the data needed to set up the experiment. To predict the effect of a pollutant on fish behavior, it might be necessary to gather data on river flow and present water quality to simulate the river.

    Sample fish may be caught and used in the laboratory experiment. Moreover, to predict dispersion of a pollutant in sea water using tracer elements, data should be collected on water movements and location of sensitive receptors to determine appropriate monitoring points.

4. Carry out the experiment, and observe and measure the relevant change in the system. For example, the effects of different pollutant concentrations on the fish should be observed and measured. The concentration of trace elements in the sea outfall should likewise be measured to determine their dispersion.

5. Extrapolate, whenever necessary, from the observations and measurements to predict the effects of the activity in the real environment. In the above two examples, this may necessitate estimating the approximate dose-effect relationships between the fish species and the pollutant, and determining dilution factors to predict the dispersion of the pollutant in sea water.

6. Interpret the results, taking into account the possible differences between experimental and actual circumstances. For instance, in the fish experiment, the absence of uptake by other organisms and the consequent reduction of dissolved oxygen in the experiment and its implication as to the accuracy of the predictions should be discussed. In the same manner, the real life contribution of such factors as decay in sunlight, different densities, and absence of biodegradation (which are controlled in the tracer experiment to predict the dispersion of a pollutant in sea water) should be accounted for and discussed in the assessment.

2.4.13.2.2.2   Resource Requirements

Experimental modeling requires substantial amounts of money, effort, time, and expertise in specialized fields.

2.4.13.3   From Conceptual Modeling to Computer Modeling

The first step in developing a predictive model is to construct a conceptual model. Most of the methods discussed in, for example, networks and impact hypotheses are based on conceptual models. To develop a quantitative predictive model, one must first represent conceptual models as mathematical equations.

Once the conceptual models are represented in mathematical language, they are amenable to computation and computerization. For example, dispersion modeling is one of the most commonly used techniques for predicting changes in air quality associated with emissions of pollutants. Relatively well established models (for example, the US Environmental Protection Agency's (EPA) computerized air quality models) are used throughout the world.

These models are based on mathematical equations that represent a simplification of basic physical processes occurring in the atmosphere. They take, as input, 1) emission of pollutants (or loadings) 2) basic meteorological data and 3) background concentrations of pollutants. They produce, as output, estimates of pollutant concentrations. These estimates are usually provided graphically as isopleths (contour lines of equal concentration) plotted around the source point.

2.4.13.3.1   Mathematical Models

Mathematical models use mathematical equations to represent the functional relationships between variables. In general, sets of equations are combined to simulate the behavior of environmental systems. The number of variables in a model and the nature of the relationships between them are determined by the complexity of the environmental system being modeled. Mathematical modeling aims to limit, as much as possible, the number of variables and thus keep the relationships between variables as simple as possible without compromising the accuracy of representation of the environmental system.

$$Cl = \frac{Q_0 C_o + Q_e C_e}{Q_0 + Q_e}$$

The above equation is an example of a mathematical model which is a simple water quality mixing model which is based on the simplest of mass balance equations. The water quality model below assumes a continuous discharge of a conservative contaminant into a stream.

where

Cl is the downstream concentration

Co is the upstream concentration

Ce is the effluent concentration

Qo is the upstream flow and

Qe is the effluent flow.

This model may be used to predict changes in downstream effluent concentrations in response to pollutant loading by changing the values of effluent concentration (Ce) and the effluent flow (Qe).

2.4.13.3.2 Types of Mathematical Models

Mathematical models can be described according to the following features:

1. *Empirical or internally descriptive:*

   - *Empirical* because they can be derived solely on the basis of statistical analysis of observations from the environment to find the "best fit" equation (empirical models are sometimes called "black box" models) or

   - *Internally descriptive* because equations are based on *a priori* understanding of the relationship between variables. The equations therefore represent some theory or assumption of how the environment works.

2. *Generalized or site-specific:*

   - *Generalized,* as they can be applicable to a range of different environmental allocations which meet certain specific characteristics or

   - *Site-specific,* as they can be developed or applied only to a specific environmental location.

3. *Stationary or dynamic:*

   - *Stationary,* if conditions in the model are fixed over the period of the prediction or

   - *Dynamic,* if the predictions are made over a period of time in which conditions in the environment change.

4. *Homogeneous or non-homogeneous:*

   - *Homogeneous,* as they can assume that conditions at the source, prevail throughout the area over which predictions are made or

- *non-homogeneous*, as environmental conditions affecting the predicted outcome vary with distance from the source.

5. *Deterministic or stochastic:*

- *Deterministic*, as input variables and relationships are fixed quantities and the predicted outcome from a given starting point is a single, unique value or

- *Stochastic*, as simple variables and parameters may be described probabilistically. These models reflect the natural variations occurring in the environment and results are presented as a frequency distribution of probable outcomes rather than as a single value.

2.4.13.3.2.1 Steps in Mathematical Modeling

There are seven steps in mathematical modeling, although not all seven must be applied in every modeling case.

1. Define the environmental system to be modeled, the system's salient features, and the effect requiring prediction (for example, the prediction of maximum concentration of a water contaminant in an area downstream from its point of discharge).

2. Select an appropriate pre-defined model or develop a new model (for the above example, a predefined model may be used to predict the downstream concentration, or in the absence of a predefined model, it may be necessary to formulate a suitable new model).

3. Collect the necessary data from existing sources or by monitoring and surveying (for the above example, data on the input variables (upstream concentration, discharge concentration, upstream flow and discharge flow) can be collected through actual monitoring and surveying).

4. If necessary, define the model parameters for the particular application, using either standard values or experimental data (calibration). For example, to predict the average annual and the maximum concentration of a pollutant emitted from a single tall stack in an open rural area, a set of atmospheric dispersion parameters should be defined for the different classes of meteorological conditions using standard empirical formulae applicable to tall stacks in open rural areas.

5. Test the validity of the model for the intended use by comparing its behavior with observations from the field.

6. Apply the model to predict the future condition of the environment.

7. Communicate the model results and assumptions to the non specialist. All relevant variables, relations, assumptions, and factors omitted from the analysis should be identified and their implications for the results discussed.

2.4.13.3.2.2   Resource Requirements

Mathematical models require varying amounts of resource inputs. A simple model, such as the river dilution model used in the above example, may require minimal input data and simple manual calculation, while a complex Gaussian plume model may require sophisticated computer techniques and demand considerable resources of input data, time, and expertise.

Assuming that an existing software program may be used, the costs of using the model may be limited to preparing the input data and labor costs for technical staff or external experts to run the model and interpret the results.

2.4.13.4 Predicting Quantitative Environmental Changes

Predictive methods for estimating quantitative changes in the environment have been commonly applied to physical systems (air, water, noise), have had some application to ecological systems, and have had limited application to social systems. Predictive models are used in EIA in two distinct ways: 1. comparison of model results with environmental standards; and 2. the evaluation of project alternatives. Where possible, experience in using the models in a developing country context is highlighted in an attempt to assess the appropriateness of applying the models in developing Asia.

In their review of EIA methods, Canter and Sadler (1997) (43) provide a listing of prediction techniques applicable to different aspects of the EIA, which are summarized in Table 2.24. A large selection of computer software is available for use in EIA. Most programs are for specific applications; many are available free of charge from government agencies and may be downloaded from the internet.

Table 2.24 Prediction techniques applicable in EIA.

Air
1. emission inventory
2. urban area statistical models
3. receptor monitoring
4. box models
5. single to multiple source dispersion models
6. monitoring from analogs
7. air quality indices

Surface Water
1. point and nonpoint waste loads
2. QUAL-IIE and many other quantitative models
3. segment box models
4. waste load allocations
5. water quality indices
6. statistical models for selected parameters
7. water usage studies

Ground Water
1. pollution source surveys
2. soil and/or ground water vulnerability indices
3. pollution source indices
4. leachate testing
5. flow and solute transport models
6. relative subsurface transport models

Noise
1. individual source propagation models plus additive model

Table 2.24 Contd...

2. statistical model of noise based on population
3. noise impact indices

Biological

1. chronic toxicity testing
2. habitat-based methods
3. species population models
4. diversity indices
5. indicators
6. biological assessments
7. ecologically based risk assessment

Historical/Archaeological

1. inventory of resources and effects
2. predictive modeling
3. prioritization of resources

Visual

1. baseline inventory
2. questionnaire checklist
3. photographic or photomontage approach
4. computer simulation modeling
5. visual impact index methods

Socioeconomic

1. demographic models
2. econometric models
3. descriptive checklists
4. multiplier factors based on population or economic changes
5. quality -of-life (QOL) indices
6. health-based risk assessment

(**Source:** adapted from Canter and Sadler, 1997) (44)

2.5 Summary

Numerous EIA methodologies have been developed in the last two decades. These methodologies are useful in identifying anticipated impacts, determining appropriate environmental factors for inclusion in a description of the affected environment, providing information on prediction and assessment of specific impacts, allowing for systematic evaluation of alternatives and the selection of a proposed action, and summarizing and communicating impact study results. The most used methodologies can be categorized as interaction matrices, networks, or checklists. Interaction matrices are of the greatest value in impact identification and the display of comparative information on alternatives. Network methodologies provide useful information on interrelationships between environmental factors and anticipated project impacts.

Checklist approaches range from simple listings of environmental factors to complex methods involving assignment of relative importance weights to environmental factors and the scaling of environmental impact factors for each of a series of alternatives. The Matrix methods are flexible and valuable tool for explaining impacts by presenting a visual display of the impacts and their causes and can be employed to identify impacts during various stages of the entire project. Overlay methods involve preparation of a set of transparent maps which represent the spatial distribution of an environmental characteristic (e.g., Extent of dense forest area). Information on a wide range of variables will be collected for standard geographic units within the study area which will be recorded on a series of maps typically one for each variable. Geographical Information Systems can have a wide application in all EIA stages, acting as an integrative framework for the entire process, from the generation, storage, and display of the thematic information relative to the vulnerability and/or sensitivity of the affected resources, to impact prediction and finally their evaluation of decision support.

Expert opinion using professional judgement and Expert systems consisting of drawing upon professional knowledge and judgement of experts as coded systems in user friendly computer software is used in may EIA which is drawing greater attention in recent days.

These methods are widely used for assessing visually the changes in the landscape before and after the activity. Cost/benefit analysis provides the nature of expense and benefit accrual from a project in monetary terms as a common practice in traditional feasibility studies and thus enables easy understanding and aids decision-making adaptive environmental assessment and management (AEAM), combines various simulation models to predict impacts. This approach broadens the potential of simulation models to evaluate the impacts of alternatives and is beneficial for project planning. Predictive models are used in EIA in two distinct ways: 1. comparison of model results with environmental standards and 2. the evaluation of project alternatives. The technical details of various above EIA methodologies and their application with specific examples are discussed in this chapter.

References

1. Canter, L.W., 1996a Environmental Impact Assessment, Second Edition, McGraw-Hill Publishing Company, Inc., New York.

2. Canter, L.W. 1986. Environmental Impact of Water Resources Projects. Chelsea, MI: Lewis.

3. Environmental Impact Assessment: Guidelines for Planners and Decision Marker, UN Publication S1/1 SCAP/351/ESCAP, 1985.

4. An introductory guide to EIA in P Wathern (ed), "Environmental Impact Assessment. Theory and Practice" Routledge, London 1988 oronto.

5. Carstea.D. 1976 Guidelines for the Environmental Impact Assessment of small structures and related activities in coastal bodies of water, HTR-6916, rev. 1, Aug.

6. Morris, P. and Therivel, R. (Eds), 2001. Methods of Environmental Impact Assessment, 2nd edition, Spon Press, London (2008 reprint).

7. Canter and Hill. L.G. 1979 Handbook of variables for Environmental Impact Assessment, Ann Arbor Science Publishers.

8. Dee, N., J. Baker, N. Drobny, K. Duke, T. Whitman, and P. Fahringer. 1972. An Environmental Evaluation System for Water Resource Planning. Water Resource Research, Vol. 9, pp. 523-535.

9. Dee.N., 1972 Environmental evaluation system for Water Resources Planning; Final Rep. Battelle-Columbus Laboratories, Columbus, Ohio.

10. Westman, W.E. 1985. Ecology, Impact Assessment and Environmental Planning. John Wiley & Sons.

11. Bisset, R, 1988 "Developments in EIA methods", in P Wathern (ed), "Environmental Impact Assessment. Theory and Practice" Routledge, London.

12. Bisset, R, "Methods for environmental impact assessment. A selective survey with case studies", in A Biswas, Q Geping (eds).

13. Environmental Impact assessment for developing countries, Tycollo International Press 1987.

14. Colombo A G 1992 "Environmental Impact Assessment" Kluver Academic Publishers, Brussels.

15. Leopold.L.B; Clarke, K. E, Harrow, B. B and Balsley, J. R. 1971 "A Procedure for evaluating Environmental Impact", Circular 645, U.S. Geological Survey, Washington. D.C,.

16. Johnson, F.L. and Bell.D.T, (1975) "Guidelines for the identification of potential Environmental Impacts in the construction and operation of a Reservoir", Forestry Research Rep. 75-6, Department of Forestry, University of Illinois Champaign.

17. Sorensen, J.C. 1971. A Framework for Identification and Control of Resource Degradation and Conflict in The Multiple Use of the Coastal Zone, Master's thesis, University of Berkeley.

18. Lohani, B.N. and N. Halim. 1983. Recommended Methodologies for Rapid Environmental Impact Assessment in Developing Countries: Experiences Derived from Case Studies in Thailand, Workshop on Environmental Impact.

19. Wathern, P. 1988. An introductory guide to EIA. *In:* P. Wathern (ed.). "Environmental Impact Assessment operation of a Reservoir", Forestry Research Rep. 75-6, Department of Forestry, University of Illinois.

20. Burrough, P A (1986). Principles of Geographical Information Systems for Land Resources Assessment. Clarendon Press, Oxford.

21. Longley, Paul A.; Goodchild, Michael F.; Maguire, David W. & Rhind (2005). Geographic Information Systems and Science (2nd ed.) England, John Wiley and sons Ltd.

22. Antunes, P., Santos, R. & Jardao, L. (2001). The Application of Geographical Information System to determine environmental impact significance", Environmental Impact Assessment Review, Vol. 21, pp 511-535.

23. Erickson, P.A., (1994). A Practical Guide to Environmental Impact Assessment (New York: Academic Press.

24. Bhatt, R.P. (2009). The Need and Use of Geographic Information System for Environmental Impact Assessment in Nepal; Journal of Water Energy and Environment, Issue No (4) PP (29-31).

25. Fedra K. (1993). GIS and environmental modelling. In Goodchild, M.F., Parks, B.O., Steyaert, L.T., editors. Environmental modelling with GIS. (Oxford: Oxford Univ. Press. pp. 35– 50).

26. Eedy, W., (1995). The use of GIS in environmental assessment, in Impact Assessment (International Association for Impact Assessment, IAIA), 13(20): 199-206.

27. Joao, E. & Fonseca, A. (1996). The role of GIS in improving environmental assessment effectiveness: theory vs practice. Impact Assess 1996; 14:371– 87.

28. Edward-Jones, G., and Gough, M., 1994 "ECOZONE: A Computerized Knowledge Management System for Sensitizing Planners to the Environmental Impacts of Development Projects," Project Appraisal, Vol. 9, No. 1, March, pp. 37-45.

29. Lein, J.K., 1988 "An Expert System Approach to Environmental Impact Assessment," International Journal of Environmental Studies, Vol. 33, pp. 13-27.

30. Coker, A., and Richards, C., Editors, 1992 Valuing the Environment: Economic Approaches to Environmental Evaluation, Belhaven Press, London, England,.

31. Pearce, D., Markandya, A., and Barbier, E., 1989 Blueprint for a Green Economy, Earth scan Publications, Ltd., London, England,. Pearce, D., and Turner, K., 1990 Economics of   Natural Resources and the Environment, Johns Hopkins University Press, Baltimore, Maryland.

32. Ranasinghe, M., 1994 "Extended Benefit-Cost Analysis: Quantifying Some Environmental Impacts in a Hydropower Project," Protect Appraisal, Vol. 9, No. 4, pp.   243-251.

33. Sadler, B., Manning, E.W., and Dendy, J.0., 1995 Editors, Balancing the Scale: Integrating   Environmental and Economic Assessment, Foundation for International Training, Toronto, Canada.

34. Asian Development Bank, 1987a. Environmental guidelines for selected agricultural and natural resources development projects. Asian Development Bank, Manila, Philippines.

35. Asian Development Bank, 1991 "Guidelines for Social Analysis of Development Projects: Operational Summary," June, Manila, Philippines.

36. Asian Development Bank, 1993a. Environmental guidelines for selected infrastructure projects. Asian Development Bank, Manila, Philippines.

37. Asian Development Bank, 1993b. Environmental Guidelines for Selected Industrial and Power Development Projects.

38. Asian Development Bank, 1991. Remote Sensing and Geographical Information Systems for Natural Resource Management. Asian Development Bank Environmental Paper No. 9. 202 pp.

39. Leopol Linstone, Mass., 197, H. A and Turnoff, M., The Delpi Method-Techniques and Applicaitons, Addison-Wesley Publishing Company, Reading 5. d,

40. L.B., F.E. Clarke, B.B. Manshaw, and J.R. Balsley. 1971. A Procedure for Evaluating Environmental Impacts, U.S. Geological Survey Circular No. 645, Government Printing Office, Washington, D.C. Assessment, Guangzhou, People's Republic of China.

41. Economopoulos, Alexander P. 1993a. Assessment of Sources of Air, Water, and Land Pollution: A Guide to Rapid Source Inventory Techniques and Their Use in Formulating Environmental Control Strategies. Part One: Rapid Inventory Techniques in Environmental Pollution. World Health Organization, Geneva.

42. Economopoulos, Alexander P. 1993b. Assessment of Sources of Air, Water, and Land Pollution: A Guide to Rapid Source Inventory Techniques and Their Use in Formulating Environmental Control Strategies. Part Two: Approaches for Consideration in Formulation of Environmental Control Strategies. World Health Organization, Geneva.

43. ERL (Environmental Resources Limited). 1984. Prediction in Environmental Impact Assessment, a summary report of a research project to identify methods of prediction for use in EIA. Prepared for the Ministry of Public Housing, Physical Planning and Environmental Affairs and the Ministry of Agriculture and Fisheries of the Government of Netherlands.

44. Canter, Larry W. and Barry Sadler. 1997. A Tool Kit for Effective EIA Practice — Review of Methods and Perspectives on their Application. A Supplementary Report of the International Study of the Effectiveness of Environmental Assessment. Environmental and Ground Water Institute, University of Oklahoma, Institute of Environmental Assessment, UK and the International Association for Impact Assessment.

Suggested Further Reading

Biswas, A.K. and S.B.C. Agarwala eds... 1992. Environmental Impact Assessment for Developing Countries. Oxford: Butterworth-Heinemann.

Bregman, J.I. and K.M. Mackenthun. 1992. Environmental Impact Statements. Chelsea, MI: Lewis.

Gilpin A. 1995. Environmental Impact Assessment EIA Cutting edge for the twenty-first century. Cambridge University Press.

Glasson J., Therivel R. and Chadwick A. 1999. Introduction to Environmental Impact Assessment. Second edition, UCL Press: London.

Goodland R. and Edmundson V. eds, 1994. Environmental Assessment and Development. The World Bank: Washington DC, USA. 159pp.

Goodman and Mercier. 1999. The Evolution of Environmental Assessment in the World Bank: from "Approval" to Results. World Bank, Washington DC.

Jain, R.V. et al. 1993. Environmental Assessment. New York: McGraw Hill.

Lee N. and George C. 2000. Environmental Assessment in Developing and Transitional Countries.

Morgan, R. K. 1998. Environmental Impact Assessment: A Methodological Perspective. Kluwer Academic Publishers: Dordrecht, The Netherlands. 307pp.

Morris, P. and Therivel, R. 1995. Methods of Environmental Impact Assessment. UCL Press. London.

O'Riordan, T. and W.R.D. Sewell eds, 1981. Project Appraisal and Policy Review. New York: John Wiley & Sons.

Petts, J. 1999. Handbook of Environmental Impact Assessment. Two Volumes. Blackwell Science, Oxford, Volumes I 484pp. and II 450pp

Sadler, B. 1996. International Study of the Effectiveness of Environmental Assessment. Environmental Assessment in a Changing World: Evaluating Practice to Improve Performance. Final Report, Ottawa, Canada: Ministry of Supply and Service Canada.

Sheate, W. 1996. Environmental Impact Assessment: Law and Policy- Making an Impact. Cameron May, London, UK. 300pp.

Smith, L.G. 1993. Impact Assessment and Sustainable Resource Management. Longman Scientific and Technical, Harlow, Essex, UK.

United Nations Environment Programme Bisset, R, 1996. Environmental Impact Assessment: Issues, Trends and Practice.

United Nations Environment Programme, Environment and Economics Unit, Nairobi, Kenya. 96pp.

United Nations Environment Programme.1988. Environmental Impact Assessment: Basic Procedures for Developing Countries. UNEP Regional Office for Asia and Pacific: Bangkok.

Wather, P. ed... 1988. Environmental Impact Assessment: Theory and Practice. London.

Unwin Hyman. Wood, C. 1995. Environmental Impact Assessment: a Comparative Review, Longman Group Ltd. Harlow, UK.

World Bank. 1991. Environmental Assessment Sourcebook. 3 vols. Washington, DC: World Bank.

Questions

1. List the various EIA methods. What are the criteria used for selecting best EIA method in a given situation?

2. What are adhoc methods? Where they are useful? What are its draw backs?

3. What are the different categories of check methods? Discuss different environmental factors to be considered in check list methods.

4. What are scaling and weighing scaling check list methods? What are different environmental factors considered in check list methods?

5. What are different types of scales used in scaling check list methods?

6. What is Battelle Environmental Evaluation System?

7. List out the conditions in which checklists are more useful or better used.

8. What are the salient features of Matrix methods? What are interaction Matrix methods. Discuss with reference to Leopoid matrix method?

9. What are stepped up matrix and net works methods?

10. Discuss what are secondary impacts that can be visualized in dredging and pulp mill projects?

11. What are overlay methods? How is GIS useful as advanced tool in overlay methods?

12. What are WHOs Rapid assessment of Pollution Load methods? Explain the terms Waste Load Factors.

13. Explain the salient features of physical models, experimental models, and mathematical models used for Rapid EIA.

14. Discuss the different steps in chronological order for i) Physical Modeling ii) Experimental Modeling and iii) Mathematical Modeling. What are their resource requirements?

15. What are predictive Models? How they are useful in EIA process? Explain the various Predictive Techniques used in different steps of EIA.

Prediction and Assessment of Impacts on Soil and Groundwater Environment

3.1 Introduction

Almost every type of action or project can produce changes in the surroundings of the land. Due to diverse planning procedures in both rural and urban areas, the soil resource is excessively consumed quantitatively and qualitatively in the current situation. Therefore the optimization of land use and its planning of preserving the natural assets of the resource soil is crucial. Some actions and projects will have a direct pollution effect, while others may induce changes or have secondary impacts. The assessment of potential land-use impacts should be comprehensive, covering characteristics of the project. The impacts may affect the soil, groundwater, leachates due to waste disposal through the sub-surface system. Soil management is also very crucial in areas where development activities are taking place. Proper soil databases and maps should be developed for strategic management of soil.

3.2 Soils and Groundwater

The integrity of soils and groundwater can be altered by a variety of physical disturbances, including the addition/removal of soil and/or water, compaction of soil, changes in use of land or ground cover, changes in water hydrology, changes in climate (temperature, rainfall, wind), and the addition or removal of substances or heat (for example, discharge of effluents into groundwater, discharge of effluents or disposal of waste onto land, leaching of contaminants into groundwater, changes in quality of surface water, and deposition of air pollutants on land) (1, 2). The effects of these vary from first order effects of leaching into soil and groundwater to changes in groundwater regime, soil structure (including erosion and subsidence), soil quality or temperature, and groundwater quality or temperature. A summary of these effects is presented in Fig. 3.1.

3.3 Methodology for the Prediction and Assessment of Impacts on Soil and Groundwater

To provide a basis for addressing soil and/or groundwater, environmental impacts, a model is suggested, which connects seven activities or steps for planning and conducting impact studies. (3-6) Fig. 3.2. In analyzing environmental impacts, both objective and subjective judgments should be taken into consideration. Objective judgments are defined as "those, which involve or use facts that are observable or verifiable especially by scientific methods and which do not depend on personal reflections, feelings, or prejudices; "subjective judgments are those which are made on the basis of values, feelings and beliefs". In the context of the environment the objective judgment describes the impact, whereas subjective judgement describes how people feel about the 'fact'.

The soil functions in terms of raw material usage, biodiversity support, biomass production, transformation of substance, storage, filtering, infrastructure is expected to be translated into either economic or ecological forms of land use.

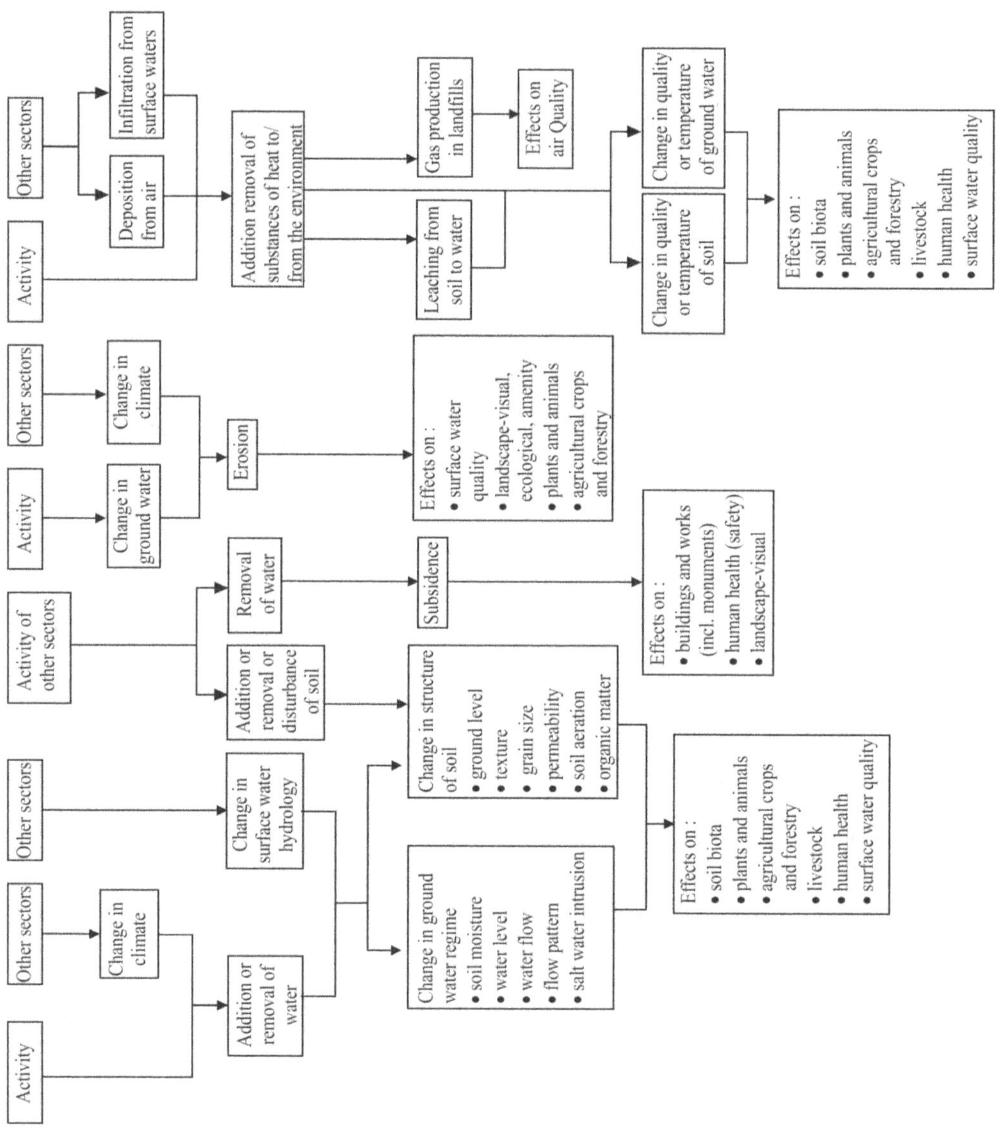

Fig. 3.1 Soil and groundwater effects

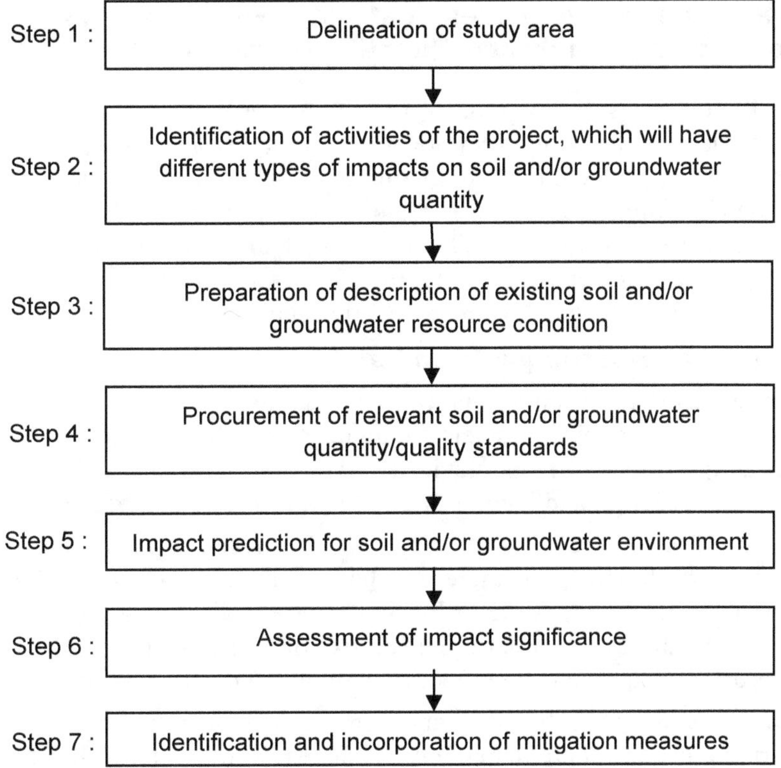

Fig. 3.2 Systematic approach for the study of impacts on soil and groundwater.

3.3.1 Delineation of Study Area

The delineation of the study area for impact assessment will be very specific based on presence of potential impacts. The study area should reflect the full reach of possible effects within the particular impact discipline that is being considered.

The proposed or future land use map along with committed land use policies, zoning, and development projects should be included in the study area.

The map should clearly distinguish between developed and undeveloped land. The categories shown on the land use map, should be

- Residential
- Commercial and industrial
- Institutional and parks or recreation
- Non-urban mixed

The map could include further divisions, such as separate commercial and industrial activity centers and public vacant lands. The categories to be used will depend largely on the type of project or action being evaluated, the characteristics of the local land area, and the geographic extent of the affected study area.

3.3.2 Identification of Activities, that will have Different Types of Impacts on Soil and/or Groundwater Quantity – Quality (7-9)

Pipeline leaks, oil spills, toxic spills from mining, airports, gas stations, factory effluents cause local contamination within their location In some cases, toxic pesticides and fertilizers contaminate the soil. Facilities for natural disasters like earthquakes, tsunamis and hurricanes hold harmful chemicals, oil and toxins, causing spills and leaks into the soil. Radioactive materials, chemicals, salts, toxic compounds and other pollutants when spilled into the soil, begin to deteriorate the texture, mineral content and quality of the soil. The pollution disturbs the biological balance of soil organisms.

3.3.2.1 Direct Land-Use Impacts on Land

1. ***Landforms***: Unique or important physical features that have special importance, as recreational or educational or scientific interests may be present in the project area. They may be unique locally or unique in a larger area. Examples are rock outcrops, river gorges, sandy beaches, and lagoons. Such features may also influence local climate.

2. ***Soil profile***: The soil profile is related to the chemical and physical nature of the soil and the prevailing climate and therefore has a direct bearing on land capability for agricultural or other purpose. Erosion is the principal process which may alter the soil profile and it can have a direct effect on existing or potential land use, and an indirect effect, through siltation on water quality, fishing, land use downstream.

3. ***Soil composition***: The chemical and mineral composition of the soil influences its engineering and agricultural capability. Changes in soil composition can occur either by, subtraction e.g., acid or alkali leaching or by adding e.g., cation exchange extraction, nitrogen fixation.

4. ***Slope stability***: Rock slopes are inherently stable. The environmental effects of slope instability are similar to those for erosion. The scale of the effects is larger in this case.

5. ***Seismicity:*** Stress, vibration due to explosions and deep well injection operations can have an effect on the stress-strain equilibria on fault planes. Renewed or increased activity can have major environmental effects for the project site.

6. ***Subsidence and compaction***: Subsidence and compaction occur naturally, but generally as a gradual and almost imperceptible process. The process can be accelerated, however, by underground excavation, vibration or loading. The major effect is on land capability, but drainage, groundwater behavior and landscape could also be affected.

7. ***Flood plains and Swamps***: Flood plains and swamps are an important part of the drainage pattern as they admit peak flows into the drainage system. Reclamation on natural flood plains or swamps may result in flooding and siltation of other areas during peak flow. The major engineering of a drainage system may either decrease the amount of agricultural land available or may destroy wetland habitats of fish, birds etc.

8. ***Land use***: The existing land use and the compatibility with existing or planned use of adjacent land are important components of the environment. Careful site selection is the principal means of controlling them, but many mitigating or abatement measures may also be available.

9. ***Mineral or engineering resources:*** The occurrence of minerals or engineering resources is of strategic and economic importance. Loss of such resources either through wasteful use or through development incompatible with subsequent mining or quarrying proposal can result in long-term economic or social impacts on the community.

10. ***Buffer zones:*** Buffer zones are spaces which provide natural environmental protection from drainage by external events. They are usually vegetated, depending on the purpose and can provide wind breaks, erosion control, sediment traps, wildlife shelter, sound insulation and visual screening.

Some projects or actions, by their very nature, have direct and obvious impacts of land use by physically destroying or clearing land and implementing a new use. Here are some examples of this kind of direct land-use revision:

(a)   A highway project with a 300 right-of-way width converts whatever the existing land use is to a transportation, land-use within that right-of-way width.

(b)   A dam constructed to create a reservoir for water supply and recreational use directly converts the previous land use to recreational use.

(c)   A regional park constructed on land previously used as pasture directly changes the number of acres of the park into a different use.

(d)   A city block of low-income housing structures is demolished to construct a shopping mall, directly converting that land to commercial use.

3.3.2.2 Examination of Existing and Future Planned Land-Use for Delineating Study Area

The first step is to get the necessary information on existing development trends, planned development projects, and especially the goals and objectives of land-use plans and policies. These existing and proposed committed projects and policies are then factored into the no-build alternative. The result is a definition of future development intensity and policy without the proposed project. The impact of the proposed project is the difference between these future conditions (no - build) and the future conditions with implementation of the proposed project or action (build). There may also be substantive differences among the various build alternatives being considered.

An initial activity is co-ordination with the regional planning, organization and with the local planning officials and zoning agencies. This early contact is valuable to:

-   Determine the existing and planned land- use and zoning for the area of the proposed project,
-   Identify any particular problem,
-   Identify goals for land- use and economic development
-   Initiate continued review and coordination throughout the project study phase.

Depending on the expected magnitude of impact of the various activities of the project, the population growth.

3.3.2.3 Environmental Impacts on Soil and Groundwater- A Typical Example: Road Construction Project (10)

Impacts and Setting

Soil is an important component of the natural environment, and is a primary medium for many biological and human activities, including agriculture. Its protection in relation to road development deserves considerable attention.

On the road itself, in borrow pits, or around rivers and streams, there are many places where damage might occur. Losses can be considerable for the road agency and others. This includes farmers losing crops and land, fishermen losing income because of sedimentation in rivers and lakes, and road users being delayed when road embankments or structures collapse. The costs of correcting these problems are often many times greater than the costs of simple preventive measures.

Loss of Productive Soil

The most immediate and obvious effect of road development on soil is the elimination of the productive capacity of the soil covered by roads. Unfortunately, the best sites for road development (flat and stable) also tend to be ideal for agriculture. The narrow, linear character of roads makes the impact of lost land seem minimal, but when the width of the right-of-way is multiplied by its length, the total area of land removed from production becomes much more significant. Soil productivity can also be reduced significantly as a result of compaction with heavy machinery during construction.

Erosion

When natural conditions are modified by the construction of a road, it marks the start of a race between the appearance of erosion and the growth of vegetation. Disturbance during construction can upset the delicate balance between stabilizing factors, such as vegetation, and others which seek to destabilize, such as running water. In some cases erosion might result in cumulative impacts far beyond the road itself, affecting slopes, streams, rivers, and dams at some distance from the initial impact.

Destabilization of Slopes

Slope stability can be upset by the creation of road cuts or embankments. Excessive incline of cut slopes, deficiency of drainage, modification of water flows, and excessive slope loading can result in landslides (Fig. 3.3). Some soils, such as shale and quick clays, are known for being difficult to drain and particularly unstable.

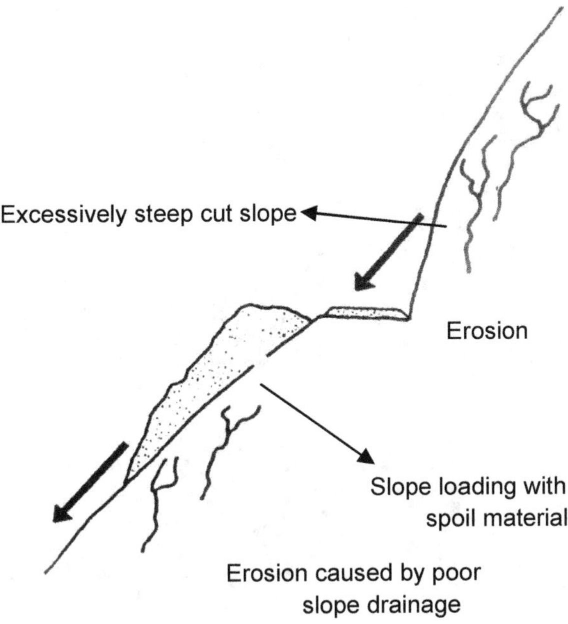

Fig. 3.3 Destabilization of slopes.

Side Tipping of Spoil Materials

Spoil material from road cuttings can kill vegetation and add to erosion and slope stability problems. Large amounts of spoil can be generated during construction in mountainous terrain. Sometimes it is difficult to design for balances between cut and fill volumes of earth at each location, and haulage to disposal sites may be expensive. This creates a need for environmental management of tipping material.

Water Flow Diversions

Diversion of natural surface water flows has been often inevitable in road projects. Diversion results in water flowing where it normally would flow.

Engineering Measures

In many cases, vegetation alone may not be enough to prevent erosive damage to slopes, and various engineering measures may be needed to complement or replace it. The use of slope retaining techniques may be necessary when

- Slopes are unstable because they are too high and steep
- Climatic conditions are such that establishment of vegetation is slow or impossible
- There is a risk of internal erosion or localized rupture because of drainage difficulties.
- It is necessary to decrease the amount of earthwork because the road width is limited.

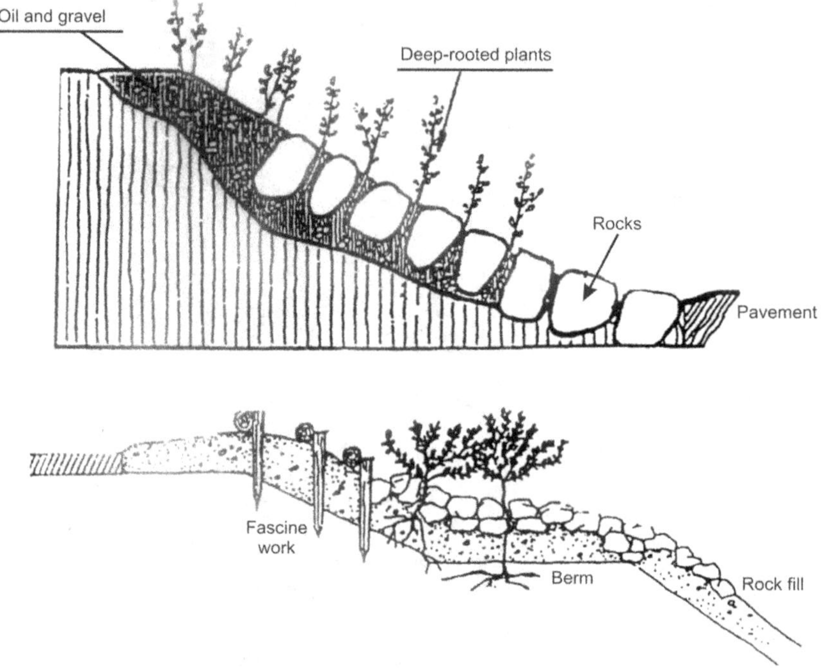

Fig. 3.4 Examples of combined techniques for slope protection.

Well established engineering measures for slope protection include:

- Intercepting ditches at the tops and bottoms of slopes. Gutters and spillways are used to control the flow of water down a slope

- Terraced or stepped slopes to reduce the steepness of a slope. A berm (or riseberm) is the level section between slope faces

- Riprap or rock material embedded in a slope face, sometimes combined with planting

- Retaining structures, such as gabions (rectangular wire baskets of rocks), cribs (interlocking grid of wood or concrete beams, filled with earth or rock), or other types of wooden barricades and grid work, usually battered back against the slope

- Retaining walls, more substantial engineering structures able to resist bending, and with a footing designed to withstand pressures at the base of the slope

- Reinforced earth, embankment walls built up as the earth fill is placed, with anchors compacted into the fill material and

- Shotcreting and Geotextiles, generally more expensive options with specific applications.

3.3.2.4 Preliminary Procedure for General Projects

An appropriate initial activity when analyzing a proposed project or activity is to consider what types of soil and/or geological disturbances might be associated with the construction and/or operational phases of the proposed project, and what quantities of potential soil contaminants are expected to occur.

Impact trees or networks can be used to delineate potential impacts on the soil and geological environments.

Regarding the identification of potential soil pollutants, a list of the materials to be utilized during the project and those materials which will require disposal could be developed. Examples of materials that may result in soil contamination include fuels and oils, bituminous products, insecticides, fertilizers, chemicals, and solid and liquid wastes. As an initial step, a simple checklist of the types and quantities of chemicals associated with each activity could be prepared and utilized. Transport and effects information on key chemicals could also be included. It may also be appropriate to consider the quality of leachates from waste materials disposed on land.

Environmental problems in land conservation in the following can be analyzed using systems analysis techniques:

1. The degradation of soil fertility due to increase in concentration of sodium, caused by water logging and application of chemical fertilizers.

2. Physical loss of soil through accelerated erosion due to the action of water and wind.

3. Impact of the conversion of good farm lands into reservoirs and dwelling areas.

3.3.2.5 Groundwater – Quantity and Quality Impacts

The consideration of groundwater quantity and quality impacts consist of identifying the types and quantities of groundwater pollutants and/or groundwater quantity changes anticipated to be associated with the construction and operational phases of the proposed project. This activity should also be performed for any alternatives to the project or proposed plan programs.

Numerous types of projects could have detrimental impacts on the soil or geological environment, or both (Table 3.1 and 3.2).

Table 3.1 Effects of developmental activities on five classic factors for soil formation.

Factors of soil Formation	Type of effect	Nature of effect
Climate	Beneficial	Adding water by irrigation; rainmaking by seeding clouds; removing water by drainage, diverting winds, etc.,
	Detrimental	Subjecting soil to excessive insolation to extended frost action, to wind etc.
Organisms	Beneficial	Introducing and controlling populations of plants and animation adding organic matter including 'nightsoil'; loosening soil by ploughing to admit more oxygen; following; removing pathogenic organisms, adding radioactive substances.
	Detrimental	Removing plants and animals; reducing organic matter content of soil through burning, ploughing, over-grazing, harvesting, etc; adding or fostering pathogenic organisms; adding radioactive substances.

Table 3.1 contd...

Factors of soil formation	Type of effect	Nature of effect
	Detrimental	Causing subsidence by drainage of wetlands and by mining; accelerating erosion; excavating
Parent material	Beneficial	Adding mineral fertilizers; accumulating shells and bones; accumulating ash, locally; removing excess amounts of substances such as salts.
	Detrimental	Removing, through harvest, more plant and animal nutrients than are replaced; adding materials in amounts toxic to plants or animals; altering soil constituents in a way to depress plant growth.
Time	Beneficial	Rejuvenating the soil through adding of fresh parent material or through exposure of local parent material by soil erosion; reclaiming land from under water.
	Detrimental	Degrading the soil by accelerated removal of nutrients from soil and vegetation cover; burying soil under solid fill or water.

Source : Goudie (3)

Table 3.2 Examples of human-induced effects on soil characteristics

Soil factor	Beneficial change	Neutral change	Adverse change
Soil chemistry	Mineral fertilizers (increased fertility) Adding trace elements Desalinize (irrigation) Increase oxidation (aeration)	Altering exchangeable ion balance Altering pH (lime) Alter via vegetation change	Chemical imbalance Toxic herbicides and herbicides Salinize Over–removal of nutrients
Soil physics	Induce crumb structure (lime and grass) Maintain texture (organic manure or conditioner) Deep plowing, after soil moisture (irrigation of drainage)	Alter structure (plowing, harrowing) Alter soil microclimate (mulches, shelter belts, heating, albedo change)	Compaction/plough pan (poor structure) Adverse structure due to chemical changes (salts) Removal perennial vegetation
Soil organisms	Organic manure Increase pH Drain/moisten Aerate	After vegetation and soil microclimate	Remove vegetation and plough (less and microorganisms Pathogens Toxic chemicals
Time (rate of change)	Rejuvenate (deep ploughing adding new soil, reclaiming land)		

3.3.2.6 Examples of Types of Projects and Associated Impacts Include:

1. Land subsidence which can occur as a result of over- pumping of ground-water resources or oil, gas resources in a given geographical area or which can occur as a result of surface or sub-surface mining activities associated with mineral extraction.
2. The impacts associated with the identification and usage of construction material for major projects, with such material coming from identified burrow areas. (There may be

changes in local surface water hydraulics and erosional patterns as a result of construction material).

3. Construction practices in general can create some concerns related to the potential for increased soil erosion in the construction area. This increase in soil erosion could lead to specific mitigation requirements, such as, the creation of sediment retention basins or the planting of rapidly growing vegetation.

4. Landslides, caused by inappropriate slope stability, which can occur as a result of over development in particular soil types within the areas having certain topographic features.

5. The potential concerns associated with constructing and operating nuclear power plants, chemical production plants, waste–disposal facilities, and/or large storage tank facilities in areas characterized by seismic instability and excessive earthquake potential. (This can influence sitting decisions and decisions associated with construction and operation activities.)

6. Strip–mining operations for coal extraction, or other mineral resource extraction wherein the land surface is to restore the original landscape, possibly in some type of alternative topographic arrangement.

7. The construction of jetties along coastal areas in order to control beach erosion and littoral drift.

8. Projects which may create acid rain in a localized area, with the acid rain, in turn, having an impact on soil chemistry and, potentially, on sub-surface groundwater resources.

9. Projects wherein the site characteristics in terms of soil and geological features are incorporated as components in the selection process; examples of such site–selection oriented projects, sludge–disposal projects, and upland locations for dredged–material disposal.

10. Projects that involve developments along the coastal areas wherein coastal erosion problems may either be increased by the project, or may influence the proposed project itself. Examples of such projects include the coastal marines and associated secondary developments, industrial development projects with associated port and boat mooring facilities, and projects, which involve the development of ports and harbors.

11. The construction and operation of surface water reservoir projects, with the purposes of the projects ranging from the single purposes of providing flood control to multiple purposes, including hydroelectric power development, provision of water supply, and so on. There are two key environmental concerns relative to soil and geological issues, the first is related to sedimentation within the reservoir and the provision of appropriate sediment – storage capacity in terms of the project lifetime; the second is related to the potential effects of such surface water reservoir projects on the subsurface environment, including changes in soil, ground water, and geological features that lie underneath the water pool of the reservoir.

12. Projects associated with permits for grazing leases or other leases related to agricultural uses, where the subsequent grazing or agricultural developments could lead to changes in soil characteristics such as erosion patterns and soil chemistry. Examples of such changes are in Table 3.2.

13. The potential effects of soil characteristics on buried pipelines, with examples, including the potential loss of the physical integrity of the pipeline as a result of acid or corroding soils.

3.3.3 Description of Existing Soil and/or Groundwater Resources Soil Characteristics

3.3.3.1 Background Information on the Soil Environment

Soil Characteristics

Soil characteristics in a given geographical area at a given point of time are a function of both natural influences and human activities.

The soil and geological environments are typically associated with the physical and chemical environment.

For example, the habitat types and associated vegetation found in an area will be a function of the soil characteristics. Additionally, cultural resources may be related to soil characteristics or, possibly, to unique geological features in an area.

The relationship between shallow, alluvial aquifers and the flow of surface streams and rivers may need to be explored. Table 3.3 summarizes the principal anthropogenic activities, which can cause groundwater pollution.

In describing quantity and quality, specific indicator parameters can be utilized. For example, the following represent some of the information, which could be compiled, and the issues which could be addressed, are:

1. Descriptions should be assembled on groundwater systems in the study area, indicating whether they are confined or unconfined, with the obvious pollution relevance being that unconfined groundwater systems tend to be more susceptible to groundwater contamination.

2. Of particular importance would be the description of karsts aquifer systems, since these areas can exhibit unique and rapid groundwater flow patterns.

3. Many areas are characterized by the presence of multiple groundwater systems. Accordingly, it would be appropriate to describe those geographical areas characterized by multiple aquifer systems.

4. If information exists on the quantitative aspects of the groundwater resource in terms of potentially useable supplies, which could be extracted, it should be summarized.

5. Information should be summarized on the uses of groundwater within the study area, with a more detailed study of this subject to be conducted later.

6. A description of the relationships between local groundwater systems and surface streams, lakes, estuaries, or coastal areas may be important, since mutual quantitative or qualitative influences can occur.

7. Groundwater pollution vulnerability is associated with the question whether or not the project area is in a recharge zone for a given groundwater system. This should be determined because there is greater pollution potential in the recharge zone. (It should be noted that for confined aquifer systems the recharge area may be located a long way from the actual segment of the groundwater system being used for purposes of water supply.)

8. Depth of groundwater is a fundamental parameter, which could be identified, with the pertinent issue that greater the depth of groundwater, the greater the degree of natural protection.

9. Unsaturated - zone permeability should be described. Here, the "unsaturated zone" refers to that segment of the subsurface environment, which is between the land surface and the water table of an unconfined aquifer system. The unsaturated zone permeability can influence the attenuation of contaminants as they move away from a source of pollution and toward the groundwater system.

10. Aquifer transmissivity should be described. This parameter represents information on the water carrying capacity of the ground water system.

11. Any existing data on groundwater quality should be summarized. If no such data exists, it may be necessary to appropriately plan and conduct a groundwater-monitoring program. In some unique cases, the quality data may need to be described in terms of aquatic ecosystems. For example, several threatened or endangered aquatic species have been found in springs associated with the Edwards aquifer in central Texas.

3.3.3.2 Unique Soil or Groundwater Problems

Many geographical areas exhibit special or unique problems that should be addressed in the description of baseline conditions for the soil or groundwater resources in the study area. Examples of these problems include saline seeps, groundwater supplies relative to existing bacteriological or other quality constituents, poor natural quality, and the presence of hazardous waste sites. Dryland farming practices involving irrigation often lead to salt accumulation in surface soils and shallow unconfined aquifers.

3.3.3.3 Pollution Sources and Groundwater Users

It is appropriate to consider which other potential and actual sources of soil and/or groundwater pollution may exist in the study area, and also to consider current and potential future usage of the groundwater resource for purposes of water supply techniques. Quantitative impact prediction is typically associated with the use of look – alike, or analogous projects for which knowledge and information are available, and/or the utilization of relevant case studies.

3.3.4 Procurement of Relevant Soil and/or Groundwater Quantity – Quality Standards

Land-use restrictions, soil quality standards, soil reclamation requirements, and groundwater quantity – quality standards, regulations, or policies are examples of institutional measures, which can be used to determine impact significance and required mitigation measures. Thus, to determine the specific requirements for a given project area will require contacting appropriate governmental agencies with jurisdiction.

The primary sources of information needed for step 3 (Figure 3.2) will be pertinent to the governmental agencies, namely, Central government, State government and/or local agencies. In addition, international environmental agencies may have information pertinent to this step.

3.3.5 Impact Prediction

The prediction of the impacts of a project – activity on the soil and/or groundwater environment(s), or conversely, the potential influence of the environment(s) on a proposed project, can be approached from three perspectives.

1. Qualitative
2. Simple quantitative, and
3. Specific quantitative

In general, efforts should be made to quantify the anticipated impacts; however, in many cases this will be impossible and reliance must be given to qualitative trend and through the spreading of excess sub-soil over the right-of-way during clean-up. In general, the mixing of sub-soil with topsoil will have an adverse impact on soil fertility and soil structure. The severity of the impact will depend on the nature of the sub-soil.

3.3.5.1 Qualitative Approaches-Groundwater Impacts (11-29)

A qualitative approach for groundwater-impact prediction involves the fundamental sub-surface environmental processes. The fundamental processes in the sub-surface environment can be examined relative to their hydrodynamic (physical), biotic (chemical), aspects. Table 3.3 summarizes processes, which may affect the constituents of ground water.

Table 3.3 Possible sources of ground water contamination.

Category I	Sources designed to discharge substances

Subsurface percolation (e.g., septic tanks and cesspools)
Injection wells
    Hazardous waste
    Non-hazardous waste (e.g., brine disposal and drainage)
Non-waste (e.g., enhanced recovery, artificial recharge, solution mining, and in-situ mining)
Land application
    Waste water (e.g., spray irrigation)
    Waste water by – products (e.g., sludge)
    Hazardous waste
    Non-hazardous waste

Category II – Sources designed to store, treat, and/or dispose of substances; discharge through unplanned release

Landfills
    Industrial hazardous waste
    Industrial non-hazardous waste
    Municipal sanitary
    Open dumps, including illegal dumping (waste)
    Residential (or local) disposal (waste)
    Surface impoundments
        Hazardous waste
        Non-hazardous waste
    Materials stockpiles (non-waste)
    Graveyards
    Animal burial
    Above ground storage tanks
        Hazardous waste
        Non-hazardous wate
    Underground storage tanks

Table 3.3 Contd...

    Hazardous waste

    Non-hazardous waste

    Non-waste

  Containers

    Hazardous waste

    Non-hazardous waste

    Non-waste

Open burning and detonation sites

Radioactive disposal sites

Category III – Sources designed to retain substances during transport or transmission

Pipelines

    Hazarous waste

    Non-hazardous waste

    Non-waste

Materials transport and transfer operations

    Hazardous and transfer operations

    Non-hazardous waste

Materials transport and transfer operations

    Hazardous waste

    Non-hazardous waste

    Non-waste

Category IV – Sources discharging substances as consequence of other planned activities

Irrigation practices (e.g., return flow)

Pesticide applications

Fertilizer applications

Animal feeding operations

Urban runoff

Percolation of atmospheric pollutants

Mining and mine drainage

    Surface mine – related

    Underground mine – related

Category V – Sources providing conduit or inducing discharge through altered flow patterns

Production wells

    Oil (and gas) wells

    Geothermal and heat recovery wells

    Water supply wells

Other wells (non-waste)

    Monitoring wells

    Exploration wells

Construction excavation

Category VI – Naturally occurring sources whose discharge is created and/or exacerbated by human activity

Ground water – surface water interactions

Natural leaching

Salt – water intrusion/brackish water upcoming (or intrusion of other poor – quality natural water)

Source : Office of Technology Assessment, 1984, p. 45.

Groundwater

1. ***Water table***: The water table elevation is an important contributory factor in engineering and agricultural land capability. It also affects the nature of habitats. A change in its seasonal fluctuation may result from a reduction in the natural recharge or from increased draw-off from the groundwater system.

2. ***Flow regime***: The groundwater flow regime, the direction and rate of flow may be altered by surface or underground engineering, especially drainage works, by draw-off and by the penetration of cap rocks of confined aquifers, any such change can have an impact on other users of the groundwater source.

3. ***Water quality***: Water quality is important for economic, ecological, aesthetic and recreational purposes. Changes in water quality may affect water treatment costs or ever deny some uses of the water. These changes can be chemical, biological or physical.

4. ***Recharge***: Impoundment, rearing or compaction of the ground surface and removal of vegetation can alter the recharge of the groundwater system. A recharge should be considered together with a water table, flow regime and water quality.

5. ***Aquifer characteristics***: Sometimes known as "Aquifer safe yield" these include all the physical parameters (porosity, permeability, etc), which govern the ability of aquifer, provide water for human use. Over pruning or waste injection can cause a decrease in the "Aquifer safe yield''.

6. ***Existing use***: The uses of groundwater system must be for engineered domestic, industrial and agricultural supply or natural agricultural and ecological dependence on the groundwater system.

Table 3.4 Characteristics of Principal Activities Potentially Causing Groundwater Pollution.

Principal characteristics of pollution				State of development[a]					Impact of water use		
Activity	Distribution	Category	Main types of pollutant	Relative hydraulic surcharge	Soil by passed	A	B	C	Drinking	Agricultural	Industrial
Urbanization											
Unsewered sanitation	ur	P-D	pno	×		××××	××	×	××××		×
Land discharge of sewage	ur	P-D	nsop	×		×	×	×	××	×	×
Stream discharge of sewage	ur	P-L	nop	××		×	×		××	×	×
Sewage oxidation Lagoons	u	P	opn	××		×	××		×	××	×
Sewer leakage	u	P-L	opn	×			××		×		×
Landfill, solid waste disposal	ur		osnh			×	×××	××	×		×
Highway drainage soakaways	ur	P-L	so	××		×	××	××	××	×	×
Wellhead contamination	ur	P	pn			××	×		×××		

Table 3.4 Contd...

Principal characteristics of pollution			State of development[a]			Impact of water use					
Activity	Distribution	Category	Main types of pollutant	Relative hydraulic surcharge	Soil by passed	A	B	C	Drinking	Agricultural	Industrial
Industrial development											
Process water/effluent	u	P	ohs	××		×	××	××	××	×	
lagoons											
Tank and pipeline leakage	u	P	oh			×	××	×××	××	××	
Accidental spillages	ur	p	oh	××		×	××	×××	×××	××	
Land discharge of effluent	u	P-D	ohs	×		×	××	××	×	×	×
Stream discharge of effluent	u	P-L	ohs	××		×	×	×	×	×	×
Landfill disposal residues											
and waste	ur	P	ohs			×	×××	×××	××		×
Well disposal of effluent	u	P	ohs	××		×	×	××			×
Aerial fallout	ur	D	a				××	×		×	×
Agricultural Development											
Cultivation with											
Agrochemicals	r	D	no		×	××	×××	×××		×	×
Irrigation	r	D	sno	×	××	××	×	×××	××××	×	
Sludge and slurry	r	D	nos			×	×	×	××	×	×
Waste water irrigation	r	D	nosp	×		××	×	××		××	
Livestock rearing/crop											
Processing :											
Unlined effluent lagoons	r	P	pno	×		×	×	××	×	×	
Land discharge of effluent	r	P-D	nsop	×		×	×	××	×	×	
Stream discharge of effluent	r	P-L	onp	×		×	×	××	×	×	

Table 3.4 *Contd…*

| Principal characteristics of pollution | | State of development[a] | | | Impact of water use | | | | | |
Activity	Distribution Category	Main types of pollutant	Relative hydraulic surcharge	Soil by passed	A	B	C	Drinking	Agricultural	Industrial
Mining development										
Mine drainage discharge	ru	P-L	sha	××		×	××	××	××	×	×
Process water/sludge										
lagoons	ru	P	has	××		×	××	××	××	×	×
Solid mine tailings	ru	P	has			×	××	××	××	×	×
Oilfield brine disposal	r	P	s	×		×	×	××	×	×	
Hydraulic disposal	ru	D	s		na	×	×	××	×	×	
Groundwater resource										
management										
Saline intrusion	ur	D-L	s		na	×	×	××	×××	×××	××
Recovering water levels	u	D	so		na		×	×		×	

Distribution

u Urban   P Point

r Rural

Category

D Diffuse

L Line

Types of pollutant

P Fecal pathogens   H Heavy metals

N Nutrients   S Salinity

O Organic micropollutants   A Acidification

× to ×××× Increasing importance or Impact

na Not applicable

3.3.5.2 Assessment of Soil and Groundwater Pollution

Leaching Into Soils and Groundwater

Water balance Mathematical models can be used to predict the extent of leachate percolating through the site.

The water balance method calculates leachate flow by balancing flows into and out of a site as follows:

$$\Delta S = I - O$$

where

  I is the inflow volume;

  O is the outflow volume;

  S is the storage volume; and

  ΔS is the change in the storage volume.

In the unsaturated zone:

$$L = P - R - Evt + Vd - Evd$$

where

  P is the precipitation volume;

  Vd is the volume of liquid disposed;

  L is the leachate volume;

  Evt is the volume lost to evapotranspiration;

  Evd is the volume of the liquid disposed lost to evaporation; and

  R is the runoff volume.

    For predicting long-term effects, the change in storage can be assumed to be zero, evapo-transpiration can be based on existing data or experiments, while run-off can be calculated using an empirical model based on surface conditions and slopes.

Darcy's Law

Darcy's Law is the basis for most models of groundwater flow in sites below the water table. The method describes the flow of groundwater through a saturated porous medium. Flow is dependent on the change in head with distance (that is, the hydraulic gradient) and the permeability of the medium. It is expressed mathematically as:

$$Q = KA(\,dH)/dL$$

where

  Q is the flow (m^3/day);

  K is the permeability (m/day);

  A is the cross-sectional area (m^2); and

  dH/dL is the hydraulic gradient (that is, the change in the water table elevation per unit change in the horizontal direction).

3.3.5.3 Changes in Groundwater Flow

The soil moisture content necessary for soil microorganisms and plants will be affected by physical disturbances and discharge of liquid effluents. This will decrease the available yield for ground water abstraction resulting either change surface hydrology or salt water intrusion into underground water resources.

Analytical or numerical solution of equations of conservations of mass in Darcy's law is the basis of many mathematical models. Based on directions of flow the ground water regime is divided into different segments and the flows into and out for each segment will be balanced using Darcy's law.

3.3.5.4 Changes in Groundwater Quality

Superimposing models of chemical conversion, biological breakdown, system process, etc., can simulate the behavior of non-conservative pollutants. Tracer experiments may be used to predict dispersion of pollutants in groundwater.

3.3.5.5 Qualitative Approaches Soil Impacts

One example of qualitative impact prediction using look – alike would be the prediction of acid rain impacts on soils as a result of a proposed project.

Another example of a qualitative approach for soil impact prediction and mitigation planning is related to pipeline construction.

There are four potential impacts of pipeline construction on drainage and soils:
1. Contamination of topsoil with excavated subsoil

2. Soil compaction

3. Soil erosion, and

4. Disruption of drainage lines or natural drainage patterns (5).

In most soils, the top several inches are relatively high in organic matter, nutrients, and soil biota. This "topsoil" provides a more fertile growing medium than the relatively inorganic and nutrient–poor sub-soil. Pipeline construction can result in the mixing of sub-soil with top-soil in several ways: through the initial grading of the right-of-way, through the excavation and backfilling of the pipeline.

Changes in Soil Structure

Changes in soil structure are caused by agricultural practices, ground conditions, surface water conditions, and by removal of subsurface soil or water. The effects of these changes can manifest on soil microorganisms, plants and animals, crops and livestock, groundwater and surface water hydrology and quality, visual landscape and amenity, and the integrity of buildings and other civil engineering works. Erosion resulting from changes in ground cover, management practices, rainfall and run-off, and wind exposure can be predicted by the universal soil loss equation.

Effects on Soil Quality

In order to determine the effects on soil quality of contaminants, it is necessary to establish the chemical composition, quality, and amount of substrate in the various soil strata; absorption and adsorption onto soil particles, uptake by plants, transport through the soil, and the chemical and biological conversion of substances.

3.3.5.5.1 Simple Quantitative Approaches – Soil Impacts

Another approach for addressing impacts on the soil environment is to use simple quantitative techniques, with a range of such techniques having been developed. One example of a simple quantitative technique is the use of "overlay mapping" which has been developed to delineate various land-use compatibilities in giving geographical areas. Overlay mapping consists of utilizing a base map of the project study area and different soil or geological features of particular impact concerns of the proposed project. Impact prediction involves identifying where overlaps of particular concerns occur.

Overlay mapping can be achieved through the development of hand drawn maps or the usage of computer- generated maps.

GIS is a database, which may contain multiple "layers" of data for the same area. Examples of possible layers are topographic data and erodibility indices are shown. All layers are referenced to common ground – datum point and orientation, allowing them to be, in essence, overlaid.

Data can be input to GIS by either analytical or digital means. An example of the former would be the use of map digitizing, and of the latter, the use of satellite imagery tapes. One of the great benefits of using GIS is its ability to collate data from diverse sources into a consistent form. Regardless of original scale and format, the data, once in the GIS, are consistent and constant. They may be output in different forms for checking, and they are available for a variety of analyses.

GIS is beginning to be used in impact studies, since it can be a valuable tool for assessing cumulative impacts. GIS can also be used to quantify rates of regional resource loss by comparing data layers representing different years. In addition, GIS can be used to develop empirical relationships between resource loss and environmental degradation.

Quantitative Models

Models have been developed to:
- simulate individual processes occurring in soil
- describe behavior of substances in soil such as nitrogen, phosphorous, and pesticides (laboratory experiments using column tests and lysimeters may also predict the behavior of substances in soil)
- predict the behavior of liquids, which are immiscible with water (for example, mathematical models for oil spills on land which simulate the behavior of oil on the surface and in the unsaturated zone and its dispersion above the groundwater table)
- simulate the behavior of gases in soil, or
- predict dispersion of heat released by pipelines or cables, or discharged in effluents.

Areas of Application

Table 3.5 lists some of the software the International Groundwater Modeling Center (IGWMC) at the Colorado School of Mines has to offer, as well as some free, public domain software. Detailed descriptions of most programs (as well as model demos) can be found on the Center's website under the IGWMC Software listing.

Table 3.5 Examples of groundwater models available from the Colorado School of Mines

Model Name	Model Description
FLOWPATH	A DOS-based two-dimensional finite difference model for state flow in confined and unconfined aquifers. It includes an elaborate, user-friendly graphic interface, and extensive graphic display of results. The model is widely used, among others, for well head protection studies.
HYDRUS-2D	A sophisticated Windows-based two-dimensional finite element model for transient unsaturated flow and solute transport. It includes an elaborate, user-friendly graphic interface, and extensive graphic display of results. The program includes modern numerical routines securing efficient and stable solutions for highly nonlinear problems.
INFINITE EXTENT	A program for aquifer test analysis with on-screen, manual curve matching, and automatic parameter evaluation. The program includes type curves for confined, leaky-confined and unconfined cases.
MICRO-FEM	A finite element model for simulation of transient quasi-three-dimensional flow in aquifer systems. This model, widely used in northern Europe, includes elaborate grid design, parameter allocation result analysis options, as well as extensive graphics.
Model IGIS	A modeling system operating in the ARC/INFO Environment under Unix. It includes MODFLOW and MODFLOWT
MODFLOWT	A three-dimensional finite difference contaminant transport modeling which fully integrates with MODFLOW.
RPTSOLV	A Windows-based finite element model for pumping test analysis in fractured rock
SUPERSLUG	A Windows-based program for slug analysis
STEPMASTER	A Windows-based program for step-drawdown tests.
THCVFIT	A simple interactive DOS program for aquifer test analysis using the HEIS method.
TWODAN	A DOS-based multi-functional analytic element model for two-dimensional steady-state flow with a user-friendly graphic interface and extensive graphic display of results. Well-suited for well-head protection studies.
The United Nations Ground Water Windows (UN-GWW)	This demo is split up in 4 self-extracting files. It shows in slide show from the hydrogeological and geochemical database options, its extensive graphic display options, as well as options for aquifer test analysis.
UNITS	A DOS-based groundwater units conversion (Shareware)
Visual MODFLOW	A MODFLOW-based modeling environment with extensive GUI and support for MT3D and other MODFLOW related programs.
ZBSoft	A three-dimensional flow and contaminant transport modeling based on the Zheng-Bennett text book on this topic.

(**Source**: Internet - www.mines.edu.igwmc)

3.3.5.5.2 Some of the most commonly used Software for Soil Assessment include

- IGEMS which have graphics and GIS capabilities for displaying environmental modeling results. The model is usually used for ambient air, surface water, soil and ground water. It also has environmental fate and transport models.

- SWAT – Soil and Water Assessment Tool – used specifically for river basin scale for quatifying impacts of land management in large and complex watershed.

- MMSOILS – Multimedia Contaminant Fate, Transport, and Exposure Modeling, which assess human exposure and health risk associated with releases of contaminated hazardous waste sites. The model addresses the transport of a chemical in groundwater, surface water, soil erosion, atmosphere, etc. The human exposure pathways considered include soil ingestion, air inhalation of volatiles and particulate, contaminated soils.

- CalTOX relates to the concentration of a chemical in soil to the risk of an adverse health effect. It computes site specific health based soil clean-up concentrations given target risk levels.

3.3.6 Assessment of Impact Significance

Several approaches can serve as a basis for interpreting the anticipated project induced changes in the soil and groundwater environments. One approach is to consider the percentage and direction of change from existing conditions for a particular soil or groundwater environmental factor. While this can be helpful, it does presume that quantitative information is available for the baseline conditions for such factors, and that anticipated changes in the factors as a result of a project can be quantified.

Another approach for impact assessment is to apply the provisions pertinent to Central, State, or Local laws and regulations related to the soil and groundwater environment to be expected with project conditions. In many cases, these institutional requirements are qualitative; however, they can be used as a "yardstick" in evaluating the project and any features the project might incorporate to minimize environmental damage.

A third approach to interpreting anticipated changes relies upon professional judgement and knowledge. The anticipated changes could be interpreted in relation to existing information on natural changes; next, the expected impacts could be placed in a historical context.

A professional-judgement-based interpretation of anticipated changes may consist of applying rules of the thumb. As an example, concerning soil erosion, the current and anticipated soil erosion patterns of a project area could be compared to regional averages or historical trends. It is generally agreed that a certain amount of soil loss is inevitable. Ideally, the loss should not exceed the rate of soil formation from parent rock and decomposed vegetation, but there is no agreement at the rate of soil formation (9).

A commonly cited, generalized upper limit of permissible or tolerate soil loss is about 11 tons/ha/yr, but the "permissibility" of such a loss depends on many local factors, such as, the fertility and drainage characteristics of the sub-soil. Many soils are vulnerable to a decline in productivity at a rate of loss from erosion considerably lower than 11 tons/ha/yr.

3.3.6.1 Environmental Analysis

After the above types of factors are considered, the resultant conclusion may not be absolute. Subjective terms, such as, the degree to which the project may induce development, may need to be used. The environmental analysis should yield the best possible prediction of environmental effects based on available information. The conclusions of the analysis of potential induced development may be that the proposed project or action will:

- Definitely cause and promote increased density of land use,
- Not cause any increase in development over what would occur in the future without the project,
- Not necessarily cause increased development, but perhaps accelerate development slated to occur anyway, and
- Not produce a development impact if local plans and policies stay unchanged, but indeed put into place the incentive for local planning bodies to change local comprehensive plans to permit higher-density land-use.

3.3.6.2 Other Secondary Effects

Secondary impacts can occur, however, due to changes in land-use or land-use plans. Many of these secondary impacts are not limited to socioeconomic effects, but can equally affect natural resources, such as, water quality or wildlife habitat.

Increased covering of the earth with impervious surface, such as parking lots or large buildings, can increase the rate and pollutant loading of surface water run-off. Secondary effects of such use can be the increased contamination of both surface water and groundwater resources. A secondary effect may then be the need to construct additional water treatment plants with associated secondary effects of the use of limited public funds.

3.3.6.3 Assessment Impacts of Induced Development

These types of possible impacts could be called secondary impacts or impacts twice removed. If induced development is predicted, the environmental impact analysis should consider, to the extent possible, the effects of this induced development. Perhaps increased density of residential or commercial and industrial land use will, in turn, create a need for additional schools, parks, public support programs and facilities, service industries, public water or power supply, solid waste and sewage disposal capacity, improvement in local roads or intersections, or increased emergency services (fire and police) and health care facilities.

Land development, resource extraction and waste – disposal projects can cause certain undesirable impacts on soil and/or groundwater resources (either quantity or quality changes).

Urban growth near a new water- supply reservoir can cause soil and/or groundwater effects as a result of urban waste disposal leachates moving through the subsurface system.

3.3.7 Identification and Incorporation of Mitigation Measures

Mitigation Measures to Prevent Soil Erosion, Compaction and Groundwater Pollution During and After Execution of Any Developmental Project.

1. Use of techniques to decrease soil erosion during either the construction or operational phase of the project: Examples of such techniques include minimization of the exposed time during the construction phase of planting rapidly growing vegetation and the use of sediment – catchment basins. Additionally, as various types of grasses and vegetation have relatively greater or less potential for minimizing soil erosion, the selection of pertinent vegetation for usage should take these characteristics into account. Remove as little vegetation as possible during the development and revegetate bare areas as soon as possible after completion of the project.

2. Where possible gentle gradients should be treated and steep slopes avoided.

3. Suitable drainage systems to direct waterways from slopes should be installed.

4. Creating large open expanses of bare soil should be avoided. These are more susceptible to wind erosion. If such large areas are created, then wind breaks may be a useful mitigation procedure.

5. For removing sediment which affects the water flows and damage the freshwater ecosystem due to any development adjacent to a water body, siltation traps need to be fixed

6. For avoiding compaction of the soil due to motor traffic, only few delineated tracks should be used for bringing the vehicles to the working area

7. To achieve natural recovery of the compacted soil, land use practices need to be continuously rotated. Rotation of military training areas, agricultural crops in giving geographical areas, and grazing patterns in areas permitted by pertinent governmental agencies are some examples

8. In earthquake prone area irrigation projects need to be redesigned with new structural designs for withstanding shocks associated with the occurrence of earthquakes.

9. Groundwater usage needs to be minimized in projects which involve large scale extraction of ground water.

10. Water conservation measures to decrease ground water usage need to be implemented in areas where land subsidence is the potential risk

11. For preventing ground water contamination site selections for solid waste disposal (10) need to be more systematic based on extensive comparative studies on different sites to avail natural attenuation capacity of certain natural environmental settings

12. Technologies involving immobilization of hazardous/pollutant constituents are to be adopted in project which generate leachates

3.3.7.1 Mitigation Measures for Major Road Construction Projects

Mitigation measures should consider how important soil properties which include (a) The soil physical characteristics of the whole profile, including - Soil texture; - Soil structure; Soil horizonation - depth (both total and of individual horizons and stoniness) (b) The soil chemistry, surface and sub soil including - organic matter content - soil pH - nutrient status - salinity parent material characteristics - soil water regime (vertical drainage and runoff characteristics) - vegetation cover, especially peat forming communities - slope gradient (c) and soil biological indicators.

3.3.7.2 Drainage Works

Roads, as linear engineering features, often modify water flow and drainage patterns over wide areas, causing rising water levels, excessive drying, erosion and vegetation die-off. An understanding of hydrogeology and drainage patterns in the watersheds to be crossed, and of the placement of drainage structures such as culverts and porous materials, plus consideration of where the cuts and fills have the least detrimental effects, can go on a long way in alleviating serious and chronic drainage problems. The cost of ignoring or reducing efforts in this area can be exceedingly high later on during the construction phase, or during the operating life of the road.

3.3.7.3 Waste Management

When dealing with major projects with workforce >1000, it is expected large quantities of solid and liquid wastes are likely to be generated causing major source for pollution resulting local health issues. For example, major construction projects are resulting serious environmental and health problems due to poor waste management, soil compaction, accidental spills etc. which can be avoided.

Measures to prevent erosion are of major importance during the work phase, and can include:

1. Planting on cleared areas and slopes immediately after equipment belonging to a specific site has been moved, and reusing stripped topsoil.
2. Temporarily covering the soil with much or fast growing vegetation
3. Intercepting and slowing water runoff, and
4. Protecting slopes by using reshaping techniques, rock fill and other methods.

3.3.7.4 Soil Remediation Methods

If the base line survey indicates the site is contaminated, then soil remediation measures have to be implemented; some which are discussed below:

1. Removal of the contamination for off-site disposal which is a commonly adopted technique, but will result in the transport of hazardous material along a public highway and the displacement of pollution to a landfill site
2. Exacavation and on site disposal which removes the need for transport, but require custom designed facility and either a waste management license or exemption for licensing
3. Adoption of on site stabilization techniques, which will remove the ability of pollutants to move off site.
4. *Insitu* bioremediation, which is effective for organic pollutants and uses natural microorganisms to breakdown organic pollutants. Even difficult materials such as halo and nitro substituted aromatics can be now be bioremediated.
5. Soil washing with acid or alkali or water is now adopted which is very effective but costly and the leachate has to be properly collected and treated or stabilized before disposal.

6. Air sparging vacuum extraction, pump and treat methods is effective in a range of contaminants from ground water.

3.4 Case Studies

3.4.1 Case Study : I

Soil and Ground Water Pollution and Remediation Project

By Intergo Enviro Tech

Study Area: Evosmos-Municipality, Greece- Covered with Industrial area, Waste and Garbage Collection Facilities

SITE FEATURES

1. Location: Island in Aegean Sea with limited water resources
2. Site Operation: Depot terminal of power plant
3. Source: Leakage of underground oil product conveyance pipeline
4. Amount of released product: 1.5 m^3 Diesel
5. Geology: Till 15-20 m unconsolidated permeable sediments (sand, gravel, silty sand)
6. Bedrock : granodlorit
7. Pizometric groundwater level: −3 m b.s.1
8. Hydr. Conductivity of aquifer: 3-8 × 10^{-5} m/s
9. Condition of the aquifer : Unconfined
10. Mean hydraulic gradient: 1-3%
11. Drinking water : Water wells located at 120 m distance (supply for 2000 people)
12. Environmental Risk : Very high
13. Groundwater impact in the drinking wells : 2, 1 mg/l TPH concentration after 15 days of the incident

Location points of the industrial units and of disposed waste
and garbage within the investigation area

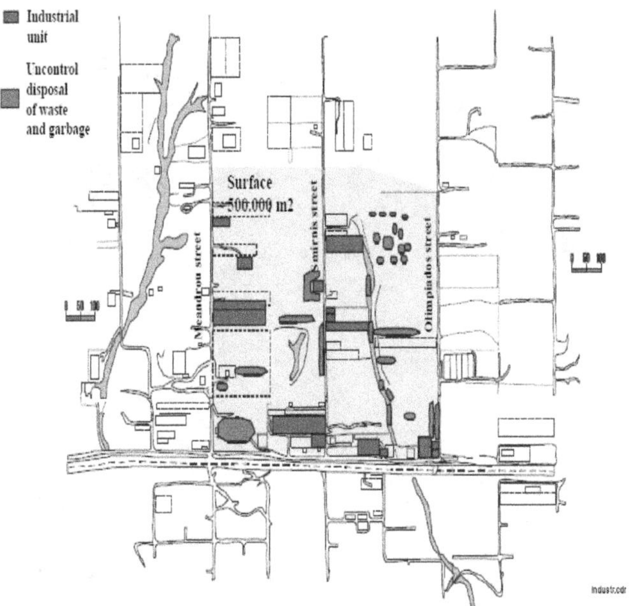

Case study Groundwater flow conditions

Groundwater table and movement direction of the groundwater under static and dynamic conditions

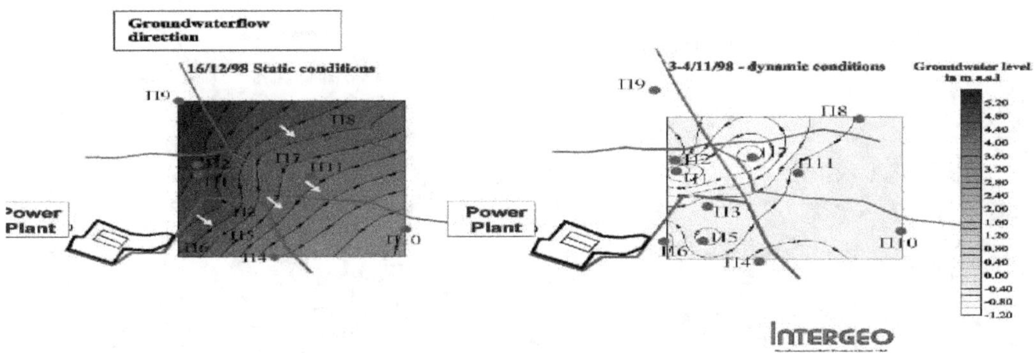

Case study Geological cross-section along the groundwater flow direction

Geological cross-section along the groundwater flow direction

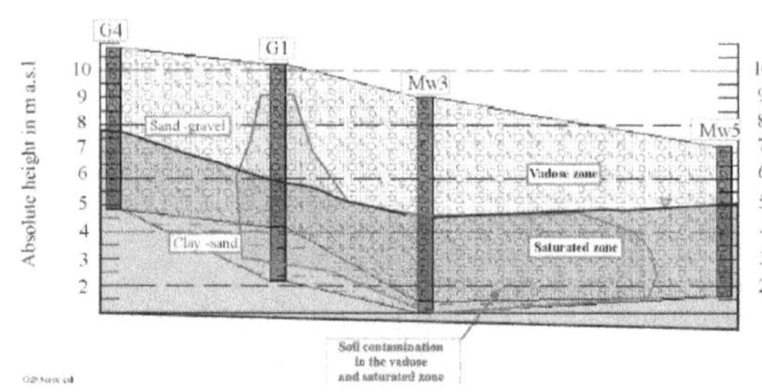

Case study Site investigation

Figure 3 : Distribution of TPH concentration in the soil in three different levels - Contaminated site - Power plant close to drinking water extractions wells

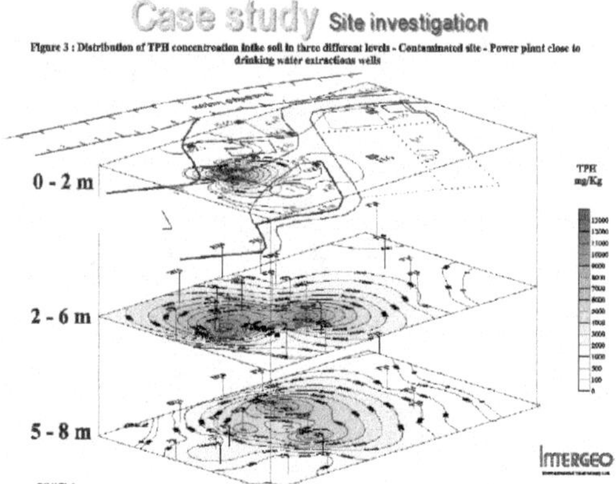

Case study Water quality in the drinking well

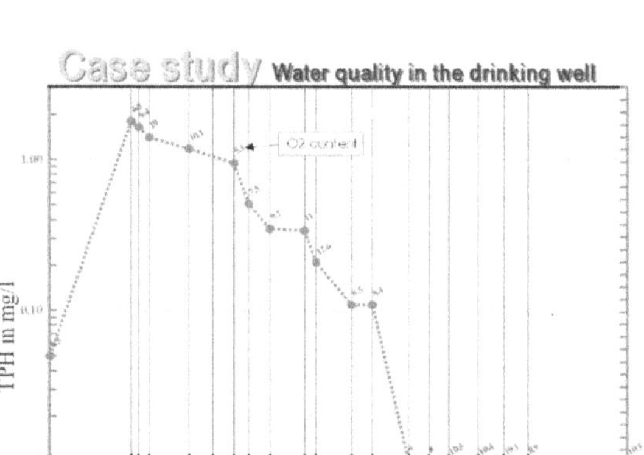

Case study TPH in the groundwater

Distribution of TPH concentration in the groundwater (date 16/12/98)

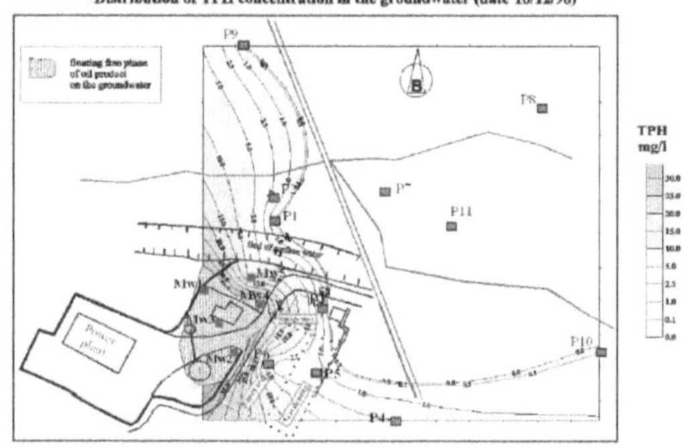

Initial concept of the proposed decontamination technique for the subsurface in the installation

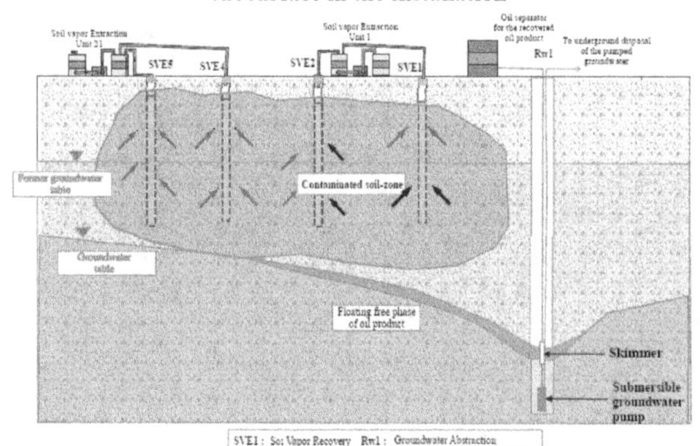

Case study Remediation Action Plan

Synoptic presentation of the remediation action plan

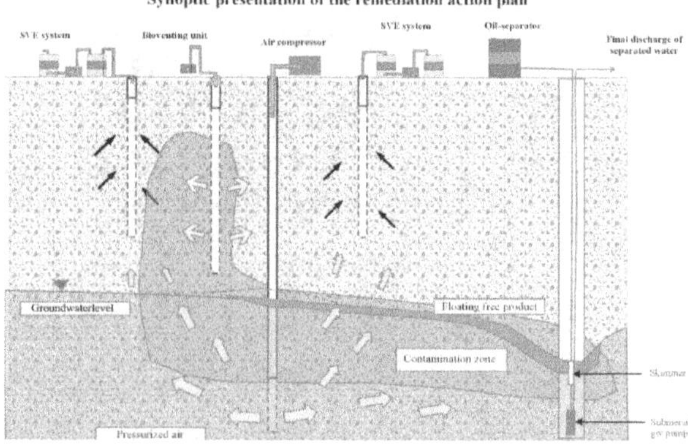

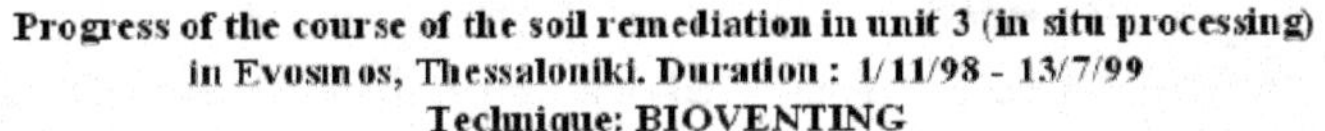

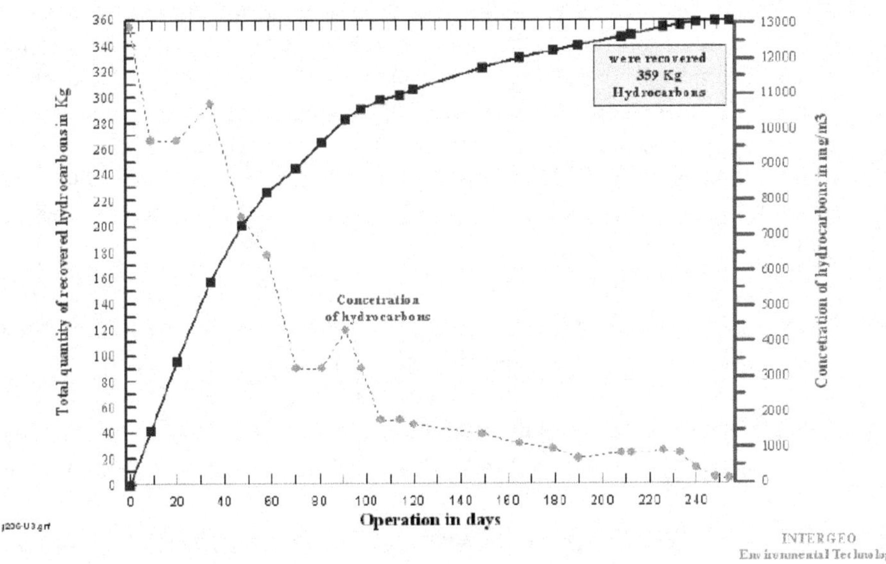

Progress of soil decontamination procedure
Operation time of SVE unit 2: 25/7/1999-17/5/2000
at the Depot terminal

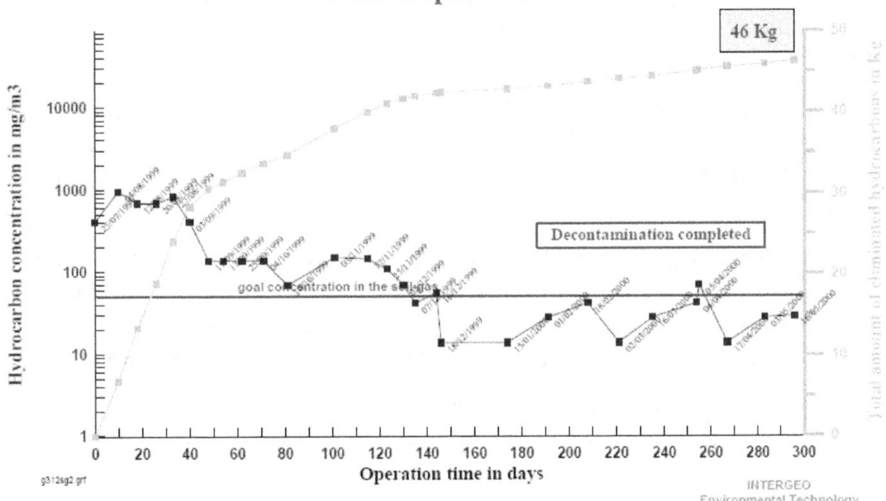

Remediation Results

- The soil remediation was completed after 20 months of implementation of the air sparging technology.

- The groundwater TPH concentration was radically reduced below the drinking water standards after 16 months of system operation.

3.4.2 Case Study 2 : Environmental Impact Assessment of Soil Quarrying from the Hills of Central Kerala – Case Study

Kerala state is one of the densely populated regions having limited land and non-renewable resource availability. Indiscriminate resource extraction due to rapid development has created serious environmental problems. This case study deals with the environmental impacts of soil quarrying from the hills of Central Kerala. These activities have reached critical levels in the peripheral areas of the major developmental centers in the state (Maya et al. 2012; Padma Lal et al. 2015). The ripple effects in the mining and quarrying sectors are often underestimated. This, in many of the occasions, has led to changes in the socio-environmental setting of the regions.

A major part of Kerala is blanketed by lateritic soil, which is a product of tropical weathering of iron rich parent rocks in which several courses of transformation takes place. In recent years, excessive quantities of lateritic soils are being quarried from the residual hillocks in the low and midlands of the State essential to meet the demands in the construction sector. In most cases, no management plans are envisaged. Studies have shown that in the Muvattupuzha river basin is under immense stress due to rampant mining and quarrying.

Environmental data on soil quarrying, quantity of resource extraction and other baseline data were collected as primary and secondary data sources covering Periyar and Muvattupuzha river basins. The data relating to weighing of the loaded vehicles moving out of the quarry locations were also collected. Administration of questionnaire with local authorities, operators and officials of Mining and Geological Departments, was collected. Rapid Environmental Impact Assessment (REIA) was carried out to evaluate the major environmental and social impacts of the soil quarrying activities in the study area.

The Rapid Environment Assessment Matrix (RIAM) proposed by Pastakia (1998) was adopted to assess the environmental impacts of soil quarrying. The purpose of the survey was to identify and address the key environmental issues in order to mitigate negative and to enhance positive impacts.

Soil quarrying, loosening, loading and transporting takes place in varying combinations, depending on the shape, size and depth of the pit, the local topography and the output. Soils that are good for building construction are characterized by good grading. Laterite soil generally gives very good results, especially if stabilized with cement or lime.

A study has shown that out of the total soil quarries, 35 are in the lowlands, 156 in the midlands and 92 in the highlands. Mechanical type of soil extraction is observed in most of the quarries.

RIAM allows both quantitative and qualitative data to be assessed. Initially the impacts of the project activities are evaluated against the environmental components/ subcomponents. For each individual environmental component a score is assigned which provides a measure of the impact expected for the component.

RIAM requires specific assessment components to be defined through a process of scoping, and these environmental components fall into four categories, Physical/ Chemical (PC), Biological/ Ecological (BE), Sociological/Cultural (SC), Economic/ Operational (EO).

From the formulae given previously, Environmental Score (ES) is calculated and recorded. To provide a more certain system of assessment, the individual ES scores are banded together into ranges where they can be compared.

The important assessment criteria fall into two groups: (1) Group A— criteria that are important to condition and, (2) Group B—criteria that are of value to the situation. Calculations are determined and expressed as:

$$(a1) \times (a2) = aT \quad (1)$$

$$(b1) + (b2) + (b3) = bT \quad (2)$$

$$(aT) \times (bT) = ES \quad (3)$$

where, a1 and a2 are the individual criteria scores for Group A, b1, b2 and b3 are the individual criteria scores for Group B, aT is the result of multiplication of all A scores, bT is the result of the summation of all B scores, and ES is the environmental score for the condition, The judgments on each component are made in accordance with the criteria and scales.

The effects of soil quarrying are directly dependent on the method of quarrying adopted in the region, geological settings, human settlement in the area and the depth of the quarry. The impact assessment performed using RIAM method for soil quarrying of hills and hillocks in the study area indicates that both manual and mechanical mining in the long term imposes significant negative impacts on landscape, land use, soil and landform features.

The mechanical quarrying of soil from hills may markedly change the landform features of the basin area, resulting in instability of the adjacent land, buildings/houses, loss of biodiversity and vegetation, accelerated erosion and caving of soil masses, etc. In addition to stability problems, uncontrolled soil quarrying operations could result in extensive modification of the landscape and/or aesthetics of the region. The land use change due to quarrying causes loss of native/agricultural vegetations. Both biotic and abiotic components of hill ecosystems operate in a balanced relationship. Obliteration of top soil could reduce the net bio-productivity of the area. The activity inevitably leads to changes in the soil profile and quality affecting the functioning of the entire ecosystem.

Further, quarrying may disturb the natural habitats of certain animals inhabiting in the affected area and lead to habitat destruction. Levelling of hill ecosystems would result in significant negative impact on the biological diversity due to habitat fragmentation.

Recommendations for impact mitigation should include:

1. Quarrying needs to be done, allowing for sufficient slope (less than 45° from horizontal level) and benches (bench width of 1 m for every 5 m height).
2. The boundary of the excavated areas should be properly protected to avoid accidents /risks
3. The fertile top soil should be collected separately and used for refilling the area after completion of the quarrying process.
4. The water spraying on the ground needs to be undertaken in dry periods in the study area and the adjoining regions of the quarry to reduce dust pollution including during transportation of the quarried soil.

5. Monitoring of the present state of hill ecosystems, optimum use of natural resources and looking for alternative materials are some of the administrative norms that need to be followed regularly.

References

- C. M. R. Pastakia, "The Rapid Impact Assessment Matrix (RIAM) - A newtool for Environmental Impact Assessment", in Environmental Impact Assessment using the Rapid Impact Assessment Matrix (RIAM), K. Jensen,Ed., Fredensborg, Olsen & Olsen, 1998, pp. 8-18.
- D. Padma Lal, K. Maya, & E. J. Shiekha, "Environmental effects of soil quarrying in Kerala: An overview", Proceedings of National Seminar on Soil Pollution and Paradigms for Sustainable Soil Management, Kerala University, Thiruvananthapuram, 2015, pp. 22-27.

3.5 Summary

Various land development projects and associated activities can cause environmental impacts on soils and groundwater by a variety of physical disturbances, including the addition/removal of soil and/or water, compaction of soil, changes in the use of land or ground cover, changes in water hydrology, changes in climate (temperature, rainfall, wind), and the addition or removal of substances or heat (for example, discharge of effluents into groundwater, discharge of effluents or disposal of waste onto land, leaching of contaminants into groundwater), changes in quality of surface water, and deposition of air pollutants on land. In analyzing environmental impacts on soil and ground water, both objective and subjective judgements should be taken into consideration.

A general approach, which connects seven important activities of any land development project, is discussed for carrying out EIA. The various technical aspects and appropriate methodologies to be adopted for implementing the seven important steps of EIA like (a) Delineation of Study Area (b) Identification of Activities, which will have different types of Impacts on Soil and/or Groundwater Quantity & Quality (c) description of the nature of existing Soil and/or Groundwater Resources (d) Background Information on the Soil Environment (e) Procurement of Relevant Soil and or Groundwater Quantity – Quantity Standards (f) Impact Prediction (g) Assessment of Impact Significance (h) Implementation of mitigation measures   for different major land development projects are discussed with specific examples. Case Studies on Soil and Ground Water Pollution and Remediation and Environmental Impact Assessment of Soil Quarrying from the Hills of Central Kerala are presented.

References

1. Toy. T. J, and Hadley, R.F. (1987 a), Chap.6, "Lands disturbed by grazing", in "Geomorphology Reclamation of disturbed lands". Academic press, Orlando, Fla, pp. 152-162.
2. Douglas A. Haith, Member, and Ethan M. Laden, (1986), Screening of Groundwater

Contamination by Travel-time Distributions, Journal of Environmental Engineering, Vol. 115, No. 3.

3. P. Maloszewski, H. Moser, W. Stichler, B. Bertleff and K. Hedin (1990), Modelling of Groundwater Pollution by River bank Filtration using Oxygen-18 data, IAHS Publ. no. 173.

4. G.S Gill and Harish Arora (2010), Determinants for Contamination Risk Zoaning of Groundwater – A Case Study of an Industrial Town of Punjab. *Water availability and management in Punjab*, 1-24.

5. Economopoulos, Alexander P. (1993a), Assessment of Sources of Air, Water, and Land Pollution: A Guide to Rapid Source Inventory Techniques and their use in Formulating Environmental Control Strategies. Part One: Rapid Inventory Techniques in Environmental Pollution. World Health Organization, Geneva Management in Punjab 1-24.

6. U. S. Department of Energy, 1988 "Energy Technologies and the Environment-Environmental Information Handbook," DOE/EH - 0077, Washington, D.C., Oct.

7. Gaoudie. A. (1984), The nature of the Environment, Basil Black-Well, Oxford, England, p. 246.

8. Drew. D. (1983), Man-Environment Processes, George Alen & Union Publishers, London, p.32.

9. ERL (Environmental Resources Limited). (1984), Prediction in Environmental Impact Assessment, a summary report of a research project to identify methods of prediction for use in EIA. Prepared for the Ministry of Public Housing, Physical Planning and Environmental Affairs and the Ministry of Agriculture and Fisheries of the Government of Netherlands.

10. Canter, Larry W. and Barry Sadler. (1997), A Tool Kit for Effective EIA Practice — Review of Methods and Perspectives on their Application. A Supplementary Report of the International Study of the Effectiveness of Environmental Assessment. Environmental and Ground Water Institute, University of Oklahoma, Institute of Environmental Assessment, UK and the International Association for Impact Assessment.

11. Chattopadhyay. S, Rani. L. A, (2005), Water quality variations as linked to land use pattern: A case study in Chalakudy river basin, Kerala, *Current Science*, **89(12):** 2163-2169.

12. Sorell. F.Y, (1982), "Air & gas pipelines in costal north Carolina: Impact and Routing considerations", CEIP Rep no, 33- North Carolina state University, Raleigh.

13. Schuknecht, M.R, and Mirels. J.K, (1986), "Hydrogeologic Impact Assessment of Proposed Urbanization Atop a Karst Aquifer," Proceedings of conference on Environmental Problems in Karst Terrians & their solutions, National Water Well Association, Dublis Ohio, pp. 435-451.

14. Pennisi. E. (1993), "Sauing Hades Creatures", *Science News*, Vol-143, PP-172-174.

15. McHarg. I.L, (1971), Design with Nature, Doubleday 1. Natural History Press, Double day & Company, Garden City, N.Y.

16. Carpenter, R.A, and Maragos, J.E. eds, (1989), "How to Assess Environmental Impact on Tropical Islands and Coastal Areas," Training Manual For South Pacific Regional Environment Programme, Environmental and Policy Institute, East West Center, Honolulu, pp-258-266.

17. Bolton K.F and Curtis, F.A., (1990), "An Environmental Assessment Procedure for Siting Solid Waste Disposal Sites", Environmental Impact Assessment Review Vol. 10, pp – 285-296.

18. Patil. A, Krishnan. S, Groundwater Situation in Urban India: Overview, Opportunities and Challenges, IWMI -Tata Water Policy Program.

19. T. Rajaram, Ashutosh Das (2008), Water Pollution by Industrial Effluents in India: Discharge scenario and case for participatory ecosystem specific local regulation. *Futures* 40, 56-59.

20. YU Weidong, DING Aizhong WU Xiaokao, (2007), Study on Relationship between Polluted River Water and Riparian Groundwater, *Journal of Water Science Research*, 296-303.

21. Mukerjee. S and Nelliyat. P, (2006), Ground Water Pollution and Emerging Environmental Challenges of Industrial Effluent Irrigation: A Case Study of Mettupalayam Taluk, Tamilnadu, IWMI-TATA Water Policy Program.

22. M. A. Momodu and C.A. Anyakora, (2010), Heavy Metal Contamination of Ground Water: The Surulere Case Study, *Research Journal Environmental and Earth Sciences* 2(1): 39-43.

23. R. Rajamanickam and S. Nagan, (2010), Groundwater Quality Modeling of Amaravathi River Basin of Karur District, Tamil Nadu, Using Visual Mod flow, *International Journal of Environmental Sciences*. Volume 1, No1.

24. Hiscock.K.M and Grischek.T, (2002), Attenuation of Groundwater Pollution by Bank Filtration, *Journal of Hydrology*, Volume 266, Issues 3-4, Pages 139-144.

25. Del Vecchio. G.M, Douglas A. Haith and Z. Member, (1993), Probabilistic screening of groundwater Contaminations, *Journal of Environmental Engineering*, Vol. 119, No. 2, Paper No. 2586.

26. Peter K. Ndiba and Lisa Axe, (2010), Risk Assessment of Metal Leaching into Groundwater from Phosphate and Thermal Treated Sediments, *Journal of Environmental Engineering*, Vol. 136, No. 4, April 1, page no. 427–434.

Questions

1. Discuss various steps to be followed for a systematic approach for the study of prediction and assessment of impacts of any developmental activity on soil and ground water.

2. Discuss the cause and effect network for assessing the impacts on soil and ground water.

3. What are the important features of land, which have to be taken into consideration for assessing the impacts of different land uses in developmental projects?

4. Discuss different activities in a major road laying project likely to have impacts on soil and ground water environment.

5. Discuss the qualitative, simple quantitative and specific quantitative methods for impact prediction on soil and ground water environment?

6. What are the impacts of induced development with reference to soil and ground water?

7. What is water balance method for calculating leachate flow through a site? What is Darcy's law for describing ground water flow through a saturated porous medium?

8. Discuss various general guidelines for implementation of mitigation measures necessary to prevent soil erosion & compaction and ground water pollution.

9. If the soil is contaminated, what are the remediation methods that can be adopted during and after the execution of any developmental project?

10. Discuss salient features of some of the most commonly used software for soil assessment.

11. Explain different approaches for Assessment of Impact Significance.

12. Discuss various Mitigation Measures for restoring soil quality to be adopted when Major Road Construction Projects are taken-up.

13. Discuss various Soil Remediation Methods.

CHAPTER 4

Prediction and Assessment of Impacts on Surface Water Environment

4.0 General Methodology for the Assessment of Impacts on Surface Water Environment

4.1 Introduction

Surface water bodies like rivers, streams, canals, ditches, ponds, reservoirs, lagoons, estuaries, coastal waters, lakes etc., play very important role in the sustainability of any ecosystem and so it is very important to assess the impacts of any developmental activity on these surface water environments. Impacts on surface waters are usually caused by physical disturbances (for example, the construction of banks, dams, dikes, and other natural or man-made drainage systems), by changes in climatic conditions, and by the addition or removal of substances, heat, or microorganisms (for example, the discharge of effluents and deposition of air pollutants into water). These activities and processes lead to first order effects as manifested by changes in surface water hydrology, changes in surface water quality, and consequently to higher order effects reflected by changes in sediment behavior, changes in salinity, and changes in aquatic ecology (1). Though developmental programs help in improving the economy of any region many times they are associated with stress on local ecosystems due to changes in land use and land cover which result in soil and ground /surface water pollution (2).

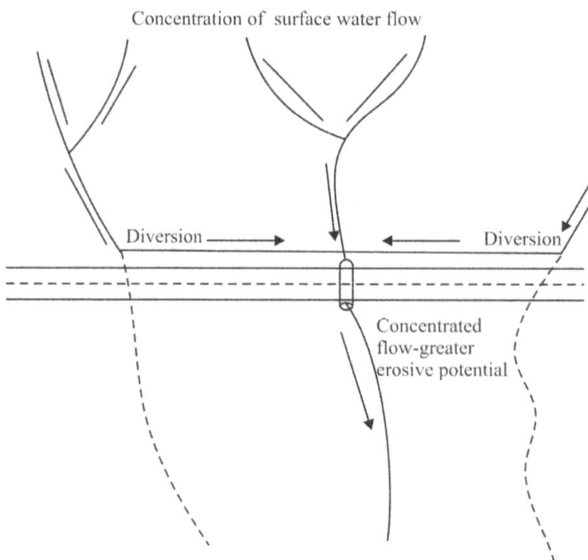

Fig. 4.1 Concentration of surface water flow.

Fig. 4.1 demonstrates the cycle of both surface-water and groundwater hydrology. Due the dynamic nature of both the quantity and quality natural variations occur in the flow and quality characteristics respectively.

4.2 Projects which Create Impact Concerns for the Surface-Water Environment (3)

Several developmental activities will result in environmental impacts on surface water bodies. The following are the list of various developmental activities, which cause significant impacts on surface water resources for which a detailed EIA is normally required:

1. Industrial power plants withdrawing surface water for cooling (this may be of particular concern during low-flow conditions).
2. Power plants discharging heated waste water from cooling cycles.
3. Industries discharging processed waste waters from either routine operations or as a result of accidents and spills.
4. Municipal waste water treatment plants discharging primary, secondary or treated waste waters.
5. Dredging projects in rivers, harbors, estuaries and or coastal area (increased turbidity and release of sediment contaminants may occur).
6. Projects involving "fill" or creation of "fast lands" along rivers, lakes, estuaries and coastal area.
7. Surface mining projects with resultant changes in surface water hydrology and non-point pollution.
8. Construction of dams for purposes of water supply, flood control or hydropower production.
9. River canalization projects for flow improvements.
10. Deforestation and agricultural development resulting in non-point source pollution associated with nutrients and pesticides and irrigation projects, leading to turn flows laden with nutrients and pesticides.
11. Commercial hazardous waste disposal sites and/or sanitary landfills, with resultant run off water and non-point-source pollution; and
12. Tourism projects adjacent to estuaries or coastal area with concerns related to bacterial pollution.

Before starting EIA on any surface water, one has to understand certain basic characteristics of qualities and quantities of surface water bodies.

4.3 Systematic Methods for Evaluation of Impacts of Various Developmental Activities on Surface Water Environment (4-6)

For assessing the environmental impacts of various human activities on surface water bodies the following six step model (Fig. 4.2) is discussed.

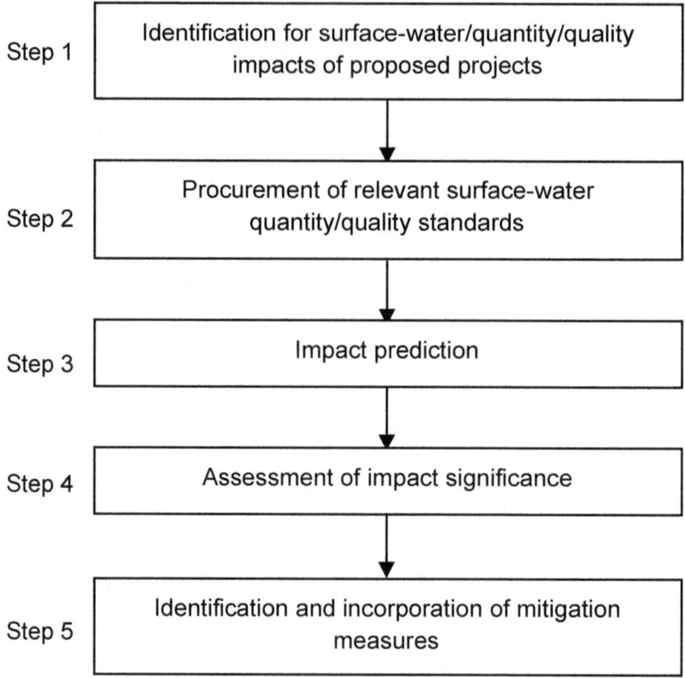

Fig. 4.2 Conceptual approach to study surface water environment impacts.

4.3.1 Step 1 Identification of Surface Water Quantity or Quality Impacts of Proposed Projects

The first activity is to determine the features of the proposed project, the need for the project, and the potential alternatives, which have already been or may now be, considered.

The key information relative to the proposed project includes such items as:

1. The type of project and how it functions or operates in a technical context, particularly with regard to water usage and waste water generation, or the creation of changes in water quality or quantity.

2. The proposed location of the project.

3. The time period required for project construction.

4. The potential environmental outputs from the project during its operational phase, including information relative to water usage and water pollutant emissions, and waste-generation and disposal needs.

5. The identified need for the proposed project in the particular location (this need could be

related to flood control, industrial development, economic development, and many other requirements; it is important to begin to consider project need because it will be addressed as part of the subsequent related environmental documentation), and

6. Any alternatives which have been considered, with generic alternatives for factors including site location, project size, project design features and pollution control measures, and project timing relative to construction and operational phases.

The focus of this step is on identifying potential impacts of the project. This early qualitative identification of anticipated impacts can help in refining subsequent steps.

For example, it can aid in describing the affected environment and in calculating potential impacts. Step 1 should also include consideration of the generic impacts related to the project type.

There is an abundance of published information generated over the past two decades which enables planners of impact studies to identify more easily the anticipated impacts of different land- use changes.

Changes in land use, land cover patterns and land management practices with increased human activity in any watershed are expected to create major impacts in the hydrological system. This leads to run off changes and significant impacts on the water quality of rivers. (7-10). Fig. 4.3 depicts relation between land use and associated water quality changes.

Many research papers are published on the effects of the land use and land cover change on the quality of surface water which reveal the linkage between land cover and the river water quality (11&12), morphological features of watersheds on the turbidity, dissolved oxygen and temperature of the river water (13), changes in the watershed scale (14). For assessing water quality issues like turbidity, harm full algal biomass Blooms (HAB), identification algal taxonomic groups over large spatial extent Remote Sensing will be very useful tool. Remote sensing data will be very useful for coastal managers for surveying on water quality issues and developing HAB monitoring strategies.

Recently a number of papers also appeared on the application of remote sensing, GIS, and multivariate analysis to explore the influence of the land cover on the suspended sediment, nutrients and ecological integrity of the stream (14-16).

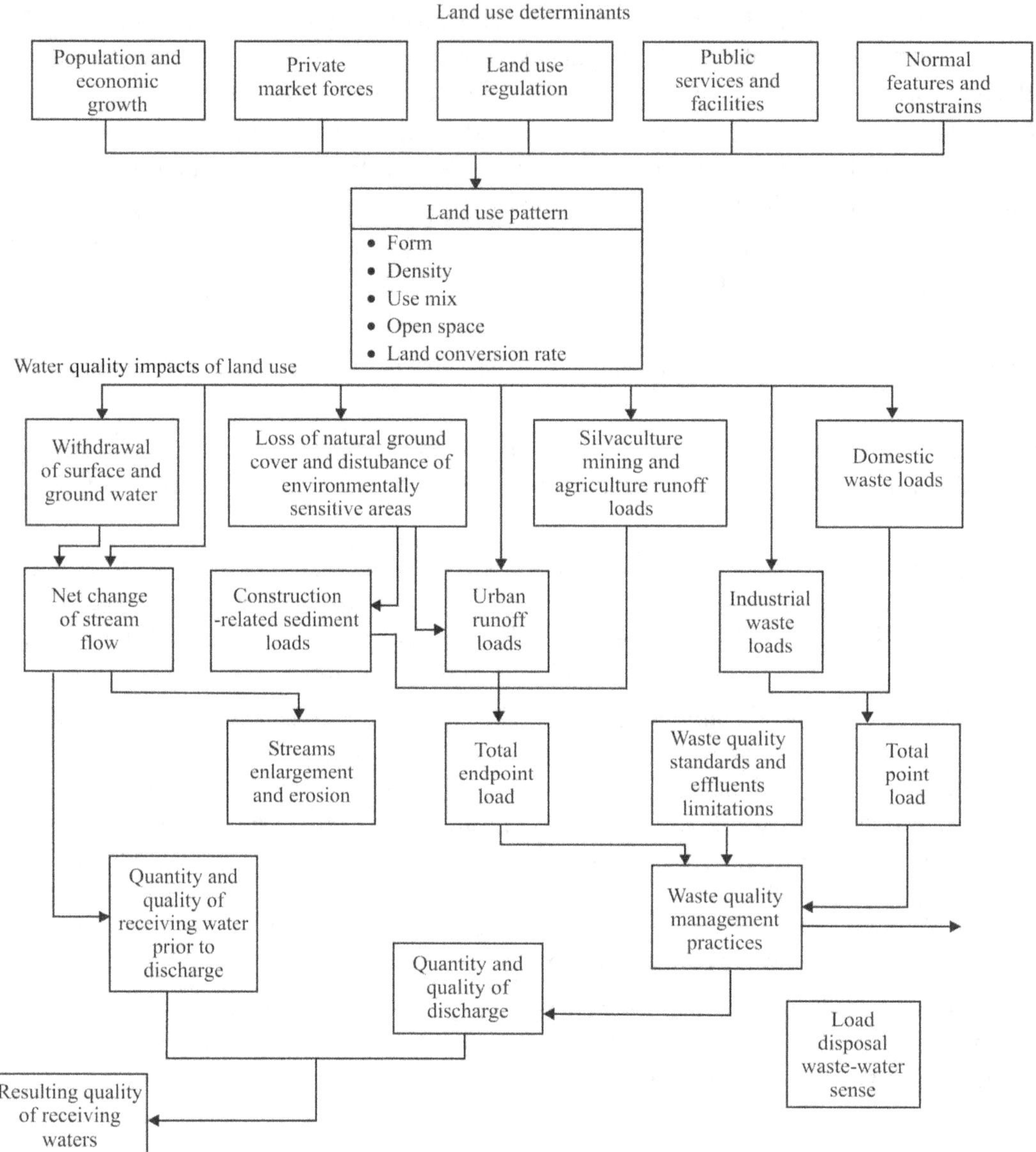

Fig. 4.3 Schematic diagram of the land-use-water-quality relationship.

(**Source** : Canter 1996 (2))

For example, rainfall in highly industrialized regions may consist of acidic precipitation which is introduced to the surface water, and may bring with it natural organics, sediments, and so on.

The summary of cause-effect network for surface waters is presented in Fig. 4.4.

Though the discharge of waste water (treated or otherwise) greatly adds to the organic loading of the surface water and clearing of land for construction, farming, etc., it can also result in increased erosion and sediment load in the surface water.

Water quality can be defined in terms of the physical, chemical, and biological characterization of the water.

Physical parameters include color, odor, temperature, solids (residues), turbidity, oil content and grease content.

Each physical parameter can be broken into sub-categories. For example, characterization of solids can be further sub-divided into suspended and dissolved solids as well as organic (volatile) and inorganic (fixed) fractions.

Chemical parameters associated with the organic content of water include biochemical oxygen demand (BOD), chemical oxygen demand (COD), total organic carbon (TOC), and total oxygen demand (TOD).

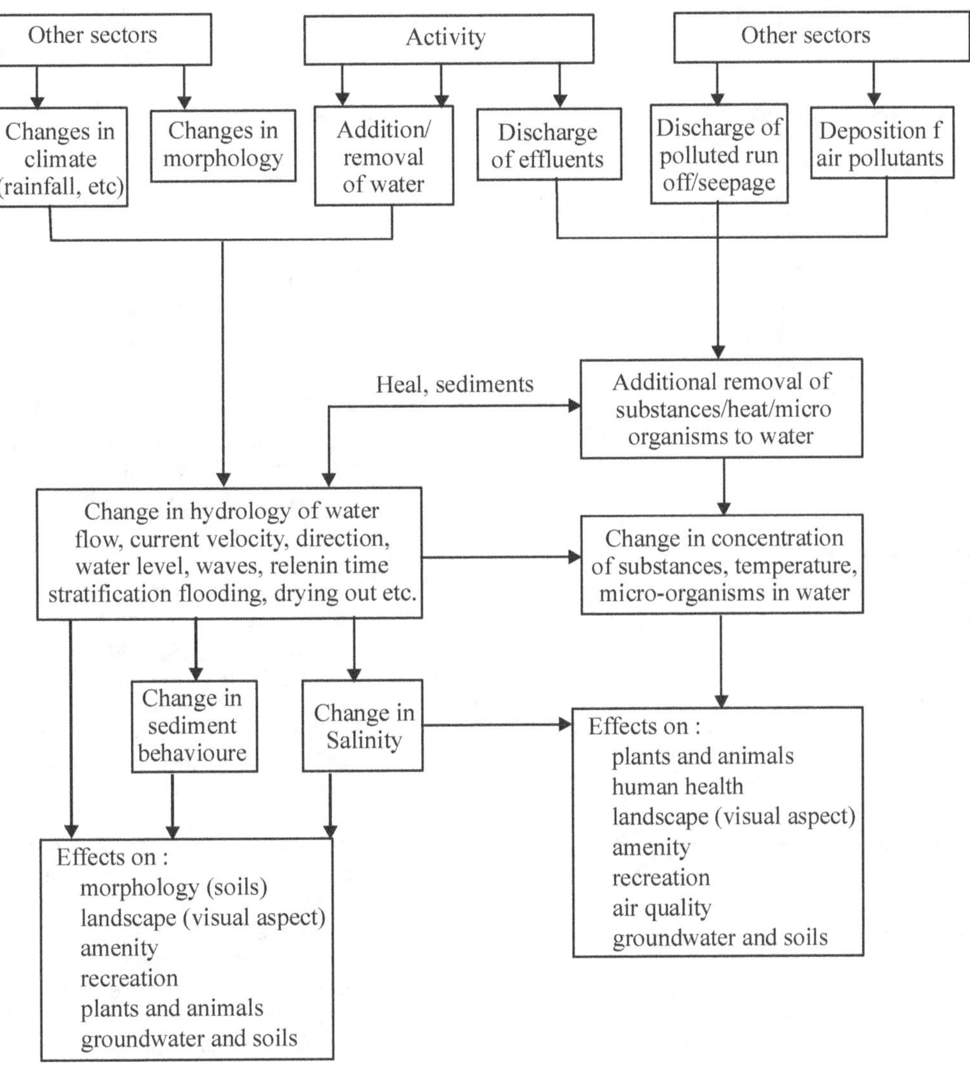

Fig. 4.4 Summary of cause-effect network for surface waters
(**Source :** ERL, 1984 (1)).

It should be noted that BOD is a measure of the organics present in the water; it is determined by measuring the oxygen necessary.

Inorganic chemical parameters include salinity, hardness, pH, acidity, and alkalinity. The presence of substances including iron, manganese, chlorides, sulfates, sulfides, heavy metals (mercury, lead, chromium, copper and zinc), nitrogen (organic, ammonia, nitrite) and phosphorus.

Biological properties include bacteriological parameters such as coliforms, fecal coliforms, specific pathogens, and viruses. Routine monitoring of biological quality of waters involve indicator groups and relies on two basic assumptions (a) that principal concern is with human faecal contamination of water and (b) that the indicators used will be present in proportion to all pathogenic species of interest. The most common organisms used are coliform bacteria (total coliforms and faecal colifoms, faecal streptococci and salmonella). Table 4.1 shows some water quality parameters assessed during impacts study.

Table 4.1 Presents some common water quality parameters surveyed in the impact assessment.

Variable	System	C	H	F	Notes
Nutrients					
Phosphorus	R	+	−	−	Several different forms. Much of load transported in sediment.
	L & P	+	−	+	Varies between hypolimnion and Epilimnion.
Nitrate	R	+	+	+	Usually higher in late autumn/winter.
	L and P	+	+	+	Levels generally increase with amount of flow through system.
Chlorophyll	OAS	+	−	+	Used as a general index of standing crop of *algae*.
Organic matter					
Biochemical oxygen	R	+	+	+	A main variable in monitoring sewage outfalls and GQRAs. Can range from demand (BOD) < 5 mg/l in clean rivers to 100,000 mg/l in industrial waste.
Chemical oxygen	R	+	−	+	Measures total organic matter which *could* use up oxygen. An alternative to BOD, e.g., where *non-labile organics* are suspected.
Metals					
Al, Cu, Cd, Hg, Pb, Zn	AS	+	+	+	Often serious pollutants of frehwaters, Toxicites usually increase with decreasing pH and water hardness.
Ca, Mg, Na, K	AS	+	+	+	Used to assess water type but not quality. Useful in conjunction with other variables to assess likely toxicity of other metals.
Others	AS	+	+	+	Industry-specific surveys may be needed (e.g., silver for electroplating, tin from old mines) but most not routinely covered.

Table 4.1 Contd…

Variable	System	C	H	F	Notes
Micro-organics	AS	+	+	+	Difficult to identify unless potential source suspected; so although potentially important rarely included in standard.
General effects	AS	+	+	+	Most are easily detected by sight/smell. Not normally a health problem as polluted water unlikely to be imbibed. Tainting can damage fisheries.
Carcinogenic Effects	AS	+	+	+	Rarely routinely done as particular carcinogen will vary with type of oil, geographic source and batch.
Others					
Ammonia	R	+	+	+	Organic decay product. Toxic to fish, and toxicity increases at high pHs.
	L & P	+	+	+	In large waterbodies, only likely to be high in intensively stocked fiheries. Small stagnant waterbodies may naturally have high levels.
Hydrogen Sulphide	R	+	+	+	Generally as for ammonia.
	L & P	+	+	+	
Cyanide	AS	+	+	+	Very toxic but occurrence limited to particular industries.
Sediment	R	+	−	+	Part of routine monitoring, especially in relation to sewage outfalls.
	L & P	−	−	+	May be of concern in fisheries and reservoirs (may block filters).
Pathogens	AS	−	+	−	Mainly for faecal contamination, especially for water-areas.
Dissolved Oxygen	R	+	−	+	A routine variable because many river animals need high levels.
	L	+	−	+	Levels vary with depth, time of day and season.
	P	−	−	+	Levels often highly variable.
PH	as	+	+	+	Interpretation is very use related. Used to qualify other data.
Alkalinity	AS	+	+	+	Used to qualify pH data.
Electrical	AS	+	+	+	Useful as an indication of the levels of other conductivity major variables.
Temperature	AS	+	+	+	Assessing thermal pollution, but mainly used to qualify other data.

System : L = lakes and reservoirs; P = ponds; R = rivers; AS = all systems (usually including groundwaters).

C,H,F = purpose : C = conservation; H = human health; F = fisheries. − = infrequently measured (but may be important in specific circumstances); + = frequently measured.

The two main sources of water pollutants to be considered are nonpoint and point sources (Table 4.2).

Non-point sources are also referred to as "area" or "diffuse" sources.

Table 4.2 Non point and point sources of pollutants. (17)

Nonpoint pollutants	Point pollutants
Pollutant from :	*Specific discharge from :*
Urban area, industrial area, or rural run-off.	Municipalities or industrial complexes
Examples: sediment, pesticides, or nitrates entering a surface water because of runoff from agricultural farms.	*Example*: Organics or metals entering a surface water as a result of waste water discharge from a manufacturing plant.

In a given body of surface water, non-point source pollution are difficult to assess and can be a significant contributor to the total pollutant loading, particularly with regard to nutrients and pesticides. Fig. 4.3 illustrates the relationship between land use changes and pollutant water quality changes in receiving waters, while Fig. 4.5 presents positive and negative effects likely to occur with different land-uses.

***Some general characteristics of non-point source pollution are as follows*:**

1. Non-point-source discharges enter surface waters in a diffuse manner and at intermittent intervals that are related mostly to the occurrence of meteorological events;

2. Pollution arises over an extensive area of land and is in transit overland before it reaches surface waters;

3. Non-point source discharges generally cannot be monitored at the point of origin, and the exact source is difficult or impossible to trace;

4. Elimination or control of these pollutants must be directed at specific sites; and

5. In general, the most effective and economical controls are land management techniques and conservation practices in rural zones and architectural or hydrological control in urban zones.

4.3.1.1 Pollution Impacts Associated with Construction and Operation of Projects

Table 4.3 provides an overview of important surface water contaminants and their impacts and Table 4.4 summarizes the impacts of certain pollutants in relation of potential impairment of water usage.

4.3.1.2 Identifying Potential Impacts

In this step various water pollutants, their quantities entering surface waters and surface water usage for various activities, runoff resulting from various precipitation episodes need to be critically examined. Materials like fuels, oils, preservatives, fertilizers, pesticides, hazardous chemicals, liquid wastes etc. which are used or disposed during project activities if not managed properly contaminate surface waters during runoff events.

Quality characteristics of industrial wastes vary considerably depending upon the type of industry.

A useful parameter in describing industrial wastes is population equivalent.

$$PE = (A) (B) (8.34)/0.17$$

where PE = population equivalent based on organic constituents in industrial wastes.

                A = industrial waste flow, mgd; B = industrial waste BOD, mg/L

                8.34 = lb / gal

                0.17 = lb BOD per person – day

A similar type of population equivalent calculation could be made for suspended solids, nutrients and other related constituents.

To express all waste loading on a similar basis, population equivalent calculations can be made for various pollutants from both point and non-point sources. Non-point sources of water pollutants have been recognized as potential major contributors.

Table 4.4 Limits of water uses due to water quality degradation.

Use / Pollutant	Drawing	Aquatic powder wildlife, Fisheries	Recreation	Irrigation	Industrial uses	cooling	Uses Transport
Pathogens	×	o	××	×	××[a]	na	na
Suspended Solids	××	××	××	×	×	×[b]	××[c]
Organic matter	××	×	××	+	××[d]	×[e]	na
Algae	×[ef]	×[g]	××	+	××[d]	×[e]	×[h]
Nitrate	××	×	na	+	××[a]	na	na
Salts	××	××	na	××	××	na	na
Trace elements	××	××	×	×	×	na	na
Organic micro-pollutants	××	××	×	×	×	na	na
Acidification	×	××	×	?	×	×	na

×× Marked impairment causing major treatment or excluding the desired use

×   Minor impairment

o   No impairment

na   Not applicable

+   Degraded water quality may be beneficial for this specific use

?   Effect not yet fully realized

a   Food industries

b   Abrasion

c   Sediment settling in channels

d   Electronic industries

e   Filter clogging

f   Odor, taste

g   In fish ponds higher algal biomass can be accepted

h   Development of water hyacinth (Eichhornia crassipers)

i   Also includes boron fluoride etc

j   Ca.Fe.Mn in testile industries etc

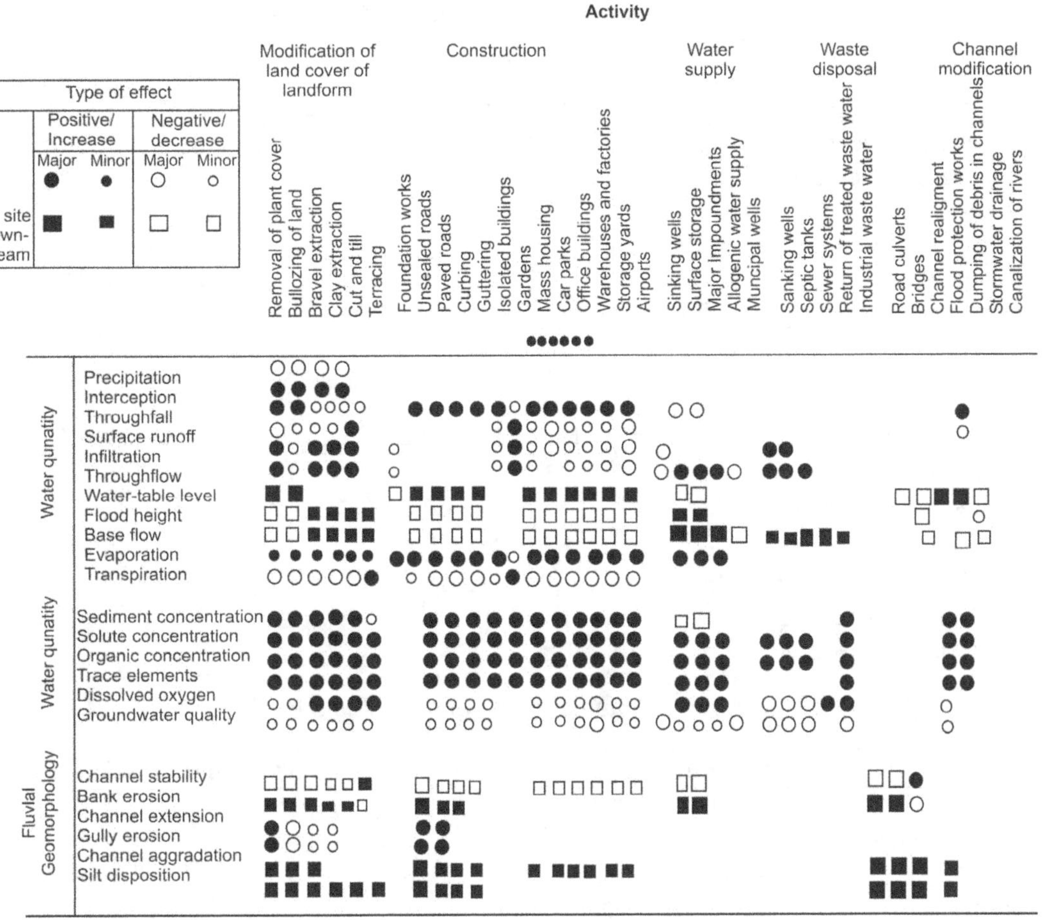

Fig. 4.5 Positive and negative effects of land use change on surface water qualities environment.

Information regarding stormwater pollution loading based on the units per acre of residential development is given in Table 4.5(a), 4.5(b) and 4.6.

Table 4.5(a) Representative rates of erosion from various land uses.

Land use	Erosion rate		
	Metric tons/cm^2-yr	Tons/ml^2-yr	Relative to forest =1
Forest	8.5	24	1
Grassland	85	240	10
Abandoned	850	2,400	100
Cropland	1,700	4,500	200
Harvested forest	4,250	12,000	500
Active surface mines	17,000	48,000	2,000
Construction	17,000	48,000	2,000

Table 4.5(a) Contd…

Stormwater pollution for selected urban uses

Residential	Density	Erosion rate			
		Nitrogen[a]	Phosphorus[a]	lead[a]	Zinc[a]
Residential large lot (1 acre)	12%	3.0	3.0	0.05	0.20
Residential small lot (0.25 acre)	25%	8.8	1.1	0.40	0.32
Townhouse apartment	40%	12.1	1.5	0.88	0.50
High-risk apartment	60%	10.3	1.2	1.42	0.71
Shopping center	90	13.2	1.2	2.5	2.06
Central Business District	95%	24.5	2.7	5.42	2.71

Table 4.5(b) Annual stormwater pollution loading for residential development.

Residential density	Phosphorus[a]	Nitrogen[a]	Lead[a]	Zinc[a]	Sediment[a]
0.5 unit/ac (1.25 person)	0.8	6.2	0.14	0.17	0.08
1.0 unit/ac (2.5 persons)	0.8	6.7	0.17	0.20	0.11
2.0 units/ac (5 persons)	0.9	7.7	0.25	0.25	0.14
10.0 units/ac (25 persons)	1.5	12.1	0.88	0.50	0.27

[a] pounds per acre per year.
[b] pounds per acre per year.

Table 4.6 Land use pollutants matrix and available loading functions.

Land use	Major pollutant	Loading functions; base
Agriculture	Sd, N, Ph, P, BOD, M	***
Irrigation return flow	TDS	**
Silviculture	Sd, N, Ph, BOD, M	*
Feedlots	Sd, N, Ph, BOD	**
Urban runoff	Sd, N, Ph, P, BOD, TDS, M, Coliform	***
Highways	Sd, N, Ph, BOD, TDS, M	*
Construction	Sd, M	*
Terrestrial disposal	N, Ph, TDS, M, Others	*
Background	Sd, N, Ph, BOD, TDS, M, radiation	**
Mining	Sd, M, radiation, acidity	*

Sd = sediment, N = nitrogen, Ph = phosphorus, P = pesticides, BOD = biochemical oxygen demand
TDS = total dissolved solids, M = heavy metals.
*** Wide range of data is available, ** Less data is available, * A little data is available

For quantitative understanding of pollution loads for protecting water quality parameters for recovered water bodies the following criteria need to be followed

1. *Load allocation (LA)*: The portion of receiving water's loading capacity that is attributed either to one of its existing or future nonpoint sources of pollution or to natural (background) sources.

2. *Waste load allocation (WLA)*: The portion of receiving water's loading capacity that is allocated to one of its existing or future point sources of pollution. WLAs constitute a type of water quality based effluent limitation.

3. *Total maximum daily load (TMDL)*: The sum of the individual WLAs for point sources and LAS for nonpoint sources and background sources. If a receiving water has only one point–source discharger, the TMDL can be expressed in terms of mass per time, toxicity, or other appropriate measures.

   If best management practices (BMPs) or other nonpoint-source pollution controls make more stringent load allocations practicable, then waste load allocations can be made less stringent. Thus, the TMDL process provides for nonpoint-source-control trade offs.
4. *Water-quality-limited segment:* Any segment of which the water quality does not meet applicable standards, and/or is not expected to meet applicable standards, even after the application of technology-based effluent limitations.
5. *Water quality management (WQM) plan*: A state or area wide waste-treatment management plan developed and updated in accordance water act.
6. *Best management practice (BMP)***:** Methods, measures or practices (or combination of practices) determined by a state or designated area wide planning agency to be the most effective practicable means (including technological, economic and institutional considerations) of preventing or reducing the amount of pollution generated by nonpoint sources to a level compatible with water quality goals that are the best means of meeting particular nonpoint-source-control needs.

"Loading functions" refer to simple mathematical expressions that have been developed to evaluate either the production and/or the transport of a given pollutant

In addition to information on pollutant types and quantities, it may also be necessary to assemble information on the transport and fate of specific pollutant materials.

For example, information may be needed on the fate of petroleum products, other organics, nutrients, metals and so on in the water environment.

It is important to know whether the pollutant will partition between the water and sediment phases or become associated with aquatic flora and fauna.

Metals can occur in surface-water systems as both dissolved and particulate constituents. Bio-geochemical partitioning of metals can yield absorbed phases and coordination complexes with dissolved organic and inorganic legends.

4.3.2 Step 2 Description of Existing Surface-Water Resource Conditions

Step 2 involves describing existing (background) conditions of the surface water resource(s) potentially impacted by the project.

Pertinent activities include assembling information on water quantity and quality, identifying unique pollution problems, key climatological information, conducting baseline monitoring, and summarizing information on point and non-point pollution sources and on water users and uses.

4.3.2.1 Compilation of Water Quantity – Quality Information

Information should be assembled on both the quantity (flow variations) and quality of the surface water in the river reach of concern, and potentially in relevant downstream.

4.3.2.2 Water Quantity

Run-off Over Land

There are a number of standard mathematical models, expert systems, and field tests using tracers are available to determine movement of the run-off on land and its appearance in surface water bodies which are important in EIA studies as they mainly cause resultant impacts on the hydrology and water quality in receiving water bodies. Run-off of pesticides, fertilizers, and other materials toxic to water bodies used for domestic, agricultural, and recreational purposes need special focus as their impacts are significant. A number of Mathematical models are available for predicting run-off for:

- permeable or impermeable surfaces;
- sewered or unsewered areas;
- short-term or long-term predictions; and
- quantity or quality, for example, pesticides, sediments, biological oxygen demand, nutrients, dissolved minerals, bacteria, etc.

The balance between hydrological inputs and outputs to surface run-off (precipitation minus evapotranspiration, infiltration, and storage equals run-off) are described by mathematical equations based on same principles in all these Runoff Models. The basic model may be manipulated to include variables describing relevant processes (for example, erosion, sedimentation, wash-off of chemicals, adsorption, biodegradation, etc.), in which case they can also be integrated to water quality models for the receiving surface waters.

Extensive calibration and verification for use in specific areas and high level of expert assistance are required for application of all these models. Further substantial information on rainfall, air temperature, drainage network configuration, soil types, ground cover, land use, and management are also essential inputs.

The following are some widely used applications where the Runoff models are used:

- prediction of traffic pollutant loads washed off road surfaces through sewers after prolonged dry periods (the accumulated load is assumed to be washed off in the first heavy rainfall and enter surface waters); and
- prediction of the run-off of a conservative pollutant applied within a catchment area (the total amount applied is assumed to be uniformly diluted in the total run-off from the catchment).

Flow Models

For several types of freshwater systems, hydrological and hydrodynamic models have been developed for use in environmental assessment for which information on water flow will be highly essential. For estimating time varying flow rates (m^3/sec) in rivers, lakes, and manmade reservoirs many hydrological models which are often constructed based on historical data collected at hydrometric monitoring stations are finding wide application. In marine systems models have been used to predict currents and water level in coastal and estuarine environments

4.3.2.3 Water Quality

The quality emphasis should be on those water pollutants expected to be emitted during the construction and operational phases of the project. If possible, consideration should be given to historical trends in surface – water quantity and quality characteristics in the study area.

Oxygen Sag Curve - Streeter Phelps Equation

The changes in dissolved oxygen resulting from increased demands for oxygen from bacteria during decomposition and supply of oxygen from natural reaeration are considered in various models for accounting organic loading.

The Streeter-Phelps equation (18) which represents the oxygen sag curve (Fig. 4.6) depicts how the oxygen concentration C changes with time and distance downstream of a discharge point. The dissolved oxygen deficit, (Cs - C) as a function of demand for oxygen and natural aeration, where Cs is the oxygen saturation concentration is described by this equation.

The basic equation

$$D_t = \frac{K_1 L_o}{K_2 - K_1} e^{-K1t - K2t} + D_0\, e^{-K2t}$$

where

Dt is the dissolved oxygen (DO) deficit at t;

Lo is the BOD concentration at the discharge point immediately after mixing (t = 0);

Do is the initial DO deficit at the point of waste discharge;

t is the time or distance downstream;

K1 is the parameter of deoxygenation; and

K2 is the reaeration parameter.

Other processes that affect BOD and resulting dissolved oxygen concentrations, and that can be integrated in this model include algae and plant respiration, benthal oxygen demand, photosynthesis, and nitrogenous oxygen demand.

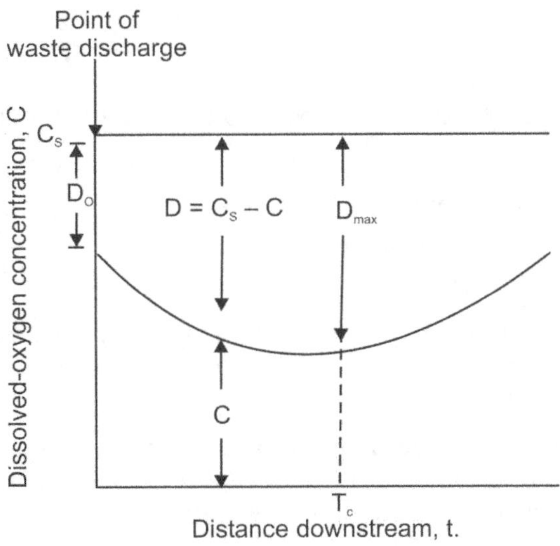

Fig. 4.6 Oxygen sag curve obtained from the Streeter-Phelps Equation
(*Source*: Canter, 1996 (2)).

4.3.2.4 Mass Balance Concept of Water Quality

Most of the water quality models are based on mass or material balance (Fig. 4.7) as described by the following equation

$$I + D + F + J = X + R + T$$

where

I is the inflow into the compartment (mass/time);

D is the discharge into the compartment (mass/time);

F is the formation due to biochemical activity in the compartment (mass/time);

J is the transfer from other compartments (mass/time);

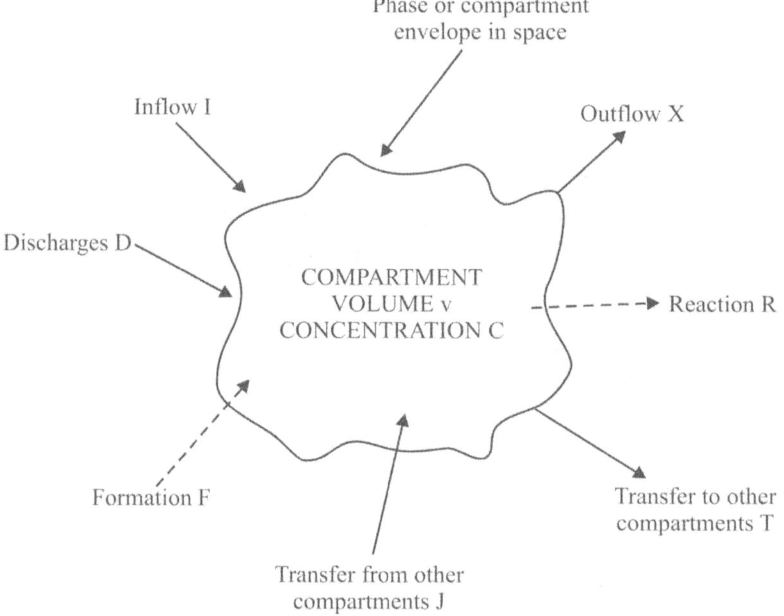

Fig. 4.7 Mass balance equation for a compartment.

(**Source**: Mckay and Peterson, 1993 (19)).

X is the outflow from the compartment (mass/time);

R is the degrading reaction (mass/time); and

T is the transfer to other compartments (mass/time)

The change in inventory (amount present in the volume of the compartment of air, water, biota) of a chemical in the identified volume is equal to inputs minus the outputs is the basis on which the mass balance equation is developed.

The simplest mass balance model is the mixing model described by Canter (1996) (3) which is the mass balance formulation of dissolved oxygen including transport exchanges with air (atmospheric exchange), and biota (production of oxygen due to photosynthesis and respiration (in water column and sediments)).

The model can be easily defined when there is only one compartment like water. In more complex models, a number of compartments may be present. McKay and Peterson (1993)

(19) describe a six compartment model of air, water, fish, soil, bottom sediments, and suspended sediment for estimating the fate of trichloroethylene (Fig. 4.8). The movement of trichloroethylene can be described by estimation of the transfer rates of some fifteen physical, chemical, and biological processes, including rates of sedimentation, run-off from soil, evaporation, reaction, and advective flows.

While the initial step involves defining the model assumptions, model equations and estimation of the various rates for the model's processes, the second step involves assessment of various discharges to the system which may come from various industrial and municipal sources, spills, and applications of chemicals (for example, pesticides). The final step is to define and estimate the transport rates between various media. The second step is very important because a chemical discharged into one media (for example, soil) may be relatively stable, but when conveyed to another media (for example, air), it may be subject to a large reaction rate since, a chemical's lifetime in the environment may then be controlled, not by how fast it will react in soil, but by how fast it can evaporate from the soil to the atmosphere McKay and Peterson, 1993 (19).

For simplification in EIA the chemical discharges associated with a specific project (for example, the effluent discharge from a pulp and paper mill) are considered. A mass balance model will then be used to estimate the changes in chemical concentrations within environmental components of concern.

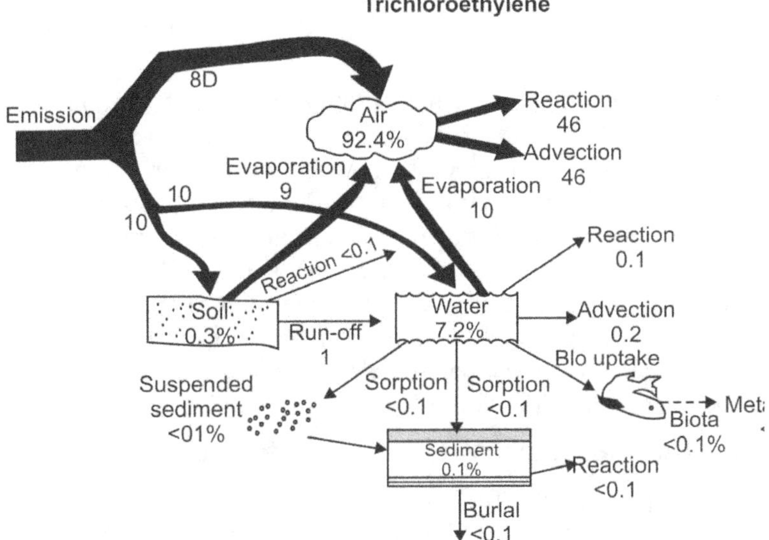

Fig. 4.8 Example mass balance diagram for the fate of trichloroethylene

(**Source:** Mackay and Peterson, 1993)(19).

Soil Erosion and Sedimentation

The underlying features of the project area determine the price of soil erosion which have been used as the basis for the Universal Loss Soil Equation (USLE). Westman (1985) (20) summarizes the work of many researchers in his presentation of the soil loss equation.

The simplest representation of the universal soil loss equation is:

$$A = 88.27 \times R \times K \times LS \times C \times P$$

where

A is the average annual soil loss (tonnes/km^2 per year);

R is a measure of rainfall intensity; and index value related to the maximum 30 minute rainfall intensity per storm (cm/hr) average over all storms;

K is the soil erodibility factor; an index from 0.001 (non erodible) to 1 (erodible) based on soil texture, structure, organic matter content, permeability (available for US soils from Soil Conservation)

LS is the slope length factor - where S is the slope angle (% of 45 degrees) and L is the length of the slope (m). The factor LS is expressed as a ratio of the erosion from that experienced on a slope of 9% and the length of 22 m. The ratios are available from standard slope charts.

C is the vegetative cover and management factor - ranges from 0.001 for a well managed woodland to1.0 for no cover;

P is the erosion control or management practice factor - ranges from 0.001 for effective contouring, terracing, and other erosion control for tilled land to 1.0 for the absence of erosion control.

88.27 is a conversion factor to convert units of A (tons/acre/year) original formulation to metric units tonnes/km^2 per year.

The USLE is often used to calculate the sediment yield or sediment input into a given water body (for example, a lake or reach of river). Models for movement of sediment within a water body concentrate on two processes: bed transport or sediment transport. The processes are the primary influence sediment behavior and can be expressed mathematically as a function of stream velocity. Bed transport is normally modeled as a function of stream bed condition and stream velocity. Sediment transport is normally modeled as a function of a sediment concentration (integrated over the full water depth) and stream velocity. Models may treat the two processes separately or simultaneously. Depending on the nature of the water body, sediment behavior may be modeled in one, two, or three dimensions. The models may also be integrated with hydraulic models to predict hydrological effects.

4.3.2.5 Development of Surface Water Quality Models

After Streeter and Phelps built the first water quality model (S-P model) to control river pollution in Ohio state of the US [18] surface water quality models have made a big progress from single factor of water quality to multi factors of water quality, from steady-state model to dynamic model, from point source model to the coupling model of point and nonpoint sources, and from zero-dimensional model to one-dimensional, two-dimensional, and three-dimensional models (21, 22). More than 100 surface water quality models have been developed up to now which are classified based on water body types, model-establishing methods, water quality coefficient, water quality components, model property, spatial dimension, and reaction kinetics by Cao and Zhang (23). After 1975, the number of state variables in the models increased greatly, and the three-dimensional models were developed at this stage, and the hydrodynamic model and the influences of sediments were introduced to water quality models (24, 25). Meanwhile, water quality models were combined with watershed models to consider nonpoint source pollution input as a variable (26, 27) due to more constraint conditions and nonpoint source pollution simulation at watershed scale. Typical models including QUAL models (28,29), MIKE11 model [30], and WASP models [31] were also developed.

For the integration of point and nonpoint sources, the US Environmental Protection Agency (USEPA) developed a multipurpose environmental analysis system (BASINS), which makes it possible to assess quickly large amounts of point and nonpoint source [32]. Meanwhile, the USEPA also listed the EFDC model as a tool for water quality management. Among the previously mentioned surface water quality models, these models including the Streeter-Phelps model, QUASAR model, QUAL model, WASP model, CE-QUALW2 model, BASINS model, MIKE model, and EFDC model were widely applied worldwide [33,34]. Recently, Kannel et al. [35] concluded that these public domain models (e.g., QUAL2EU, WASP7, and QUASAR) are the most suitable for simulating dissolved oxygen along rivers and streams. Some important surface water quality models in present use and their characteristics are summarized (Table 4.7).

Table 4.7 Main surface water quality models and their versions and characteristics

Models	Model version	Characteristics
Streeter-Phelps models	S-P model [13]; Thomas BOD-DO model [14]; O'Connor BOD-DO model [15]; Dobbins-Camp BOD-DO model [16, 17]	Streeter and Phelps established the first S-P model in 1925. S-P models focus on oxygen balance and one-order decay of BOD and they are one-dimensional steady-state models.
QUAL models	QUAL I [11]; QUAL II [18]; QUAL2E [19]; QUAL2E UNCAS [19]; QUAL 2K [20, 21]	The USEPA developed QUAL I in 1970. QUAL models are suitable for dendritic river and non-point source pollution, including one-dimensional steady-state or dynamic models.
WASP models	WASP1-7 models [22, 23]	The USEPA developed WASP model in 1983. WASP models are suitable for water quality simulation in rivers, lakes, estuaries, coastal wetlands, and reservoirs, including one-, two-, or three-dimensional models.
QUASAR model	QUASAR model [11, 24, 25]	Whitehead established this model in 1997. QUASAR model is suitable for dissolved oxygen simulation in larger rivers, and it is a one-dimensional dynamic model including PC-QUASAR, HERMES, and QUESTOR modes.
MIKE models	MIKE11 [22]; MIKE 21 [26]; MIKE 31 [27]	Denmark Hydrology Institute developed these MIKE models, which are suitable for water quality simulation in rivers, estuaries, and tidal wetlands, including one-, two-, or three-dimensional models.
BASINS models	BASINS 1 [11, 28]; BASINS 2 [11, 28]; BASINS 3 [11, 28]; BASINS 4 [28]	The USEPA developed these models in 1996. BASINS models are multipurpose environmental analysis systems, and they integrate point and nonpoint source pollution. BASINS models are suitable for water quality analysis at watershed scale.
EFDC model	EFDC model [29, 30]	Virginia Institute of Marine Science developed this model. The USEPA has listed the EFDC model as a tool for water quality management in 1997. EFDC model is suitable for water quality simulation in rivers, lakes, reservoirs, estuaries, and wetlands, including one-, two-, or three-dimensional models.

Areas of Application

Surface water models are applied to make predictions of dissolved oxygen, temperature, instream flows, suspended sediment, salinity and nutrients. Most models useful for impact assessment contain a number of basic processes. For example, the U.S. Department of Agriculture's (USDA) Simulator for Water Resources in Rural Basins-Water Quality [SWRRBWQ] was developed to simulate hydrological, sedimentation, nutrient and pesticide transport processes in a large, complex rural watershed. The SWRRBWQ model operates on a continuous time-scale and allows for subdivision of basins to account for differences in soils, land use, rainfall, etc. It can predict the effect of management decisions on water, sediment, and pesticide yield with reasonable accuracy for ungauged rural basins throughout the United States. SWRRBWQ includes five major components: weather, hydrology, sedimentation, nutrients, and pesticides. Processes considered include surface run-off, return flow, percolation, evapotranspiration, transmission losses, pond and reservoir storage, sedimentation, and crop growth. A weather generator allows precipitation, temperature, and solar radiation to be simulated when measured data is unavailable. Some widely used computer based water quality models are given in Table 4.8.

Table 4.8 Selection of parameters for river water quality monitoring surveys.

Type of survey	Physical Parameters	Inorganivcs	Organics	Nutrients	Microbiological	Hydrobiological
Proposed for inclusion in all surveys	Color pH Specific conductance suspended solids Total solids		Chemical oxygen demand total organic carbon (TOC)		Coliforms, total and fecal	
Recommended for collection of baseline data	Odor	Acidity Alkalinity Calcium (Ca) Chlorides (Cl) Dissolved oxygen Hardness Iron (Fe) Magnesium (Mg) Potassium (K) Selenium (Se) Silver (Ag) Sodium (Na)	Biochemical oxygen demand (BOD) immediate 5-day ultimate	Nitrate nitrogen NO$_3$	Total plate count	
Recommended Additional parameters where municipal and/or industrial pollution are expec	Floating solids	Arsenic (As) Barium (Ba) Beryllium (Be) Boron (B) Cadmium (Cd) Chromium (Cr) Copper (Cu) Dissolved carbon-dioxide (Co$_2$) Fluorides (F) Hydrogen sulfide (H$_2$S) Lead (Pb) Mercury (Hg) Nickel (Ni) Vanadium (V) Zinc (Zn)	Cyanide (CN) Dissolved organic carbon Methylene blue active substances (MBAS) Oil and grease Pesticides Total Phenolics	Ammonia nitrogen (NH$_3$) Nitrate nitrogen (NO$_2$) Organic nitrogen Soluble posphorus	Fecal streplococci Salmonella	Benthos Plankton counts
Optional parameters for Surveys of special purpose	Bed load Light penetration particle size Sediment concentrations	Aluminum (Al) Sulfates	Carbon alcohol extract (CAE) Carbon chloroform extract (CEE) Chlorine demand Echoviruses	Organic phosphorous Orthophosphates Polyphosphates Reactive silica Settleable solids	Shigella Viruses Coxsackie Polio Adenoviruses	Chlorophylls Fish Periphyton Taxonomic composition

The precipitation model is a first-order Markov chain model, while air temperature and solar radiation are generated from the normal distribution. Sediment yield is based on the Modified Universal Soil Loss Equation (MUSLE). Nutrient yields were taken from the EPIC model (Williams et. al., 1984) (36). The pesticide component is a modification of the CREAMS (Smith and Williams, 1980) (37) pesticide model. SWRRBWQ allows for simultaneous computations on each sub basin and routes the water, sediment, nutrients, and pesticides from the sub basin outlets to the basin outlet.

A detailed review of the current numbers of surface – water users and the quantities associated with such uses should also be assembled. The type of information that may be accumulated includes general estimates of the number of users of the surface water (private, public, industrial), types of water uses (drinking water, recreation, cooling water, etc), the location and rates of existing surface water withdrawals, and the location, quantity, and quality of existing discharges into the surface water, and so on.

4.3.3 Step 3 Procurement of Relevant Surface Water Quantity-Quality Standards

To determine the severity of the impact that may result from a project, it is necessary to make use of institutional measures for determining the impact significance.

Surface-water quantity and quality standards, regulations, or policies are examples of these measures. Thus determination of the specific requirements for a given surface water will require contacting governing agencies in one or several regions.

The intended use of the surface water with the use as drinking water supply typically results in the most stringent standards.

Effluent limitations regulating the permissible quality of discharged waste water from domestic and industrial sources may also be pertinent, along with regulations concerning non-point discharges from industrial areas. In some cases, there may be limitations on the amount and timing of water usage from a given body of water.

Water quality management policies may also be pertinent; examples of such policies include anti-degradation goals, clean-up or remediation goals, and /or goals for preservation of aquatic ecosystems and scenic beauty.

Typically, state water quality standards represent statewide goals for individual water bodies and provide a legal basis for decision making.

The standards designated by Central Pollution Control Boards (CPCB) will be based on the use or uses to be made of the water and set criteria necessary to protect the water resources and environment in general. It should be noted that most water quality standards and water-use restrictions are related to low-flow periods in the river system. For example, dissolved oxygen (DO) minima may be applicable during the 7-day, 10-yr low-flow conditions.

The maximum permissible contaminant levels in common units of water system for community supply and industrial are given in Table 4.9. and 4.10

Table 4.9 Maximum contaminant levels in community water systems.

Contaminant category	Primary standards	Maximum contaminant level
Inorganic chemicals	Arsenic	0.05mg/L
	Barium	1
	Cadmium	0.010
	Chromium	0.05
	Fluoride	4.0
	Lead	0.05
	Mercury	0.002
	Nitrate (as N)	10
	Selenium	0.01
	Silver	0.05
Organic chemicals	Chlorinated hydrocarbons	
	Endrin	0.0002mg/L
	Lindane	0.004
	Methxychlor	0.1
	Toxaphene	0.005
	2,4 – D (2,4 – dechlorophenoxyacetic acid)	
	2,4,5 – TP Silver (2,4,5 – trichlorophenoxypropionic acid) 0.1	
	Total trihalomethanes (the sum of the concentrations of bromodichloromethane, dibromochloromethane Tribromomethane (bromoform), and trichloromethane (chloroform)	0.10
Turbidity	**Turbidity**	**1.0 JTU (turbidity units)**
Radioactivity	Combined radium 226 radium 228	5 pCi/L
	Gross alpha – particle activity (including radium 226 but excluding radon and uranium	15 pCi/L
Secondary standards		
Miscellaneous	Aluminium	0.05 to 0.2 mg/L
	Chloride	250 mg/L
	Color	15 CU (color units)
	Copper	1.0 mg/L
	Corrositvity	Noncorrosive
	Fluoride	2.0 mg/L
	Foaming agents	0.5 mg/L
	Iron	0.3 mg/L
	Manganese	0.05 mg/L
	Odor	3 Ton
	pH	6.5 to 8.5
	Silver	0.1 mg/L
	Sulfate	250 mg/L
	Total dissolved solids (TDS)	500 mg/L
	Zinc	5 mg/L

4.3.4   Step 4 Impact Prediction

"Impact prediction" refers to the quantification (or, at least, the qualitative description), where possible, of the anticipated impacts of the proposed project on various surface water environment factors. Because of the complex and dynamic nature of hydrological systems, accurate prediction of impacts is often difficult and are bound to be some uncertainties which have to be recorded in the EIS.

The following considerations may be relevant to the prediction of surface water quantity-quality impacts:

1. frequency distribution of decreased quality and quantity;

2. effects of sedimentation on the stream-bottom ecosystem;

3. fate of nutrients by incorporation into biomass;

4. reconcentration of metals, pesticides, or radionuclides into the food web;

5. chemical precipitation or oxidation-reduction of inorganic chemicals; and

6. anticipated distance downstream of decreased water quality and the implications for water users and related raw-water quality requirements

Depending upon the particular impact, technically demanding mathematical models might be required for prediction which were discussed in earlier step.

Other approaches include the conduction of laboratory testing, such as, leachate testing for degraded material, and for solid or hazardous waste materials or sludges. Still other techniques might be appropriate; examples include chronic toxicity testing. The environmental effects on surface water are given below.

1. *Shoreline*: The shoreline and the river bank have special economic, ecological, aesthetic and recreational importance. Project development in the vicinity of shoreline may affect these uses. Furthermore any alteration of the shoreline may upset the land/water equilibrium and cause erosion.

2. *Bottom interface*: The river bed, lake bed and the sea floor provide habitats, determine flow regimes, influence water quality and can be a source of minerals. Then disturbance may cause shore-line erosion some distance away, create turbidity and destroy habitats.

3. *Flow variation*: Flow regulations can affect transport capacity and water quality and can have a direct effect on economic, recreational and ecological characteristics of the system and adjacent land.

4. *Water quality*: Water quality is important for economic, ecological, aesthetic and recreational purposes. Water quality changes can be physical, chemical and biological and may affect water treatment costs or even deny some uses of the water.

5. *Drainage pattern*: Any alteration to the drainage pattern can affect the capability of land and wetland habitats. It may also disrupt the natural flow variation of the catchments.

6. *Water balance*: The equilibrium between precipitation, runoff, infiltration and evapotranspiration can be upset by project development. It may also disrupt the natural flow variation of the catchments.

Table 4.10 Summary of specific quality characteristics of surface waters used as sources for industrial water supplies.

Characteristics	Boiler makeup Industrial 0-1,500 pslg	709-5,000 pslg	Utility, Once through	Cooling water Fresh Makeup recycle	Brackish Once through	Make-up recycle	Textile Industry SIC-22	Lumber Industry SIC-24	Pulp and paper industry	Chemical industry SIC-26	Petroleum industry SIC28	Primary metals industry SIC-33	Food and kindred products, SIC-20	Leather industry SIC-31
Silica (SiO$_2$)	150	150	50	150	25	25	—	—	50	—	50	—		For the above two
Aluminium (Al)	3	3	3	3	—	—	—	—	—	—	—	—		categories the quality of raw
Iron (Fe)	80	80	80	80	1.0	1.0	0.3	—	2.6	5	15	—		surface supply should be
Manganes (Mn)	10	10	2.5	10	0.02	0.02	1.0	—	—	2	—	—		that prescribed by the
Copper (Cu)	—	—	—	—	—	—	0.5	—	—	—	—	—		National Technical Advisory
Calcium (Ca)	—	—	500	500	1,200	1,200	—	—	—	200	220	—		subcommittee on Water
Mangnesium (Mg)	—	—	—	—	—	—	—	—	—	100	85	—		Quality Requirements for
Sodium and potassium (Na + K)	—	—	—	—	—	—	—	—	—	—	230	—		Public Water Supplies.
Ammonia(NH$_3$)	—	—	—	—	—	—	—	—	—	—	—			Quality Requirements for
Bicarbonate (HCO$_3$)	600	600	600	600	180	180	—	—	—	600	480	—		
Sulfate (SO$_4$)	1,400	1,400	680	680	2,700	2,700	—	—	—	850	570	—		
Chloride (Cl)	19,000	19,000	600	500	22,000	22,000	—	—	200^f	500	1,600	500		
Fluoride (F)	—	—	—	—	—	—	—	—	—	—	1.2	500		
Nitrate (NO$_3$)	—	—	30	30	—	—	—	—	—	—	8	—		
Phosphate (PO$_4$)	—	50	4	5	5	5	—	—	—	—	—	—		
Dissolved solids	35,000	35,000	1,000	1,000	35,000	35,000	150	—	1,080	2,500	3,500	1,500		
Suspended solids	15,000	15,000	5,000	15,000	250	250	1,000	—	—	10,000	5,000	3,000		
Hardness (CaCO$_3$)	5,000	5,000	850	850	7,000	7,000	120	—	475	1,000	900	1,000		
Alkalinity (CaCO$_3$)	500	500	500	500	150	150	—	—	—	500	—	200		
Acidity (CaCO$_3$)	1,000	1,000	0	200	0	0	—	—	—	—	—	75		
PH, units	—	—	5.0-8.9	3.5-9.1	5.0-8.4	5.0-8.4	6.0-8.0	5-9	4.6-9.4	5.5-9.0	6.0-9.0	3-9		
Color, units	1,200	1,200	—	1,200	—	—	—	—	360	500	25	—		
Oranics methylene blue active substances	2^6	10	1.3	1.3	—	1.3	—	—	—	—	—	—		
Carbon tetrachloride extract	100	100	^{f}	100	^{f}	100	—	—	—	—	—	30		
Chemical oxygen demand (O$_2$)	100	500	—	100	—	200	—	—	—	—	—	—		
Hydrogen Sulfide (H$_2$S)	—	—	—	—	4	4	—	—	—	—	—	—		
Temperature,°F	120	120	100	120	100	120	—	—	95^g	—	—	100		

Unless otherwise indicated, units are mg/L and values are maximums. No one water will have all the maximum values shown.

Water containing in excess of 1,000 mg/L dissolved solids
May the 1,000 for mechanical pulping operations.

7. *Flooding:* Reclamation of natural flood plains or swamps may result in flooding and siltation of other areas during peak flow.

8. *Existing use*: The use of surface waterways for new projects can deny existing uses such as for transport, recreation, water supply by creating turbidity, constructing barriers or changing the water quality.

Table 4.9 gives the quality of surface water used as source for industrial water supplies.

Some other techniques include the use of look-alike or analogous information on actual impacts from similar types of projects in other, similar geographical locations.

Finally, environmental indexing methods such as the WQI or other types of systematic techniques for relatively addressing anticipated impacts can also be considered.

It is desirable to quantify as many impacts as possible because in doing so, it has been frequently determined that the concerns related to anticipated changes are not as great as they would appear to be in the event of non-quantification. Also, if anticipated impacts are quantified, it would be appropriate to use specific numerical standards as the basis for qualitatively describing the impacts. The impact sources which involve direct utilization of hydrological systems and which involve indirect associations with hydrological systems are presented in Table 4.11.

Table 4.11 Impacts not directly associated with manipulation or utilization of hydrological system.

Sources	Potential Impacts
Roads	Changes in drainage systems, e.g., due to gradient changes, ridges, embankments, channel diversion or resectioning. Drawdown by dewatering when deep cutting. Increased runoff from impermeable surfaces, with risks of flash floods and erosion. Increased sediment loads form vehicles, road wear, and erosion of cuttings and embankments. Pollution of water courses by organic content of silt, other organics (e.g., oils, bitumen, rubber), de-icing salt (and impurities), metals (mainly vehicle corrosion), plant nutrients and pesticides from verge maintenance, and accidental spillages of toxic materials. (DoT 1993) (38.)
Urban and commercial development	Changes in drainage systems due to landscaping. Abstraction. Drawdown/changes in groundwater flow, e.g., when dewatering deep foundations. Reduced groundwater recharge, and increased runoff velocities and volumes (with flood and erosion risks from rapid storm flows) due to impermeable surfaces. Pollution of watercourses and groundwaters by a wide range of pollutants which are rapidly transported to receiving waters by increased runoff. Increased sewage treatment. (Hall 1984, Sgaw 1993, Walesh 1989) (39-41)

Table 4.11 contd...

Sources	Potential Impacts
Industrial development	As above but with greater runoff effects (from a higher proportion of hard surfaces); higher pollution levels and a wider variety of pollutants including metals and micro-organics from heavy industry and refineries, pesticides from wood treatment works, and nutrient-rich or organic effluents from breweries, creameries, etc. Thermal pollution from power plants.
Mineral extraction	**Operation phase** – Removal/realignment of watercourses. Loss of floodplain storage/flow capacity. Drawdown and reduced local streamflows caused by dewatering for dry extraction, or increased runoff from process wash water or extraction methods involving water use. Increased siltation and chemical pollution downstream e.g. from spoil heaps/vehicles/machinery/stores.
Landfill	**Restoration/aftercare phase** (Rust Consulting 1994) (42) Increased runoff from raised landforms, especially if clay-capped. Reduced groundwater recharge and river base flows if clay-sealed. Pollution of groundwater and near-surface runoff by *leachates* and by fertilizers and pesticides from restored grassland (Petts & Eduljee 1994) (43).
Forestry and deforestation	Reduced evapotranspiration and infiltration after felling – with consequent (a) decreased groundwater recharge, (b) increases in runoff, soil erosion, stream-sediment loads and siltation. Pollution by pesticides, especially herbicides used to prevent regrowth after clear felling.
Intensive agriculture	Enhanced runoff and erosion from bare soils. Drainage or irrigation impacts. Pollution of surface and groundwaters by: fertilizers, pesticides, organics from soil erosion, silage clamps and muck spreading, heavy metals from slurry runoff, and pathogens in animal wastes.
River engineering/manipulation Resectioning/channelization widening, deepening, realigning/straightening), e.g. to increase channel capacity for flood defence or drainage, or to facilitate project layout.	Brooke 1992, Brookes 1988, 1999 (4-6) Loss of channel and bank habitats. Enhance erosion and hence silt production (especially during construction, when pollution risks also increase). Increased flood risk and siltation downstream. Lowering of floodplain water table caused by deepening.
Embanking and bank protection (e.g., with concrete) usually for reasons as above.	Floodplain inundation and siltation prevented, with consequent risk of soil drought and loss of wetlands. Drainage from floodplain inhibited (unless sluices installed) with consequent waterlogging.

Table 4.11 *contd...*

Sources	Potential Impacts
Clearing bank vegetation	Loss of wildlife habitats and visual/amenity value.
Fluvial dredgings, and deposition of e.g., to maintain/enhance flood capacity or navigation.	Damage to channel habitats and biota at dredging sites. Increased sediment load and hence turbidity and smothering of downstream benthic and marginal ecosystems.
Diversion, e.g., to increase water supply. receptor area, or as a flood relief channel	Decreased supply in donor area. Channelisation and to evaporative loss from open channels. Risk to habitats in main river corridor.
Development on river floodplains	DETR 2000, EA 1997, Smith & Ward 1998 (44-46)
Use of floodplain area Construction of flood defences Laying impermeable surfaces	Increased flood risk upstream and downstream. Reduced groundwater recharge and river baseflows. Loss of ecological, heritage and visual/amenity/recreational features.
Reservoirs and dams : General On-stream dams : above dam On-stream dams : below dam On-stream dams : barrier effects Off-stream dams (not on a main	Petts 1984 (47) Loss of terrestrial habitats/farmland/settlements. Local climate change and rise in water table. Visual impacts of retaining walls. Water-borne pathogens. earthquake/landslip/ failure risks. Loss of river section; changes in flow regime; siltation. Reduced flows, oxygen levels and floodplain siltation. Migration of fish and invertebrates blocked. Changes in groundwater recharge, levels and flow directions.
Irrigation	Water abstraction (often from rivers). Increased evapotranspiration and local runoff. Risk of waterlogging and salination
Drainage schemes Water abstraction Sewage treatment works	May involve channelisation. Increased soil drought risk and oxidation of organic soils. Water table lowered and wetlands lost. Increased flood/erosion risk downstream. Water resources depleted. Water table lowered. Risks of river Lowflows, loss of wetlands, soil droughts and subsidence. Petts & Eduljee 1994 (43)

Increases in silts, nutrients (especially if treatment is poor), heavy metals, organics, and pathogens, e.g., fecal coliforms.

It is necessary for professionals to use their best judgement. Impact prediction involves the question whether the pollutants are conservative, non-conservative, bacterial or thermal.

Conservative pollutants are not biologically degraded in a stream, nor are they lost from the phase as a result of precipitation, sedimentation or volatilization.

The basic approach for prediction of downstream concentration of conservative pollutants is to consider the dilution capacity of the stream and use a mass-balance approach with appropriate assumptions.

Non-conservative pollutants refer to organic materials that can be biologically decomposed by bacteria in aqueous systems. Nutrients are also non-conservative, since they can be involved in biochemical cycling and plant uptake.

Predication of impacts resulting from non-concervative and bacterial pollutants and thermal discharges require mathematical-modelling.

Aquatic Ecosystem Modeling Approaches

Several methods (Table 4.12) have been developed to quantify and assess biological impacts on aquatic resources (3). One example is the "instream flow incremental methodology".

This approach is based on the concept that a particular species can be correlated with a set of particular habitat requirements, such as, specific water quality, velocity, depth, substrata, temperature, and cover conditions.

4.3.5   Step 5 : Interpretation of Impact Significance

For protection and assessment of a significant impact by public opinion collection, there are a number of specific numerical standards or criteria. For example, a number efficient discharge standards are prescribed for discharging into lakes or land with professional judgement.

The application of the professional judgement in the context of assessing impacts related to the biological environment; for example, the biological scientist in the study team would render judgements as to the applicability of various laws and the potential significance of the loss of particular habitats.

4.3.6   Step 6 : Identification and Incorporation of Mitigation Measures

The next activity is that of identifying and evaluating potential impact mitigation measures. Mitigation measures may need to be added to the project proposal to make it acceptable.

These mitigation measures might consist of decreasing the magnitude of the surface – water impacts or including the features that will compensate for the surface water impacts.

Table 4.12 Computer-based water quality model

Model name	Description	Major features	Data requirements	Output
CE-THERM-R1	1-D vertical reservoir model for temperature	Temperature, total dissolved solids (TDS), suspended solids (SS) coupled to density; Specify outflow ports or ports based on temperature objective; Reregulation pool, pumped-storage, and/or peaking hydropower options targets if using outflow-port	Inflow rates and constituent values; outflow rates, operations; Structural configuration and hydraulic constraints of outlets; Initial constituent profiles; Morphometric data Meteorological data Process and rate coefficients Release flow and temperature configurations decision routine	Vertical profiles and outflow values for constituents over time (printed and/or plotted); Statistics of predicted and observed values Flux information Operations schedules for multilevel outlet
CE-QUAL-R1	1-D vertical reservoir model for water quality	All CE-THERM-R1 features Allows simulation of most major physical, chemicals, and biological processes and associated water quality constituents Simulates anaerobic processes Monte Carlo simulations	Same as CE-THERM-R1 plus additional water quality data and coefficients	Same forms of CE-THERM –R1
CE-QUAL-W2	2-D longitudinal vertical hydro-dynamic and water quality model for reservoir, estuarine and other 2-D waterbodies	Solves 2-D hydrodynamics Head of flow boundary conditions Allows multiple branches. Simulates temperature, salinity and up to 19 other water quality Variables	Basically same as CE-QUAL (THERM)-R1 Tidal boundary conditions for estuarine applications Morphometric data, including widths for each cell	Velocities and water quality constituents at all points on 2-D vector plots and 2-D constituent concentration contour or shading plots Time-series data and plots Statistical output Restart files for subsequent hot restart simulations

(**Source:** Canter, 1996) (2).

Model name	Description	Major features	Data requirements	Output
SELECT	1-D vertical steady-state model of selective withdrawal from a reservoir	Computation of withdrawal zone distribution from a density-stratified reservoir Release temperature, density, conservative constituents computed Multiple outflow types (spillway, Water quality gate, flood-control Outlet, etc.) handled internally Users specifies ports operating or Selects ports internally based on Quality objective (e.g., temperature) Reacration of hydropower and flood-Control releases	Reservoir profiles for temperature (density) and conservative constituents Outflow rate, operation Structural configuration and hydraulic constraints or outlet (s) Quality targets if deciding port operation	Vertical profile of withdrawl zone Release qualities Appropriate port operations to meet quality targets
STEADY	1-D longitudinal steady-State stream temperature and DO model	1-D steady state Steady flow Allows branches, loops, and lateral inflows and withdrawals Flow can be piecewise nonuniform	Flows, depths, and velocities Average equilibrium temperature and heat-exchange coefficient Inflow temperature and DO Rate coefficients	Printed output for predicted temperature and DO at each node
CE-QUAL1-D RIV 1	dynamic flow, time-varying stream Hydraulic (RIV 1H) Models	Simulates dynamic (highly unsteady) flows Simulates up to 10 time-varying water quality constituents Allows branching systems Allows multiple control structures Stream, structural, and wind reaeration options Direct energy balance or equilibrium Temperature approach for temperatures	Physical data, cross-section geometry, elevations, and locations of nodes; lateral inflows and tributaries; control structures Initial conditions Boundary conditions for flow and quality Rate coefficients and other parameters Meteorological data or equilibrium Temperatures and exchange Coefficients	Hydraulic information and water quality constituent values printed for all nodes at specific print intervals Time-series plots of selected variables at selected nodes
HEC-5Q	Reservoir system Simulation/optimiza-tion model for multiple water-resources purposes including water quality, water supply hydropower and flood control	Balanced reservoir system regulation determination Optimum gate regulation for multiple water quality constituents	Inflow quantity and quality Initial water quality conditions System configuration and physical description Reservoir regulation manual time operation criteria System diversions Water quantity and quality targets at system control points	Reservoir and river water quality profiles Reservoir and river discharge rates, elevations and travel

The specific mitigation measures will be dependent upon the particular project type and location. Some typical mitigation measures with reference to certain water impact issues are given in Table 4.13.

Table 4.13 Some typical mitigation and enhancement measures relating to water-impact issues.

Damage to riparian features and/or change in channel morphology caused by river works, etc.

Use project management and restoration techniques to minimize and repair damage. Create new features such as pools and riffles. Use dredging's positively, e.g., for landscaping or habitat creation. (Brooke 1992, Brookes 1988) (4, 5).

Increased sediment loads and turbidity caused by river channel works

Select appropriate equipment and timing, e.g., construct new channels in the dry and allow vegetation to establish before water is diverted back in.

Impacts of development on floodplains

If development is permitted : (a) steer away from wetlands and high-flood-risk areas. (b) ensure that new flood defenses do not increase flood risk elsewhere. (c) take compensatory measures, e.g., floodways and flood storage areas/reservoirs to provide flood storage and flow capacity. (d) allow for failure/overtopping of defenses, e.g., by creating flood routes to assist flood water discharge. (e) take opportunities for enhancement in redevelopment, especially where (as in many urban sites) existing conditions are poor, e.g., use river corridor works to restore floodplain (by removing inappropriate existing structures), enhance amenity and wildlife value, and create new floodplain wetlands. (EA 1997, Smith & Ward 1998) (45, 46).

Impacts of mineral workings, especially on floodplains

Operational phase – Carefully manage the use and storage of materials/spoil, and runoff from spoil heaps/earthworks. Use siltation lagoons. Route dewatering water into (a) lagoons, wells or ditches to recharge groundwater, (b) watercourses to augment streamflows. **Restoration phase** – Careful backfill and aftercare management. Enhancement, e.g., of amenity/wildlife value (Rust Consulting 1994) (42)

Impact of new roads and bridges, or road improvement schemes

Use : careful routing, designs to minimize impacts on river corridors (not just channels), and measures to control runoff, e.g., routed to detention basins or sewage works, and not into high-quality still waters. If construction imposes river realignment, create new meandering channel with vegetated banks.

Impacts of dams and reservoirs

Adjust size of location (avoid sensitive areas). Minimise height and slope of embankments, and plant trees.

Water depletion by abstraction

Promote infiltration and hence groundwater recharge in urban areas (next page). Minimise water use, e.g., by metering and the installation of water-efficient equipment/appliances.

Table 4.13 Contd...

Increased runoff from urban and industrial developments

Use sustainable urban drainage schemes with (a) efficient piped drainage and sewer measures i.e., at or near the point of rainfall, to e.g., porous surfaces (car parks, pavements, etc.), soakaways (gravel trenches, vegetated areas); flow detention measures (grass swales, vegetated channels, stepped spillways, detention/balancing ponds/storm reserevoirs, and project layout/landscaping to increase runoff route). (Ferguson 1998, Hall 1984 Schwab *et. al.* 1993, Walesh 1989) (39-41).

Increased runoff and pollution (including sediments) from construction sites

Minimise soil compaction and erosion. Ensure careful storage and use of chemicals, fuel, etc., Install adequate sanitation. Guard against accidental spillage, vandalism and unauthorized use.

Chemical pollution from built environments, e.g., roads, urban/industrial areas

Control runoff (as above). Use oil traps, siltation traps/ponds/lagoons, vegetated **buffer zones** and wetlands, e.g., constructed reed beds.

Increased sewage and/or sewage-pollutant content

Increased sewage and/or **sewage treatment level**, e.g., from primary to secondary or secondary to tertiary.

Chemical pollution from an accidental spillage

Effective contingency plans. Use booms and dispersants.

Groundwater pollution

Avoid contamination from leaking storage tanks, etc., by appropriate bounding of tanks and improved site management. Use buffer zones (EA 1998b). (45).

(Source : Methods of EIA by Peter Morris and Riki Therivel, Spon Press NY 2001) (48)

Examples of certain actions, things which could be considered for mitigation or control measures, depending on the type of project, are listed below :

1. Decrease surface-water usage and waste water generation through the promotion of water conservation and waste water treatment and re-use. Pre-treat waste waters prior to discharge into receptor.

2. Minimize erosion during the construction and operational phases of the project; this could be facilitated by the use of on-site sediment-retention basins and by planting rapidly growing vegetation.

3. In projects involving the use of agricultural chemicals, consider measures that could be used to plan better the timing of chemical applications, the rate of application, and the extent of such applications in an effort to minimize erosion and chemical transport to surface-water systems. "Integrated Pest Management" (IPM) could also be used to decrease the pesticide loading from agricultural areas. IPM is an approach which combines biological, chemical, cultural, physical, and/or mechanical means, as appropriate to deal with unwanted insects, weeds and other pests a decrease in the non-point-source-pollution contribution to the surface-water environment.

4. Manage non-point-source pollution through the application of Best Management Practices (BMPs) as determined by a state or a designated area-wide planning agency to be the most effective practicable means of achieving pollutant levels compatible with water quality goals (Novotny and Chesters, 1981) (49). This determination should be made after a process of problem assessment, examination of alternative practices and appropriate public participation.

5. Develop a non-point-pollution-control program for coastal waters: information is available on management measures for agricultural sources, forestry, urban areas, marines, recreational boating, hydro-modification projects (channelization and channel modification dams, and stream-bank and shoreline erosion), wetlands, riparian areas, and vegetated treatment systems (U.S.EPA, 1990, 1993).

6. Use constructed wetlands to control non-point-source pollution involving nutrients, pesticides and sediments. As an example, a constructed system might include in hydraulic order, a sediment basin, grassy filter, wetland and deep pond.

7. Consider alternative wastewater treatment schemes to achieve treatment goals in a cost-effective manner. For point sources, the treatment schemes could include primary, secondary and/or tertiary processes involving physical, biological and/or chemical principles of pollutant removal. For thermal effluents the use of cooling ponds or towers might be appropriate.

8. Use discharge credit trading within watersheds to enable the trading of permitted pollution credits between parties responsible for both point and non point-source discharges.

9. Consider project operational modes that minimize detrimental impacts. One example is related to operating dam-reservoir projects in a recent survey of water resources projects operated by the U.S.

10. Use techniques such as sediment removal and macrophyte (weed) harvesting for restoring lakes and reservoirs from water quality deterioration and eutrophication. These techniques have been described in terms of their scientific basis, method of application, effectiveness, beneficial and detrimental impacts and costs in Cooke et. al., (1986) (50).

Summary

Several developmental activities will result in environmental impacts on surface water bodies. Impacts on surface waters are usually caused by physical disturbances, changes in climatic conditions, and the addition or removal of substances, heat, or microorganisms etc. These activities and processes lead to first order effects as manifested by changes in surface water hydrology, changes in surface water quality, and consequently to higher order effects reflected by changes in sediment behavior, salinity and aquatic ecology. For evaluating the environmental impact by any project activity on surface water bodies systematically a six step model is discussed. The technical details of these steps i.e., (a) Identifying Potential Impacts of proposed project (b) description of Existing Surface – Water Resource conditions-compilation of Water Quantity – Quality Information (c) Procurement of Relevant Surface–

water Quantity-Quality Standards (d) Impact Prediction (e) Interpretation of Impact Significance and (f) Identification and Incorporation of Mitigation Measures are presented with appropriate examples.

Case Study 1: Evaluation of the Impacts of Land Use on Water Quality: A Case Study in The Chaohu Lake Basin

Juan Huang, Jinyan Zhan, Haiming Yan, Feng Wu, and Xiangzheng Deng

Hindawi Publishing Corporation The Scientific World Journal Volume 2013, Article ID 329187, 7 pages http://dx.doi.org/10.1155/2013/329187

It has been widely accepted that there is a close relationship between the land use type and water quality. This study aims to analyze the influence of various land use types on the water quality within the Chaohu Lake Basin based on the water quality monitoring data and RS data from 2000 to 2008, with the small watershed as the basic unit of analysis.

Study Area: The Chaohu Lake Basin (Figure 4.9) belongs to the drainage system in the lower reaches of the Yangtze River, and it is the fifth largest freshwater lake in China, with a total watershed area of 13350 km^2. The total annual inflow from 33 rivers is 4.8 .109 m^3 year−1 and the total outflow is 3.4. 109 m^3 year−1. A large portion of the inflow is from Nanfei River, Hangbu River, and Yuxi River. The average annual temperature in the Chaohu Lake Basin ranges between 15∘C and 16∘C, with a mean annual rainfall of 1100 mm. The Chaohu Lake Basin is one of the most densely populated regions in Anhui Province [23], with a population of more than 9.65 million. The Chaohu Lake Basin also plays an important role in the local economic development, accounting for 24.65% of GDP in Anhui Province.

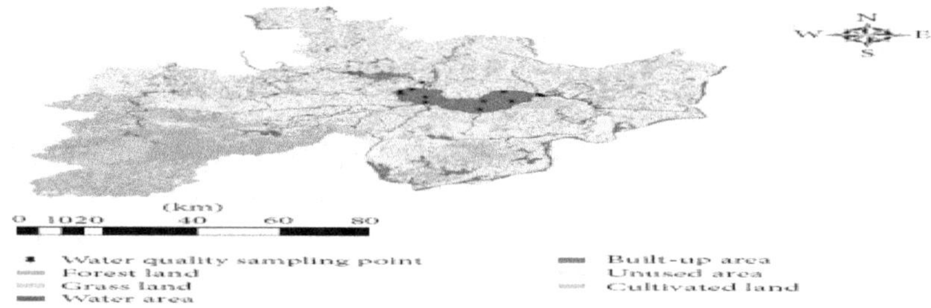

Fig. 4.9 The Chaohu Lake Basin. Water quality points are shown in the figure. Upstream catchment of each water quality sampling point and land use types were delineated.

During 1996-1999, the data from water quality monitoring points around Chaohu Lake showed that the percentage of water exceeding the level V had decreased from 80% to 60%. Since 2000, water quality of the main tributaries of Chaohu Lake, including Nanfei River, Shiwuli River, Pai River, and Shuangqiao River, is always exceeding the level V, with ammonia nitrogen as the key pollutant. But the water quality of other tributaries is better, generally fluctuating between level III and level IV (Table 4.14). In 2003, Chaohu Lake reached the meso-eutrophication the whole, with the total phosphorus (TP) and total nitrogen (TN) as the main pollutants. The water quality of Chaohu Lake improved slightly in 2005, reaching the meso-eutrophic state on the whole (Fig.4.10). The mean value of

CODmn and TN succeeded in achieving the goal of the tenth Five-year Plan in Chaohu Lake, but the mean value of TP failed. For the moment, Chaohu Lake is known as one of the three most polluted freshwater lakes in China [29]. Since Chaohu Lake serves as the primary drinking water source of Hefei City, it has been ranked as the key lake to be managed.

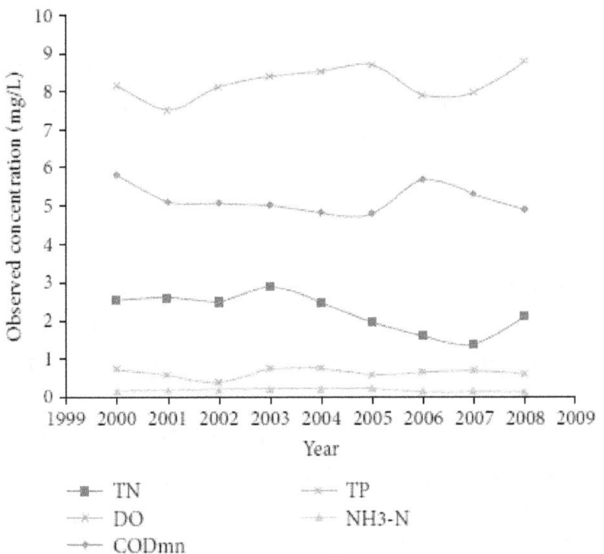

Fig. 4.10 Water quality change from 2000-2008 in the study area.

Table 4.14 The correlation coefficients between different land uses and water quality variables at the scale of the whole Watershed (TN=total Nitrogen; TP=Total phosphorous).

Variables (%)	ln (TN) coef.	ln (TP) coef.	ln (CODmn) coef.	ln (NH3-N) coef.	ln (DO) coef.
Cultivated land	−0.46	−0.42	−0.08	0.31	0.06
Forest land	−4.11**	−9.59***	−5.47***	−9.26***	0.31
Grassland	−6.52***	−7.87***	−2.93***	−8.83***	0.54
Water area	−0.30	0.26	0.29*	1.22**	0.0029
Built-up area	0.86	1.54**	0.70**	2.60***	−0.27*
Constant	1.20***	−1.44***	1.69***	−1.04***	2.08***
R-squared	0.49	0.75	0.80	0.69	0.36

*** $P < 0.01$, ** $P < 0.05$, * $P < 0.1$.

In this study the authors reported significant negative relation between water pollution and forest land and grass land, negative impacts occur on water quality while the influence of cultivated land on water quality is very complex. Significant negative relationship is observed between water quality indicators and landscape diversity.

References

1. ERL (Environmental Resources Limited). (1984), Prediction in Environmental Impact Assessment, a summary report of a research project to identify methods of prediction for use in EIA. Prepared for the Ministry of Public Housing, Physical Planning and Environmental Affairs and the Ministry of Agriculture and Fisheries of the Government of Netherlands.

2. Stone ML. (2016), Land Use Practice and Water Quality, Blue Crab Population Dynamics and Fish Biodiversity in Blackbird Creek, Delaware. Master's Thesis. Natural Resources Program, Department of Agriculture and Natural Resources, Delaware State University, Dover, Delaware. p. 103.

3. Canter, Larry W. (1996), Environmental Impact Assessment. Second edition. McGraw Hill Publishing Company, Inc., New York.

4. Brookes JS (1992), River and Coastal Engineering. In Environmental Assessment: A Guide to the Identification, Evaluation and Mitigation of Environmental Issues in Construction Schemes CIRA Research Project 424 chapter 4 Birmingham; CIRA.

5. Brookes A (1988), Channelised Rivers: Perspective of Environmental Management Chchester; WILEY

6. Brookes A (1999), Environmental Impact Assessment for Water Projects. A Hand Book of Environmental Impact Assessment, Vol. 2 *J. Petts* (ed) 404-430 Oxford Blackwell Science.

7. E. Ngoye and J. F. Machiwa (2004), "The influence of land-use patterns in the Ruvu river watershed on water quality in the river system," Physics and Chemistry of the Earth A, B, C, vol. 29, no. 15–18, pp.1161–1166.

8. L. Sliva and D. D. Williams (2001), "Buffer Zone versus Whole Catchment Approaches to Studying Land Use Impact on River Water Quality," *Water Research*, Vol. 35, No. 14, pp. 3462–3472.

9. S. T. Y. Yong and W. Chen (2002), "Modeling the Relationship between Land Use and Surface Water Quality," *Journal of Environmental Management*, Vol. 66, No. 4, pp. 377–393.

10. J. Bai, H. Ouyang, R. Xiao et al. (2010), "Spatial Variability of Soil Carbon, Nitrogen, and Phosphorus Content and Storage in an Alpine Wetland in the Qinghai-Tibet Plateau, China," *Australian Journal of Soil Research*, Vol. 48, No. 8, pp. 730–736.

11. G. Yang and J. Wang (2003), *Economic Development of Taihu Watershed*, Science Press, Beijing, China.

12. U. S. Tim and R. Jolly (1994), "Evaluating Agricultural Nonpoint-source Pollution using Integrated Geographic Information Systems and Hydrologic/water Quality Model," *Journal of Environmental Quality*, Vol. 23, No. 1, pp. 25–35.

13. R. C. Harrel and T. C. Dorris (1968), "Stream order, Morphometry, Physico-chemical Conditions, and Community Structure of Benthic Macro-Invertebrates in an Intermittent Stream System," *American Midland Naturalist*, vol. 80, no. 1, pp. 220–251.

14. N. E. Roth, J. D. Allan, and D. L. Erickson (1996), "Landscape influences on stream biotic integrity assessed at multiple spatial scales," *Landscape Ecology*, Vol. 11, No. 3, pp. 141–156.

15. F. H. Bormann, G. E. Likens, and J. S. Eaton (1969), "Biotic regulation of particulate and solution losses from a forest ecosystem," *Bioscience*, Vol. 19, No. 7, pp. 600–610.

16. K. B. Jones, A. C. Neale, M. S. Nash et al. (2001), "Predicting nutrient and sediment loadings to streams from landscape metrics: a multiple watershed study from the United States Mid-Atlantic Region," *Landscape Ecology*, Vol. 16, No. 4, pp. 301–312.

17. Metcalf and Eddy, Inc. (1991), Wastewater Engineering: Treatment, Disposal, and Reuse. 3rd Edition, McGraw-Hill, Inc., Singapore.

18. H.W. Streeter and E. B. Phelps (1925), A Study of the Pollution and Natural Purification of the Ohio River, United States Public Health Service, U.S. Department of Health, Education and Welfare.

19. Mckay, Donald, and Sally Peterson (1993), Mathematical Models of Transport and Fate. in Ecological Risk Assessment. ed. Glenn W. Suter II. Lewis Publishers, Ann Arbor, MI. 538 pp.

20. Westman, W.E. (1985), Ecology, Impact Assessment and Environmental Planning. John Wiley & Sons.

21. Q. G. Wang, X. H. Zhao, M. S. Yang, Y. Zhao, K. Liu, and Q, Ma (2011), "Water quality model establishment for middle and lower reaches of Hanshui river, China," *Chinese Geographical Sciences*, Vol. 21, No. 6, pp. 647–655.

22. Z.-X. Xu and S.-Q. Lu (2003), "Research on hydrodynamic and water quality model for tidal river networks," *Journal of Hydrodynamics*, Vol. 15, No. 2, pp. 64–70.

23. X. J. Cao and H. Zhang (2006), "Commentary on study of surface water quality model," *Journal of Water Resources and Architectural Engineering*, Vol. 4, No. 4, pp. 18–21, (Russian).

24. E. Wolanski, Y. Mazda, and P. Ridd (1992), "Mangrove hydrodynamics", *Coastal and Estuarine Studies*, Vol. 41, pp. 43–62.

25. M. J. Zheleznyak, R. I. Demchenko, S. L. Khursin, Y. I. Kuzmenko,P. V. Tkalich, and N. Y. Vitiuk (1992), "Mathematical modeling of radionuclide dispersion in the Pripyat-Dnieper aquatic system after the Chernobyl accident," *Science of the Total Environment*, Vol. 112, No. 1, pp. 89–114.

26. C. T. Hunsaker and D. A. Levine (1995), "Hierarchical approaches to the study of water quality in rivers spatial scale and terrestrial processes are important in developing models to translate research results to management practices," *BioScience*, Vol. 45, No. 3, pp. 193–203.

27. W. J. Grenney, M. C. Teuscher, and L. S. Dixon (1978), "Characteristics of the solution algorithms for the QUAL II river model," *Journal of the Water Pollution Control Federation*, Vol. 50, No. 1, pp. 151–157.

28. L. C. Brown and T. O. Barnwell Jr. (1987), The Enhanced Stream Water Quality Models QUAL2E and QUAL2E UNCAD: Documentation and User Manual, US

Environmental Protection Agency, Environmental Research Laboratory, Athens, Ga, USA.

29. Danish Hydraulics Institute (1993), MIKE11, User Guide & Reference Manual, Danish Hydraulics Institute, Horsholm, Denmark.

30. R. B. Ambrose, T. A. Wool, and J. P. Connolly (1988), WASP4, A Hydrodynamic and Water Quality Model-Model Theory, User's Manual and Programmer's Guide, US Environmental Protection Agency, Athens, Ga, USA.

31. "The US Environmental Protection Agency," http://water.epa .gov/scitech/datait/models/basins/fs-basins4.cfm.

32. S. F. Fan, M. Q. Feng, and Z. Liu (2009), "Simulation of water temperature distribution in Fenhe reservoir," *Water Science and Engineering*, Vol. 2, No. 2, pp. 32–42.

33. N. J. Morley (2007), "Anthropogenic effects of reservoir construction on the parasite fauna of aquatic wildlife," *Eco Health*, Vol. 4, No.4, pp. 374–383.

34. P. R. Kannel, S. R. Kanel, S. Lee, Y.-S. Lee, and T. Y. Gan (2011), "A review of public domain water quality models for simulating dissolved oxygen in rivers and streams," *Environmental Modeling and Assessment*, Vol. 16, No. 2, pp. 183–204.

35. Williams, J. R., Jones, C. A. & Dyke, P. T. (1984). "A modeling approach to determining the relationship between erosion and soil productivity". *Transactions of the American Society of Agricultural Engineers* 27, 129–144.

36. Smith, R. E. and J. R. Williams. (1980), Simulation of the surface water hydrology. JjH CREAMS Field Scale Model for Chemicals, Runoff, and Erosion from Agricultural Management Systems. Vol. I: Model Documentation. Conserv. Research Report No. 26. USDA-Sci. and Educ. Admin, pp. 13-35.

37. DoT (1993), "Design manual of roads and bridges", *Environmental Assessment* Vol 11, London, HMSO.

38. Hall MJ (1984), *Urban Hydrology* London E & FN Spon.

39. Shaw EM (1993), *Hydrology in Practice* 3rd edition Chapman & Hall.

40. Walesh SG (1989), Urban surface water management, New York, Wiley.

41. Rust Consulting (1994), *Hydrology and Mineral Workings-effects on nature conservation*; guidelines; Technical Annex English Nature Research Reports 106 &107 Peterborough, EN.

42. Petts J & G Eduljee (1994), *Environmental Impact Assessment for waste treatment and disposal facilities*, Chichester; Wiley.

43. DETR (2000), Planning policy guidance note 25 (PPG25) *Development and Flood Risk*, consultation paper

44. EA (1997), Our policy and practice for the protection of food plains. Bristol: Environmental agency (www.environment-agency.uk).

45. Smith K & R Ward (1998), Floods: Physical processes and human impacts; Chichester Wiley.

46. SO (1995), National Planning Policy Guidance 7(NPPG7). *Planning and Flooding*. Edingburgh: SO.

47. Petts GE (1984), *Impounded rivers; perspectives for ecological management*, Chichester: Wiley.

48. Methods of EIA by Peter Morris And Riki Therivel, Spon Press NY 2001.

49. Novotny, V and Chesters, G. (1981), Hand book of Non-Point pollution. Van Nostrand Reinhold Company, New York.

50. Cooke G.D Welch, E.B. Petersons, S.A, and Newroth, P.R. (1986), Lake & Reservoir Restoration, Butterworth, Publishers Stoneham, Mass.

Further Reading

1. Bird, S.L, and Hall. R, (1988), "Environmental Impact Research program; coupling Hydrodynamics to a multiple – Box water quality model", WES/TR/EC 88-7, U.S army Engineer waterways Experiment station, Vicksbcug, Miss.

2. Olson R.K and Marshall K. (1991), Workshop Proceedings: The role of created and Natural Wetlands in controlling non-point resource pollution. EPA 600/9-9/-042 Mantech Environmental Technology, Corvallis, ore.

3. Wengrzynsk, R.L (1991), "Constructed wetlands to Central Non-point source pollution, "PAT – APPL – 7, 764 924/WEP, U.S, Dept of Agriculture; Agricultural Research, Service, Washington D.C.

4. Frank, I and Brownstone, D. (1992), The Green Encyclopedia, Prentice-Hall General Reference, New York, pp 167-168.

5. Foster, G. R. (1981), Modeling the soil erosion process. In: Hydrologic Modeling of Small Watersheds. C. T. Haan, ed. Amer. Soc. of Agric. Engr.

6. Foster, G. R., L. J. Lane, and J. D. Nowlin (1980b), Model to estimate sediment yield from field-sized areas: Selection of parameter values. IH1 CREAMS Field Scale Model for Chemicals, Runoff, and Erosion from Agricultural Management Systems. 118 Vol. II: User Manual. Conserv. Research Report No. 26. USDA-Sci. and Educ. Admin, pp. 193-281.

7. Foster, G. R., L. J. Lane, J. D. Nowlin, J. M. Laflen, and R. A. Young (1980c), Model to estimate sediment yield from field-sized areas: Development of model. Im_ CREAMS Field Scale Model for Chemicals, Runoff, and Erosion from Agricultural Management Systems. Vol. I: Model Documentation. Conserv. Research Report No. 26. USDA-Sci. and Educ. Admin, pp. 36-64.

8. Foster, G. R. and L. D. Meyer (1972), Transport of soil particles by shallow flow. *Trans, of the ASAE* 15:99-102.

9. Foster G. R., L. D. Meyer, and C. A. Onstad (1977), Runoff erosivity factor and variable slope length exponents for soil loss estimates. *Trans, of the ASAE* 20:683-687.

10. Foster, G. R., W. H. Neibling, S. S. Davis, and E. E. Alberts (1980d), Modeling particle segregation during deposition by overland flow. In: Proc. of Hydrologic Transport Modeling Symposium. Amer. Soc. of AgricT~Engr. St. Joseph, MI. pp. 184-195.

11. Frere, M. H., J. D. Ross, and L. J. Lane (1980), The nutrient sub-model. DH CREAMS Field Scale Model for Chemicals, Runoff, and Erosion from Agricultural Management Systems. Vol. I: Model Documentation. Conserv. Research Report No. 26. USDA Sci. and Educ. Admin, pp. 65-87.

12. Leonard, R. A. and R. D. Wauchope (1980), The pesticide submodel. Inj. CREAMS-A Field Scale Model for Chemicals, Runoff, and Erosion from Agricultural Management Systems. Vol. I: Model Documentation. Conserv. Research Report No. 26. USDA-Sci. and Educ. Admin, pp. 88-112.

13. Smith, R. E. and J. R. Williams. 1980. Simulation of the surface water hydrology. JjH CREAMS Field Scale Model for Chemicals, Runoff, and Erosion from Agricultural Management Systems. Vol. I: Model Documentation. Conserv. Research Report No. 26. USDA-Sci. and Educ. Admin, pp. 13-35.

14. U. S. Department of Agriculture (USDA). 1980. CREAMS Field Scale Model for Chemicals, Runoff, and Erosion from Agricultural Management Systems. Conserv. Research Report No. 26. Sci. and Educ. Admin. 643 pp.

15. Wischmeier, W. H. and D. D. Smith. (1978), Predicting Rainfall Erosion Losses. Agriculture Handbook No. 537. USDA-Sci. and Educ. Admin. 58 pp.

16. Yalin, Y. S. (1963), An expression for bed-load transportation. *Journal of the Hydraulics Division*, Proc. of the ASCE 89:221-250.

Questions

1. Discuss the general methodology for the assessment of impacts of developmental activities on surface water environment.
2. What key information is required of any proposed project for identification of water quantity and quality impacts?
3. What are point and non point source of pollution? Discuss their relation with land use/ land cover changes.
4. Discuss the relevant factors for the prediction and assessment of surface quality and quantity impacts.
5. Discuss the loading factors to be considered for protecting the quality of receiving water bodies.
6. What is meant of impact prediction with reference to Surface Water Environment? Discuss the application of various mathematical models for impact prediction.
7. Discuss different potential impacts that can be predicted to occur in various types of projects where hydrological systems are directly utilised or manipulated.
8. Discuss typical mitigation measures that should be adopted to various water related issues.

CHAPTER 5

Prediction and Assessment of Impacts on Biological Environment

5.0 Introduction

Biodiversity refers to the wealth of species and ecosystems in a given area and of genetic information within populations. It is of great importance at global and local levels. Areas of high biodiversity are prized as store houses of genetic material, which form the basis of untold numbers and quantities of foods, drugs, and other useful products. The more species there are, the greater the resource available for adaptation and use by mankind. Species, which are pushed to extinction, are gone forever; they are never again available for use. Preservation of biodiversity is of global concern, but the causes of loss and their solutions are very often local in scale.

Our Life support systems need for sustainability a wide range of biological diversity at levels of genetic species, habitat and ecosystems. The Bio diversity of any ecosystem cannot be duplicated or replicated by any technological intervention and so is a non renewable resource/treasure which need to be protected (1). At present developmental programs involving activities like habitat destruction/fragmentation, industrial agriculture, forestry, over exploitation of plant and animal species/introduction of new species, soil, water, air pollution etc., are eroding the biodiversity of the ecosystems at (2) at unprecedented rate. Many developmental activities are likely to play a major role in the overall reduction of biodiversity, and proper planning at the project level can go a long way in limiting the loss, while still serving the needs of the people for which the project is started.

Thus the causes of biodiversity loss are "*embedded in the way we live*" (WRI, 1992) (2) and a shift in the path of human development is required to stop this loss which needs integrating biodiversity concerns into our planning.

Some development activities have direct impacts on biological systems. For example, clearing of land for infrastructure will destroy vegetation and displace animals. Introduction of contaminants may cause direct mortality of plants and animals. However, in many cases the changes in the physical environment caused by development that often lead to secondary or high order changes in plants and animals. For example, changes in downstream flow as a result of an upstream dam on a river may change the productivity of fish population. Alternatively, industrial pollution may be transported downstream and move through the food chain and ultimately contaminate the fish and wildlife populations that depend on the river.

The issue of impacts on flora and fauna is much broader than a concern for individual specimens and any useful discussion in this area must be considered in the larger context of biodiversity conservation.

Biodiversity is very useful for species adaptation for changing ecosystem conditions due to human activity. Diverse systems as they possess high degree of species redundancy allows substitutions and quickly facilitate the return of equilibrium state. Genetically diverse populations are less susceptible for extinction than less diverse populations as they can cope with changing environmental conditions

For decision making on proposed developmental programs/projects/plans/policies biodiversity in terms of its ecosystem services provides a measuring tool for impact assessment.

The Eco system services of bio-diversity can be mainly classified into following four categories

- Provisioning services: Harvestable goods such as fish, timber, bush meat, fruits, genetic material.

- Regulating services: Maintaining biological diversity itself, including natural processes and dynamics.

- Cultural services: Source of artistic, aesthetic, recreational or scientific enrichment, or non-material benefits.

- Supporting services: Necessary for the production of all other ecosystem services, such as soil formation, nutrients cycling and primary production.

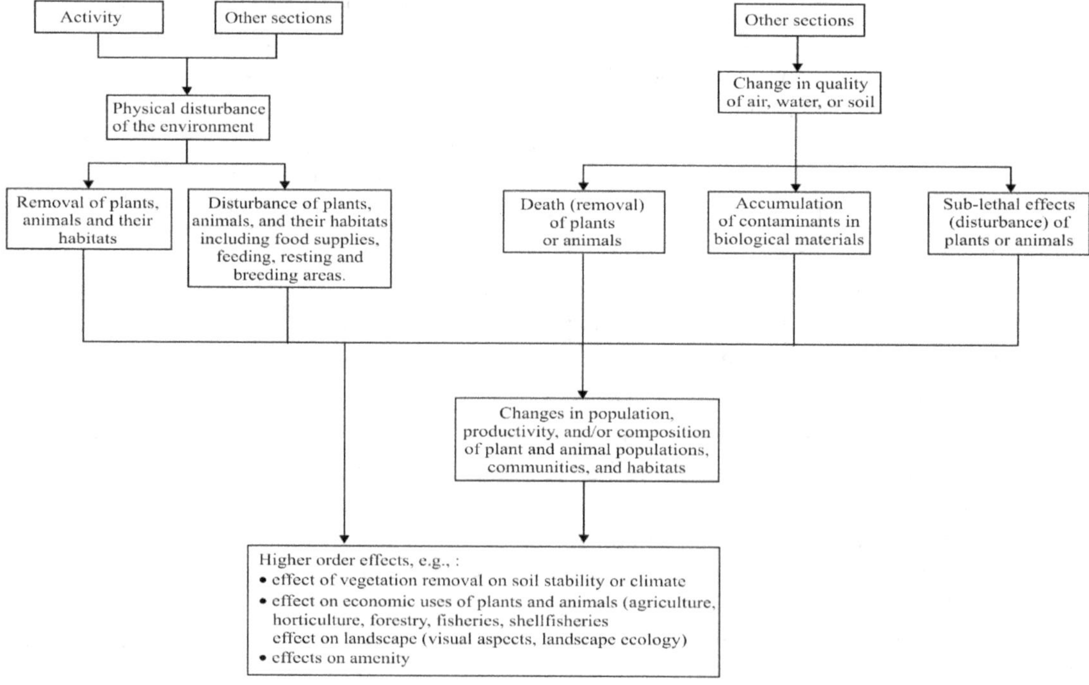

Fig. 5.1 Summary of cause-effect network for biota (source; ERL 1984).(3).

Decision making in Biological impact Assessment (BIA) is part of the Impact assessment which is about weighing these changes against each other, including those of alternative initiatives.

A simplified conceptual model of potential effects on biota is presented in **Fig. 5.1.** The complex and dynamic nature of ecological systems imposes difficulties in obtaining adequate baseline data for making accurate impact predictions and formulating dependable impact predictions.

Increased pace in implementation of developmental programs in recent times adopting new technological interventions resulted in the rapid depletion of biodiversity raising concern by 170 countries (4) around the globe and come to an agreement on biodiversity to conserve, sustainably use, and share equitably all biological resources with a shift in development path/programs.

For evaluating new projects/policies/programs a number of EIA methods (5) like environmental impact assessment (EIA), strategic environmental assessment (SEA), health impact assessment (HIA), risk assessment, and social impact assessment (SIA) are in practice

To address biodiversity concerns occurring due the implementation of many developmental programs, existing EIA methods by making few adaptations/changes can amend the projects/policies/plans. Associated EIA techniques include Integrated Environmental Management (IEM), Environmental Assessment (EA), Environmental Management and Audit Schemes (EMAS), and others specific to demands of countries and organizations.

EIA and SEA are the two techniques the process details of which are discussed in Chapter 1 are also relevant to biodiversity because they address environmental impacts. Procedurally EIA and SEA are similar but they differ in scope with EIA acting on a project level and SEA working on a more 'strategic' program and policy level. However these methods due to structural constraints cannot infuse biodiversity constraints in the early stages of project planning. A shift in planning developmental programs incorporating conservation of Biodiversity (CBD) concepts is highly essential

5.1 General Methodology for the Assessment of Impacts on Biological Environment (6-8)

The main objectives for any biological impact assessment are:

1. Preventing rapid depletion of biological resources due to any project/policy/plan and conserve bio diversity i.e., judiciously maintaining biological resources for Earths life support systems, keeping in view future requirements for human development.

2. Utilization of biological resources in a sustainable way i.e., Without hindering future development, planning utilization bio resources for supporting livelihood of people who traditionally depend on them .

3. Planning for equitable distribution of genetic resources for fair sharing of their benefits

Prediction and assessment of impacts on the biological environment involve a number technical and professional considerations related to both the predictive aspects and the interpretation of the significance of anticipated changes.

5.1.1 Biological Impact Assessment

The biological assessment of the impact of any proposed project or action, may include:

- Results of on-site inspections or surveys
- Views of recognized experts
- Review of literature and other information
- Analysis of effects of the proposed project or action on the species and habitat
- Analysis of alternative actions considered.

The biological assessment should be conducted at a level of detail suitable to the project or action characteristics and the biological requirements of the listed species. This will usually encompass a very large geographic area, sometimes even all the species known, in the entire state or country even though the particular proposed project or action may affect only a very small area. Such a comprehensive approach is appropriate. It remains the ultimate responsibility of the Central/State government not to assist or sponsor any activity that may adversely affect an endangered species in compliance with the Endangered Species Act. The agency must therefore assume a proper share of accountability in identification of the presence of a listed species or critical habitat within the area of likely project effect.

Variously, the biological assessment will be simple and identifies the following factors that cause changes in biodiversity.

Direct drivers of change

(i) Land use and land cover changes

(ii) Disruption. isolation and isolation of colonies

(iii) Destruction/harvesting of species

(iv) Adding Pollution due to emissions, toxic effluents ,chemicals into ecosystem

(v) Ecosystem disturbance due to addition of invasive/alien/genetically modified species

Indirect drivers of change

(i) demographic,

(ii) economic,

(iii) socio-political,

(iv) cultural an

(v) technological processes or interventions.

Differentiation between endogenous drivers and exogenous drivers need to be made. The temporal, spatial and organizational scales need to be addressed.

5.1.2 Biological Environment

The biological environment includes plants and animals, the distribution and abundance of the various species and the habitats of communities. Species forming a community are often inter dependent so that a direct environmental effect on one species is likely to have indirect effect on other species. This interference acts primarily through food chains but can also act through one species providing a habitat for another species.

(A) *Terrestrial Species*

1. ***Terrestrial vegetation:*** It includes in its broadest sense to include agricultural crops, pasture, the introduction, proliferation or control of noxious weeds as well as the native species.

2. ***Terrestrial wild life:*** Included in this group are native mammals, birds, reptiles, amphibians and invertebrates. Migration routes, resting areas, feeding grounds and water sources concentrate wildlife so that such places are particularly sensitive to developmental project activity.

3. ***Other terrestrial fauna:*** Included in this group are domestic and farm animals. Human dependence on such animals extends beyond the food chain to include economics and companionship. Insect and snails are especially important as carriers of parasitic diseases, which afflict the human community.

4. ***Aquatic/marine flora:*** These are important because they provide an important habitat and food for other aquatic marine life and sustain our fresh water or marine fisheries. Mangrove forests, various species of sea weeds and kelp are important. The proliferation of fresh water species can have an impact on the economic use of inland waterways.

5. ***Fish:*** They are considered separately because they provide an important source of animal proteins. In addition to fresh water and marine fish, invertebrates such as prawns, shellfish, crabs and squid should be considered. Species in the brackish-water estuarine environment are of particular importance to man's food chain.

Other aquatic marine fauna: Other species that are not of direct economic importance may form a part of the food chain. Any project, which has a major impact on species populations, can have an equally major indirect impact on the economically important varieties of marine life.

(B) *Habitats and communities*

In considering the environmental effects of development on habitats and communities the special features of terrestrial, aquatic estuarine and marine ecosystems should be considered separately. Special consideration should be given to bird life in considering wetland habitats.

1. ***Terrestrial habitats***: Swamps, wet lands, bird nesting areas, grazing areas, watering places and migration routes should be considered.

2. ***Terrestrial communities***: Special plant communities at high altitudes or those that are residual in an otherwise altered environment should be considered.

3. ***Aquatic, estuarine or marine habitats***: Nursery and breeding areas near shoreline are considered. Wetlands are important. Damage may occur by siltation. Chemical,

physical and biological pollutants may each have major impacts. Oil spills are important in marine environment. Gravel beds are important for spawning.

4. ***Aquatic, estuarine or marine communities***: The food chain relationships involving fish and invertebrates and vegetation are important and should be understood. Siltation and chemical pollution can severely disrupt the balance or the very existence of the community. Project design should aim to leave communities intact and at the very least protect key community components such as invertebrates.

5.1.3 Species Population

The viability of population depends on the presence of a suitable environment with adequate resources. All organisms are constantly affected by and interaction with a complex set of environmental factors including abiotic (physico-chemical factors like water, temperature light, oxygen, nutrients, toxins, pH etc) and biotic factors (which involve interactions between species i.e., competition, predation, parasitism and mutualism). Species can tolerate normal short term environmental variations while populations undergo marked temporary fluctuations, they tend to remain stable in long term. Species also may be capable of responding to slow progressive environmental changes by evolving or changing their geographical range. However, their adaptations have evolved in response to gradual past environmental conditions and may be unable to adjust quickly enough to rapid environmental changes. One of the greatest threats to most species is habitat loss, together with associated habitat fragmentation due to urbanization. The key issue which causes irreversible population loss is the ability of the species populations to survive in and move between small isolated habitat patches scattered within an urban or agricultural matrix.

5.2 Systematic Approach for Evaluating Biological Impacts

To provide a basis for evaluating biological environment impacts, a six-step protocol was formulated for planning and conducting impact studies. This protocol is flexible and can be adapted to various project types by modification as needed to enable the addressing of concerns of specific projects in unique locations.

The various phases associated with the evaluation of biological environment impacts are:

1. Identification of the potential biological impacts of the construction and/or operation of the proposed project of activity, including habitat changes or loss of chemical cycling and toxic events, and disruptions to ecological succession;

2. Description of the environmental setting in terms of habitat types, selected floral and faunal species, management practices, endangered or threatened species, and special features (such as wetlands);

3. Procurement of relevant laws, regulations or criteria related to biological resources and protection of habitat or species;

4. Conducting of impact prediction activities including the use of analogies (case studies), physical modeling and/or mathematical modeling, as based on professional judgment;

5. Use of pertinent information from step 3, along with professional judgment and public input, to assess the significance of anticipated beneficial and detrimental impacts; and

6. Identification, development and incorporation of appropriate mitigation measures for the adverse impacts.

Fig. 5.2 gives the relationship among the six steps or activities in the protocol.

The six steps can be used to plan a study focused on biological environment impacts, to develop the scope of work for such study, and/or to review biological-impact information in EAs or EISs.

In assessing biological impacts the main focus should be on review or redesign of various alternatives for conservation of biodiversity, development of mitigation measures to improve bio resources and enhance biodiversity index and compensate for the project residual effects

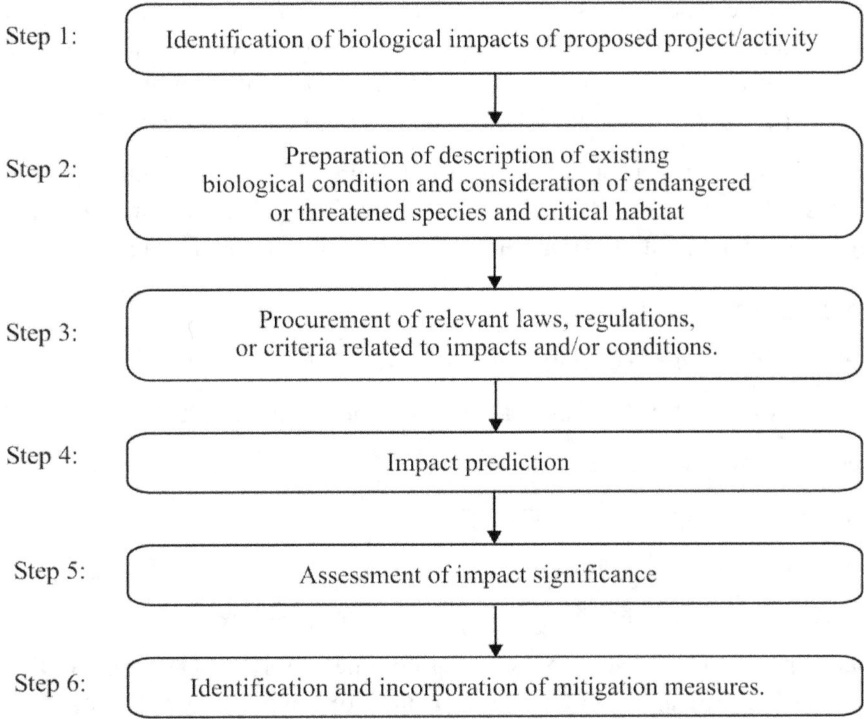

Fig. 5.2 Six- step protocol for evaluation of Biological Environment Impacts.

5.2.1 Step 1: Identification of Biological Impacts

The first step is to qualitatively identify the potential impacts of the proposed project (or activity) on biological resources, including habitats and species. Many projects can cause terrestrial habitat loss. The loss and degradation of terrestrial environments could be classified into eight casual categories: 1. Land conversion for industrial and residential use, 2. Land conversion for agricultural use, 3. Land conversion for transportation use, 4. Timbering practices, 5. Grazing practices, 6. Mining practices, 7. Water management practices, and 8. Military, recreational, and other activities. These causal activities contribute to the degradation and loss of ecological values, including animal and plant species, ecosystem structure (abundance biomass, community composition, species richness, species diversity, trophic organization, and spatial structure), and ecosystem function (energy flow, nutrient cycling, and water retention).

5.2.2   Step 2: Description of Existing Biological – Environment Conditions

The second step in the methodology involves the preparation of the description of the flora and fauna – and other natural resources and habitats constituting the biological – environment setting. This description should primarily focus on community types (habitat types) and their geographical distribution. It may be desirable to identify certain selected species and to include descriptions of those selected species for each community type. There are several options for achieving this step, like 1. use of species lists with qualitative descriptions, 2. use of structured data presentations with qualitative – quantitative descriptions, 3. use habitat – based methods, and 4. use of energy system diagrams.

Available biodiversity information is usually limited and descriptive, and cannot be used as a basis for numerical predictions. There is a need to have enough data on biodiversity to develop and understand the criteria for measurable impact.

5.2.2.1 Floral Components

A five- level classification of vegetation is shown in Table.5.1.

General vegetation patterns of entire area, plant species in upland forests

Tree species in lowland forests

Shrubs and vines of high and lowland forests

Shrubs and vines of floodplain forest

Herbaceous component of the forest category under consideration.

Rare plant species in entire area as per guidelines of IUCN (International Union for Conservation of Nature) and CITES (Consortium for International Trade of Endangered Species).

Table.5.1 A Five level classification of Vegetation

Level 1 (vegetative structure)	Level II (dominant plant types)	Level III (size and density)	Level IV (site or habitat or associated use)	Level V (special plant species)
Forest (trees with average height greater than 15 ft with at least 60%	e.g., oak, hickory, willow, cotton wood, elm, basswood maple, beach,	Tree size (diameter at breast height) Density (number of average stems per acre)	e.g., upland (i.e., well-drained terrain), floodplain, slope face, woodlot, greenbelt, parkland, residential land	Rare and endangered species, often ground plants associated with certain forest types
Woodland (trees with average height greater than15 ft with 20 – 60%	e.g., pine, spruce, balsam fir, hemlock, douglas fir, cedar	Size range (difference between largest a smallest stems)	e.g., upland (I.e., well - drained terrain), floodplain, slope face, woodlot, greenbelt, parkland,	Rare and endangered species, often ground plants associated with certain forest types
Orchard or Plantation (same as woodland or forest but with regular spacing)	e.g., apple, peach, cherry, spruce, Pine	Tree size; density	e.g., active farmland, abandoned farmland	residential land Species with potential in landscaping for proposed development

Table 5.1 *Contd…*

Level 1 (vegetative)	Level II (dominant plant)	Level III (size and density)	Level IV (site or habitat or structure)	Level V (special plant types)
Bush (trees and hrubs generally less than 15 ft high with high density)	e.g., sumac, willow, lilac, hawthorn, tag alder, pin, cherry, scrub oak, juniper	Density	e. g., vacant farmland, landfill, disturbed terrain (e.g., former construction	Species of significance to landscaping for proposed development site
trees and shrubs of forms along borders such as road, yields, yards, play-grounds	Any trees or shrubs	Tree size; density	E.g., active farmland, road right-of-way, yards, playgrounds	Species of value as animal habitat and utility in screening
Wetland (generally low, dense plant cover in wet areas)	e.g., cattail, tag, alter, cedar, cranberry, reeds	Percent cover	e.g., floodplain, bog, tidal, marsh, reservoir backwater, river	Species and plant communities of epial importance ecologically and hydrologically; rare and endange red species
Grossland (herbs, with grasses)	e.g., big blue stem, bunch grass, dune grass	Percent cover	e.g., prairie tundra, pasture, vacant farmland	Species and communities of special ecologi cal significance; rare and endan gered species
Field (tilled or recently tilled farmland)	e.g., corn, spyabean wheat, also weeds	Field size	e.g., sloping or flat, ditched and drained, muckland, irrigated	Special and unique crops; exceptional levels of productivity in standard crops

5.2.2.2 Faunal Components of Entire Area

Amphibians – Frogs, toads, salamanders; Reptiles – Turtles, lizards, snakes

Naiads (freshwater mussels or freshwater clams); Fishes- Sport fisheries

Birds - Mammals Rare faunal species

To serve as another illustration of a structured data presentation, the vegetation in a study area can be classified according to several schemes such as

1. the floristic/taxonomic pattern which classifies individual plants assigning to species, genera, families, and so on, using the universally recognized system of Linnaean botanical nomenclature.

2. the form and structure (or physiognomic) schemes which classify vegetation according to overall expression (for example, forest/grassland) with special attention to dominant plants (quantitatively most abundant). and

3. various ecological/ecosystem schemes, which classify plants according to their habitat (for example, sand dunes, wetlands, lake shores), or some critical parameter of the environment, such as, soil moisture or seasonal air temperatures (9).

A second approach, which could be used for structured data presentations, involves the assemblage of specific information on floral and faunal indicator species within the study area.

Another approach to a structured data presentation of flora and fauna is further augmented by including food, web relationships for the individual ecosystem of concern in the study area. Food-web relationships represent an attempt to show the interdependence of various floral and faunal components within the ecosystem. The system presentation aids in understanding that changes in a certain aspect of the biological environment lead to changes in other interacting features of the biological system. An illustration of a food-web relationship is shown in Fig. 5.3

Species diversity indices could also be used in step 2. Included in this group are various types of ecological – sensitivity ranges for aquatic or terrestrial ecosystems; such ratings often focus on system resiliency or sensitivity to various environmental perturbations.

A third option for describing the biological setting is to utilize only the quantitative habitat- based methods, such as, Habitat Evaluation system (HES) or Habitat Evaluation Procedure (HEP). Use of such methods requires considerable information and the development of numerical indices of habitat quality as a part of the evaluation process. The quantitative use of habitat-based methods are likely to be limited to larger-scale projects, which have significant concerns relative to their anticipated biological impacts.

The final suggested option for addressing the biological environmental setting involves the use of energy- system diagrams Fig. 5.4.

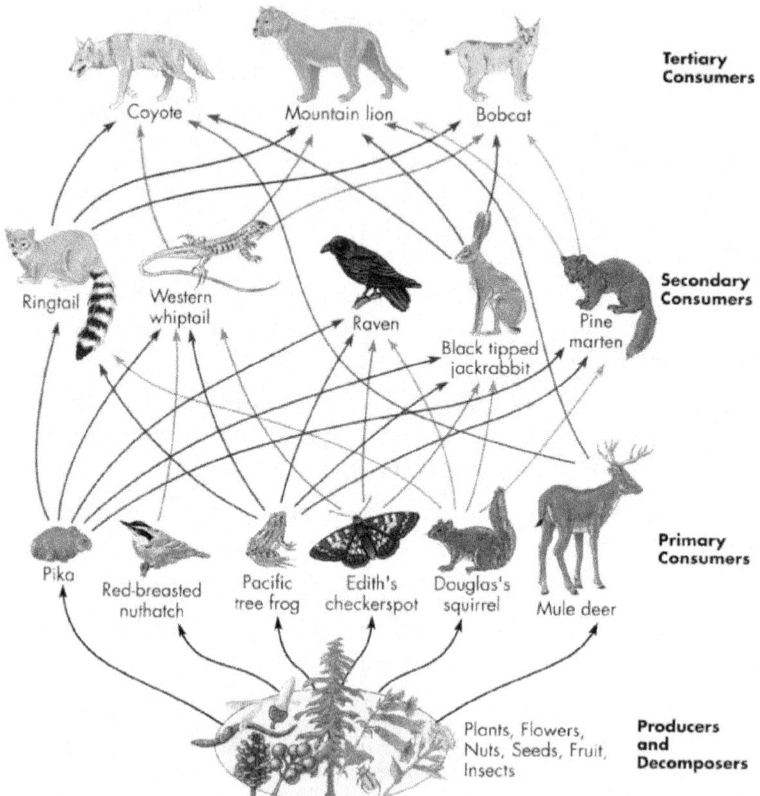

Fig. 5.3 Food Web Relationship.

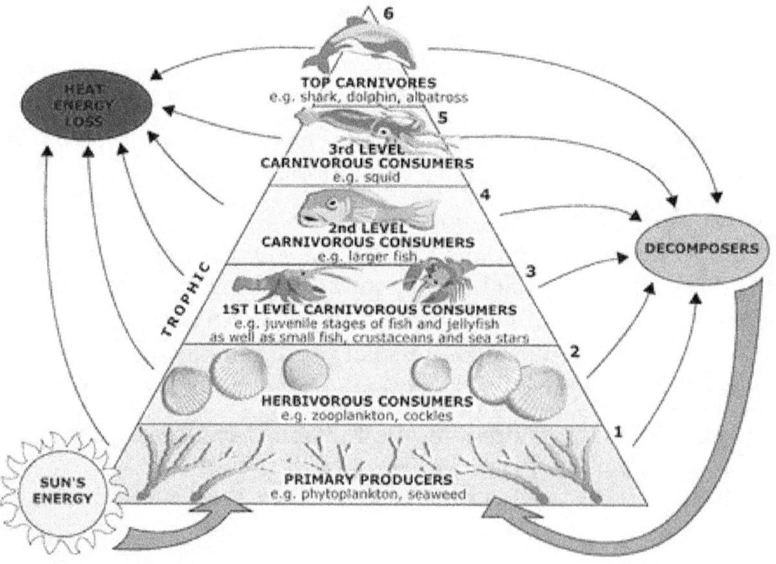

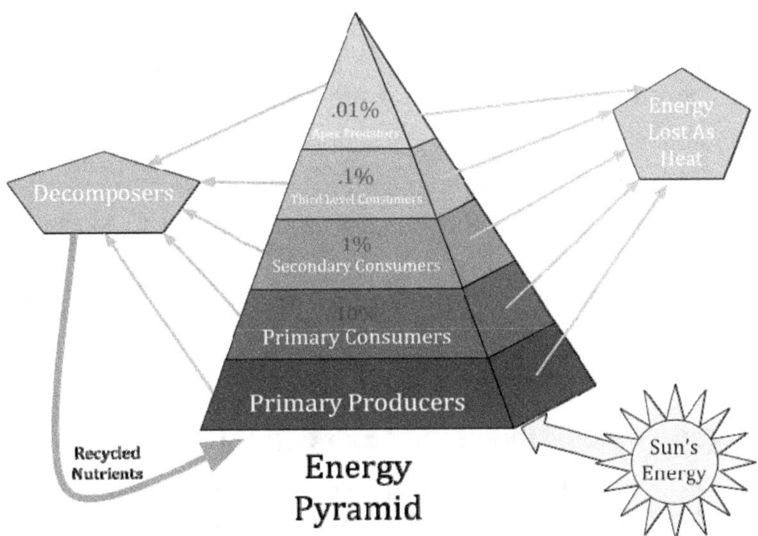

Fig. 5.4 Energy system diagrams in Biological Environmental Setting.

Such diagrams are based on accounting for the flows of all energies in the biological system, including the main components of the system such as plants, animals, and chemical processes, as well as outside actions that may cause changes. The energy-system pictorial approach involves the development of mathematical equations to describe energy flow from the sun to primary producers, from primary producers to secondary producers, and so on. It is necessary to identify the energy flow scenario within a system and to build up data on the rates of energy transition.

The following questions have to be answered after getting full information on the existing biological setting in the project area:

1. Are there any historic sightings of the species within the project area? Review the literature and check it with local universities and experts.

2. Does designated critical habitat exist in the project area? If not, does the habitat required, or suitable for use by the species, for nesting, feeding, or resting (animals) or survival (plants) exist in the project area? In some cases, the answer may be obvious. In other circumstances, a field view of the area of potential project effect may be required.

3. Will the characteristics of the project or action cause any disturbance or other adverse effects on such species or habitat known or assumed to or existing in the project area?

All parties should agree, in writing, to the approach, duration, and level of analysis detail, of studies for each affected species before any work commences.

Previously prepared biological assessment for other projects in the area may be used if the information is still currently applicable. The biological assessment should be completed within 180 days after its initiation.

Formal consultation

If it is assessed that the proposed project or action will affect a listed endangered or threatened species or critical habitat, the following information has to be obtained:

- Description of the proposed project or action,
- Description of any listed species or critical habitat that may be affected,
- Description of the effects on the species or habitat, including an analysis of any cumulative effects,
- Relevant reports including the Environmental Impact Statement of biological assessment, and
- Any other relevant information.

5.2.2.3 Description of the Biodiversity Existing Environment

"Biodiversity" (or the variety of life and its processes,) is the basic property of nature that provides enormous ecological, economic, and aesthetic benefits. It is considered as an index of a nation's wealth. Its loss is recognized as a major national as well as global concern, with profound ecological and economic consequences (10,11). Table 5.2 summarizes some of the components of biological diversity or biodiversity. The hierarchy is given in Fig. 5.5.

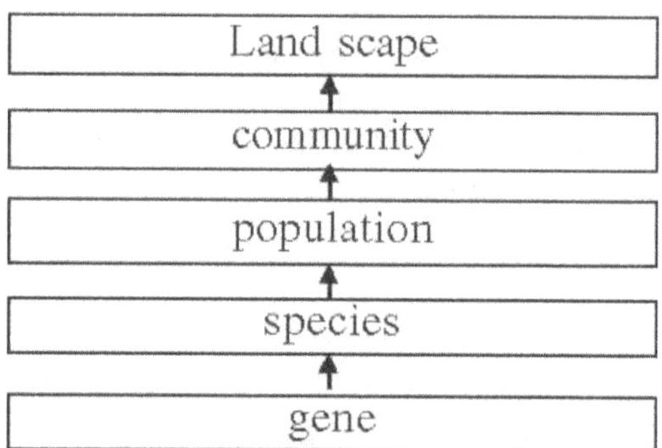

Fig. 5.5 Hierarchy of ecosystem.

Fig. 5.5 Emphasizes the diversity at molecular, micro and macro levels and hence needs finer approach to the environmental impact at all levels.

For detailed information on biodiveristy and its erosion, pl refer to Heywood and Watson (12) and Decastre and Young (13).

Table 5.2

Components of Biological Diversity

- **Regional ecosystem diversity:** The pattern of local ecosystems across the landscape, sometimes referred to as "landscape diversity" or "large ecosystem diversity".

- **Local ecosystem diversity:** The diversity of all living and non-living components within a given area and their interrelationships. Ecosystems are the critical biological/ecological operating units in nature. A related term is "community diversity" which refers to the variety of unique assemblages of plants and animals (communities) individuals species and plant communities exist as elements of local ecosystem, linked by processes such as succession and predation.

- **Species diversity:** The variety of individual species, including animals, plants, fungi, and microorganisms.

- **Genetic diversity:** Variation within species. Genetic diversity enables species to survive in a variety of different environments, and allows them to evolve in response to changing environmental conditions.

- The **hierarchical nature** of these components is an important concept. Regional ecosystem patterns form the basic matrix for, and thus have important influences on, local ecosystems, which, in turn, form the matrix for species and genetic diversity, which can in turn affect ecosystem and regional patterns.

- **Relationships and interactions** are critical components: Plants and animal communities, and other elements exist in complex webs, which determine their ecological significance.

Source : Council on environmental quality, (14).

Thus Biodiversity considerations are very important in environmental management. The basic goal of biodiversity conservation is to maintain naturally occurring ecosystems, communities, and native species and identify activities in less sensitive areas, to minimize the impacts of such activities where possible, and to restore lost diversity, where practical (14). Certain principles (not rules) can be enumerated for incorporating consideration of biodiversity into environmental management. These principles include the following (14):

1. Take a "big picture," or ecosystem view.
2. Protect communities and ecosystems.
3. Minimize fragmentation and promote the natural pattern and connectivity of habitats.
4. Promote native species, and avoid introducing nonnative species.
5. Protect rare and ecologically important species.
6. Protect unique or sensitive environments.
7. Maintain or mimic naturally occurring structural diversity.
8. Protect genetic diversity,
9. Restore ecosystems, communities, species, and
10. Monitor biodiversity impacts.

5.2.3   STEP 3 : Procurement of Relevant Legislation and Regulations

The primary sources of information on pertinent legislation, regulations, criteria, or guidelines related to the biological environment include environmental and/or natural resource agencies of the central and state levels. Local agencies and/or conservation groups may also provide pertinent information. Procurement of this information will facilitate the evaluation of baseline conditions and the data obtained can serve as a basis for impact-significance determination (step 5).

Most of the biological-environment legislation, regulations, criteria, or guidelines are qualitative in terms of specific requirements. This is in contrast to the substantive areas of air quality, surface water and groundwater quality, soil quality, and environmental noise for which numerical standards are available. Sound professional judgment must be exercised in applying the qualitative requirements for the biological environment in this step.

5.2.4   STEP 4: Impact Prediction

The most technically demanding step in addressing the biological environment is the prediction of the impacts of the project-activity and various alternatives, on the biological environmental setting. As a general principle, the impacts should be quantified where possible with qualitative descriptions provided for those impacts which cannot be quantified. From historical perspective, impact prediction for the biological environment has focused on land-use or habitat changes and the associated implications of those factors relative to the biological system. Several options are available for impact prediction approaches, including qualitative descriptions of impacts, the use of habitude methods or ecosystem models, and the use of physical models or simulations. Broader impact issues of increasing importance are biological diversity and sustainable development.

It is necessary to examine the relative merits and demerits of impacts of alternatives with reference to existing base line data and then compare against threshold values/objectives for biodiversity achievement in the project/statutory standards. The proposed action plan for mitigation should be examined in the context of national biodiversity standards and international commitments. For industries like mining, cumulative threats and impacts likely to come due to repeated project activities of the same or different nature over space and time need to be considered

5.2.4.1 Qualitative Approaches

Qualitative descriptions could be associated with a discussion on of land- use or habitat changes. One tool which can be helpful in identifying the types of impacts (effects) that might take place on the biological system is the list of 52 effects found in Table.5.3. The general approach would be to consider each of these factors and determine its applicability to the project and the environmental setting. If deemed applicable, then either specific qualitative information could be assembled, or, at least, qualitative discussions prepared, on the implications of the project relative to the particular biological items identified. In using this approach, the considerable exercise of professional judgement would be required.

Additional columns should address the following issues.

1. The likelihood of impact, shown in terms of a relative scale of high, medium, and low.

2. The duration of the impact in terms of whether it would be associated with the short-term construction phase of the project versus the long term operational phase. In addition, this column could include information on the actual anticipated duration of the impact.

3. The reversibility of the impacts with two codes, one denoting items that are irreversible and the other denoting those particular impacts that might be recoverable. This column could also relate to the possibility of success in trying to reduce the impact and to potentially reverse it through various developed programs.

4. The relative resiliency of individual plant or animal species within the study area. (It is quite well-known that some species are more tolerant of change than others.)

5. Potential mitigation measures for a given project type.

Table 5.3 List of potential effects on the biological system.

1. Resiliency and fitness of ecosystem types; for example, lowland forest, upland forest grassland, marsh, bog and streams.
2. Total standing crop of organic matter.
3. Annual plant productivity.
4. Mulch or litter removal as related to top-soil stripping.
5. Animal production.
6. Sediment load carried by streams.
7. Aquatic macroinvertebrate populations.
8. Drift rate of aquatic macroinvertebrates.
9. Population density of fish.
10. Sediment-load effects on fish growth.
11. Sediment-load effects on fish spawning.
12. Species diversity of the aquatic biota.
13. Undesirable proliferations of biota.
14. Localized survival of rare plant and animal species.
15. Habitat carrying capacity of both aquatic and terrestrial systems.
16. Abandonment of habitat.
17. Endemic populations of plants and animals.
18. Wildlife breeding and nesting sites.
19. Endangered plant and animal species.
20. Vegetation communities of denuded areas.
21. Wildlife refuges and sanctuaries.
22. Scientific and educational areas of biological interest.
23. Vegetation recovery rates.
24. Forage areas for both upland and lowland game species.
25. Migratory game bird species.
26. Terrestrial microbial communities.
27. Amount of forest removed.
28. Population density of past species.
29. Domestic animal species.
30. Amount of grassland removed.
31. Natural drainage systems.
32. Natural animal corridors.
33. Eutrophication.
34. Expansion of population range for both plant and animal species.
35. Cropland removal.
36. Potential for wildlife management.
37. Food-web index, including herbivores, omnivores, and carnivores.
38. Species diversity of the terrestrial biota.
39. Nutrient supply available to terrestrial biota.
40. Sport fishing and hunting.
41. Resultant air pollution effects on crop yield.
42. Relict vegetation types.
43. Responses of sensitive native plants to air pollutants, both particulates and gases.
44. Unnatural dispersion and subsequent overutilization of habitats.
45. Noise level effects on reproductive inhibition of small mammals.
46. Air pollutant effects on tree canopy.
47. Noise level effects on broodliness of upland and lowland game birds.
48. Water-temperature stability.
49. Areas of high brush-fire potential.
50. Water quality and dependent biota.
51. Noise level effects on insect maturation and reproduction, and
52. Natural biological character loss.

Source: Adapted from Hill, (5).

If need be we need to recognize that biodiversity is influenced by cultural, social, economic and biophysical factors. In some important large infrastructure/mining projects it may be also essential to provide insight into cause effect chains. If possible, quantify the changes in biodiversity composition, structure and key processes, as well as ecosystem services may help the EIA report in terms of understanding the consequences of the loss of biodiversity associated.

5.2.4.2 Modeling Approaches

5.2.4.2.1 Mathematical Models and Mapping

Physical Disturbance

Simple qualitative maps presenting spatial distribution of plants and animals biota can be prepared for identifying impacts of physical disturbance caused by project activities. By overlaying plans of project facilities like buildings, machinery roads etc. over existing environment the impact of physical disturbance of project activities can be depicted using GIS and Remote sensing data (14a). GIS provides for plants the area last while for animals /communities the zone of influence (Fig 5.6) which may extend beyond physical disturbance area. In case of noise arising from traffic on a road/machine facility the zone of influence can go upto ~100meters from the source

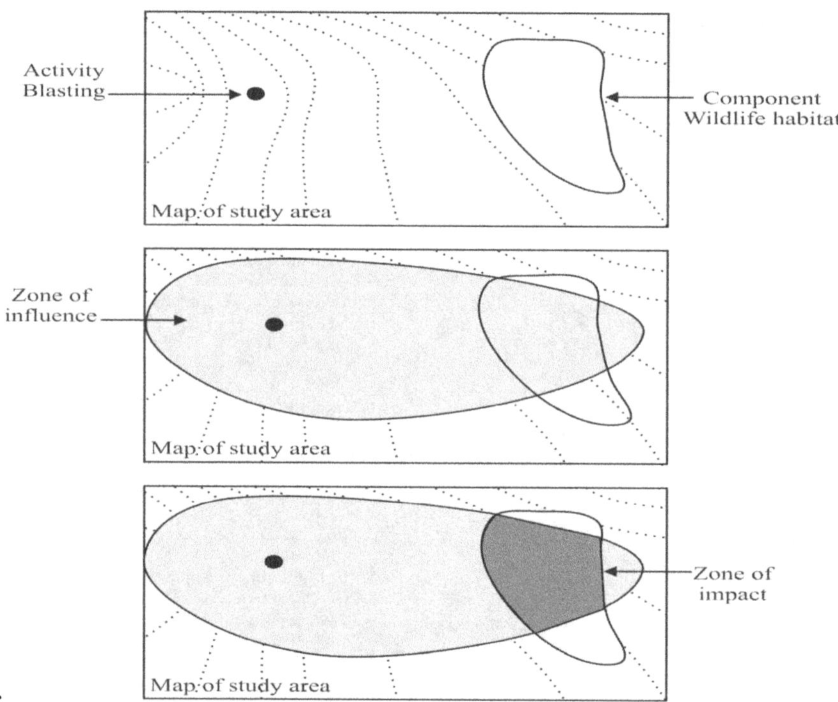

Fig. 5.6 Simple illustration of zone of influence.(14a))Multivariate forest structure modelling and mapping using high resolution airborne imagery and topographic information ; Jon Pasher *, Douglas J. King; Remote Sensing of Environment 114 (2010) 1718–1732.

5.2.4.2.2 Habitat Alienation

In the terrestrial environment, the destruction of vegetation and loss of soil usually results in reduction of habitat for animals. The relationship between the extent of the physical disturbance on the area and the amount of habitat lost or degraded is non-linear. Small changes in critical areas can make large areas unsuitable for animal habitat. This is because animal habitat is usually a combination of the basic necessities for an animal: food, water, cover, and other resources. Some habitats are critical for survival, for example, wetlands that act as staging areas or wildlife migration, or mangroves ecosystems that provide breeding areas for aquatic organisms.

Canter (1996) (9) describes two habitat-based methods used for prediction of biological impacts: the Habitat Evaluation System (HES) used by the US Army Corps of Engineers in the evaluation of water resource project in the lower Mississippi; and the Habitat Evaluation Procedure (HEP) developed by the US Fish and Wildlife Service. HEP, originally developed for use in evaluating water resource projects has been applied in many other contexts. Most habitat based methods involve the use of expert judgement to construct simple indices of habitat quality based on key ecosystem variables. In most cases, the habitat quality indices must be redeveloped for each assessment. These methods are best used to compare the merits of alternatives. They provide neither an absolute measure of impact, nor the degree of significance of the impact.

5.2.4.2.3 Changes in Animal Populations

In many cases, the primary concern is with impacts on fish and wildlife populations. This is because these populations often have economic and social importance or are protected by national legislation or international treaties. Population dynamics models are often developed to predict changes in animal populations. The basic model equation (Walters, 1986) (15) is:

$$Nt + 1 = sat\ Nt + sjt\ Rt$$

where

Nt is the population size at specific time t in the annual cycle;

Rt is the recruitment to the population during the time cycle between t and t+1;

sat is the survival rate of animals (N t) from t to t+1; and

sjt is the survival rate of new recruits (Rt) during the time cycle between t and t+1.

This simple equation allows for the projection of how the population will change over time. Each of the basic components of the equation (that is, recruitment and survival rates) are usually modeled as functions of other ecological parameters and outside interventions. For example, one model of recruitment in a fish population might have recruitment as a function of the population size, fecundity (eggs/female), available spawning habitat, net migration, and water quality. Similarly the survival rates may be a function of population size, harvesting, habitat, and water quality. In conducting an EIA, one first predicts the changes in those factors upon which recruitment and survival are dependent. Once estimates of recruitment and survival parameters are calculated, the model may be applied to predict changes in population. Walters (1986) (15) provides an excellent description of how to develop the basic model, estimate parameters, and test its accuracy in prediction.

5.2.4.2.4   Habitat Evaluation Procedures Software
(source: Internet www.mesc.usgs.gov/hep/hep.htm/)

The philosophy behind the Habitat Evaluation Procedures is that an area can have various habitats, and that these habitats have different suitabilities for species that may occur in that area. Further, we assume that the suitabilities can be quantified (via Habitat Suitability Indexes) and that the different habitats have measurable areal extents. The overall suitability of an area for a species we postulate can be represented as a product of the areal extents of each habitat and the suitability of those habitats for the species. If this is true, we may further postulate that as habitat changes through time, by natural or human-induced processes, we can quantify the overall suitability through time by integrating the areal extent-suitability product function over time. Thus, we can quantitatively compare two or more alternative management practices of an area with regard to those practices affecting species in that area. For example, we can judge the effects of logging, mining, cattle grazing, versus no use. Furthermore, HEP allows us to quantify the effects of mitigation (not so great a negative impact) or compensation (improve another like area to make up for lost habitat in the impacted area). This is an important tool for land use managers, as they can quantify the effects of alternative management plans over time, and provide for mitigation and compensation that can allow fair use of the land and maintain healthy habitats for affected species. The HEP accounting program uses the area of available habitat and Habitat Suitability Index (HSI) to compute the values needed for Habitat Evaluation Procedures (HEP) as described in the Ecological Services Manual (ESM 102) and the HEP training course NR561 [Habitat Evaluation Procedures]. The compiled program requires two floppy disk drives or a hard disk, and 64 kilobytes of RAM.

5.2.4.2.5 Areas of Application

Table 5.4 provides examples of computer models available for prediction of impacts on habitat. The transport fate exposure model is given in Fig. 5.7.

Table 5.4 Software for programs for habitat evaluation from United States Geological Survey Midcontinent

Software	Description
HEP Habitat Evaluation Procedures (HEP)	The HEP accounting program uses the area of available habitat and Habitat Suitability Index (HIS) to compute the values needed for Habitat as described Ecological services Manual (ESM 102) and the HEP training course HEP500 [Habitat Evaluation Procedures (HEP) as described in the Ecological Services Manual (ESM 102) and the HEP training course HEP500 [Habitat Evaluation Procedures]
HIS Suitability Index	The HIS software is a system of programs that uses mathematical models to compute an HIS value for selected species from field Habitat measurements of habitat variables. The development and use of HIS models are described in the Ecological Services Manual (ESM 103) and the HEP Training Course HEP 500 Habitat Evaluation Procedures.
PHABSIM	The Physical Habitat Simulation System. This extensive set of programs is designed to predict microhabitat conditions in rivers as a function of streamflow and the relative suitability of those microhabitat conditions to aquatic life. The appropriate use of this set of programs is taught in IF 130, [Using the Computer-based Physical Habitat Simulation System (PHABSIM)].
TSLIB The Time Series	TSLIB programs provide for data entry, analysis, and display of daily or monthly flow or habitat values. Some programs are useful for integrating microhabitat and macrohabitat, and some are of value in the analysis of water operations systems. Many of the concepts of time Library series analysis are taught in IF250 [Theory and Concepts of the Instream Flow Incremental Methodology].
SNTEMP Stream Network Temperature Model	SNTEMP predicts the water temperature in streams and rivers from data describing the stream's geometry, meteorology, and hydrology. It handles a dendritic netw ork of streams through time and space. SNTEMP is taught in IF312 [Stream Temperature Modeling]. See More About SNTEMP and SSTEMP.
SSTEMP Stream Segment Temperature Model	SSTEMP is a scaled down version of SNTEMP suitable for single (to a few) reaches and single (to a few) time periods. SSTEMP is taught in IF312 [Stream Temperature Modeling]

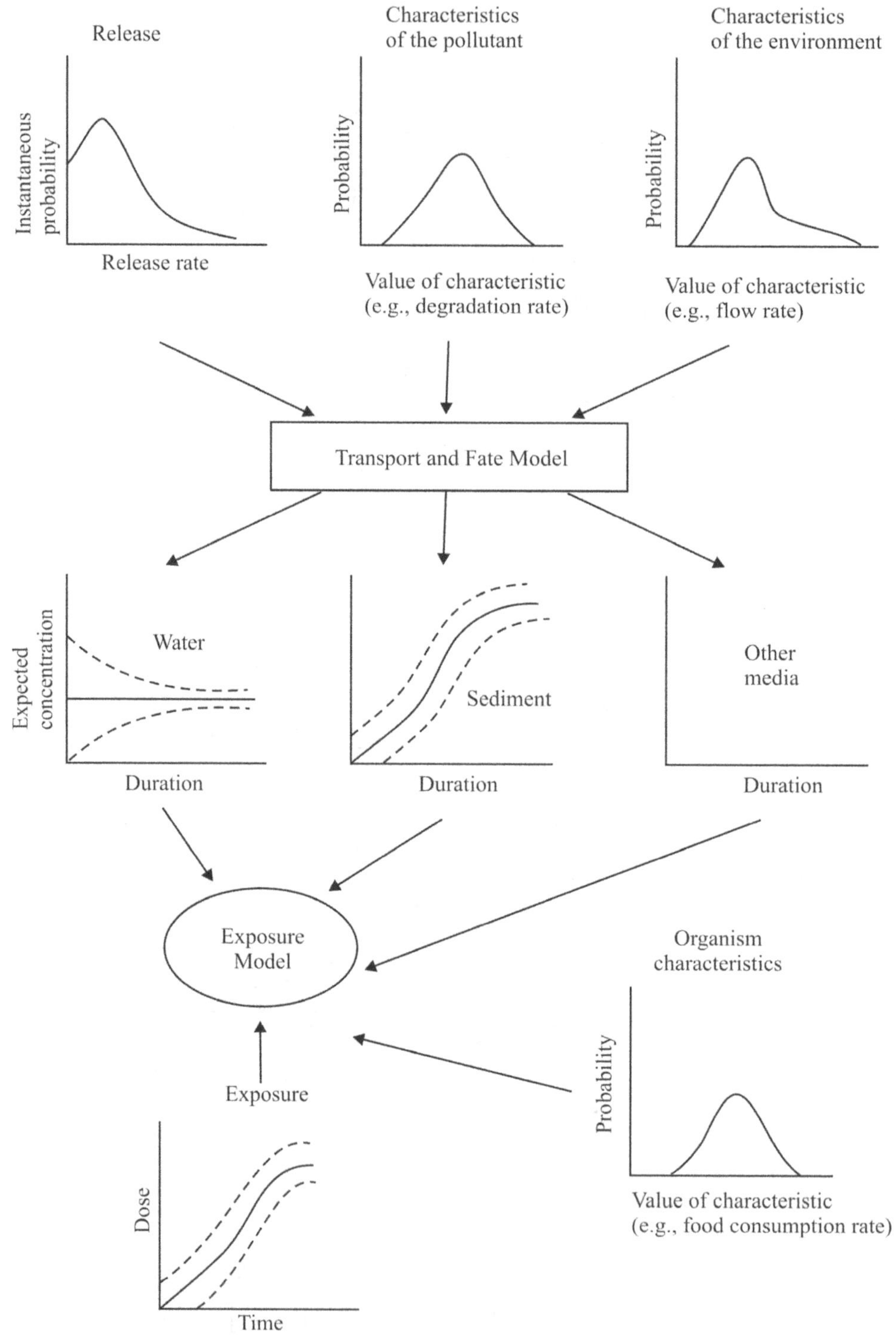

Fig. 5.7 The exposure assessment process.

5.2.4.2.6 Physical Models

In addition to mathematical models, physical models could also be used for biological impact prediction and assessment. Examples include bio-assay, chronic-toxicity testing, microcosms and scaled ecosystem models. Bio-assays and chromic-toxicity testing are focused on the potential toxic effects on terrestrial or aquatic plant or animal species of discharges or releases of residuals by the proposed project-activity.

5.2.4.2.7 Ecotoxicology - Impacts of Pollutants on Biota

The dose of a chemical to an organism is a function of both the concentration of the chemical in the immediate environment and the duration of exposure of the population to that concentration. The two factors interact in multiplicative way; hence the dose of a chemical received by an organism is defined (Westman, 1985) (16) as: *dose* is equal to the concentration of chemical times duration of exposure at concentration.

5.2.4.2.7.1 Exposure

Exposure has been defined as contact with a chemical or physical agent. It is the process by which an organism acquires a dose (Suter, 1993) (17). The estimation of exposure of a target organism requires an exposure scenario that answers to four questions (Suter, 1993) (17)

1. given the output of fate models, which media (ecosystem components) are significantly contaminated,
2. to which contaminated media are the target organisms exposed,
3. how are they exposed (pathways and rates of exposure), and
4. given an initial exposure, will the organism modify its behavior to modify exposure pathways or rates (attraction or avoidance).

Table 5.5 identifies some of the major exposure pathways, while Fig. 5.7 provides an example of exposure pathways for two target species: mink, a small carnivorous mammal and the great blue heron, a large piscivorus bird.

Behavioral responses of organisms may modify subsequent exposure. Animals commonly avoid contaminated food or media, however there are cases where animals are attracted. Due to lack of behavioral information, most assessors normally assume that behavior does not modify exposure (Suter, 1993) (17). Because of the complexity involved, most EIA practitioners will have to rely on existing computer software models to provide estimates of exposure.

EXAMS provides a means of rapidly evaluating the fate, transport and exposure concentrations of synthetic organic chemicals–pesticides in aquatic ecosystems. To date, there has been little usage of exposure models in EIA. Three possible reasons can be suggested: 1. there has been little emphasis on assessing the exposure of biotato pollutants 2. EIA practitioners are unaware or unskilled in the use of the tools and techniques for exposure assessment and 3. the basic baseline data to parameterize the models is unavailable and too costly to obtain.

5.2.4.2.7.2 Effects

Effects assessment is the process of determining the relationship between exposure and its effects on the target organism. Most effects assessments are based on toxicity tests. Suter (1993) (17) outlines the basic steps in the effects assessment Fig. 5.8.

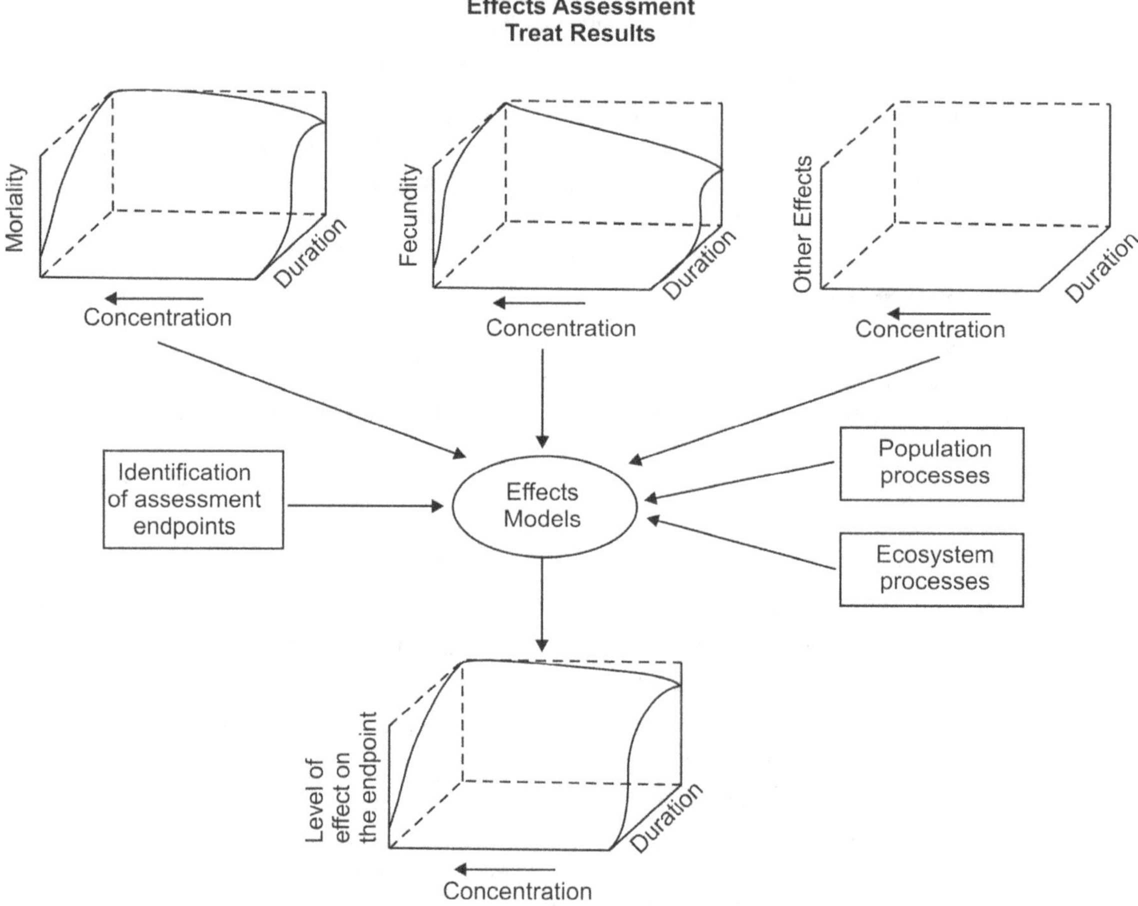

Fig. 5.8 Process of effects assessment (*source:* Suter, 1993) (17).

1. toxicity tests are conducted to determine the effects of various combinations of exposure concentrations and the duration on the frequency or severity of the responses of concern, such as increased mortality and decreased fecundity;

2. statistical models are fit to test data and an exposure to the response model is selected to represent the toxicological responses in the effects models.

3. effects models are generated that represent the relationship between the test results and the target organisms, and

4. the test results and data concerning relevant population and ecosystem processes are used to parameterize the effects model which is then used to derive a function relating to the level of effects on the target organism to the exposure.

The example of the exposure pathways are shown in Table 5.5 and Fig.5.9

Table 5.5 Exposure pathway.

Media	Pathways	Comment
Air – gases and aerosols	Respiration	Assuming accurate fate model estimates, exposure is relatively predictable based on assumptions of homogenous distribution in air.
Water – soluble chemicals	Respiration	Assuming accurate fate model estimates, exposure is relatively predictable based on assumptions of homogenous distribution in water.
Sediment (solids and pore water)	Benthic animals absorb chemicals, respire pore water or free water, and ingest sediments, sediment associated food or food from the water column. Plants rooted in the sediment may take up material from sediments, surface water and air	Processes are very complicated and usually simplifying assumptions are required
Soil (solids, pore water and pore air)	Organisms in soils may absorb material from soil, pore water, pore air, ingest soil, soil-associated food.	Processes are very complicated and usually simplifying assumptions are required
Ingested Food and Water	Consumption by fish and wildlife	Assume that test animal consumption rates in laboratory for a given availability of food or water are the same as those occurring naturally in the environment
Multi Media	More than one of above pathways	It is often possible to assume one pathway is dominant. In some cases, it will be necessary to estimate the combined dosage.
MULTIMED model Version 1.01-Dec 92	The Multimedia Exposure Assessment Model (MULTIMED) for exposure assessment simulates the movement of contaminants leaching from a waste disposal facility. The model consists of a number of modules which predict concentrations to a receptor due to transport in the subsurface, surface air, or air. To enhances the user-friendly nature of the model, separate interactive pre-(PREMED) and post-processing (POSTMED) programs allow the user to create and edit input and plot model output.	
MUTIMDP model Version 1.00 – Oct 96	The Multimedia Exposure Assessment Model (MULTIMED) for exposure assessment simulates the movement of contaminants leaching from a waste disposal facility. The MULTIMED model has been modified (MULTIMDP) to simulate the transport and fate of first and second-generation transformation (daughter) products that migrate form a waste source through the unsaturated and saturated zones to a downgradient receptor well.	

(**Source :** Suter, 1993). (17).

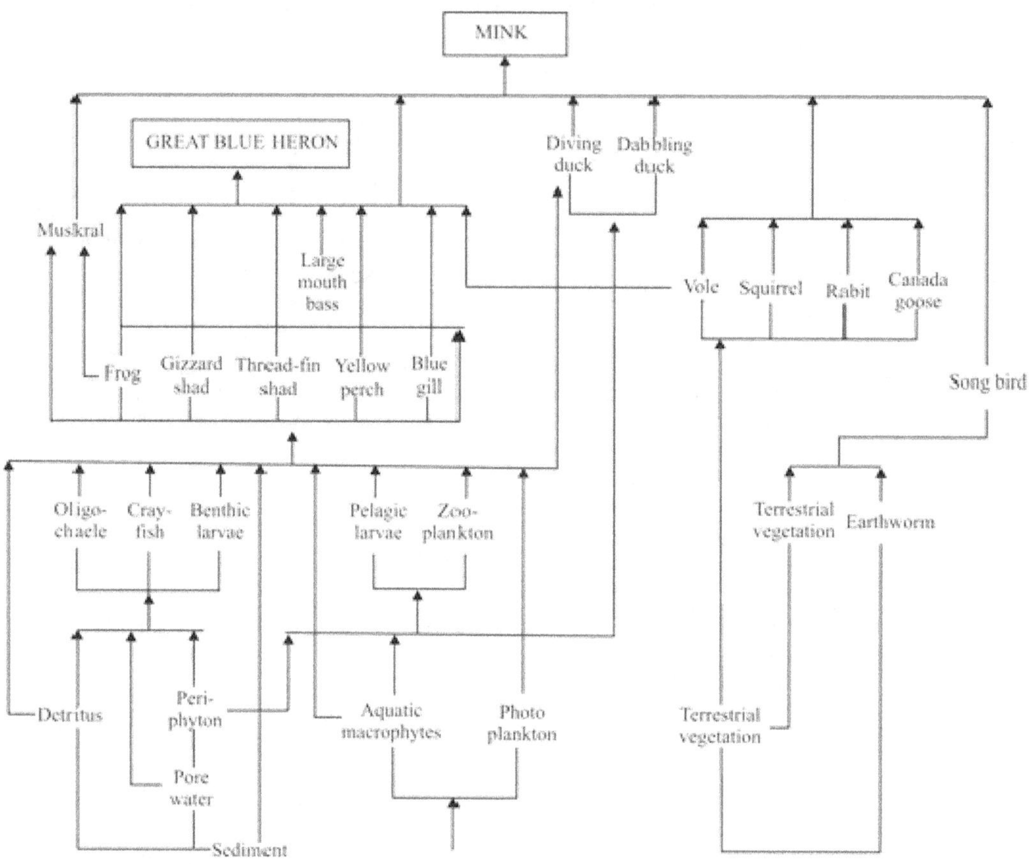

Fig. 5.9 Example of exposure pathways
(**Source :** Suter, 1993) (17).

The EXAMS Model System (source: Center for Exposure Assessment Modeling (CEAM)).

The Exposure Analysis Modeling System, first published in 1982 (EPA-600/3-82-023), provides interactive computer software for formulating aquatic ecosystem models and rapidly evaluating the fate, transport, and exposure concentrations of synthetic organic chemicals – pesticides, industrial materials and leachates from disposal sites. EXAMS contains an integrated Database Management System specifically designed for storage and management of project databases required by the software. User interaction is provided by a full-featured Command Line Interface, context-sensitive help menus, an on-line data dictionary and Command Line Interface users' guide, and plotting capabilities for review of output data. EXAMS provides 20 output tables which document the input data sets and provide integrated results summaries for aid in ecological risk assessments.

EXAMS' core is a set of process modules that link fundamental chemical properties to the limnological parameters that control the kinetics and transport in aquatic systems. The chemical properties are measurable by conventional laboratory methods; most are required under various regulatory authority. When run under the EPA's GEMS or pcGEMS systems, EXAMS accepts direct output from qsar software. EXAMS' limnological data are composed of elements historically of interest to aquatic scientists worldwide, so generation of suitable environmental datasets can generally be accomplished with minimal project-specific field investigations. EXAMS provides facilities for long-term (steady-state) analysis of chronic chemical discharges, initial-value approaches for study of short-term chemical releases, and full kinetic simulations that allow for monthly variation in mean climatological parameters and alteration of chemical loadings on daily time scales. EXAMS has been written in generalized (N-dimensional) form in its implementation of algorithms for representing spatial detail and chemical degradation pathways.

EXAMS provides analyses of: Exposure: the expected (96-hour acute, 21-day and long-term chronic) environmental concentrations of synthetic chemicals and their transformation products.

Fate: the spatial distribution of chemicals in the aquatic ecosystem, and the relative importance of each transformation and transport process (important in establishing the acceptable uncertainty in chemical laboratory data), and Persistence: the time required for natural purification of the ecosystem (via export and degradation processes) once chemical releases end.

5.2.4.2.8   Dose - Response Functions

The most common model to test results is the dose-response function. The pattern of response with increasing dose is assumed to be S-shaped (Fig. 5.10). This function assumes that there is no threshold below which there is no response.

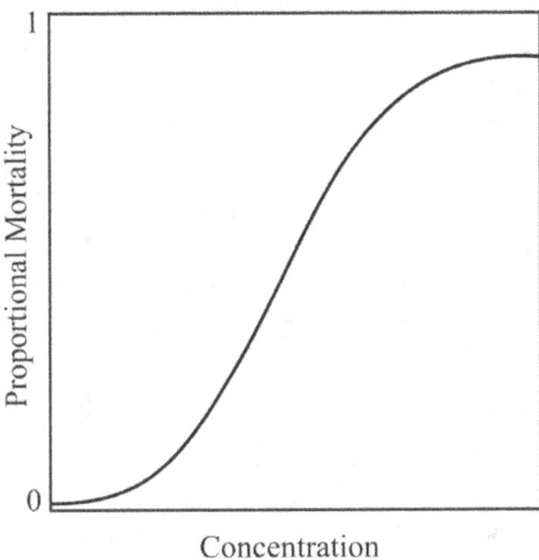

Fig. 5.10 A typical S-shaped dose-response curve.

(**Source** : Suter, 1993) (17).

5.2.4.2.9 Impact Comparison with reference to Environmental Standards

The quantitative and qualitative significances of expected impact on the organism under examination will be evaluated in the final step by exposure assessments and effect assessments and compared with published standards.

Due to limitations of models and getting practical data in predicting biochemical impacts in the target organisms in environment,   many times EIA practitioners ignore these aspects. Predictions based on transportation and dilution of chemical concentrations of pollutants using simple physical models provide fairly realistic data of actual effects on target organisms and compared against published environmental standards to categorize the observed/predicted environmental impact is significant or not.

5.2.4.2.10 Areas of Application

Table 5.6 provides examples of models available for exposure assessment in aquatic systems.

Table 5.6 Center of Exposure Assessment Software. DOS release of selected CEAM software are available through the World Wide Web.

Model Name	Description
CEAM information system Version 3.21 – May 95	The Center for Exposure Assessment Modeling (CEAM) of the U.S. EPA serves as the focal po1int for ORD's multimedia exposure assessment modeling and ecological risk assessment activities. The CEAM Information System (CEAMINFO) is a collection of reference and information documents and/or files that summarize CEAM mission, activities, documentation, software products, assistance, support, and software product distribution.
CORMIX model Version 3.20 – Dec 96	Comell Mixing Zone Expert System (CORMIX) can be used for the analysis, prediction, and design of aqueous toxic or conventional pollutant discharges into diverse water bodies. The major emphasis is on the geometry and dilution characteristics of the initial mixing zone – including compliance with regulatory constraints – but the system also predicts the behavior of the discharge plume at larger distances. The system consists of three subsystems: CORMIX1 for submerged single port discharges, CORMIX2 for submerged multiport diffuser discharges, and CORMIX3 for buoyant surface discharges. Further information is available from Comell University concerning technical support available to users for the CORMIX model system, version 3.20, dated December 1996.
PRZM2 model Version 2.00 – Oct 94	Pesticide Root Zone Model – 2 (PRZM2) links two subordinate models, PRZM and Vadose Zone Flow and Transport (VADPFT) to provide a deterministic simulation of the fate of agricultural pesticides in the crop root and underlying unsaturated zone. PRZM2 can simulate multiple pesticides or pesticide parent-daughter product relationships, and estimate probabilities of concentrations of fluxes in or from various media to perform exposure assessments. PRZM/VADOFT codes are linked in PRZM2 using an execution supervisor that can build loading modules tailored to site-specific situations.

Table 5.6 *Contd...*

Model Name	Description
PLUMES model Version 3.00 – Dec 94	PLUMES includes two initial dilution plume models (RSB and UM) and a model interface manager for preparing common input and running the models. Two farfield algorithms are automatically initiated beyond the zone of initial dilution. PLUMES also incorporates the flow classification scheme of the Cornell Mixing Zone Model (CORMIX) with recommendations for model usage, thereby providing a linkage between the systems. PULMES models are intended for use with plumes discharged to marine and some freshwater bodies. Both buoyant and dense plumes, single sources, and many diffuser outfall configurations can be modeled.
PATRIOT model Version 5.10 – Oct 93	Pesticide Assessment Tool for Rating Investigations of Transport (PATRIOT) provides rapid analyses of ground water vulnerability to pesticides on a regional, state, or local level. PATRIOT assesses ground water vulnerability by quantifying pesticide leaching potential in terms of pesticide mass transported to the water table. It integrates a tool that enables analysis of pesticide leaching potential with data required for area-specific analysis anywhere in the U.S. PATRIOT is composed of: 1. pesticides fate and transport model (PRZM2). 2. comprehensive database, 3. interface that facilities database exploration, 4. directed sequence of interaction that guides user on providing necessary information to perform alternative model analysis model analyses, and 5. user-selected methods of summarizing and visualizing results.
QUAK2EU model Version 3.22 – May 96	The Enhanced Stream Water Quality Model (QUAL2E) is a steady state model for conventional pollutants in branching streams and well mixed lakes. It can be operated either as a steady-state or dynamic model and is intended for use as a water quality planning tool. The model can be used to study impact of waste loads on instream water quality and identify magnitude and quality characteristics of non-point waste loads. The Enhanced Stream Water Quality Model with Uncertainty Analysis (QUAL2EU) is an enhancement to the QUAL2E model that allows the user to perform uncertainty analysis.
SWMM model Version 4.30 – May 94	Storm Water Management Model (SWMM) – comprehensive computer model for analysis of quantity and quantity problems associated with urban runoff. Both single-event and continuous simulation can be performed on catchments having storm sewers, or combined sewers and natural drainage, for prediction of flows, stages and pollutant concentrations. Extran Block solves complete dynamic flow routing equations (St. Venant equations) for accurate simulation of backwater, lopped connections, surcharging and pressure flow. Modeler can simulate all aspects of the urban hydrologic and quality cycles, including rainfall, snowmelt, surface and subsurface runoff, flow routing through drainage network, storage and treatment.

Table 5.6 *Contd...*

Model Name	Description
SMPTOX3 model Version 2.01-Feb 93	U.S. EPA regulatory programs have sponsored development of an interactive computer program for performing waste load at locations for toxics-Simplified Method Program – Variable Complexity Stream Toxics Model (SMPTOX3). It predicts pollutant concentrations in dissolved and particulate phases for water column and bed sediments and total suspended solid. Separate simulation routines are provided for model calibration, waste load allocation, and sensitivity analysis.
WASP model Version 5.10 – Oct 93	The Water Quality Analysis Simulation Program (WASP) is a generalized framework for modeling contaminant fate and transport in surface waters. Problems studied using WASP framework include biochemical oxygen demand and dissolved oxygen dynamics nutrients and eutrophication, bacterial contamination, and organic chemical and heavy
QUAK2EU model Version 3.22 – May 96	The Enhanced Stream Water Quality Model (QUAL2E) is a steady state model for conventional pollutants in branching streams and well mixed lakes. It can be operated either as a steady-state or dynamic model and is intended for use as a water quality planning tool. The model can be used to study impact of waste loads on instream water quality and identify magnitude and quality characteristics of non-point waste loads. The Enhanced Stream Water Quality Model with Uncertainty Analysis (QUAL2EU) is an enhancement to the QUAL2E model that allows the user to perform uncertainty analysis.
SWMM model Version 4.30 – May 94	Storm Water Management Model (SWMM) – comprehensive computer model for analysis of quantity and quantity problems associated with urban runoff. Both single-event and continuous simulation can be performed on catchments having storm sewers, or combined sewers and natural drainage, for prediction of flows, stages and pollutant concentrations. Extran Block solves complete dynamic flow routing equations (St. Venant equations) for accurate simulation of backwater, lopped connections, surcharging and pressure flow. Modeler can simulate all aspects of the urban hydrologic and quality cycles, including rainfall, snowmelt, surface and subsurface runoff, flow routing through drainage network, storage and treatment.
SMPTOX3 model Version 2.01-Feb 93	U.S. EPA regulatory programs have sponsored development of an interactive computer program for performing waste load a locations for toxics-Simplified Method Program – Variable Complexity Stream Toxics Model (SMPTOX3). It predicts pollutant concentrations in dissolved and particulate phases for water column and bed sediments and total suspended solid. Separate simulation routines are provided for model calibration, waste load allocation, and sensitivity analysis.
WASP model Version 5.10 – Oct 93	The Water Quality Analysis Simulation Program (WASP) is a generalized framework for modeling contaminant fate and transport in surface waters. Problems studied using WASP framework include biochemical oxygen demand and dissolved oxygen dynamics nutrients and eutrophication, bacterial contamination, and organic chemical and heavy

(**Source** : Internet – *ftp.epa,giv/epa_ceam/wwwhtml/software,htm*).

Monitoring and auditing are extremely essential in biological impact assessment studies. It also serves to verify that the proponent is compliant with the environmental management plan (EMP).

Management plans including clear management targets for afforestation, biodiversity protection, enhancement of greenery etc., and the roles and responsibilities of various organization involved (forest department etc) with appropriate monitoring should be established to ensure that mitigation is effectively implemented.

Negative effects or trends if identified during the implementation of biological methods need to be addressed immediately. The EMP should define responsibilities, budgets and any necessary training for monitoring biological management. Monitoring should essentially focus on those components of biodiversity most likely to change as a result of the project. The use of indicator organisms or ecosystems that are most sensitive to the predicted impacts needs to be identified such that earliest possible indication of undesirable change is recorded.

In the area of forest management, models are developed for studying.

1. Tropical deforestation
2. Erosion control
3. Agroclimatic analysis

Some of the environmental planning problems analyzed by systems analysis techniques include.

1. Planning to avoid damage to the environment due to floods and storage.
2. Impact on environment due to wastewater treatment and disposal.
3. Impact on environment due to land filling method of solid wastes disposal.
4. Environmental planning of urban and rural settlements.
5. Ecological modelling.

5.2.5   STEP 5: Assessment of Impact Significance

Impact significance is a function of impact magnitude and the value, sensitivity and recoverability (resilence) of ecological receptors.

Interpretation of the anticipated impacts of a proposed project (or activity) should be considered not only in terms of individual species, but also relative to the general characteristics of the affected habitat(s) and overall ecosystem. One basis for significance determination is to apply the institutional information described earlier, including relevant laws, regulations, criteria and guidelines.

Another basis for impact interpretation is the biological science of professional interpretation approach. This involves the application of professional judgement and knowledge of biological-ecological principles, and it demonstrates why it is necessary for a biological scientist to be a part of an interdisciplinary study team. Examples of some biological-ecological principles and considerations which could be applied in impacts interpretation include the following:

1. The role of the individual species in the food-web relationship, with this interpretation based on recognizing the biological environment as a system,

2. An analysis of the carrying capacity of the biological setting relative to individual species of concern within the project area,

3. An evaluation of the resiliency of plant and animal species, and the interpretation of that resiliency relative to the anticipated changes caused by the project,

4. An evaluation of the implications of the project relative to species diversity within the terrestrial and aquatic habitats in the study area. In general, there is less ability to resist change when the species diversity is lower, therefore, this evaluation could also serve as a basis for interpreting the overall fragility of the biological environmental setting,

5. Consideration of natural succession and the implications of the project in terms of disruptions that might occur in this successional process,

6. A review of species that exhibit the ability to reconcentrate particular chemical constituents through natural environmental processes,

7. An evaluation of the implications of the proposed project on species of economic importance within the study area (these include species that might be of interest from the perspective of hunting or fishing activities), and

8. Any anticipated changes that might occur in threatened or endangered species or critical habitat within the study area.

5.2.6 STEP 6 : Identification and Incorporation of Mitigation Measures

Mitigation measures for biological impacts can include avoidance, minimization, rectification, preservation, and/or compensation and are associated with project location alignment design or construction and operating procedures. The location of a project can be a key factor. It is usually determined.

Fresh water ecosystems are almost profoundly influenced by adjacent terrestrial ecosystems and mitigation frequently involves maintaining these areas. Some of the measures for the mitigation of various biological impacts relating to fresh water ecosystem are given in Table 5.7.

Table 5.7 Mitigation measures relating to impacts on freshwater ecosystems.

Impact	Mitigation
Sediments/silt	Collect in **siltation traps, French drains,** or siltation basins/ponds/lagoons (maintenance is essential). Use vegetated **buffer zones** (30 – 100 m), including wetlands, as filters. Phase major construction periods to avoid wet seasons. Minimise disturbance during construction or operation, e.g., reduce bare areas by zoning, and install fences to protect adjacent areas. Avoid vegetation removal where possible. Revegetate bare areas rapidly, using temporary cover crops or mulches where necessary. Minimise dredging disturbance and erosion associated with bare areas, e.g., grade spoil heaps, and cover with tarpaulins.

Table 5.7 contd...

Impact	Mitigation
Organic matter, nutrients and salt	Reduce silt inputs as above (P is primarily carried with silt). Reduce N inputs by minimizing soil disturbance. Encourage formation of wet organic soils (i.e., create wetlands, extensive waterbody margin habitats, and wet woodland) to promote denitrification. In sewage treatment use nutrient stripping, tertiary treatments, separation of effluents, storm overflows.
Heavy metals, Micro organics, and other toxic materials	Treat or recycle industrial pollutants at source, and monitor effluents. Reduce silt inputs (as above). Reed beds may remove or manage many industrial and domestic effluents but proper design and maintenance is essential. Buffer zones (30 – 100 m) may give a reprieve from diffuse pollutants but can lead to long-term accumulation and/or release if these are not degradable. Minimise surface drainage from polluted areas. Reduce use where possible (e.g., of *Pesticides*). Test any fill material placed in surface waters during the construction phase. Ensure isolation of waste-storage facilities and landfill sites from surface and groundwater bodies, and monitor for *leachates*. Discharge vehicle and other wash waters to foul sewers rather than surface-water drains. Guard against accidental pollution by: effective safety systems (with back-up), security systems against fire or vandalism where potential pollutants are stored or delivered; contingency plans; and education/training of personnel.
Oils	Install silt/petrol traps (gully traps) in road or parking areas and ensure a proper maintenance. Bund or dike around temporary fuel/oil storage areas during construction. Vegetated buffer zones may retain petroleum products while they degrade. Guard against accidental pollution.
Acidification	Strip power station flue gases. Control afforestation and modify forestry practices. Avoid use of liming to increase the pH of waterbodies because of adverse effects on the ecosystem.
Heat	Re-circulate and/or use to heat local buildings
Changes in flow regime and aquifer recharge	Procedures are outlined in Table. It is difficult to reproduce natural flow conditions using physical structures; so where possible, mimic natural processes by encouraging infiltration, e.g., use vegetated areas, porous artificial surfaces, or detention basins.
River engineering	Where possible, maintain natural river depths and course, bottom sediments, and floodplain/flood regimes. Use natural materials for bank protection/stabilization, e.g., vegetation fringes and bankside trees instead of concrete or steel reinforcements. Limit damage by working from one bank and retaining vegetated areas, etc., Make new channels sinuous (not straight), and create new features such as pools, *roffles* and islands. Use dredgings for landscaping, etc.

Table 5.7 *contd…*

Impact	Mitigation
Physical loss or other damage	Destruction or degradation of long-established semi-natural habitats should be strongly resisted, since current technology and understanding are not sufficient to allow full recreation. Whenever possible, the development should be relocated or rezoned. For other habitats, loss or damage may sometimes be minimized by retaining key areas and protecting specific species migration routes, shelter and refuge zones. Consider habitat creation or enhancement to ameliorate loss.
Disturbance of wild life	Create/maintain buffer zones. During construction: restrict working/access/service areas and extent of temporary roads, physically protect habitat/wildlife areas including food areas, and plan activities around critical periods (e.g., breeding, nesting). During operational phase restrict access to valuable wildlife areas, and provide other focuses to reduce public pressure.

(**Source:** Methods of EIA By Peter Morris and Riki Therivel Spon Press NY 2003) (18)

5.3 Biodiversity Impact Assessment (BIA) as a Planning Tool

5.3.1 What is Biodiversity Impact Assessment (BIA)?

Biodiversity Impact Assessment (BIA) is a new technique which helps existing techniques to achieve the Conservation of Bio Diversity. CBD's with three objectives viz. biodiversity conservation, sustainable use, and equitable sharing. Introducing biodiversity concerns into conceptual stages of planning, BIA achieves the integration needed to spur innovative solutions which place all the three above objectives at the core of planning processes.

5.3.2 BIA- A Tool for the Planning Process

Biodiversity Impact Assessment (BIA) is a new planning tool to achieve the Conservation of Bio Diversity (CBD)'s with three objectives viz. biodiversity conservation, sustainable use, and equitable sharing. Introducing biodiversity concerns into conceptual stages of planning BIA achieves the integration needed to provide innovative solutions which place all the three above objectives at the core of planning processes. *BIA should have the following functions to achieve above objectives viz:* i) BIA should be integrated with the planning process even from staring phases of the project ii) BIA approach should have positive and not adverse iii) BIA solutions should be easy to adopt and focused and iv) BIA should provide alternatives with dual approach.

5.3.2.1 Integration into the Planning Process

Brundtland Commission defines Sustainable development as a development process which meets the needs of the present without compromising the ability of future generations to meet their own needs (WCED, 1987) (19). For achieving sustainable development, biodiversity need to be conserved adopting innovative projects, programs, policies necessitating integration of BIA process from the initial stages of project planning which facilitates sustainable development paths. Though EIA helps in some way towards integrating biodiversity concerns into decision making processes, as its application to the middle of the planning process it is unable to take full advantage of the CBD's mandate for sustainability.

To achieve above objectives BIA should have the following functions

- BIA should be integrated with the planning process even from staring phases of the project
- BIA approach should positive and not adverse
- BIA solutions should be easy to adopt and focused
- BIA should provide alternatives with dual approach

Thus BIA helps as a supplement for EIA and SEA for achieving sustainable development with focus on CBD and its success depends on its ability to affect changes in the early stages of planning.

5.3.2.2 A Non-adversarial Technique

Establishing a place in the early stages of decision making processes requires a technique palatable to developers, government officials, and local citizens. BIA must achieve a non-adversarial means of supporting biodiversity. It is important to embrace a position which recognizes that development *per se* is not destructive and emphasizes that BIA is a tool for making development sustainable, not for stopping development.

5.3.2.3 Adaptive in Application and Resolute in Purpose

Achieving a position in the early stages of planning also requires a technique is easily adapted to the unique situation posed by every project, program, and policy. BIA must be adaptive in application with several alternate methodologies and opportunity for independent developments and ideas. This adaptive nature will spur innovation needed to change the course of development. This does not mean that BIA forfeits its purpose of integrating biodiversity concerns. There should be a built-in system of checking that biodiversity conservation, sustainable use, and equitable sharing remain at the core of the BIA process. BIA must remain resolute in its purpose.

Biodiversity Impact Assessment (BIA) is a new planning tool to achieve the Conservation of Bio Diversity (CBD)'s with three objectives viz. biodiversity conservation, sustainable use, and equitable sharing. Introducing biodiversity concerns into conceptual stages of planning, BIA achieves the integration needed to provide innovative solutions which place all the three above objectives at the core of planning processes. *BIA should have the following functions to achieve above objectives viz:* i) BIA should be integrated with the planning process even from staring phases of the project ii) BIA approach should have positive and not adverse iii) BIA solutions should be easy to adopt and focused and iv) BIA should provide alternatives with dual approach

5.3.2.4 A Dual Approach

An additional factor in developing BIA is that many projects, programs, and policies are intended for purposes other than biodiversity's conservation, sustainable use, and equitable sharing. Often these are the very projects, programs, and policies that adversely effect biodiversity. BIA must be a tool not only for developing biodiversity policies but also for integrating biodiversity concerns into other sectoral projects, programs, and policies.

This means BIA methodology must apply to two avenues of planning:

- planning for conserving, sustainably using, and equitably sharing biodiversity resources
- integrating biodiversity concerns into planning for other sectors.

5.3.2.5 BIA Methodology

The methodology depicted in Figure 5.11 is adapted from a framework which developed out of discussions by an international expert group of ecologists and economists brought together by IUCN in 1996 (20) to discuss the role of economics in addressing biodiversity issues (IUCN, 1996) (20).

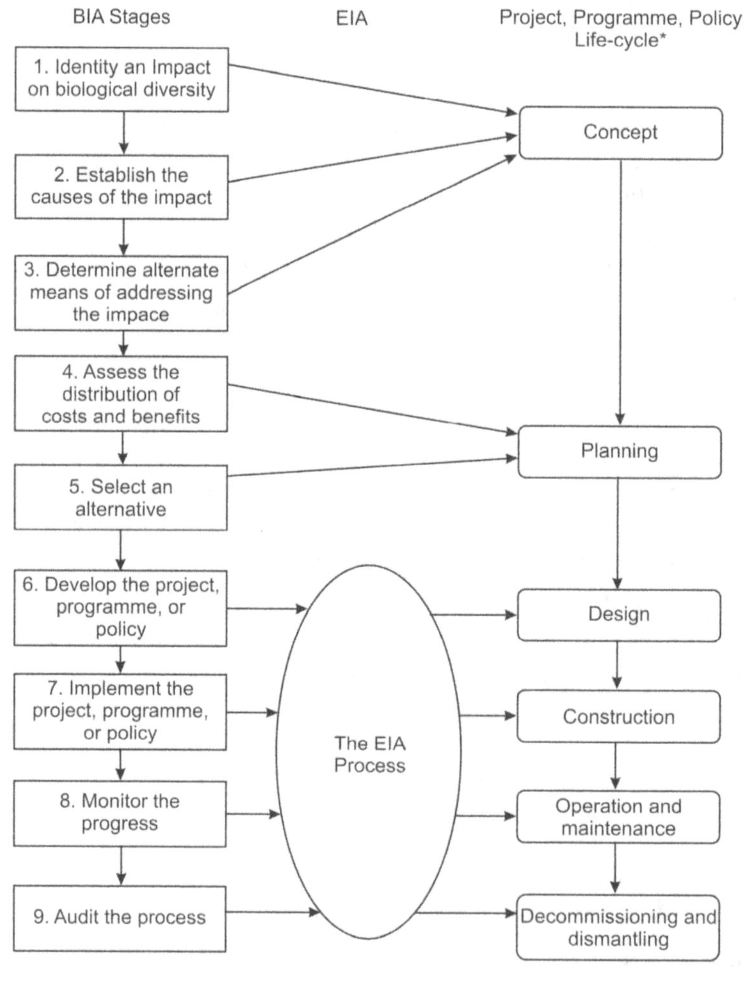

Fig. 5.11 BIA Process and its relation to the project programs.

This methodology builds on conventional impact assessment techniques and carries biodiversity concerns to the early stages of planning. The methodology presented is not meant as an end solution but is instead intended to spur ideas about possible approaches.

5.3.2.6 Identify an Impact on Biological Diversity

The purpose of BIA is to bring the objectives of the Convention into initial stages of planning, so the first step is to think about a project, program, or policy from the biophysical perspective. This entails looking at the existing state of biodiversity and asking if there is a loss or threat (an impact) that needs to be addressed. Information about the state of biodiversity can be obtained through existing studies, by using tools such as checklists and matrices, and through interviews and discussions with the public. The first step is to ask the questions:

- Which biodiversity is significant in the region?
- What is happening to that biodiversity?
- Are there species, communities, or ecosystems which are threatened in the region?

To give an example, a paper company executive wishing to develop a new product determines from asking these questions that a member of the region's waterflow, the Wood duck, is declining in population. At a different level, a federal policy maker realizes a decline in the Wood duck population across its range.

5.3.2.7 Establish the Causes of the Impact

Having identified a loss or threat, the next stage is to establish proximate and underlying causes of this loss or threat. This is not a straight forward task and the process is prone to bias if the assessment group is too homogeneous, just as with the scoping stage of EIA. A broader perspective can be obtained by using a team with diverse backgrounds and by opening the process to public participation.

A useful guideline is to ask if the loss is caused by proximate factors such as:
- habitat loss and fragmentation
- introduced species
- over-exploitation of plant and animal species
- pollution of soil, water, and atmosphere
- global climate change or
- industrial agriculture and forestry.

It is important to ask if the proximate causes are related to underlying causes such as:
- economic incentives or disincentives (subsidies or taxes, for example) or
- social conduct laws (zoning legislation, for example)

The temporal and spatial parameters of the project, program, or policy is important as these determine which causes the proponent is able to address.

5.3.2.8 Determine Alternate Means of Addressing the Impact

After knowing the impact and its causes, the next step is to determine what can be done about it. The proponent identifies several means of addressing the impact and its causes including a 'do-nothing' alternative. Brainstorming sessions, open dialogues with stakeholders, and surveys of similar impacts and responses can help this process. Innovative solutions should be encouraged and considered at this stage.

5.3.2.9 Assess the Costs and Benefits of Each Alternative

Each alternative is subjected to an analysis of the social, economic, and environmental costs and benefits to determine which alternative is a 'best response'. This analysis also determines the distribution of those costs and benefits which will help proponents meet the fair and equitable sharing objective of the CBD.

5.3.2.10 Select an Alternative

From the above analysis one alternative is selected. Recognizing the constraints the country and region's social, cultural, and political values the selection should reflect an equitable sharing of the distribution of benefits derived from the use of biological resources. Making the selection and supporting reasons explicit and public helps to ensure equitable distribution. Having identified the costs and benefits of each alternative and the distribution of those costs and benefits, the policy maker decides to repeal the existing subsidies as this is clearly the best option for the stakeholders.

However, if the political situation was such that the farmers had a particularly strong lobbying position this option may not be feasible. At this stage the policy proponent may go back to the drawing board and discover that a policy of replacing the perverse incentives with more benign subsidies would help the farmers and make this option politically feasible.

5.3.2.11 Implement the Project, Program or Policy

Implementing the project, program or policy is a lengthy process and a strategy should be developed to ensure it goes smoothly. Strategies also ensure biodiversity considerations and mitigatory measures developed throughout if the BIA are properly administered. Continued interest in biodiversity at this stage helps the group identify impacts which eluded them before and address those impacts as appropriately as possible. The new legislation will likely take time to pass through legislative procedures. The policy maker should follow the process throughout, ensuring that the *raison d'être* for the policy remains intact.

5.3.2.12 Monitor the Progress

Monitoring the progress of the project, programme, or policy throughout the implementation and operative stages is crucial to ensure it is conducted appropriately. Also monitoring identifies problems as they arise and in time to be corrected. Data collected through the monitoring exercises can add to information available through the clearinghouse mechanism and BCIS and thereby help decision makers in other countries or organisations. In monitoring the subsidy repeal, the policy maker may find an unintended consequence of the policy is to clear forests instead of wetlands. He then returns to the first stages, determine what is happening to the state of biodiversity, find proximate and underlying causes, etc.

5.3.2.13 Audit the Process

Auditing is crucial for improving the BIA process. Audits review the entire process and determine if appropriate impacts were identified, their causes established accurately, the alternatives given appropriate considerations, and costs and benefits analyzed correctly. An audit also determines if BIA meets the CBD's objectives and if the process is capable of instituting change in the path of human development.

5.3.2.14 Conclusions and Recommendations

Procedurally, existing impact assessment techniques, with a few adaptations, amend projects and policies so that biodiversity concerns are addressed. But the structural constraints of traditional impact assessment techniques render them unable to infuse the early stages of planning with consideration for biodiversity. EIA and SEA alone are unable to affect the shift in the development path called for in the CBD. BIA is a new technique of impact assessment that holds potential to help EIA and SEA achieve the three mandated objectives of the Convention. By introducing biodiversity concerns into the conceptual stages of the planning process BIA can achieve the full integration needed to spur innovative solutions which place biodiversity conservation, sustainable use and equitable sharing.

5.4 Specific procedure -Typical example of Assessment of Impacts of Developmental Activities on Vegetation and Wildlife

5.4.1   Introduction

The impacts of vegetation and wildlife will be most likely intense in rural areas and for proposed projects or actions covering large geographic areas or setting future management policies. In urban areas, however, small tracts of natural vegetation and habitat may be extremely important if there is an absence of similar habitat in the area or region. Wetlands are a significant habitat for numerous species of plants and animals. Vegetation and wildlife studies often focus on threatened or endangered species.

The environmental analyst should assess the possible project or action effects on vegetative ecosystems and wildlife species that are protected by law. Game species and other unprotected species and the systematic approach for this involve mostly (a) assigning the existing biological resumes (b) impacts analysis of project activities and (c) mitigation. The typical regulatory mitigation measures for the mitigation of biological impacts in various developmental activities are summarized in Table 5.8.

Table 5.8 Regulatory Mitigation measures for the mitigation of Biological Impact.

Biological impact	Possible mitigation measures and regulatory program requirement
Loss of wildlife and application. wildlife habitat	A wildlife-protection plan is required as part of any mining permit. Wildlife agencies must be consulted.
	Timing, shaping, and sizing operations must be conducted to avoid breeding or nesting season and trees, protecting key food, cover, and water resources.
	Fencing will keep large mammals from direct contact with toxic chemicals in sedimentation ponds and from roadways to reduce the number of road kills.
	Revegetation must use species with high nutritional or cover value.
	Topsoil handling and replacement prior to revegetation must be conducive to wildlife.
	Topsoil storage must be covered with vegetation, thus providing cover for wildlife.
	A 30-m buffer zone on each side of streams must be undisturbed.

Disturbance of aquatic habitat	A regulatory program designed for restoration, protection, organisms enhancement, and maintenance of aquatic life must be habitats implemented.
	Surface and underground mine openings must be cased and sealed to prevent escape of acid and toxic discharge.
	Buffer strips must be left between mining operations and waterways.
	All streams restoration is to include alternating patterns of riffles, pools, and drops.
	All diversions must be removed.
Erosion and sedimentation	Surface runoff must be collected in sediment ponds.
	Disturbed soils must be revegetated.
Destruction of vegetation	Affected land must be restored to premining productive capacity.
	Topsoil must be removed, segregated, stored, and redistributed with minimum loss or contamination.
	Topsoil and subsoil may be removed separately and replaced in sequence.
	Native vegetation or appropriate substitutes after mining must be established.
	Agricultural lands must be returned to the same or greater productive capacity obtained under remaining conditions.

Source : Developed from data in U.S. Department of the Interior (21).

5.4.2 Describing Existing Resources

Vegetation and wildlife studies begin like most other studies with co-ordination with central, state and local agencies for information on the presence of any special species or particularly valuable vegetation types in the project area. Goals and objectives for the area should be reviewed particularly if the project involves a large geographic area.

For small, simple projects or actions, it will be sufficient to verbally describe the existing resources. At the level expected impact increases, photographs and vegetation (habitat) mapping will most likely be required. Detailed studies are usually contained within a supporting technical report to the Draft Environmental Impact Statement or the Environmental Assessment. Habitat mapping can begin with a review of aerial photographs of the project area. Much preliminary work can be accomplished prior to doing any field surveys. Field surveys will then verify the habitat mapping and finalize the classification of vegetation communities.

Vegetative communities can be described generally or in terms of dominant species. Significant secondary species and understory species complete the descriptions. Any special wildlife habitat features, such as, feeding or nesting sites, water supplies, cover, or travel corridors, should be individually identified and emphasized. Unique or rare habitats or vegetative communities, relative to the presence of similar habitat types in the area or region, should be noted.

These are examples of the types of general vegetative community or wildlife habitat descriptors that may be used:

  – Hardwood forest – areas where greater than 50 percent of the area is dominated by trees,

– Abandoned field Scrub – areas not subject to moving for at least the current growing season and subject to invasion of woody plants,

– Agricultural – areas maintained for annual crop production or pasturing include hedgerows and drainage ways, and

– Human-dominated – moved aprons, lawns, and residential land-scaping and gardens.

For each of these general descriptors, supporting text would further describe the resources, including representative plant species. For example, the description of hardwood forest should include dominant species, understory species, and a discussion of tree, size and forest successional maturity. For all natural areas, the extent of evidence of disturbance or intervention by humans may be important to note.

Examples of more detailed habitat descriptors may include:

– Rivers, streams, floodplains and wetlands

– Open water

– Marine and coastal areas

– Aquatic bed

– Riparian (streamside)

– Wetlands, by classification

– Sage scrub

– Scrub and shrub

– Annual grassland

Oak-hickory forest

– Southern sycamore woodland

– Willow forest

– Conifer forest

– Bottomland hardwoods

– Ruderal (disturbed by humans)

– Ornamental and agricultural

– Developed or urban

The environmental analyst should become knowledgeable about the communities of fish and wildlife, include reptiles and amphibians that are expected to be present within particular vegetative ecosystems or habitat types in the project area and region. It is not necessary to try to list all possible faunal (animal) species within the technical report or draft environmental document. Most often, such an attempt will not be complete. Examples of common species should be given, however, and any special species of concern should definitely be emphasized.

If the environmental impact assessment is being conducted on a large management plan, as for a national forest or as a part of land management resource area, significantly more details will be involved in the description of existing vegetation and wildlife resources, the assessment of impacts of various management practices, and the selection of indicator species.

Characteristic plant species are listed for each of the 25 identified natural communities. Detailed vegetative diversity and wildlife analyses can be conducted using computer models for incorporation of suitable habitat and population indices information.

A major tool in national forest management is the Management Indicator Species (MIS). These species can be selected to estimate the effects of forest management activities on wildlife communities and on the forest ecosystem as a whole. Each selected species is representative of a group (guild) of many other species that have the same general habitat requirements. Effects of management activities on the indicator species are assumed to represent the effects on other species in the guild. A forest plan will include a list of indicator species and calculation of acres of suitable habitat for each species.

5.4.3 Impact Analysis

The level of detail required for impact analysis for vegetation and wildlife will depend on the specific characteristics of the proposed project alternatives and the expected degree of effects. Examples of the types of impacts that may be applicable are loss of unique vegetative communities, direct loss of wildlife habitat and species, deterioration of remaining habitat, barriers to wildlife travel corridors, and effects on recreational activities and land use.

5.4.4 Loss of Valuable Vegetative Community Types

The analysis of degree of impact of direct loss of vegetation will depend heavily on the value of the vegetative community to be destroyed. If the vegetation is common and unremarkable, the effects can be quantified by the amount of each type of community to be destroyed for each proposed alternative.

The key to ensuring an efficient analysis is the identification and quantification of any special or unique natural communities to be destroyed. Special areas would meet such rare and unique criteria as virgin or mature forests of high-quality functional value, namely wildlife habitat, erosion control, recreational use, or visual quality.

5.4.5 Direct Loss of Wildlife and Habitat

For projects or actions requiring land clearance and removal of natural vegetation, the most obvious impact of wildlife will be loss of habitat and individual animals. Species with small home ranges will be most affected. Larger species may immigrate to adjacent areas, but the wildlife biologist should be cautious in assuming that adjacent areas can support any individuals that may invade. Often, the community will already be at its carrying capacity for the particular species, that is, at its maximum ability to support a particular number of individuals without causing stress or imbalance to the species population as a whole.

The amount of habitat, by type, destroyed by each proposed alternative should be quantified. Any special functions provided by the habitat, such as food supply, water supply, and nesting or resting resources, should be identified. Represented species that would incur loss of individuals should also be identified. There may be more emphasis placed on game species as a result of indicated agency, organization, or public interest and concern during the scoping process.

As with wetlands, methodologies exist for the assessment of the functional value of wildlife habitat, based on the number of functions provided and the quality and rarity of similar areas in the region. The analysis of vegetation and wildlife impacts should identify the functional attributes to be lost.

The determination whether a biological change constitutes an adverse impact depends on the predicted future biological conditions with and without the proposed project or action.

5.4.6 Barriers to Travel Corridors

The analyst must consider whether the proposed project action would cause removal of connecting travel corridors between areas of wildlife habitat. Such corridors may cover large areas or may be very small but very important. This type of impact can often occur in sub-urban or partially rural areas where farmland or naturally vegetated land is being converted to a developed use in a piecemeal approach. Often large natural areas may remain, but travel ways for wildlife among the remaining large areas may be limited to narrow strips of woodland, fence rows, or riparian areas along the streams,

Linear projects, such as, highways, railroads, power lines, pipelines, or artificial drainage channels, are particularly likely to produce barriers to wildlife travel. The effect on wildlife populations can be particularly adverse if feeding or watering areas are separated from nesting or resting areas. Species requiring large home ranges are most affected. The analysis must consider not only the proportion of habitat lost, which in some cases may be a small percentage, but also whether that portion would split and render useless the remaining habitat because travel is restricted. Some species are more sensitive than the others. For example, even a small two-lane roadway may present a genuine barrier if the particular species cannot cross the paved area or is particularly sensitive to any human disturbance whatsoever. Some species require very remote areas.

5.4.7   Recreational Use and Enjoyment

Direct or indirect impacts to vegetation can produce secondary effects on recreational resource values. Vegetation is a major amenity in both expansive natural settings and smaller urban parks and open areas in urban environments can support recreational activities such as birding, picnicking, walking, bicycling, and general high quality visual resources. Larger natural areas support fishing, camping, hunting, hiking and research studies. To the degree possible, the impact of the proposed project of action alternatives on both active and passive recreational activities and qualities should be comparatively assessed and quantified.

5.4.8   Mitigation

Mitigation for potential impacts on vegetation and wild life may be very site-specific, such as, replacing landscaping or creating open space and parks in more urban areas, or geographically expansive in scope, such as, implementing particular management techniques in national forests. In some cases, rare plants or particular animals may actually be transplanted or trapped and moved to other safe locations.

Mitigation measures include 1. avoiding, 2. minimizing, 3. rectifying, 4. reducing and 5. compensating.

If particularly sensitive or valuable natural areas were destroyed, the first mitigation technique should be development and feasibility analysis of avoidance alternatives. If total avoidance is not possible, design refinements may reduce the quantity of exceptional natural area affected.

Techniques to improve the productivity and functional value of the remaining habitat can be used to offset adverse impacts. Such measures may include installation of nesting boxes or trees, creation of waterholes and open spaces, planting of food supply vegetation, or increasing the overall vegetation diversity.

Habitat impacts can also be mitigated through compensatory preservation or created replacement habitats. Depending on the value of the lost habitat, the required replacement ratio may be as high as 5:1, and the replacement should be functional in kind to that lost one. As with wetland mitigation programs, the habitat replacement plan should include detailed plans for physical construction and for planting of various plant species. Sometimes trees and vegetation removed by the proposed project or action can be saved and used to replant the created habitat.

Barriers to wildlife travel corridors can sometimes be mitigated through provision of wildlife underpasses in highway or railroad fills or similar types of protected travel corridors for power lines or artificial drainage channels. Wildlife losses through road kills can be further minimized by installation of fencing to prevent wildlife from crossing the highway and to direct wildlife movement to the provided underpasses.

Proposed mitigation measures should be coordinated with appropriate federal, state, and local agencies. The mitigation plan should include documentation of designated funding, responsible parties, performance criteria, monitoring methods, and schedule.

5.5 Specific procedure -Typical Example of Assessment of Impacts of Road development on Flora and Fauna

5.5.1 Direct Impacts

5.5.1.1 Habitat Loss

The consumption of land and consequent loss of natural habitat is inherent in road development. Where new roads intersect habitat, the area occupied by the road itself, burrow pits, and quarries is subtracted from the total habitat area available to flora and fauna.

5.5.1.2 Habitat Fragmentation

When a road cuts through an ecosystem, the sum of the two parts created by the cut is less than the value of the initial whole, even when the habitat loss is ignored. Ecosystems are characterized by complex, interdependent relations between component species and their physical environment, and the integrity of the ecosystem relies on the maintenance of those interactions. Roads tend to fragment an area into weaker ecological sub-units, thus making the whole area more vulnerable to invasions and degradation.

5.5.1.3 Corridor Restrictions

Most animal species tend to follow established patterns in their daily and seasonal movements. The areas, through which they travel on their way to and from feeding, breeding and birthing grounds, and between their seasonal ranges, are known as corridors. When a road intersects or blocks a wildlife corridor, the result is either cessation of use of the corridor because animals are reluctant to cross the road, an increase in mortality because of collisions with vehicles, or a delay in migration which may result in the weakening or disappearance of an entire generation of the population. Some animals are attracted to roads for various reasons, including protection from predators, good food supplies, better travel conditions, and so forth.

5.5.1.4 Aquatic Habitat Damage

Road development has perhaps its most serious effects on aquatic ecosystems. Erosion from poorly constructed and rehabilitated sites can lead to downstream siltation, ruining spawning beds for fish. Constriction of flows at water crossings can make the current too fast for some species. Alterations of flood cycles, tidal flows, and water levels can upset tropic dynamics by affecting the life cycle of plankton, and have corresponding effects on the rest of the food chain.

In the process of rechanneling, natural streambeds are dug up and useful obstructions, including large boulders are removed (Fig, 5.12). The same applies to shade trees on the banks. Frequently, the result is a straight, featureless channel, which may be an efficient evacuator of water, but has little in the way of the eddies, shaded areas, sheltering ledges, and turbulence essential to the health and existence of so many aquatic species.

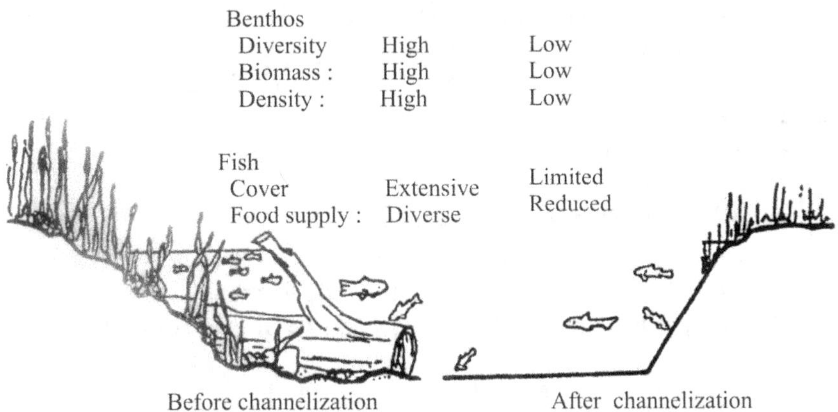

Fig. 5.12 Effects of Stream Re-Channelization.

The issue of blockage or restriction of fish migration is extremely important and needs to be assessed for each relevant project. This is critical in areas of the world where streams are dry for part of the year, but during the monsoon season are active fish spawning waters.

5.5.1.5 Interruption of Biogeochemical Cycle

The flow of nutrients and materials is a major determinant in ecosystem structure and function, and road development can easily disrupt it through alteration of flows of surface and groundwater, removal of biomass, and relocation of topsoil. Also, human activity can be a major source of nutrients (sewage, animal dung, and eroded topsoil) which, provided can raise turbidity and biological oxygen demand (BOD) of the water to the point where certain aquatic species simply cannot survive.

Based on typical ecosystem characters like soil erodibility, soil fertility and consequential human activity the nature of alteration likely to happen for an ecosystem due to proposed project activities can be established which then can be used to predict the potential impacts that alter the biogeochemical cycle of the ecosystem

5.5.2 Indirect Impacts

In many cases, indirect impacts are more damaging than direct ones, and their effects can be felt farther, sometimes several dozen kilometers, from the road. Where the road provides access to areas which were previously relatively untouched by human activities, the environmental assessment should take into account these frequently long effects. Some indirect impacts encountered commonly are:

5.5.2.1 Accessibility

Roads increase contact between humans and the natural environment, which in most cases leads to ecosystem modification. Penetration of previously unmodified areas makes them available for a host of human activities of varying effect, from recreation, forest and mineral exploitation to colonization and urbanization. Upgrading of existing roads generally facilitates an increase in the number of people having access and is accompanied by an increase in the likelihood of impacts. A classic example of the accessibility impact is the widespread land degradation occurring in Brazilian Amazonian, which has been induced in large part by road-building initiatives.

5.5.2.2 Ecological Disequilibrium

The establishment of new plant and animal species along the right-of-way can upset the dynamic balance which exits in ecosystems. Native species face competition for resources from new arrivals, and predator-prey relationships can be altered, often to the detriment of the native species. Non-native species can gain a competitive advantage because of a lack of natural controls and become dominant. The result is usually a simplified ecosystem that is more vulnerable to further impacts.

In some cases, road development may actually alter the ecological equilibrium in a positive way by providing for the creation of new ecotones, which tend to be relatively biodiverse. This will only apply if the total area of the existing system is relatively large compared to the newly created ecotone.

5.5.2.3 Contamination of the Biota

The presence of motor vehicles introduces the potential for contamination of the soil, air, and water adjacent to the road, and in the case of surface water, well beyond the immediate surroundings. Chronic contamination can become a serious problem for animal species, especially those at the top of the food chain, because of bioaccumulation of pollutants (Fig. 5.13).

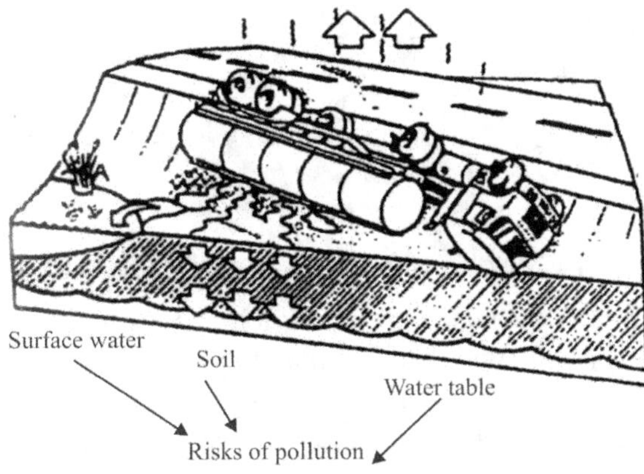

Fig. 5.13 Risk of Contamination from accidents.

5.5.2.4 Fires

Due to increased pace in human developmental activities the frequency of fires also increase leading to wide range of impacts sometimes sudden and severe.

5.5.2.5 Transmission of Disease

Increased road net works facilitate transmission of disease carrying bacteria leading to significant impacts on plants and animal species (Fig. 5.14) as disease carriers for both floral and faunal (like livestock/plant products)can get access to wilderness areas along with road corridors spreading diseases

5.5.2.6 Ecosystem types and Sensitivity

Based on ecosystem factors like biodiversity, climate, soil type, adjacent ecosystems and their size the biophysical environment consisting of different ecosystems, experience varying levels impacts and variable resilience under different growth scenarios.

For example the structure of forest ecosystems is highly variable and depends climate and altitude.

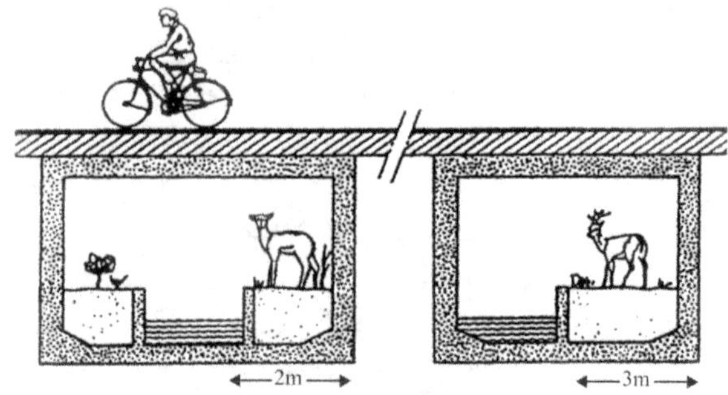

Fig. 5.14 Wildlife underpass and Hydraulic structure.

Water crossings: Aquatic ecosystems are particularly sensitive to road development, and there are a number of ways in which the impacts can be lessened. Standing water can be bridged instead of filled. Stream rechaneling should be avoided as much as possible, but where it must be done, efforts should be made to recreate lost channel diversity.

5.5.2.7 Traffic Control Measures

Reduction of the speed limit may reduce the rate of collisions between vehicles and animals. Some jurisdictions apply lower speed limits, particularly at night in areas of frequent animal crossings. Signs warning motorists of the presence of animals in places where animal corridors cross the road may also help to reduce collisions. Roadside reflectors may be used to scare animals away from the roadway when vehicles approach at night. Grassland ecosystems are dominated by herbaceous species and generally occur in areas experienced relatively low rainfall, large amounts of sunshine, and plentiful winds. Grasslands, are major carbon dioxide processors, and are thus important to global climate regulation.

Cave, limestone, and karst ecosystems often harbor rare species and display a high degree of endemism. Because of their relative inaccessibility, cave ecosystems have not been explored very extensively, and it is thought that they may contain many unknown species; this is especially true of water-filled cave systems. The systems provide habitat for highly specialized species which often have very limited distributions.

In some areas, cave bats are essential pollinators of economically important tree crops, while other species consume enormous quantities of pest insects. Cave ecosystems are particularly vulnerable to interruptions in groundwater flow, which can result from deep road cuts.

Tundra ecosystem occurs at high latitudes and is characterized by permanently frozen ground and highly adapted, very fragile plant species. Although annual precipitation is low, the frozen subsurface means that surface water tends to accumulate. Road building in tundra areas invite a host of problems related to the melting of the permanently frozen ground and sinking of the roadbed. Most roads are only negotiable during the lengthy winters. As with desert ecosystems, climatic extremes make recovery from disturbances very slow.

5.5.3   Determining the Nature and Scale of Impacts

Assessment of potential environmental impact should take into account (a) The extent of the proposed road development, (b) The duration of the construction period, and (c) The characteristics of the local natural environment through which the road will pass.

5.5.3.1 Extent of the Project

As the extent of project area increases, the significance of the biological impact also increases. It is necessary to provide the design specifications of any road project like the width of the road and right-of-way, amount of cutting and filling, number and location of water crossings, type of water crossing structures, and degree of expected groundwater flow etc.

5.5.3.2 Duration of the Construction Period

Generally road construction periods that do not exceed the annual reproduction cycle of key organisms in the ecosystem and hence will cause low impacts, while the construction activities which are focused, intense and spread over longer periods tend to have greater impact.

5.5.3.3 Evaluation of the effected systems

An evaluation of the ecosystem or ecosystems to be intersected by a road should have two objectives: (a) to take stock of the existing organisms in relation to how their function might be changed by a road, and (b) to determine the area's sensitivity to the magnitude and types of change that are expected. Whenever possible, these evaluations should be completed in the context of the local watershed or drainage basin in which the project is to be located.

5.5.3.4 Characterization

The descriptive component of the evaluation should comprise of:

- An inventory of biotic and abiotic resources, as well as their geographic distribution. This can be aided by working with local natural scientists and scientific institutions, and by making use of computer databases and biological inventories that may be available from national and international agencies.
- An estimation of productivity
- A description of species associations, relationships, keystone species, biodiversity, and the food chain
- a listing of rare or vulnerable species
- an estimation of ecological significance, which implies importance in the workings of nature on a grander scale- an ecosystems and
- a description of the resource needs of species-biogeochemical cycling and the food chain should be considered.

5.5.3.5 Sensitivity

The proposed changes can be evaluated by the following:

- environmental variables which are likely to experience changes of a magnitude greater than that of natural variations

- previous experience with change (evidence of soil erosion, invasion by non-native species, ecosystems simplification) and

- Likely effects on species which are instrumental in the formation and maintenance of habitat, offer crucial links in the food chain, are particularly vulnerable, whose corridors are intersected by the proposed road, and whose resource use will be affected by the development.

5.5.3.6 Use of Indicator Species or Groups

Physical, chemical, biological attributes can be used as overall indicators for evaluating the health of any ecosystem and understanding its susceptibility to damage from a road laying project while it will be extremely difficult and time consuming if one aims to evaluate the impacts at species level due the variety of plant and animal species and the complexity of their relationship

Indicators fall into four groups.

(i) Response indicators which provide evidence of the biological condition at the organism, population, community, ecosystem, or landscape level, e.g., biodiversity

(ii) Exposure indicators which indicate the presence of a stressor, e.g., algal blooms

(iii) Habitat indicators which are used to characterize conditions necessary to support an organism, population, community, or ecosystem and

(iv) Stressor indicators which are natural processes, environmental hazards, or management actions that produce changes in exposure and habitat, e.g., water quality.

Monitoring of indicators from each group is required for evaluating ecosystems and should also be selected to represent several levels of organization within an ecosystem. Indicators are of limited use in predicting impacts directly, but can be used to describe conditions as well as show trends and environmental response; they can therefore contribute to baseline studies and monitoring.

5.5.3.7 Rapid Appraisal

Rapid appraisal is a useful tool for assessments in which the complexity of the environment is so great, or the time available so limited, that a full-scale study is not feasible. Rapid appraisal allows a great deal of varied information to be brought together and synthesized in a blend of modern scientific with traditional knowledge instead of relying solely on quantitative research and empirical results.

5.5.4 Remedial Measures

5.5.4.1 Prevention

When planning new roads or changes in width or alignment, sensitive natural environments should be identified early in the planning process so that alternate routes and designs may be considered. Wherever possible, road developments should be located more than one kilometer away from sensitive areas to avoid severe impacts on flora and fauna. Water crossings should be minimized, and buffer zones of undisturbed vegetation should be left between roads and watercourses. Groundwater recharge areas should be avoided, and major roads should not be constructed through national parks or other protected areas.

5.5.4.2 Planting

Planting in road rights-of-way and adjacent areas can help to support local flora and fauna. In some cases, planting may provide additional habitats and migration routes for local animals, while also guarding against erosion. Border plant species may need to be chosen for resistance to wind or fire in some areas. Planting should be done wherever possible with native species, which are likely to require little maintenance and may prove beneficial in maintaining ecosystem integrity.

5.5.4.3 Animal Crossings

Animal crossing points like tunnels, culverts, hydraulic structures, bridges can be used to facilitate migration of animals and to reduce collision rates, especially for protected or endangered species.

5.5.4.4 Fencing

Fencing or plant barriers can reduce the risk of collisions between animals and vehicles. In some cases, semi-permeable fencing is used, which excludes species that are more likely to be involved in collisions while letting less problematic species through. Fences may interfere with the migratory patterns conflict with traffic patterns of animals, or may simply shift the points where migratory patterns conflict with traffic patterns along the route. Fencing may also, in some cases, interfere in predator-prey relationships, allowing predators to gain significant advantage because prey escape routes are restricted.

5.5.4.5 Compensation

One common compensatory measure is to replace damaged or lost biotopes with others of equal or similar characteristics and ecological significance. Environments damaged by a road project can be restored, and nearby biotypes of the same significance can be protected as parks or reserves. This is only feasible when the affected area is not unique.

The more important steps in the EA process relative to the incorporation of concerns about conserving biodiversity in the road development process are highlighted below.

5.5.4.6 Collect Relevant Data

Baseline data should identify areas of ecological interest within the study area. The identification criteria; adapted to the scope of the investigations, will be those commonly used in ecology: biodiversity, rarity and vulnerability of species, wildlife corridors, and so forth.

Identify potential impacts of road development proposals and carry out a comparative analysis of the various route alternatives in terms of their respective consequences for the natural environment.

5.5.4.7 Select Preferred Design

Select the design that interferes the least with wildlife movements and creates the least disturbance to nutrient cycling, especially as related to water movement.

5.5.4.8 Prepare Mitigation Plan

Mitigation plans should be suited to the scope of the project, the extent of environmental impacts. All measures proposed should balance cost with effectiveness.

5.5.4.9 Legislation and Regulations

Laws pertaining to plant and animal species, protected areas, hunting, fishing, and forestry should be used where available and develop as a more permanent means of impact minimization.

Aquatic ecosystems, such as swamps, ponds, marshes, lakes, rivers and streams, are habitats for important food sources and are characterized by a great wealth of flora and fauna, and high productivity. In general, these ecosystems are important because of their role in regulating the flow in waterways, in filtering water, and in serving as habitats for migratory birds and fish.

Island ecosystems, depending on their size and distance from the mainland, tend not to be especially biodiverse, and generally have a high incidence of endemic species. Interaction between island ecosystems and other terrestrial systems is very limited, or even non-existent, because of the expanse of open water between them.

Island ecosystems are particularly vulnerable to changes that reduce habitat area and population size, because these populations have few choices once their habitat is degraded or lost, or their food sources have become extinct. They are also vulnerable to the introduction of non-native species, which normally do not have any local predators, and quickly get out of control.

Mountain and alpine ecosystem, because of their relatively high altitudes and extreme weather conditions, tend not to be especially rich in species (often highly endemic). They are characterized by steep slopes and are therefore prone to erosion when disturbed. Alpine vegetation, in particular, tends to be very fragile, and recovery of damaged areas can take decades.

Desert ecosystems are characterized by extreme temperature fluctuations, low annual rainfall, and high evaporation. As a result, their species diversity tends to be low (also endemic) and vegetation is usually sparse. What rainfall they do receive often comes in brief but very intense episodes; these have tremendous erosive potential, given that the soils are generally sparsely covered and low in organic matter. For climatic reasons, recovery or recolonization of damaged areas tend to be slow.

Coastal and riparian ecosystem is found at the boundaries between aquatic and terrestrial, diverse, and productive. This applies more to wet climates that it does to dry ones. These systems usually exhibit a large number of species because they contain species from both bordering systems. Many species which inhabit these systems are living at the extremes of their ranges and are therefore especially vulnerable to changes in environmental conditions. Coastal ecosystems include mangrove swamps, salt marshes, dunes, beaches, and near shore islands, while riparian zones are found at the intersection of terrestrial and fresh water systems such as swamps, rivers, lakes, and estuaries. Coastal and riparian ecosystems are also preferred human settlement areas and are being lost rapidly to development.

Savannah ecosystems occur at a variety of latitudes and are characterized by semiarid climatic conditions. Their vegetation consists mainly of widely spaced drought resistant tree species, interspersed with herbaceous plants. Seasonal fluctuations in rainfall are very great, and erosion of disturbed soil can be a serious threat during the wet season.

Summary

Changes in the physical environment caused by development that often lead to secondary or high order changes in plants and animals and overall biodiversity. Biodiversity which is the wealth of species and ecosystems in a given area and gives genetic information within populations is of great importance for preservation at both global and local levels. At the ecosystem level, biodiversity provides flexibility for adaptation to changing conditions, such as those induced by human activity. Many developmental activities are likely to play a major role in the overall reduction of biodiversity, and proper planning at the project level based on full scale EIA can go a long way in limiting the loss, while still serving the needs of the people for which the project is started. To provide a basis for evaluating biological environment impacts, a six-step protocol is discussed for planning and conducting impact studies. This protocol is flexible and can be adapted to various project types by modification as needed to enable the addressing of concerns of specific projects in unique locations. The technical details of various phases associated with the evaluation of biological environment impacts i.e., 1. identification of the potential biological impacts of the construction and/or operation of the proposed project of activity, including habitat changes or loss of chemical cycling and toxic events, and disruptions to ecological succession; 2. description of the environmental setting in terms of habitat types, selected floral and faunal species, management practices, endangered or threatened species, and special features (such as wetlands); 3. procurement of relevant laws, regulations or criteria related to biological resources and protection of habitat or species; 4. conducting of impact prediction activities including the use of analogies (case studies), physical modeling and/or mathematical modeling, as based on professional judgment; 5. use of pertinent information from step 3, along with professional judgment and public input, to assess the significance of anticipated beneficial and detrimental impacts; and 6. identification, development and incorporation of appropriate mitigation measures for the adverse impacts are discussed with examples.

Biodiversity Impact Assessment (BIA) is a new planning tool to achieve the Conservation of Bio Diversity (CBD)'s with three objectives viz. biodiversity conservation, sustainable use, and equitable sharing. Introducing biodiversity concerns into conceptual stages of planning, BIA achieves the integration needed to provide innovative solutions which place all the three above objectives at the core of planning processes. *BIA should have the following functions to achieve above objectives viz:* i) BIA should integrated with the planning process even from staring phases of the project ii) BIA approach should have positive and not adverse iii) BIA solutions should be easy to adopt and focused and iv) BIA should provide alternatives with dual approach. These aspects are discussed in detail in this chapter

As the complex and dynamic nature of ecological systems impose difficulties in obtaining adequate baseline data making accurate impact predictions and formulating dependable impact predictions, the specific approaches to be adopted are presented with two examples (a) Impacts of Developmental Activities on Vegetation and Wildlife and (b) Impacts of Road development on Flora and Fauna.

References

1. Swanson, T. (1997). Global Action for Biodiversity. Earthscan, London.

2. WRI (World Resources Institute). (1992), (Global Biodiversity Strategy: Guidelines for action to save, study, and use earth's biotic wealth sustainably and equitably). WRI, IUCN, and UNEP.

3. ERL (Environmental Resources Limited). (1984), Prediction in Environmental Impact Assessment, a summary report of a research project to identify methods of prediction for use in EIA. Prepared for the Ministry of Public Housing, Physical Planning and Environmental Affairs and the Ministry of Agriculture and Fisheries of the Government of Netherlands.

4. Department of the Environment (1995), Preparation of Environmental Statements for Planning Projects that Require Environmental Assessment. HMSO, London.

5. ISEA (International Summit on Environmental Assessment). (1994), Final Report for the International Summit on Environmental Assessment, Quebec City, Canada, 12-14.

6. Gontier, M., Balfors, B., Mörtberg, U.M. Biodiversity in environmental assessment – current practice and tools for prediction. Submitted March 2005 to Environmental Impact Assessment Review.

7. Gontier, M., (2005), Integrating landscape ecology in environmental impact assessment using GIS and ecological modelling. *In:* Tress, B., Tress, G., Fry, G., Opdam, P. (eds.) 2005. From landscape research to landscape planning: Aspects of integration, education and application. Springer. *In press.*

8. Balfors, B., Mörtberg, U.M., Brokking, P., Gontier, M. Impacts of region-wide urban development on biodiversity in strategic environmental assessment. Under revision, submitted November 2004 to *Journal of Environmental Assessment Policy and Management.*

9. Canter, Larry W. (1996), Environmental Impact Assessment. Second edition. McGraw Hill Publishing Company, Inc., New York, NY. 660 pp.

10. Hill, L.G. (1975), Personal Communication to author, Oklahoma Biological station, University of Oklahoma, Norman.

11. Glowka, L., F. Burhenne-Guilmin, H. Synge, J. McNeely, and L. Gundling. (1994), A Guide to the Convention on Biological Diversity. IUCN, Gland, Switzerland.

12. Heywood, V. H and Watson, R. D. ed. (1993), Global biodiversity assessment. Cambridge University Press.

13. De Castri, F. and Young T. ed. (1996), Biodiversity, Science and Development. Towards a new partnership. CAB Int. IUBS, UK Cambridge University Press.

14. Council on Environmental Quality, (1993), "Incorporating Biodiversity considerations into Environmental Impact analysis under the National Environmental Policy act, CEQ, Washington D.C

15. Walter, Carl. (1986), Adaptive Management of Renewable Resources. Macmillan. New York. 374 pp.

16. Westman, Walter E. (1985), Ecology, Impact Assessment, and Environmental Planning. John Wiley & Sons, New York, NY. 532pp

17. Suter, Glenn II. (1993), Exposure. in Ecological Risk Assessment. ed. Glenn W. Suter II. Lewis Publishers, Ann Arbor, MI. 538 pp

18. Methods of EIA by Peter Morris and Riki Therivel, Spon Press NY 2001.

19. WCED (World Commission on Environment and Development). (1987), Our Common Future. Oxford University Press, Oxford.

20. IUCN. (1996), Using Economics to Attack Biodiversity Loss. Paper for the IUCN workshop on the development of a framework for biodiversity loss and assessment Gland, Switzerland, 22-24 April.

21. U.S. Dept of the Interior; (1979), "Final Environmental Impact Statement; Permanent Regulatory Program Implementing Sec 501 (b) of the surface mining control & Reclamation Act of 1977", Washington D.C, PP B III – 71 – B III – 81.

Questions

1. What is meant by biodiversity. How important is it in assessing biological impacts of any developmental activity? Discuss through a conceptual model of potential effects on biota by any developmental activity.

2. What are the key elements in the assessment of biological impacts of any project activity?

3. Discuss various phases involved using six step conceptual model for the study of biological impacts of any developmental activity.

4. Discuss the physical and mathematical models approach for impact prediction. What is meant by impact significance? How do you assess it with reference to biological and ecological effects?

5. Discuss the principles and main aims of any impact mitigation. Program for mitigating biological impacts of any developmental project.

6. What is Bio Diversity Impact assessment ((BIA)? To support the objectives of Conservation of Bio Diversity (CBD) how should BIA be planned?

7. Discuss the salient features of Bio Diversity Impact Assessment (BIA) Process as a planning tool.

8. Discuss the important aspects of assessment of impacts of any developmental activity on vegetation and wild life.

9. Discuss the salient features of impact assessment of a major road project on flora and fauna.

CHAPTER 6

Prediction and Assessment of Impacts on the Air Environment

6.0 Introduction

Many developmental activities are expected to add air pollutants to the atmosphere or alter the local weather and in a broader perspective the climate of the region which in turn may result in adverse effects on people, plants, animals, materials buildings etc. These effects can occur at local, regional or even global scale. The scientific and methodological approaches for evaluating the impacts of any new project activity on the air environment are discussed in this chapter. For evaluation of the air quality impacts of any project a six step methodological approach is generally adopted. Before applying this method one has to acquire the basic information on the following aspects of air pollution:

6.1 Basic Information on Air Pollution

For systematically evaluating the impacts of potential projects or activities on air quality, the basic information on air pollution sources and the effects of specific air pollutants have to be examined. Several sources of air pollution are envisaged from different activities like:

- Large point sources in thermal power plants
- Small to medium industrial stacks from chemical, fertilizer, refinery, steel, etc
- Area sources from mining of coal/minerals, solid waste dumping grounds
- Mobile sources in town planning, highways, ports and harbours.

6.2 Air Pollution Sources

The sources of air pollution can be classified based on the type of activity, the frequency of occurrence and spatial distribution, and the types of emissions and can be delineated as arising from natural sources or man-made sources. Sources like windblown dust, volcanic eruptions, lightning-generated forest fires and biological activity etc., can be termed as natural sources while transportation vehicles, industrial processes, power plants, construction activities, and military training activities, etc., can be termed as man-made sources.

Based on the number and spatial distribution, air pollution sources can be classified as single or point sources (stationary), area or multiple sources (stationary or mobile), and line sources. Pollutant emissions from industrial process stacks, as well as stacks for different fuel combustion processes are typical examples of point sources while vehicular traffic, fugitive dust emissions from resource or material stockpiles or construction, or military training activities over large geographical areas, etc., are examples of area sources. Figure 6.1 presents various types of sources which can be used for analyzing air pollutant sources in a given geographical area.

Based on the preliminary regulations and standards prescribed by various statutory and regulating agencies, the impacts of various activities on the air quality have to be examined.

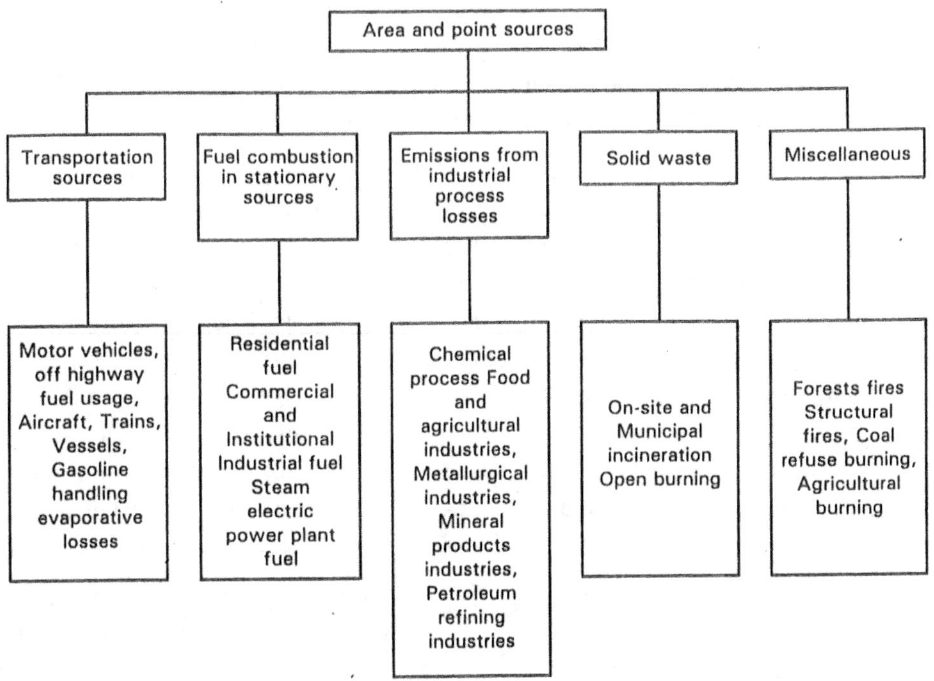

Fig. 6.1 Types of air pollution sources [I]

6.3 Effects of Air Pollution (Human Health, Material Damage, Climatic and Aesthetic Effects)

The air pollution effects can be grouped into human health, material damage, climatic and aesthetics.

6.3.1   Effects on Human and Animal Health

Include eye irritation, headaches, and aggravation of respiratory difficulties. Plants and crops have been subjected to abnormal growth patterns, leaf discoloration or spotting and death. Animals such as cattle have been subjected to the undesirable consequences of atmospheric fluorides.

6.3.2 Materials Damage

Examples are property devaluation because of odours, deterioration of materials such as concrete statuary, and discoloration of painted surfaces on cars, buildings and bridge structures.

6.3.3 Climatic Changes: Greenhouse Effects, Ozone Depletion, etc.

6.3.4 Aesthetic Effects

Include reduction in visibility, discoloration of air, photochemical smog-related traffic disruptions at airports, and the general nuisance aspects of odors and dust.

Activities of major concern are the burning of waste, the emission of dust and smoke, and the emission of chemical impurities such as heavy metals, acid or other toxic gases. Principal effects are on human health, aesthetic value (sight and smell) adjacent land uses, temperature modifications and humidity changes. Closely related to the subject of air quality is that of atmospheric visibility which is of both economic and aesthetic importance. Poor visibility due to gas, vapor, smoke or dust emission can have major impacts. Excessive heat emission at ground level can create katabolic winds and give rise to conditions favoring thermal inversions. Inversion layers can concentrate impurities in the atmosphere at, or close to ground level. Some localities are more susceptible to temperature inversion than others, the topographic character of the area and local wind patterns are important contributing factors.

Highways, tall buildings and major earth works (e.g., contouring) can modify wind patterns locally. Firebreaks in forests can produce a wind tunneling effect. In addition large paved areas or bodies of water can generate thermal updrafts.

Environmental effects on human communities are of major importance because they have always direct impacts. The human characteristics of the environment however are perhaps the most difficult to summarize or predict because of their complexity and the apparent inconsistency of human responses. However an E.I.A that does not take into account the human response to a development project would have omitted the most fundamental consideration in the human environment namely the human community itself. Effect of air and water pollution on human health and safety include physical safety, aspects of psychological well-being, parasitic diseases, communicable diseases and physiological diseases.

Based on the preliminary regulations and standards prescribed by various statutory and regulating agencies, the impacts of various activities on the air quality have to be examined. The typical standards of WHO and CPCB are given in Table 6.1 & 6.2.

Table 6.1 Ambient air quality standards World Health Organization guidelines.

Pollutant	Time weighted average	Averaging time
Sulphur dioxide (SO$_2$)	500 (μg/cum)	10 minutes
	350 (μg/cum)	1 hour
	100-50(μg/cum) (a)	24 hours
	40-60 (μg/cum) (a)	1 year
Carbon monoxide (CO)	30 (μg/cum)	1 hour
	10 (μg/cum)	8 hours
Nitrogen dioxide (NO$_2$)	400 (μg/cum)	1 hour
	150 (μg/cum)	24 hours
Ozone	150-200 (μg/cum)	1 hour
	100-200 (μg/cum)	8 hours
Total suspended particulates	150-230 (μg/cum) (a)	24 hours
	60-90 (μg/cum) (a)	1 year
Thoracic particles (PM$_{10}$)	70 (μg/cum) (a)	1 year
Lead (Pb)	0.5-1 (μg/cum)	1 year

Table 6.2 Indian ambient air quality standards.

Pollutant	Time weighted average	Concentration in ambient air (μg/cum) (a)		
		Sensitive areas	Industrial areas	Residential, rural and other areas
Sulphur dioxide	Annual (a)	15	80	60
	24 hours (b)	30	120	80
Nitrogen dioxide	Annual	15	80	60
	24 hours	30	120	80
Suspended particulate matter	Annual	70	360	140
	24 hours	100	500	200
Respirable particulate matter size less than 100 mm	Annual	50	120	60
Lead	Annual	0.50	1.0	0.75
Carbon monoxide	24 hours	0.75	1.5	1.00
	8 hours	1.0 (mg/cum)	5.0 (mg/cum)	2.0 (mg/cum)
	1 hour	2.0 (mg/cum)	10.0 (mg/cum)	4.0 (mg/cum)

6.4 Air Quality

Atmospheric changes are generally caused by the release of reactive substances into air by stationary or mobile sources, and by changes in surface morphology (for example, the construction of large buildings, clearance of vegetation, forestation, and creation of water impoundments). Possible environmental changes range from first order (immediate impact) effects of changes in concentration of substances in the air to higher order longer-range and secondary impacts) effects of physical and chemical changes on climate (for example, turbulence effects, haze, microclimates over water, heat emission effects, greenhouse effect), to the deposition of substances on soils, water, materials and vegetation, to effects of deposited substances on materials (for example, soiling, corrosion), to effects of changes in climate and air quality on visibility in the atmosphere). Fig. 6.2 summarizes the cause-effect network for atmospheric effects on air pollution.

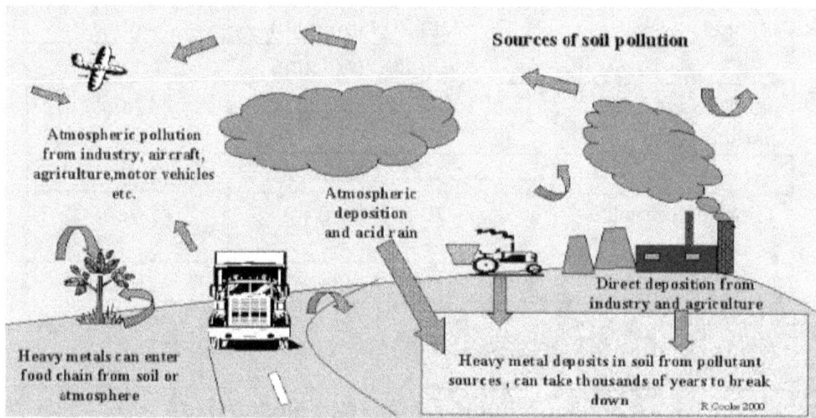

Fig. 6.2 Cause-effect network for atmospheric effects on air pollution

Source: Y. Anjaneyulu; Air Pollution and Control Technologies, B.S Publications (1999).

6.4.1 Predicting Changes in Pollutant Concentrations

The assessment of air quality impacts usually focuses on determining concentrations of air pollutants. Predicted concentrations are often compared against national or local air quality standards or objectives. Much of the pre-project air quality data collection is directed at determining pollutant concentrations at different times, at different locations, and the variations in concentration in time and in space. This information not only determines a baseline for comparison against changes, but also provides background information for predictive models. In cases where there is concern for higher order effects, predictions of pollutant concentrations are necessary inputs into predictions of deposition rates; exposure to flora, fauna, and man-made changes to local climate and visibility.

The normally followed impact assessment methodology to be used to determine the air quality impact pathways and risk ratings for any project are:

1. Determine the 'impact pathway' (how the Project impacts on a given air quality value or issue).
2. Describe the 'consequences' of the impact pathway.
3. Determine the maximum 'consequence' associated with the impact.

4. Determine the likelihood of the consequence occurring to the level assigned in step.

5. Using the Consequence and Likelihood in the matrix to determine the risk rating.

6.5 Generalized Approach for Assessment of the Impact of Air Pollution

To evaluate the impacts on air quality by any project activity, a six-step activity model proposed for planning and assessment. It is highly flexible and can be modified for adoption to different types of pollutants and for addressing specific problems in various types of locations.

6.5.1 Six main steps in the proposed Model for EIA Study on Air Environment

1. Evaluation and identification of sources and quantity of air pollutant emissions of different phases of the proposed activity like construction, operation and development.

2. Detailed evaluation of the project area for the existing ambient air quality, emission inventory, and meteorological data.

3. Enumeration of appropriate laws, regulations, or criteria to be followed for maintaining ambient air quality and/or pollutant emission standards.

4. Carrying out impact assessment studies adopting mass balances, dispersion calculations, comprehensive mathematical models, and/or qualitative predictions based on case studies and professional judgement.

5. Assessment of significance of anticipated beneficial and detrimental impacts. and

6. Development of appropriate mitigation measures for the adverse impacts.

6.5.2 Step 1 Evaluation and Identification of Sources and Quantity of Air Pollutant Emissions of Different Phases of the Proposed Activity

In the first step one has to examine what types and quantities of air pollutants are likely to be emitted during the construction and/or operational phases of the proposed project. The typical air pollution sources of some projects are given in Table 6.3. One can use the emission factor information based on the project type or activity. An emission factor is the average rate at which a pollutant is released into the atmosphere as a result of some activity, such as combustion or industrial production, divided by the level of that activity [2]. Emission factors relate the types and quantities of pollutants emitted to indicators such as production capacity, quality of fuel burned, or vehicle-km traveled by an automobile. The emission factors for different air pollutants from various types of refuse incinerators and automobiles are presented in Table 6.4 and 6.5 respectively.

Table 6.3 Air pollution sources of some activities

Activity	Air pollution sources
Solid waste disposal	Refuse incineration, open burning, sewage sludge and incineration
Combustion activity	Coal combustion as fuel oil combustion, Natural gas combustion,
	Wood waste combustion in boiling lignite carbon
	Coal cleaning, Sand and gravel process, Stone quarrying process
Industrial mineral products	a) Boiler coal burning

Table 6.3 contd....

production	b) Chemical reactor gaseous effluents
Chemical industrial process	c) Waste water treatment plants
	d) Acidification and neutralization reaction.
	e) Solvent evaporation
	f) Solvent evaporation in reaction process

Table 6.4 Emission factors for refuse incinerators without controls-emission factor rating:**A**

Incinerator type	Particulates kg/MT	Sulphur oxides[b] kg/MT	Carbon Monoxide kg/MT	Hydrocarbons[c] kg/MT	Nitrogen oxides[d] kg/MT
Municipal					
Multiple chamber					
uncontrolled	15	0.75	17.5	0.75	1
With settling chamber					
and water-spray system	7	0.75	17.5	0.75	1
Industrial-commercial					
Multiple chamber	3.5	0.75	5	1.5	1.5
Single chamber	7.5	0.75	10	7.5	1
Controlled air	0.7	0.75	–ve	–ve	5
Flue-fed	15	0.25	10	7.5	1.5
Flue-fed (modified)	3	0.25	5	1.5	5
Domestic single chamber					
Without primary burner	17.5	0.25	150	50	0.5
With primary burner	3.5	0.25	–ve	1	1
Pathological	4	–ve	–ve	–ve	1.5

a–average factors given based on EPA procedures for incinerator stack testing. Using high side of particulate, HC and CO emission range; when operation is intermittent and combustion conditions are poor b-Expressed as SO_2: c-Expressed as methane: d-Expressed as NO_2

Source: Adapted from US Environmental Protection Agency, 1973. pp. 2.1-3. [3]

Table 6.5   Air pollution emission factors for passenger vehicles

Transport mode	CO_2 per km	Organic compounds	CO	NO$_x$ Grams per kilometer	SO$_x$
Truck (petrol)					
Single occupancy	5.29	11.28	96.86	7.23	0.811
Average occupancy	2.85	5.92	49.91	3.80	0.42
Car					
Single occupancy	3.95	9.06	71.81	5.67	0.49
Average occupancy	2.39	5.32	42.25	3.35	0.21
Vehicle ride share					
Three-person car pool	1.30	3.03	23.95	1.90	0.17
Four –person car pool	0.987	2.25	17.95	1.41	0.10
Nine-person car pool	0.599	1.26	10.75	0.81	0.10
Bus (diesel)					
Transit	1.37	0.88	4.26	6.41	N.A.

Table 6.5 contd….

Rail intercity					
Diesel	1.51	3.95	2.11	3.17	1.79
Electric	0.917	neg	0.17	3.88	7.30
Commuter (diesel)	1.869	3.66	5.07	14.46	2.22
Transit (electric)	1.30	neg	0.21	5.22	10.19
Aircraft	2.01	1.76	1.83	3.80	0.28
Bicycle	0	0	0	0	0
Walk	0	0	0	0	0

6.5.3 STEP 2: Detailed Evaluation of existing Ambient Air Quality, Meteorological conditions and Natural Quality existing in the Project Area

6.5.3.1 Compilation Air Quality Information

All the information on the existing air quality should be compiled, particularly for the pollutants likely to be emitted from various project activities as identified in Step 1. To utilize this information appropriately, one must carefully describe the characteristics of each sampling site, including any unique factors about the site, such as surrounding land usage, height of the sampling device above the ground surface, and the type and calibration history of the sampling equipment. Graphical presentation of air quality information may be useful, particularly if there appear to be trends, either upwards or downward, in the air quality levels of any of the air pollutants.

The collected raw data should be compared in terms of the existing air data and presented in accordance with the averaging times in pertinent ambient air quality standards. Evaluation of annual average concentrations, along with the pertinent statistical distributions should be done. Data distributions may be required for arriving at 8 h or 24 h averaging data.

If the baseline air quality in the proposed project or activity area is not available, data from nearby areas with similar characteristics in terms of land usage and climatological features be utilized.

Baseline ambient-air quality data will be useful for assessing whether the air quality before the project activity exceeds attains or does not comply with relevant standards. Some relevant toxic air pollutants also have to be examined in addition to normal air pollutants. Greater significance should be given to, and greater attention should be given to, those pollutants which do not meet or barely meet the allowable ambient air concentrations. If one or more pollutants are in a non-attainment area some additional air quality management methods may be needed.

6.5.3.2 Procurement or Development of Emission Inventory

For evaluating the potential air quality impacts of a proposed project or activity, it is necessary to identify the study area (potential area or region of influence) associated with the air pollution emissions. The delineation of the study area can be made using the boundaries of the land associated with the project, or the delineation can include larger area by considering the atmospheric dispersion patterns within the vicinity of the proposed project or activity. So an emission inventory gives the overall scenario of air pollutants emissions in the existing meteorological conditions of the project area.

If the appropriate emission inventory suitable for the study area is not available, a comprehensive emission inventory of the study areas has to be prepared taking into account the following paints.

a) All pollutants and sources of emissions in the project area are to be classified and considered.

b) Information on emission factors for each of the identified pollutants and sources have to be collected with average values.

c) Daily quality and quantity of materials handled, processed, or burned, or other unit production information, of the individual identified sources have to be estimated.

d) The rate which each pollutant is emitted to the atmosphere and annual averages have to be estimated.

e) Specific pollutant emissions from each of the identified sources have to be added up.

6.5.3.3 Key Meteorological Data

Categorization of meteorological parameters which hinder dispersion of pollutants emitted to the atmosphere have to be made on the following lines.

a) Meteorological data which describe the general air pollution dispersion characteristics of the study area.

b) Meteorological data useful to describe the atmospheric dispersion of air pollutants from a project activity quantitatively.

c) Meteorological data necessary for air pollution dispersion modeling.

The general atmospheric dispersion conditions will provide a fundamental understanding of atmospheric transport. More importantly, limiting times, months, or seasons can be identified during this process and this information can be used in construction - phase planning and operational-phase decision making. Data indicative of the general characteristics of the area with regard to air pollution dispersion include mixing height, inversion height and mean annual wind speeds. Wind rose information can be used to qualitatively disclose the atmospheric dispersion of air pollutants from an activity.

6.5.3.4 Baseline Monitoring

To establish the concentration of specific pollutants, ambient air monitoring has to be carried out to verify the experienced changes in air quality concentrations for those pollutants which are identified as potential problems in the earlier steps.

6.5.4 Step 3 Examination of Appropriate Air Quality Emission Regulation Laws and Air Quality Standards to be maintained as per Local, State and Central Government Notifications

The primary sources of information on air quality standards, criteria, and policies of the basic local, state, and central government agencies like state PCBs and central PCBs which have the statutory authority to maintain the air resources have to be collected.

Documentation of this information will allow the determination of the significance of air quality impacts incurred during projects or activities and will aid in deciding between alternative actions or in assessing the need for mitigation measures for a given alternative. Specific air quality management policies or requirements may be in existence for particular areas, and the particular requirements of such policies have to be ascertained.

6.5.5 Step 4 Carrying out Impact Assessment using Mass-Balance, Mathematic Modeling and other Qualitative Approaches

Impact assessment can be carried out using a number of scientific methods like mass balances, ambient air dispersion models, and plume dispersion models.

6.5.5.1 Mass Balance Approach

With reference to the existing air environment inventory of the project area, air pollutant emissions from the construction and/or operational phase of a project have to be considered for which an inventory based on mass balance approach has to be prepared. The preparation of emission inventory for a proposed project or activity involves the following steps:

1. In the project area the pollutants likely to be emitted from different phases of the proposed project or activity like both the construction and the operational phases have to be classified and the sources have to be identified.

2. Information on the emission factors for each of the identified sources for each pollutant and their annual averages and aggregates have to be completed.

3. Procuring the specific unit-production information based upon the source and its type which may be either number of kilometers of vehicle traveled or tons of coal consumed or the extent of the area under construction, which multiplied with the relevant emission factor gives the overall mass balance value.

4. Computation of the rate at which each pollutant is emitted into the atmosphere, with this rate typically being extended to an annual basis. Systematic comparison of the emission from the proposed project or activity with the existing emission inventory for the area has to be carried out on an annual basis which can be computed by first assessing the rate at which each pollutant is emitted into the atmosphere from different project activities.

5. The pollutant emissions from each of the identified source categories associated with the proposed project or activity are to be added to arrive at the total value.

6.5.5.2 Amesoscale Impact Calculation has to be carried out based on the expected increase in the existing emission inventory for one or more pollutants as a result of the construction and or operational phase of the proposed project or activity using the relationship

Percentage increase in inventory

$$= \frac{\text{Project} - \text{activity emmission inventory information (100)}}{\text{Existing emission inverntory information}} \quad \ldots\ldots(6.1)$$

The increase in the percentage inventory for each pollutant for each project activity has to be assessed and the total percentage increase has to be arrived at by adding all the individual values in the inventory.

The percentage change in the current inventory for one or more air pollutants can be used to interpret and assess the impact on the basis of following criteria:

a) The existing air quality for the pollutants of interest.

b) The quantity of emission and magnitude of the percentage change.

c) The time period of the expected percentage change.

d) The potential for visibility reduction.

e) Any local, sensitive receptors to damage from the pollutants.

Further in addition to the above factors expected emissions from the proposed activity with reference to applicable emission standards have to be examined, Though it is expected that the proposed activities will be in compliance with pertinent emission standards, the extent of compliance involved in the proposed activity should be thoroughly examined.

6.5.5.3 Box Model Approaches

The box model is a simple atmospheric dispersion model, which can be adopted to compute the ground level concentrations of specific air pollutants of concern from the project activity [4]. In the box model it is assumed that the pollutants emitted to the atmosphere are uniformly mixed in a volume, or "box" of air of fixed dimensions (5) and the downwind, crosswind and vertical dimensions of the box and the time period over which pollutant emissions to be considered must be established. It is also assumed that the emissions, wind speed, and characteristics of air available for dilution will not vary over time, i.e., the box is under steady state conditions [6]. Further is this model it is also assumed that discharges mix completely and instantaneously with the air available for dilution and the released material is chemically stable and remains in the air.

The average concentrations of the pollutant using the box model can be mathematically expressed as

$$C = \frac{Qt}{xyz} \qquad\qquad .....(6.2)$$

where C = average concentration of gas or particulate <20 μm in size, throughout the box, including at ground level, μg/m^3

Q = release rate of gas, or particulates <20 μm in size, from source type(s), μg/s

t = time period over which assumption of uniform mixing in box is valid(typicalperiod, lh)
x = downwind dimension of box, m
y = crosswind dimension of box, m
z = vertical dimension of box, m

Adopting mass balance approach the important pollutants of interest can be examined for their ground level concentrations. One method of establishing the dimensions' of the box and the time period for the emissions is to use the data on limiting meteorological conditions corresponding to the worst case conditions.

The data from the box model can be examined and evaluated on a pollutant to pollutant basis, with reference to existing ambient air quality and corresponding standards, comparing existing pollutant concentrations and the concentration from the proposed project activity.

For more precise estimation of the dispersion of pollutants from a point source, the following air emission models are used.

6.5.5.4 Air Quality Dispersion Models

Based on the types of sources [elevated point (stacks), ground level point, ground level area, or line], pollutant type (gases or particulates), averaging times (short-term. 24-h, monthly, or annual), and atmospheric reactions (deposition, photochemical smog formation, or acid rain formation) the air pollutant dispersion models can be classified as:

a) Manual calculation for calculator models and

b) Computer models

6.5.5.4.1 Manual Calculation Models

There are three basic manual calculation models: (a) Pasqual-Gifford (b) Ground level point source model and (c) Area sources model which are widely used for calculating the impact on air quality of projects or activities. They are useful for calculating short term, average concentrations of air pollutants at specific locations.

6.5.5.4.1.1 Pasqual Model as Modified by Gifford

Elevated point source model. This is one of the important manual calculation models and can be indirectly used to analyze the impact on air quality of single, elevated point sources [7]. It is particularly suitable for various industrial gaseous emissions as many industries like chemical plants, heat and steam generation facilities have elevated stacks.

$$C_{x,y,0} = \frac{Q}{\pi \sigma_y \sigma_z u} \exp\left[-\left(\frac{H^2}{2\sigma_z^2} + \frac{y^2}{2\sigma_y^2}\right)\right] \qquad .....(6.3)$$

where $C_{x,y,0}$ = ground level concentration of gas, or particulates <20 μm in size, at a distance

    x, in m downwind from source, and distance y in m crosswind (90° from wind direction) from source, in μg/m³.

Q = release rate of gas, or particulates <20 μm in size, from elevated point source, μg/sσ_y = horizontal dispersion coefficient which represents amount of plume spreading in crosswind direction from source, and under a given atmospheric stability condition, m.

α_z= vertical, dispersion coefficient which represents amount of plume spreading in vertical direction at a distance downwind from source, and under a given atmospheric stability condition, m

    u = mean wind speed, m/sec

H = effective stack height (actual physical height plus any rise of plume as it leaves the stack), m; plume rise is a result of momentum effect caused by vertical velocity.

6.5.5 4.1.2 Ground Level Point Source Model

This is another manual calculation model that can be used for ground level point sources. The ground level concentration is given by:

$$C_{x,y,0} = \frac{Q}{\pi \sigma_y \sigma_z u} \exp - \left[\frac{y^2}{2\sigma_y^2} \right] \qquad .....(6.4)$$

6.5.5.4.1.3 Area Source Model

The other air pollution sources associated with many projects and activities are area sources (examples include air pollutants from agricultural operations, open burning, wind erosion, and pesticide applications) and line sources (examples include unpaved roads and vehicular traffic) for which the following equation can be used for the calculation of the ground level concentrations.

$$C_{x,0,0} = \frac{Q}{\pi (\sigma_y^2 + \sigma_{y0}^2)^{1/2} \sigma_z u} \qquad .....(6.5)$$

where $C_{x,y,0}$ = ground level concentration of gas, or particulates less than 20 um in size, directly downwind and at a distance x in m downwind from the source, $\mu g/m^3$

σ_{y0} = one-fourth of emission width of area or line source along axis which coincides with wind direction, m

6.5.5.4.1.4 Computer Models

A number of computer-based air quality simulation models which are a part of the U.S.EPA's UNAMAP (Users Network for Applied Modeling of Air Pollution) program are available.

Using these models not only the impacts of the project or activity of the project on air quality can be assessed but also various modifications of the proposed project or activity can be evaluated to assess the effectiveness of the efforts to minimize the impacts of the project activity. Based on both the technical capabilities of models and model-related managerial issues such as economic considerations, and necessary training and experience of model users, the appropriate model can be selected for a given condition.

APC compatible **SCREEN model** was developed by the USEPA to facilitate an initial evaluation of the air quality impacts of stationary sources [8]. It can be used in a **"Screening mode"** to calculate ground level concentrations under limited dispersion conditions.

A number of models specific for computation of fugitive-dust emissions, emissions at military airbases, heavy gases evaporation from toxic spills and hazardous waste sites are also available. The **Fugitive-Dust Model** (FDM) is a computerized air-model specifically designed for computing concentration and deposition impacts from fugitive dust sources [9].

Emissions for each source are apportioned by the user into a series of particle size classes. A gravitational setting velocity and a deposition velocity are calculated by FDM for each class.

A user friendly model, **Emissions and Dispersion Modeling System (EDMS)** was developed by Segal:

a) To compute emission inventory of all sources at an airport or air base.

b) The concentrations produced by these sources at four airport locations.

Atmospheric dispersion of denser than air vapors can be computed by a package called **SLAB** [10] which can be applied to area sources at ground level or an elevated horizontal jet. Further dispersion of these released vapors can be modelled either as a steady state plume, a transient puff or a combination of both based on the release time.

A model was developed for estimating evaporation which can occur from a toxic-chemical spill. The model is based on calculating the temperature of the spilled chemical Pool based on the net energy input into the pool from all possible sources. The pool temperature is then used to calculate the evaporation rate under steady state conditions.

A model to address site specific area source emissions hazardous wastes sites also was developed [1 1]. The details of some selected USEPA recommended plume models are given in Table 6.6.

Table 6.6 Selected USEP are commended plume models and appropriate applications

Model	Averaging period	Source Type	Terrain	Land use
SCREEN	Hourly, daily	Point, area	Simple, complex	Rural, urban
ISCST2	Hourly to annual	Point, area, volume	Simple	Rural urban
ISCLT2	Monthly, seasonal annual	Point, area, volume	Simple	Rural, urban
MPTER	Hourly to annual	Point	Simple	Rural
COMPLEX 1	Hourly to annual	Point	Complex	Rural
SHORTZ	Hourly to annual	Point, area	Complex	Urban
LONGZ	Seasonal to annual	Point, area	Complex	Urban

The choice of model for a given application will be based on:
- need for accuracy in prediction.
- type of emission point, area, line, module, hot or cold, intermittent, or continuous.
- meteorological conditions in the receiving area (wind turbulence, stability, inversions, mixing layer height, rainfall).
- topographical conditions in the receiving area.
- location of receptors.
- nature of effects on receptors. and
- nature of substance emitted.

Table 6.7 summarizes the most commonly used models in India

Table 6.7 List of Commonly used models in India

Model	Type	Source	Duration
ISCST3	Gaussian	Point(P), Area (A), Line(L), Volume(V)	Short
ISCLT3	Gaussian	P, A, L, V	Long
PAL	Gaussian	P, A, L	Short
CDM	Gaussian	P, A	Long
ADMS	Gaussian	L	Short
HIWAY2	Gaussian	L	Short
CALINE3 CAL3QHC CAL3QHCR	Gaussian	L	Short
ROADWAYS	Numerical	L	Short
UAM	Numerical	P, A, L	Short
AERMOD	Gaussian	P, A, L, V	Short
Research	Gaussian	P, A, L, V	Short /Long

6.5.6　Step 5 Assessment of Significance of Impacts of Predicted Beneficial and Detrimental Effects

Evaluation of the significance of anticipated changes related to the proposed project should be carried out through public meetings and public participation programs. Professional judgement is required based on the percentage changes from baseline conditions in terms of air pollutant emissions levels and/or exposed human population, or the PSI. These changes should be considered during both the construction and operational phases of a project.

For certain type of projects or air-pollutant prediction methods, there are numerical standards or criteria which can be used as a basis for interpretation. A final impact significance can be assessed based on the specific-effects of the types of air pollutants from a proposed project-activity and identification of sensitive receptors in the study area [12, 13].

6.5.7　Step 6 Development of Appropriate Mitigation or Remediation Plans for Reducing Adverse Impacts

Remediation or mitigation measures involve project-activity design or operational features that can be used to minimize the magnitude of the impact on air quality. Mainly the design should be revised to reduce the air pollutants expected to be emitted from the project. The revised project or activity can then be reassessed to determine whether other remediation or mitigation measures will help in eliminating or sufficiently minimizing the deleterious impact on air quality.

The following are some examples when mitigation measures can help in minimizing detrimental effects of air pollution.

1. Regulatory control on the practice of open burning of agricultural crop residues, like delineation of specified times for burning, and the establishment of distance requirements between residences and open burning areas. This will help in reducing air pollutant concentrations in sensitive areas [5].

2. Development of vegetation cover and watering or use of wind breaks and chemical stabilizers. Watering is the most common method for temporary dust control (nearly 0%). Planting of rapid growing vegetation in construction areas will reduce dust generation [14].

3. For reducing air pollutant emissions from unpaved roads, paving the surface, treating with penetration chemicals, working soil-stabilization chemicals into the road bed, watering, and traffic control regulations are some of the remediation methods [15]:

4. In the case of open waste piles and staging areas, dry surface impoundments, landfills, land treatment systems, and waste stabilization measures are adopted as Fugitive dust control techniques [15].

5. For controlling or reducing airborne pesticide residues resulting from spraying, pesticide and other materials, low pressure spray nozzles can be used to minimize the generation of fine particles. Further by spraying the pesticide in periods of low wind velocities so that its dispersion can be reduced.

6. Several alternative fuels, including methanol, ethanol, compressed natural gas liquefied petroleum gas, electricity, and reformulated gasoline [16] are recommended by USEPA for reducing air pollution from various combustion process in automobiles and processes.

7. Exhaust emission control technologies for petrol and diesel-fueled cars, trucks,and buses are available to minimize pollutants from the transportation sector.

8. A number of control systems like cyclones, scrubbers, fabric filters, and electrostatic precipitators for control of particulate emissions are now available. Flue-gas desulphurization and carbon adsorption can be used for SO_2 control. Thermal incineration, flares, carbon adsorption, absorption, condensers, fabric filters, electrostatic precipitators, and venturi scrubbers are some of the control systems using for reducing pollutant levels from various point source air pollutants [17, 18].

9. Plantations: Trees remove gaseous pollution by uptake via leaf stomata, though some gases are removed by the plant surface by intercepting airborne particles. However these particles often resuspended into atmosphere with leaf fall.

10. Transport of air pollutants and their concentrations, either from road ways or any air pollution source is influenced by high rise buildings, noise barriers etc. Solid structures exhibit lower removal rates for air pollution compared to vegetation

6.6 Detailed Methodology for Air Pollution Dispersion Modelling Approaches for Assessment of Air Quality Impacts (AQI) for a Specific Study Area (19-91)

6.6.1 Various Key Steps in Air Pollution Dispersion Modelling Approach for Air Quality Impact Assessment for a Specific Study Area

Air Quality Impact assessment methodology using Air Pollution Dispersion modeling approach involves the following five key steps:

1. Input data collection
2. Dispersion modelling
3. Processing dispersion model output data
4. Interpretation of dispersion modelling results
5. Preparation of an impact assessment report

6.6.2 Input Data Collection

The first stage in the Air Quality impact assessment is the collection of all the information required to complete the dispersion modelling. This includes development of a) air emissions inventory b) compilation of meteorological data c). background air quality data and   d) terrain data.

6.6.2.1 Emissions Inventory

The emissions inventory is the foundation of the air quality impact assessment. Developing a sound emissions inventory should be a priority task and requires the collation of a significant amount of data. A thorough air emissions inventory for a study area identifies all sources of air pollution, the air pollutants emitted from each source, and estimates the emission concentration and rate of all air pollutants emitted.

The details of each step of the development of an emissions inventory are given below:

6.6.2.1.1 Identify all sources of Air Pollution and Potential Emissions

A thorough understanding of the premises is essential in developing an emissions inventory. Undertaking a site visit of the existing premises or a detailed review of the engineering drawings for the proposed premises is necessary to identify all sources of air pollution and gain an understanding of the process and industry. This knowledge can be supplemented with a literature review on the industry and its most prevalent air pollution issues. For all sources of air pollution at a premises, identify the following:

- release type and source release parameters
- location (in metres relative to fixed origin, elevation and discharge geometry)
- potential air pollutants emitted.

Release type

Source configuration may be one of the following types:

Point sources

For a point source, emissions emanate from a very small opening such as a stack or vent. Stacks usually emit hot gases forcefully into the atmosphere at a fixed height above ground level.

Tall point sources: The term 'tall' point source usually refers to sources that protrude out of the surface boundary layer (e.g., over 30 to 50 m tall).

Wake-affected point sources: Where nearby buildings interfere with the trajectory and growth of the plume, the source is called a wake-affected point source. A point source is wake-affected if stack height is less than or equal to 2.5 times the height of buildings located within a distance of 5L (where L is the lesser of the height or width of the building) from each release point.

Wake-free point sources: Wake-free point sources are more than 2.5 times the height of the largest nearby building, so that surrounding buildings do not influence the stack top airflow.

Area sources

An area source has a more realistic two-dimensional structure but only a limited vertical extent. It is a source with a large surface area such as a liquid surface (pond, lagoon) or a landfill surface.

Line sources

A line source is a special case of a long, thin area source. In practice, these sources are taken to be at ground level and thin. A line source becomes an area source if the breadth exceeds 20% of the length.

Volume sources

A volume source is an essentially three-dimensional structure. Usually there are a sufficient number of emission points to consider an uniform emission rate over the full source structure. They are diffuse sources, such as emissions from within a building.

Source release parameters: For proposed study area this information is obtained from the engineering drawings and plans.

Point: stack height, stack diameter, temperature, discharge velocity, moisture, pressure, carbon dioxide and oxygen concentration.

Diffuse area: surface area, side length and release height.

Diffuse volume: side length, release height, and initial horizontal and vertical plume spread (σ_y and σ_z).

Estimation of Emission Rates

There are a number of methods that can be used to estimate the emission rate from each source. The EPA's preferred methods are direct measurement for existing sources and manufacturers' design specifications for proposed sources. Emission factors are generally used when there is no other information available or when emissions can reasonably be demonstrated to be negligible.

Manufacturers' Design Specifications or Performance Guarantees

Manufacturers' design specifications or performance guarantees can be used to estimate the emission rate of air pollutants from proposed sources. Such specifications provide a reliable means of determining the upper limit to the emission rate or concentration of air pollutants for sources that are maintained and operated in a proper and efficient manner.

Post-commissioning testing may be required to establish that sources comply with the manufacturers' design specifications and/or performance guarantees.

Direct measurement

For sources where manufacturers' design specifications or performance guarantees are unknown, emission rates and source release parameters should normally be established from the results of source emission sampling and analysis.

Emission Factors

An emission factor is usually an equation that relates the quantity of a pollutant released to process throughout. These factors are generally averages of all available data of acceptable quality, and are generally assumed to be representative of long-term averages for all facilities in the source category. As stated above, emission factors are generally used when there is no other information available or when emissions can reasonably be demonstrated to be negligible. Some databases of emission factors include: • US EPA's AP-42 Emissions Factors (www.epa.gov/ttn/chief)

• National Pollutant Inventory Emissions Estimation Technique Manuals (www.npi.gov.au/handbooks/approved_handbooks/sector-manuals.html).

Calculate emission concentration for point sources

The concentration of a pollutant emitted from a source is calculated using equation :

$$Ci = ERi_{FR} \qquad \qquad .....(6.6)$$

where

C_i = the concentration of pollutant i emitted from a source in mg/m^3

ER_i = the rate pollutant i is emitted from the source in mg/s

FR = the gaseous volumetric flow rate in m^3/s

The inventory should contain two emission concentrations:

1. Actual concentration of a pollutant emitted from a source (mg/Am^3) calculated using the actual gaseous volumetric flow rate (Am^3/s) and measured emission rate in Eq.6.6

2. Concentration of a pollutant emitted from a source corrected to the reference conditions as specified in the Regulation (mg/Nm^3 @ O_2%). This is calculated using the gaseous volumetric flow rate corrected to normal conditions (dry, 273K, 101.3 kPa) and the measured emission rate in Equation 6.6.

Assess Compliance with the Regulation

The inventory must be used to demonstrate compliance with the Regulation. All sources of air emissions must comply with the requirements of the Regulation. If a source does not comply, the emissions inventory must be revised to reflect the implementation of new technology and/or pollution control equipment that will comply with the Regulation.

Presentation of Emissions Inventory

The results of the emissions inventory must be presented to include the following information:

- all release parameters of stack and fugitive sources (e.g., temperature, exit velocity, stack dimensions, flow rate, moisture content, pressure, carbon dioxide and oxygen concentration).

- pollutant emission concentrations and a comparison against the relevant requirements of the Regulation.

A suggested format for summarizing and presenting the results of the emissions inventory in the impact assessment report is provided in Tables 6.8 and 6.9. The additional data that should be included in the impact assessment report for complex mixtures of odour and hydrogen sulfide is included in Table 6.10

Table 6.8 Stack source Release Parameters

Source	Release type	Stack height (m)	Exit temp (C)	Diameter Exit (m)	Velocity Exit (m/s)	Oxygen Conc. (%)	Moisture Content (%)	Flow rate (Am/s)	Flow rate (Nm/s)
Boiler No. 1	Wake-affected	20	150	4	15	10	15	188.5	103.4

Table 6.9 Stack emission concentrations and regulation limits

Pollutant	Emission rate (g/s)	Emission concentration 3) (mg/Am)	Corrected emission concentration 3 at stack (mg/Nm) reference conditions	Regulation emission concentration 3 at limit (mg/Nm stack reference conditions)
Sulphur dioxide	40	212.2	N/A	N/A
Solid particles	2	10.6	31.6	100
Oxides of nitrogen	15	79.6	237.4	350

Table 6.10 Peak Odor Emission Rates

Source	Source type	Odour emission rate (OUm) 3/s)	Stability class	Peak odour emission rate 3/s) (OUm)	
				Near-field	Far-field
Lagoon No. 1	Area	20,000	A, B, C, D	50,000	46,000
			E, F	46,000	38,000

6.6.3   Compilation of Data

The meteorological data used in the dispersion model is of fundamental importance as it drives the transport and dispersion of the air pollutants in the atmosphere. The most critical parameters are wind direction, which determines the initial direction of transport of pollutants from their sources, wind speed, which dilutes the plume in the direction of transport and determines the travel time from source to receptor, and atmospheric turbulence, which indicates the dispersive ability of the atmosphere.

6.6.3.1 Minimum Data Requirements

The meteorological data used in the dispersion modelling is one factor that determines the level of assessment.

Level 1 Impact assessments are conducted using 'synthetic' worst-case meteorological data. Table 6.11 lists the wind speed and stability class combinations that need to be included in the synthetic worst-case meteorological data file.

Level 2 Impact assessments are conducted using at least one year of site-specific meteorological data. The meteorological data must be 90% complete in order to be acceptable for use in Level 2 impact assessments (i.e., for one year, there can be no more than 876 hours of data missing). If site-specific meteorological data are not available for a Level 2 impact assessment, at least one year of site-representative meteorological data must be used. The site representative data should be:

- preferably collected at a meteorological monitoring station. Where measured data is unavailable or of insufficient quality for dispersion modeling purposes, a meteorological data file may be generated using a prognostic meteorological model such as TAPM.

- correlated against a longer-duration site-representative meteorological database of at least five years (preferably five consecutive years) to be deemed acceptable. It must be clearly established that the data adequately describes the expected meteorological patterns at the site under investigation (e.g., wind speed, wind direction, ambient temperature, atmospheric stability class, inversion conditions and katabatic drift).

6.6.3.2   Preparation of Level 1 Meteorological Data

Wind Speed and Stability Class

Gaussian plume dispersion models use stability categories as indicators of atmospheric turbulence and the dispersive properties of the atmosphere. Based on the work of Pasquill and Gifford, seven stability categories have been defined: A – very unstable. B – unstable.

C – slightly unstable. D – neutral. E – slightly stable. F – stable. and G – very stable conditions. In most dispersion models, stability classes F and G are combined into one class, F.

The stability class at any given time depends on:
- static stability (vertical temperature profile of the atmosphere, i.e., migrating high and low air-pressure masses)

- convective or thermal turbulence (caused by the rising of air heated at ground level)
- mechanical turbulence (a function of wind speed and surface roughness, i.e., wind flow over rough terrain, trees or buildings).

Table 6.11 lists the minimum wind speed and stability class combinations that must be included in a Level 1 meteorological data file.

Table 6.11 Wind speed and stability class combinations for a Level 1 data file

Stability class	Wind speed (m/s)																			
	0.5		1	1.5	2	2.5	3	3.5	4	4.5	5	6	7	8	10	12	14	16	18	20
A	*		*	*	*	*	*														
B	*		*	*	*	*	*	*	*	*	*										
C	*		*	*	*	*	*	*	*	*	*	*	*	*	*						
D	*		*	*	*	*	*	*	*	*	*	*	*	*	*	*	*	*	*	*
E	*		*	*	*	*	*	*	*	*	*										
F	*		*	*	*	*	*														

Ambient Temperature

For Level 1 impact assessments, the maximum and minimum ambient temperatures that are representative of the site must be included in the Level 1 meteorological data file to account for the range in possible plume rise. Higher ambient temperatures will result in the lowest plume rise and hence the largest impacts.

Mixing Height

For Level 1 impact assessments, the mixing height for neutral and unstable conditions (classes A–D) can be calculated using an estimate of the mechanically driven mixing height. The mechanical mixing height, h, can be calculated as follows:

Mechanical mixing Height for Stability Classes A–D

$$h = 0.3 \times u^* / f \qquad \qquad .....(6.7)$$

where

h = mixing height (m); u^* = friction velocity (m/s); f = coriolis parameter

For Level 1 impact assessments, the mixing height, h, for stable conditions (classes E and F) can either be set at an unlimited value (e.g., 5000 m) or calculated as follows:

Mechanical mixing height for stability classes E and F

$$h = 0.4 \times (u^* L / f)^{0.5} \qquad \qquad .....(6.8)$$

where

h = mixing height (m)

u^* = friction velocity (m/s)

L = Monin–Obukhov length (m)

f = coriolis parameter

Monin–Obukhov length

The Monin–Obukhov length, L, characterises the stability of the surface layer. The surface layer is defined as the layer above the ground in which the vertical variation of heat and momentum flux is negligible. The surface layer is typically 10% the height of the mixed layer. The parameter, L, can be calculated using the linear approximation to Golder's plot (Golder 1972) as follows:

Monin–Obukhov length

$$1/L = X + Y \times \log_{10}(Z_o) \qquad .....(6.9)$$

where

L = Monin–Obukhov length (m)

X & Y = parameters dependent on the Pasquill–Gifford stability class (Table 6.11)
Z_o = surface roughness height (m) (Table 6.12)

Table 6.12 Parameterisation of Golder's plot

Parameter	Pasquill–Gifford stability class					
	A	B	C	D	E	F
X	−0.096	−0.037	−0.002	0.000	0.004	0.035
Y	0.029	0.025	0.018	0.000	−0.018	−0.0365

In Equation 6.9

- the value of Z_o is the surface roughness height, unless the surface roughness height is outside the range Z_{omin} to Z_{omax} presented in Table 6.13
- if the surface roughness height $<Z_{omin}$ use the value of Z_{omin} for Z_o
- if the surface roughness $>Z_{omax}$ use the value of Z_{omax} for Z_o.

Table 6.13 Upper and lower limits for surface roughness heights for each Pasquill– Gifford stability class

Parameter	Pasquill–Gifford stability class					
	A	B	C	D	E	F
$Z_{o\,min}$	0.001	0.001	0.001	0.001	0.001	0.001
$Z_{o\,max}$	18.0	30.0	1.25	50.0	1.6	9.0

Table 6.14 presents typical values of surface roughness height for various land uses.

Table 6.14 Typical values of surface roughness height for various land-use categories (AUSPLUME version 6.0)

Land-use category	Roughness height Z_o (m)	Land-use category	Roughness height Z_o (m)
Hill	2.0	High-rise	1.0
Industrial area	0.8	Commercial	0.8
Forest	0.8	Residential	0.4
Rolling rural	0.4	Flat rural	0.1
Flat desert	0.01	Water	0.0001

Surface Friction Velocity

The surface friction velocity, u*, is a measure of mechanical turbulence and is directly related to the surface roughness. The parameter, u*, can be calculated using the procedure given in (Businger and Fleagle 1980; McRae 1981).

Coriolis Parameter

The Coriolis parameter accounts for variation in wind direction with height (wind shear) at different latitudes and can be calculated in accordance with well-established techniques. The coriolis parameter, f, can be calculated as follows:

$$F = 2\Omega\sin(\varphi)$$

where

Ω = Earth's rotation rate ($2\pi/86400$ or $7.29 \times 10^{-5} \text{rad·s}^{-1}$)

π = pi or 3.1416 radians (rad)

86400 = number of seconds in the day (s/day)

φ = latitude in radians (rad)

Table 6.15 lists an example of typical mixing heights for a location with a similar latitude to Sydney (34°) and in a rural location (surface roughness height of 0.3 m) to be included in the Level 1 meteorological data file.

Table 6.15 Typical mixing heights for a rural location (km)

Stability class	Wind speed (m/s)																		
	0.5	1	1.5	2	2.5	3	3.5	4	4.5	5	6	7	8	10	12	14	16	18	20
A	0.2	0.4	0.6	0.8	1.0	1.2													
B	0.2	0.4	0.5	0.7	0.9	1.0	1.2	1.4	1.6	1.8									
C	0.2	0.3	0.5	0.6	0.8	0.9	1.1	1.2	1.4	1.5	1.8	2.1	2.4	3.1					
D	0.2	0.3	0.4	0.6	0.7	0.8	1.0	1.1	1.3	1.4	1.7	2.0	2.2	2.8	3.3	3.9	4.5	5.0	5.0
E	5.0	5.0	5.0	5.0	5.0	5.0	5.0	5.0	5.0	5.0									
F	5.0	5.0	5.0	5.0	5.0	5.0	5.0												

Table 6.16 lists an example of typical mixing heights for a location with a similar latitude to Sydney (34°) and in an urban location (surface roughness height of 1.0 m) to be included in the Level 1 meteorological data file.

Table 6.16 Typical mixing heights for an urban location (km)

Stability class	Wind speed (m/s)																		
	0.5	1	1.5	2	2.5	3	3.5	4	4.5	5	6	7	8	10	12	14	16	18	20
A	0.3	0.6	1.0	1.3	1.6	2.0													
B	0.3	0.5	0.8	1.1	1.4	1.7	1.9	2.2	2.5	2.7									
C	0.2	0.4	0.7	0.9	1.1	1.3	1.5	1.8	2.0	2.2	2.6	3.1	3.5	4.4					
D	0.2	0.4	0.6	0.8	1.1	1.3	1.5	1.7	1.9	2.1	2.6	2.9	3.4	4.3	5.0	5.0	5.0	5.0	5.0
E	5.0	5.0	5.0	5.0	5.0	5.0	5.0	5.0	5.0	5.0									
F	5.0	5.0	5.0	5.0	5.0	5.0	5.0												

6.6.3.3 Preparation of Level 2 Meteorological Data

For guidance on processing meteorological data for dispersion modelling purposes, the USEPA guide (USEPA 2000) and USEPA processor (USEPA 1996) should be used.

Stability Class

In order of preference, the following methods outlined in USEPA (2000) should be used to determine stability class:

Turner's 1964 method: This requires information on solar altitude or zenith angle, cloud cover, cloud ceiling height and wind speed. Solar altitude can easily be calculated, but cloud cover and ceiling height are generally determined through visual observations.

Solar radiation–delta temperature method: This retains the basic structure and rationale of Turner's 1964 method but eliminates the need for observations of cloud cover and ceiling height. The method uses the surface-layer wind speed (measured at 10 m) in combination with measurements of total solar radiation during the day and a low-level vertical temperature difference (i.e., at 2 m and 10 m) at night.

Sigma theta method (the standard deviation of the horizontal wind direction fluctuation): All modern meteorological data loggers include software to determine sigma theta.

For Level 2 impact assessments, hourly stability class should be estimated using the USEPA meteorological pre-processor for regulatory models (USEPA 1996) or a processor that includes similar techniques.

Mixing Height

Mixing height is the depth through which pollutants released to the atmosphere are typically mixed by dispersive processes. Dispersion of pollutants in the lower atmosphere is greatly aided by the convective and turbulent mixing that takes place. Mixing height determines the vertical extent of dispersion for releases occurring below that height. Releases occurring above that height are assumed to have no ground-level impact (with the exception of fumigation episodes). Therefore, the greater the vertical extent of the mixed layer, the larger the volume available to dilute pollutant emissions.

Morning and afternoon mixing heights are estimated using vertical temperature profiles (otherwise known as 'upper air data') and surface temperature measurements. For Level 2 impact assessments, hourly mixing heights should be estimated from the twice-daily mixing height values, sunrise and sunset times, and hourly stability categories by using the USEPA meteorological pre-processor for regulatory models (USEPA 1996) or a processor which includes similar techniques.

6.6.4 Background Air Quality, Terrain, Sensitive Receptors and Building Wake Effects

6.6.4.1 Background Air Quality Data

Including background concentrations of pollutants in the assessment enables the total impact of the proposal (i.e., impact of emissions on existing air quality) to be assessed. The background concentrations of air pollutants are ideally obtained from ambient monitoring data collected at the proposed site. As this is extremely rare, data is typically obtained from a monitoring site as close as possible to the proposed location where the sources of air pollution resemble the existing sources at the proposal site.

Accounting for Background Concentrations

For impact assessments of sulfur dioxide (SO_2), nitrogen dioxide (NO_2), ozone (O_3), PM_{10}, total suspended particulates (TSP), deposited dust, lead (Pb), carbon monoxide (CO) and hydrogen fluoride (HF), the existing background concentrations of the pollutants in the study area should be assessed.

Level 1 Assessments

- Obtain ambient monitoring data that includes at least one year of continuous measurements.
- Determine the maximum background concentration of the pollutant being assessed for each relevant averaging period.

Level 2 Assessments

- Obtain ambient monitoring data that includes at least one year of continuous measurements and is contemporaneous with the meteorological data used in the dispersion modelling.
- At each receptor, add each individual dispersion model prediction to the corresponding measured background concentration (e.g., add the first hourly average dispersion model prediction to the first hourly average background concentration) to obtain hourly predictions of total impact.

6.6.4.2 Terrain and Sensitive Receptors

The dispersion modelling input file requires information regarding the surrounding terrain and sensitive receptors. Terrain and receptor files to be developed include the location and height in metres relative to a fixed origin. The location of any particularly sensitive receptors (and likely future sensitive receptors) such as residences, schools and hospitals can also be specifically included in the receptor file.

Building wake Effects: PRIME is the EPA's preferred building wake algorithm. AUSPLUME v. 6.0 includes the PRIME building wake algorithm and the Building Profile Input Program (BPIP) for entering the location and dimension of buildings. The location and dimensions of buildings located within a distance of 5L (where L is the lesser of the height or width of the building) from each release point for buildings with a height greater than 0.4 times the stack height should be entered in BPIP.

6.6.4.3 Dispersion Modelling

6.6.4.3.1 Dispersion Models

Dispersion models provide the ability to mathematically simulate atmospheric conditions and behavior. They are used to calculate spatial and temporal sets of concentrations and particle deposition due to emissions from various sources. Dispersion models can be used to determine the affected zone around an emitter by producing results that can be compared against impact assessment criteria.

Dispersion models can provide concentration or deposition estimates over an almost unlimited grid of user-specified locations, and can be used to evaluate both existing and forecast emission scenarios. In this capacity, air dispersion modelling is a useful tool in assessing the air quality impacts associated with existing and proposed emission sources. The results of the dispersion modelling analysis can be used to develop control strategies that should ensure compliance with the assessment criteria. Dispersion models can also be used to estimate the cumulative impacts of various industries that are located close to one another.

Dispersion models are widely used by environmental regulators in Australia, New Zealand, the United States, the United Kingdom and Europe, and industry well understands their limitations. The results have been shown, through numerous model evaluation studies, to be sufficiently robust to be relied on to calculate concentration limits for point-source stack emissions.

6.6.5   Approved Dispersion Models for Air Quality Assessment for Typical urban Areas

AUSPLUME v. 6.0 or later is the approved dispersion model for use in most simple, near field applications in NSW, where coastal effects and complex terrain are of no concern.

AUSPLUME is a Gaussian plume model, based on the assumption that cross-sections through elevated plumes from point sources of pollution have a Gaussian (or normal) distribution of concentration. AUSPLUME is also a steady-state model, which assumes the atmosphere is in a state of uniform flow, and wind velocity is a function of height alone and does not vary with direction. The mathematical basis of AUSPLUME is the Victorian EPA's Plume Calculation Procedure (EPA Victoria 1985), which itself is an extension of the ISC model (USEPA 1995). AUSPLUME v. 6.0 or later is specifically *not* approved for use in the following applications:

- **complex terrain, non-steady-state conditions:** AUSPLUME is a steady-state model and is unable to adjust the winds to reflect the effects of terrain. The straight-line trajectory assumption of the plume model is unable to handle the curved flow associated with terrain-induced deflection of channelling. AUSPLUME should not be used for terrain where the height of any receptor exceeds the lowest release height.

- **buoyant line plumes** e.g., discharges from the roof vents of aluminium smelters

- **coastal effects such as fumigation:** AUSPLUME is unable to consider large changes in meteorological conditions which can occur over short distances across a coastline.

- **high frequency of stable calm night-time conditions:** Pollutants can accumulate under such conditions or will flow downwind with the drainage flow. AUSPLUME has no memory of the previous hour's weather conditions as each hour is treated independently of the next and material is carried away instantaneously, to the edge of the grid, even if only light winds are prevailing.

- **high frequency of calm conditions:** AUSPLUME cannot handle calm conditions because of the inverse wind speed dependence plume equation. AUSPLUME assumes a minimum wind speed, which shoots the plume out to the edge of the grid even though the plume may not have moved at all.

- **inversion break-up fumigation conditions.**

- There are also other situations where another dispersion model may be more scientifically sound than AUSPLUME. In these instances, CALPUFF or TAPM (Section 6.3) may be appropriate. The two key factors that should be considered in evaluating whether to use a conventional plume model, such as AUSPLUME, or a more sophisticated approach are:

  1. Is the steady-state assumption in the plume model valid?

  2. Do the technical parameterizations in the plume model adequately treat the situation to be modelled?

6.6.5.1 Advanced Dispersion Models for Specialist Application

In circumstances where the AUSPLUME dispersion model is not approved or suitable for use, other dispersion models may be appropriate. Guidance on choosing appropriate alternative dispersion models can be found in the USEPA publication *Guideline on Air Quality Models* (USEPA 1999). CALPUFF and TAPM are the most commonly used alternative dispersion models for regulatory dispersion modelling applications in NSW. However only experienced and appropriately trained professionals should use them.

CALPUFF

CALPUFF is a multi-layer, multi-species, non-steady-state Gaussian puff dispersion model that is able to simulate the effects of time and space-varying meteorological conditions on pollutant transport. This enables the model to account for a variety of effects such as spatial variability of meteorological conditions, causality effects, dry deposition and dispersion over a variety of spatially varying land surfaces, plume fumigation, low wind speed dispersion, pollutant transformation and wet removal. CALPUFF has various algorithms for parameterizing dispersion processes, including the use of turbulence-based dispersion coefficients derived from similarity theory or observations.

CALPUFF has been accepted by the USEPA as a guideline model to be used in regulatory applications involving the long-range transport of pollutants (> 50 km). It can also be used on a case-by-case basis in situations involving complex flow and non-steady-state cases up to 50 kilometers from the source.

CALPUFF v. 5.7, CALMET v. 5.5 and CALPOST v. 5.4 or later should be used.

TAPM

The Air Pollution Model (TAPM) was developed by the CSIRO to simulate three dimensional meteorology and pollution dispersion in areas where meteorological data are sparse or non-existent. The modelling system contains a number of dispersion modules. These include a particle/puff dispersion model for dispersion from point, line, area and volume sources, and a three-dimensional grid-point model for urban air pollution studies. The dispersion models allow for plume rise and building wake effects, wet and dry deposition and photochemistry for urban airshed applications. TAPM v. 2.0 or later should be used.

Dispersion modelling methodology using AUSPLUME

Unless otherwise stated, the default options specified in the *Technical Users Manual* (EPA Victoria 2000) must be used with AUSPLUME. These options are the most appropriate for most impact assessment applications.

Terrain effects

- Use the Egan half-height approach to account for terrain effects.

Building wake effects

- All building dimensions must be entered in 10-degree increments. Use the USEPA utility program BPIP (USEPA 1995) within AUSPLUME to calculate the 36 wind-direction dependent building dimensions.
- Use the PRIME method to account for building wake effects.
- The USEPA's guidance document on good engineering practice (USEPA 1985) must be taken into account when designing new stacks to avoid building wake effects.

Horizontal dispersion curves

- For stacks < 100 m high, use sigma theta values or Pasquill–Gifford curves.
- For stacks > 100 m high, use Briggs rural curves.

Vertical dispersion curves

- For stacks < 100 m high, use Pasquill–Gifford curves.
- For stacks > 100 m high, use Briggs rural curves.

Enhance plume spreads for buoyancy

- Enable this option for both the horizontal and vertical dimensions. **'Adjust Pasquill–Gifford formulae for roughness height'** • Use this option.

Plume rise parameters

- Use the AUSPLUME defaults.

Wind speed categories

- Use the AUSPLUME defaults.

Wind profile exponents

- Use Irwin rural wind profile exponents for rural areas.
- Use Irwin urban wind profile exponents for urban areas.

Dispersion modelling methodology using CALPUFF

CALPUFF includes an option to automatically set all the options to the USEPA default values. These include:

- Wind speed profile: ISC Rural
- Transitional plume rise modelled
- Stacktip downwash
- Partial plume penetration
- Dispersion curves: Pasquill–Gifford or dispersion coefficients using turbulence-based micro-meteorology
- No adjustment of dispersion curves for roughness
- Terrain: partial plume adjustment method

6.6.6 Processing Dispersion Model Output Data

6.6.6.1 Calculate Peak Concentrations

The evaluation of odor impacts requires the estimation of short or peak concentrations on the time scale of less than one second. Dispersion model predictions are typically valid for averaging periods of one hour and longer. Dispersion models, such as AUSPLUME, therefore need to be supplemented to accurately simulate atmospheric dispersion of odors and the instantaneous perception of odors by the human nose.

The prediction of peak concentrations from estimates of ensemble means can be obtained from a ratio between extreme short-term concentration and longer-term averages. Properly defined peak-to-mean ratios depend upon the type of source, atmospheric stability and distance downwind. Table 6.17 shows the EPA-recommended factors for estimating peak concentrations for different source types, stabilities and distances as developed by Katestone Scientific (1995 and 1998).

The P/M60 ratios in Table 6.17 are for an idealised situation for one source in flat terrain where the receptor is located along the centreline of the single plume. The ratios do not consider fluctuations away from the centre line, terrain influences or plume interaction from multiple sources.

The EPA requires peak ground-level concentrations to be calculated for the following pollutants:

- hydrogen sulfide
- complex mixtures of odor.

A screening level assessment of peak GLCS can be undertaken by applying the ratios in Table 6.17 to multiple sources at a premises. These ratios can be applied to the emission rates entered into the dispersion model as follows:

1. Determine the source type, stability class and if the receptors are near-field or far field or both.
2. Select the appropriate P/M60 ratios from Table 6.17
3. For wake-affected point sources, determine the meteorological conditions (i.e. wind speed and stability class) under which the source is wake-affected and wake-free.
4. Apply P/M60 ratios to odour and hydrogen sulfide emission rates so they vary with wind speed and stability class.

Table 6.17 Factors for estimating peak concentrations in flat terrain (Katestone Scientific 1995 and 1998)

Source type	Pasquill–Giffordstability class	Near-field P/M60*	Far-field P/M60*
Area	A, B, C, D	2.5	2.3
	E, F	2.3	1.9
Line	A–F	6	6
Surface wake-free point	A, B, C	12	4
	D, E, F	25	7
Tall wake-freepoint	A, B, C	17	3
	D, E, F	35	6
Wake-affected point	A–F	2.3	2.3
Volume	A–F	2.3	2.3

* Ratio of peak 1-second average concentrations to mean 1-hour average concentrations

Availability of dispersion modelling software and guidance documents:

Windows-based **AUSPLUME** v.6.0 can be purchased by writing to: Environment Protection Authority of Victoria, 27 Francis Street, Melbourne Victoria 3000.

The **BPIP PRIME** (BPIPPRM) user's manual and software can be electronically downloaded, free of charge, from the USEPA website at

www.epa.gov/scram001/tt22.htm#bpipprm

The **CALPUFF** dispersion modelling package and guidance documents can be electronically downloaded, free of charge, from the Earth Tech Incorporated website at earthtec.vwh.net/download/download.htm

The **TAPM** software can be purchased by writing to:

Dr Peter Hurley, CSIRO Atmospheric Research PMB 1, Aspendale Victoria 3195

Guidance documents and information about TAPM can be obtained from the CSIRO website at www.dar.csiro.au/tapm/

Other dispersion modelling software and guidance documents can be electronically downloaded free of charge from the USEPA website at www.epa.gov/ttn/scram/

6.6.6.2 Interpretation of Dispersion Modelling Results

The primary purpose of an air quality impact assessment is to determine whether emissions from a premises will comply with the appropriate environmental outcomes. The assessment criteria outlined below reflect the environmental outcomes adopted by the EPA.

To ensure that the environmental outcomes are achieved, emissions from a premises must be assessed against the assessment criteria. The cumulative impact of emissions from several facilities also needs to be considered. Impacts of sulphur dioxide (SO_2), nitrogen dioxide (NO_2), ozone (O_3), Lead (Pb), particles (PM_{10}), total suspended particulates (TSP), deposited dust, carbon monoxide (CO) and hydrogen fluoride (HF) must be combined with existing background levels before comparison with the relevant impact assessment criteria.

Impact assessment criteria for SO_2, NO_2, O_3, Pb, PM_{10}, TSP, deposited dust, CO and HF are given in Table 6.18.

Table 6.18 Impact Assessment Criteria for SO_2, NO_2, O_3, Pb, PM_{10}, TSP, Deposited Dust, CO and HF

Pollutant	Averaging Period	Concentration		Source
		Pphm	**µg/m³**	
Sulphur dioxide (SO_2)	10 minutes	25	712	NHMRC (1996)
	1 hour	20	570	NEPC (1998)
	24 hours	8	228	NEPC (1998)
	Annual	2	60	NEPC (1998)
Nitrogen dioxide (NO_2)	1 hour	12	246	NEPC (1998)
	Annual	3	62	NEPC (1998)
Photochemical oxidants (as ozone)	1 hour	10	214	NEPC (1998)
	4 hours	8	171	NEPC (1998)
		g/m²/month[a]	**g/m²/month**[b]	NEPC (1998)
	Annual	-	30	EPA (1998)
Deposited dust[2]	Annual	2	4	NERDDC (1988)
		Ppm	**mg/m³**	
Carbon monoxide (CO)	15 minutes	87	100	WHO (2000)
	1 hour	25	30	WHO (2000)
	8 hours	9	10	NEPC (1993)
		µg/m³[d]	**µg/m³**[e]	
Hydrogen fluoride	90 days	0.5	0.25	ANZECC (1990)
	30 days	0.84	0.4	ANZECC (1990)
	7 days	**1.7**	**0.8**	ANZECC (1990)
	24 hours	**2.9**	**1.5**	ANZECC (1990)

a   Maximum increase in deposited dust level.

b   Maximum total deposited dust level. c Dust is assessed as insoluble solids as defined by AS 3580.10.1–1991 (AM-19). d General land use, which includes all areas other than specialized land use. e. Specialized land use, which includes all areas with vegetation sensitive to fluoride, such as grape vines and stone fruits.

6.6.6.3 Presentation of Air Quality Impact Assessment Results

The results of an impact assessment should be presented as follows:

1. Concentration, hazard index and/or risk contours (isopleths) to define potential affected zones

2. Concentration, hazard index and/or risk predictions in tabular form for each of the following:

   a. existing and likely future sensitive receptors

   b. maximum exposed off-site receptor

   c. maximum outside the boundary of the premises.

6.6.6.3.1   What if Impact Assessment Criteria are Exceeded?

If the EPA's impact assessment criteria are exceeded, the dispersion modelling must be revised to include various pollution control strategies until compliance is achieved. To determine incremental increases in the cost of air pollution abatement, a sensitivity analysis can be carried out by varying:

- source release parameters
- separation distance
- level of management practice.

The results can be used to select the most cost-effective and environmentally effective control strategy.

6.6.7   Application of GIS for Presentation of Air Quality Impact

Environmental impact assessment (EIA) of air pollution diffusion of different air pollutants on GIS platform can provide a decision support system with enhanced microscale spatial air pollution analysis for   sustainable environmental planning.

For processing spatial information on air pollution Geographic Information System (GIS) will be very useful. GIS a combination of geography, cartography and surveying integrated with various computation process is a very useful tool for managing geographical information resources. GIS with its capability to store large geographical data bases and ability to carryout spatial analysis finding huge applications in EIA.

ArcGIS Engine is a collection of GIS components and developer resources that can be embedded, allowing the developers to add dynamic mapping and GIS capabilities to existing applications or build new custom mapping applications. (Shengjun Zhong, Legang Zhou, ZhufangWang."Software for Environmental Impact Assessment, of Air Pollution Dispersion Based on ArcGIS", Procedia Environmental Sciences, 2011)

New Software's are being developed in which ArcEngine is used to deploy GIS data and maps, and visualization of atmospheric pollution diffusion using application programming interfaces (APIs) for COM and .NET(**Figure 6.3**).

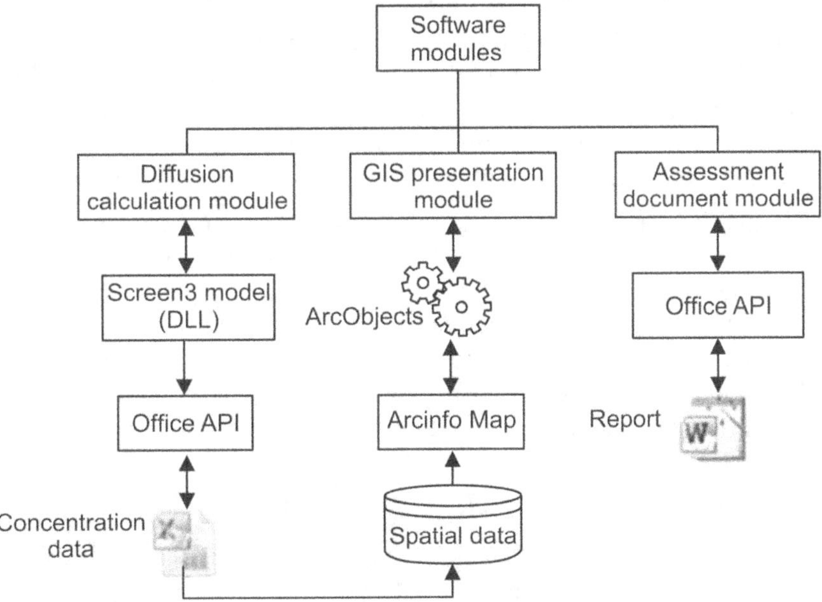

Fig. 6.3 Integrated Air quality Models and GIS Software Architecture.(Zhong, Shengjun, Legang Zhou, and ZhufangWang. "Software for Environmental Impact Assessment of Air Pollution Dispersion Based on ArcGIS", Procedia Environmental Sciences, 2011.

Spatial data will be the base of maps, and it stores the necessary space information of the system and provides data support for spatial analysis and graphics operation. The spatial data is managed by the internal layer database of the GIS platform. The features of real world can be represented as different layers in a map. A new map layer will be built to load the concentration information into an existing map. Then the embedded interpolation algorithm of the ArcGIS engine was used to constructing a raster layer and the interpolation results would be used to create a concentration contour lines layer. The raster values represent the concentrations of the regional area that one could use different colors to differ the pollution situation by using the embedded function in ArcGIS engine to render the raster layer. Finally, the raster layer and the contour layer will be loaded into the map, so that the distribution of the pollutant is shown on the map with good visualization effect.

For enhancing microscale air pollution transport and air quality analysis GIS is very useful because of its excellent storage, manipulation, modeling, and mapping tool for spatial data. Spatial information such as street coordinates and accompanying attributes can be exported and manipulated as input to air quality models such as CALINE3 and CAL3QHC. Output from air quality models in the form of pollution concentrations at specified receptor locations can be input to GIS for hot-spot identification, estimation of contributions of off-road mobile sources, and impact analysis. GIS tools applied to air quality analysis include contour generation, classification, thematic analysis, point-in-polygon analysis, and polygon overlay. Several case studies demonstrating these capabilities using TRANSCAD, a transportation-based GIS package, are available for microscale air quality analysis.

6.6.7.1 Case study of surface ozone over MS Gulf Coast®

(R) (Anjaneyulu Yerramilli, Venkata Srinivas C, Venkata Bhaskar Rao Dodla, LaToya Myles, William R Pendergrass, Christoph A Vogel, Hari Prasad Dasari, Francis Tuluri, Julius M Baham; Robert Hughes, Chuck Patrick, John Young, Shelton Swanier; 2010, A simulation study of surface Ozone pollution in the Central Gulf Coastal region using WRF/Chem; Atmospheric Environment; ATMENV-D-10-00784, Elsevier Publications)

The spatial distribution of simulated ozone, SO_2 and NO_2 at 1000 CST 9 June'06 representing daytime is shown in Figure 6.4. The precursors SO_2 and NO_2 have higher concentrations in the northern and northeastern Mississippi and northwestern Alabama. The peak ozone concentration at this time is about 45-55 ppbv and it is spread over northern and southeast Louisiana, and northwestern and southern parts of Mississippi. The concentration patterns of NO_2 and SO_2 species indicate large plume like peaks due to sources like coal fired power plants, manufacturing and other fuel combustion units distributed in the northern, central parts and along the Gulf coast in the domain.

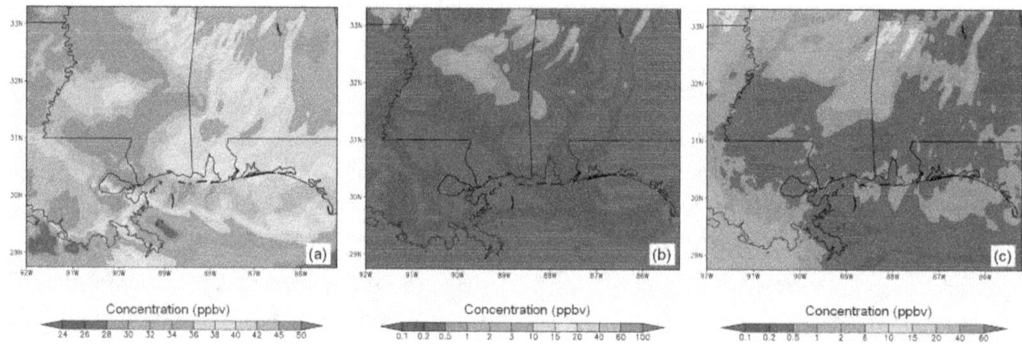

Fig. 6.4 Spatial distribution of concentrations of simulated a) Ozone b) SO_2 and c) NO_2 from the model lowest layer at 16 UTC 9 June. The concentration units are in ppbv.

6.6.7.2 Presentation of Air quality impacts of a Transportation Corridor Using Integrated Air Quality Dispersion model SCREEN3 and ArcGIS modules (Zhong, Shengjun, Legang Zhou, and Zhufang

Wang. "Software for Environmental Impact Assessment of Air Pollution Dispersion Based on ArcGIS", Procedia Environmental Sciences, 2011.)

Figure 6.5 presents the output of an integrated air quality dispersion model SCREEN3 and ArcGIS modules for a typical transportation corridor showing spatial presentation of the source information, the pollutant concentration distribution of the plume and a concentration curve at the plume centerline

The parameters of the emission source are as follow:

⌘ Source type = POINT

⌘ Emission rate (g/S) = 100.0

⌘ Stack height (m) = 100.0

⌘ Stack inner diameter (m) = 2.5

⌘ Stack exit velocity (m/s) = 25.0

⌘ Stack gas temperature (K) = 450.0

⌘ Ambient air temperature (K) = 293.0

⌘ Receptor height (m) = 0.0

⌘ Urban/Rural option = Urban

⌘ Wind direction = Northeastern

The model examines a range of stability classes and wind speeds to identify the "worstcase" meteorological conditions intended to estimate maximum ground-level concentrations.

Fig.6.5 Contour Representation of air pollutant concentration of a plume arising from a transportation corridor.

Reference: Software for Environmental Impact Assessment of Air Pollution Dispersion Based on ArcGIS; Shengjun Zhonga*, Legang Zhoua, Zhufang Wangb; Procedia Environmental Sciences 10 (2011) 2792 – 2797 (Elsievier).

Summary

In this chapter the basic information required on the air pollution sources and the effects of specific air pollutant for predicting the impacts of various project activities on the air quality of an area are discussed. Based on the preliminary regulations and standards prescribed by various statutory and regulating agencies the impacts of various project activities on the air quality are discussed comparing with typical standards of WHO and CPCB. The systematic methodology for air quality impact assessment and the air quality impact pathways and risk ratings for different types project are discussed in detail. A six-step activity model is proposed for planning and assessment which includes i.e., valuation and identification of sources and quantity of air pollutant emissions of different phases of the proposed activity like construction, operation and development, ii. detailed evaluation of the project area for the existing ambient air quality, emission Inventory, and meteorological data, iii. enumeration of appropriate laws, regulations, or criteria to be followed for maintaining ambient air quality and/or pollutant emission standards, iv. how to carry out impact assessment studies adopting mass balances, dispersion calculations, comprehensive mathematical models, and/or qualitative predictions based on case studies and professional judgement, v. assessment of significance of anticipated beneficial and detrimental impacts and vi. Development of appropriate mitigation measures for the adverse impacts.

Detailed Methodology for Air Pollution Dispersion Modeling Approaches for Assessment of Air Quality Impacts (AQI) for any specific study area are discussed in detail with latest developments. How to Present Air Quality Impact Assessment Results in terms of concentration, hazard index and/or risk contours (isopleths) to define potential affected zones are described. How air pollution diffusion of different air pollutants on GIS platform can provide a decision support system with enhanced microscale spatial air pollution analysis for sustainable environmental planning is presented in detail. Case studies on surface ozone over MS Gulf Coast and Air quality impacts of a Transportation corridor using integrated air quality dispersion model SCREEN3 and ArcGIS modules are described.

References

1. USEPA. (1992), Air Quality Atlas. EPA 4001 K – 92 - 002. Office of Air Quality Planning and standards. Research Triangle Park, NC.

2. USEPA. (1993), Research Triangle Park, NC. 8-1. *Compilation of air pollutant emission factors.* 2ndEdition. AP- 42.

3. USEPA 1972. Guidelines on Air Quality Models. EPA 450/2 –78 –027. Research TrianglePark, NC Appr. 1928. 16-27. A-12-A-24.

4. L.W. Canter. (1985), Environmental Impacts of Agricultural Production Activities. Lewis Chelsea Publishers, MI. 169-209.

5. L. Ortolano, (1985), Estimating Air Quality Impacts. *Environmental Impact Assessment Review.* 5:9-35.

6. D. B. Turner, (1979), Atmospheric Dispersion Modelling a Critical Review. *Journal of the Air Pollution Control Assessment.* **29(5):**502-519.

7. R. W. Brole, (1988), Screening Procedures for Estimating the Air Quality Impact of Stationary sources. EPA 450/4 – 88 - 010 USEPA. Research Triangle Park, NC.

8. P. Carry, (1990), Fugitive Dust Model (FDM) for Micro Computers. EPA SW / DK 90 – 94 USEPA Seattle.

9. D. L. Ermark, (1988), *SLAB.* Denser than Air Atmospheric Dispersion Model UCRL-99882, Lawrence Livermore National Laboratory. Livermore, Calif.

10. J. S. Touma, (1989), Review and Evaluation of a New Source Dispersion Algorithms for Mission Sources of Super fuel Sites. EPA 450/4 - 89 - 020 USEPA. Research Triangle Park, NC.

11. J. R. Barker and D. T. Tingey, (Eds), (1991), Air Pollution Effects on Biodiversity. Van Nostrand Reinhold, New York.

12. D. A. Cataldo, (1990), Evaluation, Actualization of Mechanisms Controlling Facilities Effects of arising smokes transport, transformations, rate and terrestrial ecological effects of brass obstructions. AD – A 227 BY / Y/ WEP. Battle Pacific Northwest Laboratory. Richland Wash.

13. U.S. Army Construction Engineering Research Lab. (1989), Environmental Review Guide from USA Rev. 5 volumes.

14. C. Conherd, P. Engle Client, G. E. Muleski, J. S. Kimmey and K. D. Resbury, (1990), Control of Fugitive and Hazardous Dust. Arryes Data Corporation Park, Ridge NJ.

15. A. J. Buonicose and W. T. Davis, (1992), Air pollution engineering manual. Van Nostrand Reinhold New York.

16. M. K. Sink. (1991),Handbook : control technologies for hazardous air pollutants. EPA 625/6-9-041, US. Environmental Protection Agency, Cincinnati.

17. Earth Tech (2000), A User's Guide for the CALPUFF Dispersion Model (Version 5), Earth Tech Incorporated, Long Beach CA, USA.

18. EPA Victoria (1985), Plume Calculation Procedure: An Approved Procedure under Schedule E of State Environment Protection Policy (The Air Environment), Publication 210, Environment Protection Authority of Victoria, Melbourne.

19. EPA Victoria (1986), The AUSPLUME Gaussian Plume Dispersion Model, First Edition, Publication 264, Environment Protection Authority of Victoria, Melbourne.

20. EPA Victoria (1999), AUSPLUME Gaussian Plume Dispersion Model: Technical User Manual, Publication 671, Environment Protection Authority of Victoria, Melbourne.

21. EPA Victoria (2000), AUSPLUME Gaussian Plume Dispersion Model: Technical User Manual, Environment Protection Authority of Victoria, Melbourne.

22. Hurley, P. (2005), The Air Pollution Model (TAPM) Version 3: User Manual, CSIRO Atmospheric Research Technical Paper No. 31, CSIRO Division of Atmospheric Research, Melbourne.

23. USEPA (1999), Guideline on Air Quality Models, 40 CFR, Chapter I, Part 51, Appendix W, United States Environmental Protection Agency, Washington DC, USA.

24. DEC (2005), Approved Methods for the Sampling and Analysis of Air Pollutants in New South Wales, Department of Environment and Conservation NSW, Sydney.

25. EPA (2001), Draft Policy: Assessment and Management of Odour from Stationary Sources in NSW, NSW Environment Protection Authority, Sydney.

26. EPA (2001), Technical Notes – Draft Policy: Assessment and Management of Odour from Stationary Sources in NSW, NSW Environment Protection Authority, Sydney.

27. EPA Victoria (1990), The AUSPLUME Gaussian Plume Dispersion Model, First Edition, Publication 264, Environment Protection Authority of Victoria, Melbourne.

29. EPA Victoria 1999, AUSPLUME Gaussian Plume Dispersion Model: Technical User Manual, Publication 671, Environment Protection Authority of Victoria, Melbourne.

28. EPA Victoria (2000), AUSPLUME Gaussian Plume Dispersion Model: Technical User Manual, Environment Protection Authority of Victoria, Melbourne.

29. Department of Environment and Heritage, National Pollutant Inventory Emissions Estimation Technique Manuals (www.npi.gov.au/handbooks/approved_handbooks/sector-manuals.html)

30. USEPA (1995), AP 42, Fifth Edition Compilation of Air Pollutant Emission Factors, Volume1: Stationary Point and Area Sources (www.epa.gov/ttn/chief/ap42/index.html)

31. Businger, J. and Fleagle, R. (1980), An Introduction to Atmospheric Physics, Academic Press, New York.

32. Golder, D. (1972), 'Relations among stability parameters in the surface layer', Boundary Layer Meteorology 3, 47–58.

33. Hurley, P. (2005a), The Air Pollution Model (TAPM) Version 3, Part 1: Technical Description, CSIRO Atmospheric Research Technical Paper No. 71, CSIRO Division of Atmospheric Research, Melbourne.

34. Hurley, P. (2005b), The Air Pollution Model (TAPM) Version 3: User Manual, CSIRO Atmospheric Research Technical Paper No. 31, CSIRO Division of Atmospheric Research, Melbourne.

35. Hurley, P., Physick, W., Luhar, A. and Edwards, M. (2005), The Air Pollution Model (TAPM) Version 3, Part 2: Summary of Some Verification Studies, CSIRO Atmospheric Research Technical Paper No. 72, CSIRO Division of Atmospheric Research, Melbourne.

36. McRae, G.J. (1981), Mathematical Modelling of Photochemical Air Pollution, Chapter 4, 'Turbulent Diffusion Coefficients', PhD Thesis, California Institute of Technology, Los Angeles.

37. Standards Australia (1987a), AS 2922–1987, Ambient Air – Guide for the Siting of Sampling Units, Standards Australia, Sydney.

38. Standards Australia (1987b), AS 2923–1987, Ambient Air – Guide for the Measurement of Horizontal Wind for Air Quality Applications, Standards Australia, Sydney.

39. USEPA (1996), Meteorological Processor for Regulatory Model (MPRM), User's Guide, EPA454/B-96-002, United States Environmental Protection Agency, Washington DC, USA.

40. USEPA (2000), Meteorological Monitoring Guidance for Regulatory Modelling Applications, EPA-450/R-99-005, United States Environmental Protection Agency, Washington DC, USA.

41. Zanetti, P. (1990), Air Pollution Modelling: Theories, Computational Methods and Available Software, Van Nostrand Reinhold, New York.

42. DEC 4th Quarter (2003), present, Quarterly Air Quality Monitoring Report Part A: DEC Data, Department of Environment and Conservation NSW, Sydney.

43. DEC 4th Quarter (2003), present, Quarterly Air Quality Monitoring Report Part B: Industry Data, Department of Environment and Conservation NSW, Sydney.

44. EPA 1992–3rd Quarter (2003), Quarterly Air Quality Monitoring Report Part A: EPA Data, NSW Environment Protection Authority, Sydney.

45. EPA 1992–3rd Quarter (2003), Quarterly Air Quality Monitoring Report Part B: Industry Data, NSW Environment Protection Authority, Sydney.

46. EPA Victoria (2000), AUSPLUME Gaussian Plume Dispersion Model: Technical User Manual, Environment Protection Authority of Victoria, Melbourne.

47. Schulman, L.L., Strimaitis, D.G. and Scire, J.S. (1998), Development and Evaluation of the PRIME Plume Rise and Building Downwash Model, Preprint Volume for the Tenth Joint Conference on Applications of Air Pollution Meteorology with A&WMA, American Meteorological Society, Phoenix, Arizona.

48. SPCC (1975-1991), Quarterly Air Quality Monitoring Reports, State Pollution Control Commission, Sydney.

49. USEPA (1995), User's Guide to the Building Profile Input Program, Revised February 1995, PA-454/R-93-038, United States Environmental Protection Agency, Washington DC, USA.

50. Earth Tech (2000), A User's Guide for the CALPUFF Dispersion Model (Version 5), Earth Tech Incorporated, Long Beach CA, USA.

51. Earth Tech (2000), A User's Guide for the CALMET Meteorological Model (Version 5), Earth Tech Incorporated, Long Beach CA, USA.

52. EPA Victoria (1985), Plume Calculation Procedure, Environment Protection Authority of Victoria, Melbourne.

53. EPA Victoria (1986), The AUSPLUME Gaussian Plume Dispersion Model, First Edition, Publication 264, Environment Protection Authority of Victoria, Melbourne.

55. EPA Victoria (1999), AUSPLUME Gaussian Plume Dispersion Model: Technical User Manual, Publication 671, Environment Protection Authority of Victoria, Melbourne.

54. EPA Victoria (2000), AUSPLUME Gaussian Plume Dispersion Model: Technical User Manual, Environment Protection Authority of Victoria, Melbourne.

55. Katestone Scientific (1995), The Evaluation of Peak-to-Mean Ratios for Odour Assessments, volumes I and II, Katestone Scientific Pty Ltd, Brisbane.

56. Katestone Scientific (1998), Peak-to-Mean Concentration Ratios for Odour Assessments, Katestone Scientific Pty Ltd, Brisbane.

57. USEPA (1985), Guideline for Determination of Good Engineering Practice Stack Height (Technical Support Document for the Stack Height Regulations), Revised, EPA-450/4-80023R, United States Environmental Protection Agency, Washington DC, USA.

58. USEPA (1995), User's Guide for the Industrial Source Complex (ISC3) Dispersion Models, Volume 1 and Volume 2, US EPA Office of Air Quality Planning and Standards Emissions, Monitoring, and Analysis Division Research Triangle Park, North Carolina.

59. USEPA (1997), Addendum to ISC3 User's Guide, The PRIME Plume Rise and Building Downwash Model, United States Environmental Protection Agency, Washington DC, USA.

60. ANZECC (1990), National Goals for Fluoride in Ambient Air and Forage, Australian and New Zealand Environment and Conservation Council, Canberra.

61. AWT (2001), Literature Review – Australian and Overseas Odour Threshold Data and Ambient Air Quality Criteria for Hydrogen Sulphide, A Report to the NSW EPA, Australian Water Technologies, Sydney.

62. Carson, P. and Round, J. (1989), Feedlot Odours, Department of Primary Industry, Brisbane.

63. CARB (2003a), Hot Spots Analysis and Reporting Program (HARP), California Environmental Protection Agency, Air Resources Board, www.arb.ca.gov/toxics/harp/harp.htm

64. CARB (2003b), Hot Spots Analysis and Reporting Program, User Guide Version 1.0,

65. California Environmental Protection Agency (2001), Air Resources Board, Sacramento.Health, Exposure Scenarios and Exposure Settings.

66. Department of Health and Ageing and Health Council (2002), Canberra.

67. Environmental Health Risk Assessment: Guidelines for Assessing Human Health Risks from Environmental Hazards, 2002,

68. Department of Health and Ageing and Health Council, Canberra. Health 2003,

69. Australian Exposure Assessment Handbook, Consultation Draft, December 2003, Department of Health and Ageing and Health Council, Canberra.

70. EPA (1998), Action for Air: The NSW Government's 25-Year Air Quality Management Plan, NSW Environment Protection Authority, Sydney.

71. NEPC (1998), Ambient Air – National Environment Protection Measure for Ambient Air Quality, National Environment Protection Council, Canberra.

72. NERDDC (1988), Air Pollution from Surface Coal Mining: Measurement, Modelling and Community Perception, Project No. 921, National Energy Research Development and Demonstration Council, Canberra.

73. NHMRC (1996), Ambient Air Quality Goals Recommended by the National Health and Medical Research Council, National Health and Medical Research Council, Canberra.

74. OEHHA (1994), Benzo[a]pyrene as a Toxic Air Contaminant, Executive Summary, California Air Resources Board and the Office of Environmental Health Hazard Assessment, Sacramento.

75. OEHHA (2003), The Air Toxics Hot Spots Program Guidance Manual for Preparation of Health Risk Assessments, Office of Environmental Health Hazard Assessment, California Environmental Protection Agency, Sacramento.

76. Streeton, J.A. (1990), Air Pollution, Health Effects and Air Quality Objectives in Victoria, Environment Protection Authority of Victoria, Melbourne.

77. SPCC (1983), Air Pollution from Coal Mining and Related Developments, State Pollution Control Commission, Sydney.

78. Victorian Government Gazette, (2001), 'State Environment Protection Policy (Air Quality Management)', No. S 240, Government of Victoria, Melbourne.

79. WHO (2000), WHO Air Quality Guidelines for Europe, 2nd Edition, World Health Organization, Geneva.

Questions

1. What are different sources of air pollution? Explain what is meant by air quality and how is predicted. Discuss the effects on various environmental components and human beings.

2. How do you identify different sources of air pollution and their potential emissions arising out of a proposed project?

3. Discuss the six main steps in the proposed model for EIA study on air environment.

4. Discuss the Air Quality impact assessment using mass-balance, mathematic modelling and other qualitative approaches.

5. How do you assess significance of impacts of predicted beneficial and detrimental effects?

6. Explain various mitigation or remediation measures for reducing adverse impacts of air pollution arising out in any project.

7. What are the various key steps to be followed in Air Pollution Dispersion Modeling?

8. Discuss salient features of Approved dispersion models for Air Quality Assessment for typical urban areas.

9. Discuss how to process meteorological data for dispersion modelling purposes as per the USEPA guide (USEPA 2000) and USEPA processor (USEPA 1996). What is Pasquill model?

10. How do you account for background air quality, terrain, sensitive receptors and building wake effects in any air pollution dispersion modelling?

11. What are the various Advanced dispersion models for specialist application in air pollution dispersion modeling?

12. You have been assigned to carryout Environmental Impact Assessment on Air Pollution generating project activities in a town. Discuss in detail the various key steps to be carried-out for an EIA.

13. Discuss in detail various parameters to be studied in air pollution dispersion modelling.

14. Discuss various dispersion modeling approaches.

15. How do you Process dispersion model output data? How do you calculate peak concentrations?

16. Explain the various key points in the Presentation of Air Quality Impact Assessment results.

17. Explain the salient features of GIS and discuss how advantageous will the Application of GIS be for presentation of Air Quality Impact results.

CHAPTER 7

Prediction and Assessment of Impacts of Noise on the Environment

7.0 Introduction

Noise is defined as any sound independent of loudness that can produce an undesirable psychological effect in an individual or a group. Thus noise is an unwanted sound energy and is also considered as a pollutant when it exceeds certain limits. Noise has a short residence and decay time and hence does not remain in the environment for long periods like air or water pollutants. By the time the average individual is spurred to action to reduce, or control, or at least, complain about sporadic environmental noise, the noise may no longer exist, either to notice or to measure.

In scientific terms, noise or sound is a pressure oscillation in the air or water or any medium, which conducts and travels (radiates) away from the source. If noise can be controlled at the source, there is a saving in energy and energy costs, which will have an impact on production and production costs.

Noise as a pollutant produces contamination in environment becoming a nuisance and a cause for annoyance, and affects the health of a person, his activities and mental abilities. Environmental noise pollution has not been an entirely new phenomenon, but rather a problem that has grown steadily worse with time due to similar factors responsible for air and water pollution almost in crisis proportions, namely increasing population, urbanization, industrialization, technological change and usual relegation of environmental considerations to a position of secondary importance relative to economic ones.

The intensity of noise can have a direct effect on biological and human communities. The intensity of a noise determines the distance over which it can be heard. The acceptability and therefore the impact of a new noise depends on existing noise levels.

The period over which a noise is likely to occur is a factor, which contributes to the impact of the noise. Noise during sleeping hours will have a major effect on the human community. Noise during a mating season may have a major effect on wild life. Short periods of noise may have less effect than persistent noise.

How often a noise is repeated is a further factor, which governs the environmental effect of noise. Between communities there may be considerable variation in tolerance of frequency of repetition of noise.

Fig.7.1 and Fig.7.2 show typical ambient day and night noise levels as a function of population density in an industrially developed society and in different environments respectively (1,.2).

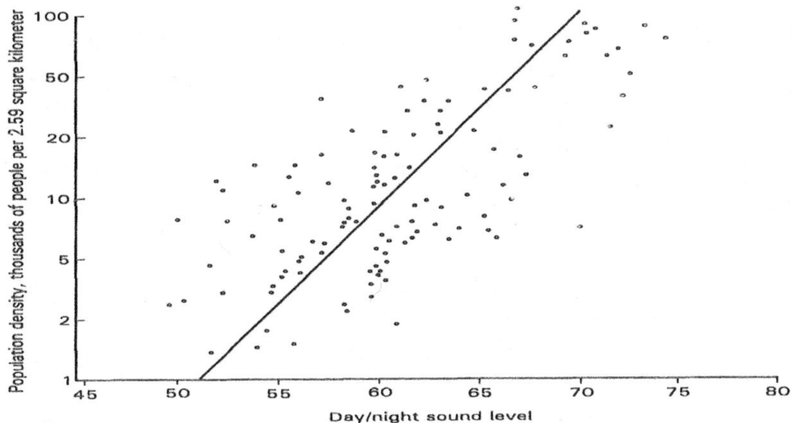

Fig. 7.1 Day/night sound level as a function of population density.

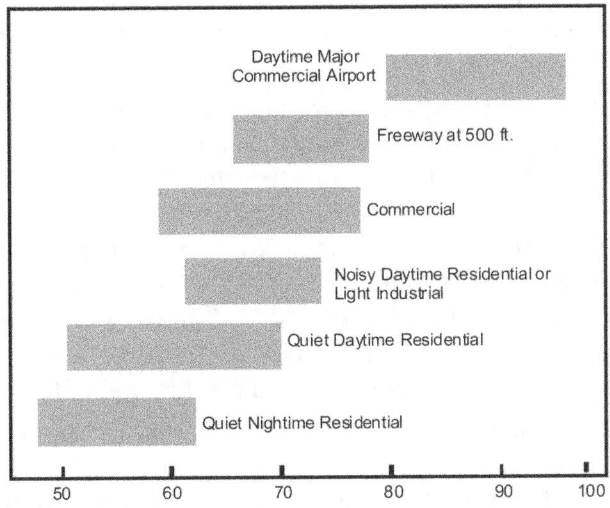

Fig. 7.2 Ambient noise levels at different land uses.

Noise annoys, distracts, disturbs and with sufficient exposure causes physiological effects leading to deafness. Annoyance results from interference with concentrated work, rest or sleep or with individual communication or speech. Noise in the work place reduces productivity, efficiency, accuracy and safety. The noise from industries is called industrial noise, from transport as transport noise or ambient noise present in the neighborhood due to natural and other causes.

As noise travels through air and under goes changes it is also considered as one form of air pollution and noise regulation laws were enacted as a part of US Clean Air Act. The 1990 Clean Air Act Amendments added a new title IV, relating to noise pollution. The U.S. Code designates the original title IV (noise pollution) as subchapter IV a code 7641.

The Noise Pollution Clearing house in US is a national non-profit organization with extensive online noise related resources. The Noise Pollution Clearing house seeks to: Raise awareness about noise pollution create, collect, and distribute information and resources regarding noise pollution, strengthen laws and governmental efforts to control noise pollution; Establish networks among environmental, professional, medical, governmental, and activist groups working on noise pollution issues and assist activists working against noise pollution.

The extent to which noise contributes to the deterioration of our environment could not be easily determined like that of pollution from other sources. It has been possible to assess impurities in the air or in water quantitatively but all persons are not affected to the same degree by the same noise. There occurs a vast variation in individual sensitivity to noise and people are affected differently when they are at home or outside or at work.

Not all sounds at the same decibel levels are perceived to be equally loud as the human ear is less sensitive to some frequencies as others. Also people perceive noise to be more intrusive the longer it persists, the more often it is heard, and the time of day it is heard. A number of descriptors for noise have been developed to account for these factors, (US Department of Transportation, 1978):

1. Equivalent Sound Level (Leq) - the constant sound level which, in a given situation and time period, conveys the same sound energy as does the actual time varying sound in the same period. The equivalent sound level is the same as the average sound level.

2. Perceived Noise Levels (PNL) - a sound level in decibels by adjusting 1/3 octave band measured levels to correspond to a subjective impression of noisiness.

3. Effective Perceived Noise Level (EPNdB) - a measure to estimate noisiness of a particular sound. It is derived from instantaneous perceived noise level values by applying correction factors for pure tones and the duration of the noise.

4. Sound Exposure Levels (SEL) - the level (dB) of sound experienced at a given time period.

5. Single Event Noise Exposure Levels (SENEL) - the level (dB) of sound experienced during a single event (for example, passage of an aircraft).

6. A–Weighted Sound Levels the measurement of sound approximating the auditory sensitivity of the human ear (that is, efficient at medium and speech range frequencies). It is measured by an electric weighting network that is progressively less sensitive to sounds below 1000 hertz and is the human.

7.1 Basic Information on Noise

"Sound" is mechanical energy from a vibrating surface and is transmitted by a cycling series of compressions and rarefaction of the molecules of the materials through which it passes sound can be transmitted through gases, liquids, and solids. A vibrating source, which produces sound has a "total power output," and the sound results in a sound pressure wave that alternatively rises to a maximum level (compression) and drops to a minimum level (rarefaction). Noise level is related to total power output. The number of compressions and rarefaction's of the air molecules in a unit of time is referred to as its "frequency." Frequency is expressed in hertz (Hz), which is the same as the number of cycles per second. Humans can detect sounds with frequencies ranging from about 16 to 20,000 Hz (2).

Sound power (total power output or sound pressure) does not provide practical units for sound or noise measurement for two basic reasons (2). First, a tremendous range of sound power (or sound pressure) can be produced. Expressed in microbars (mbar, one-millionth of 1 atm pressure), the range is from 0.0002 to 10,000 mbar for peak noises within 100 ft of large jet and rocket-propulsion devices. Second, the human ear does not respond linearly to increases in sound pressure. The human response is essentially logarithmic. Therefore noise measurements expressed by the term "sound-pressure level" (SPL), which is the logarithmic ratio of the sound pressure to a reference pressure and is expressed as a dimensionless unit of power, the decibel (dB). The reference level is 0.0002 mbar, the threshold of human hearing. The equation for sound-pressure level is as follows:

$$SPL = 20 \log_{10}(P/P_0)$$

where   SPL = sound pressure level, dB

         P = sound pressure, mbar

         P_0 = reference pressure, 0.0002 mbar

Table 7.1 contains a summary of various sound pressure and corresponding A-weighted decibel levels, with examples of recognized noise sources cited. Fig. 7.3 lists some common and easily recognized sounds along with a subjective evaluation scale.

Table 7.1 Spl, sound pressure, and recognized sources of noise in our daily experiences.

Sound pressure, μbar	SPL, dBA	Example
0.0002	0	Threshold of hearing
0.00063	10	
0.002	20	Studio for sound pictures
0.0063	30	Studio for speech broadcasting
0.02	40	Very quiet room
0.063	50	Residence
0.2	60	Conventional speech
0.63	70	Street traffic at 100 ft
1.0	74	Passing automobile at 20 ft
2.0	80	Light trucks at 20 ft
6.3	90	Subway at 20 ft
20	100	Looms in textile mill
63	110	Loud motorcycle at 20 ft
200	120	Peak level from rock and roll band
2,000	140	Jet plane on the ground at 20 ft

Source : Chanlett (3)

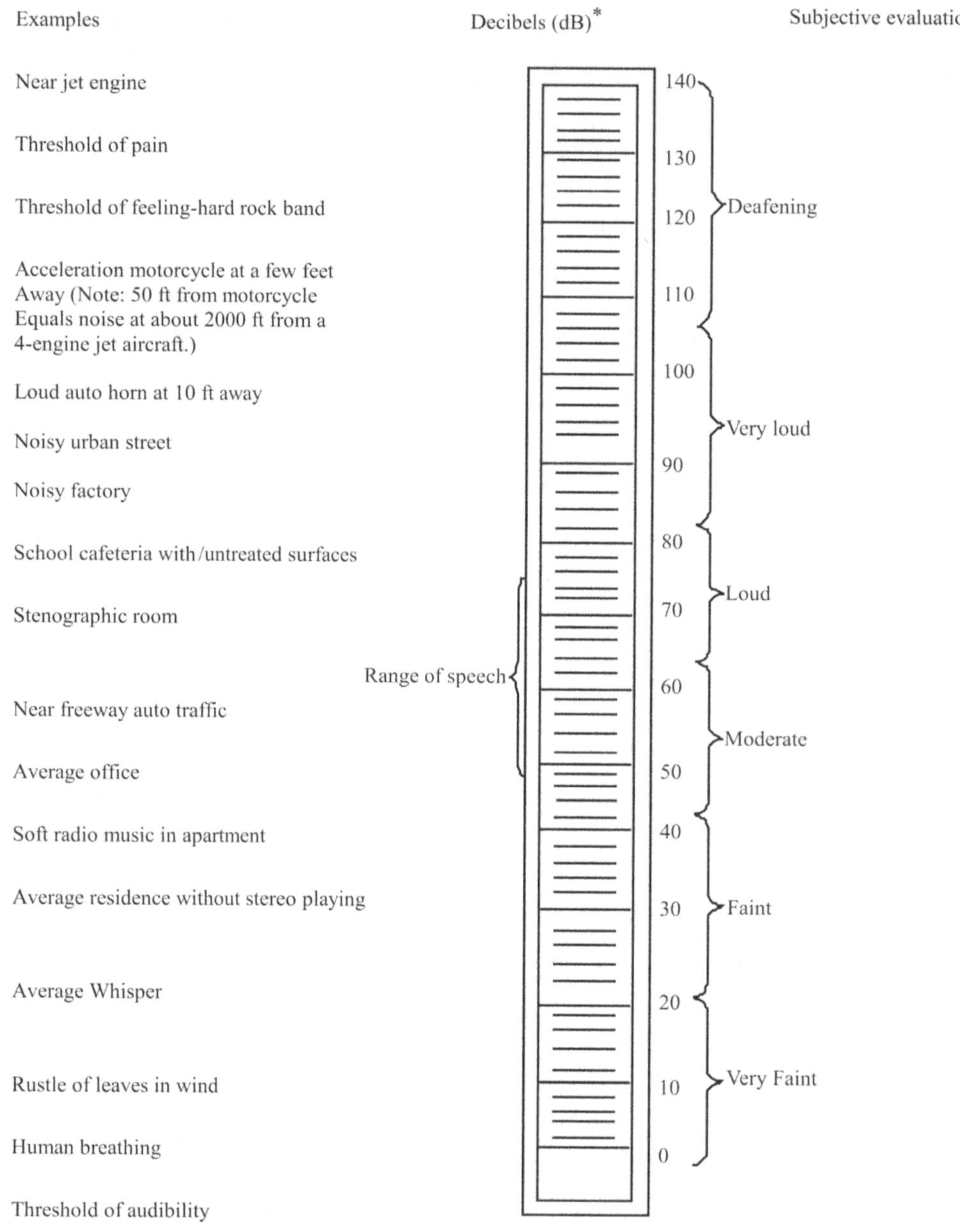

* dB are "average" values as measured on the A-scale of a sound-level meter

Fig. 7.3 Examples of common sounds in decibels.

In most noise considerations, the "a-weighted sound-level" scale is used. This scale is appropriate because the human ear does not respond uniformly to sounds of all frequencies, being less efficient in detecting sounds at low and high frequencies than at medium, or speech, frequencies (1). To obtain a single number representing a sound level containing a wide range of frequencies and yet representative of the human response, it is necessary to weigh the low and high frequencies with respect to average, or "a," frequencies. Thus, the resultant SPL is "a-weighted," and the units are a-weighted decibels (dBA). The a-weighted sound level is also called the "noise level." Sound-level meters have an a-weighting network, thus yielding a-weighting network, dB, or dBA, readings.

7.2 Noise Sources

7.2.1 Traffic Noise

Problems due to noise pollution are chiefly confined to urban areas. In the rural areas noise is not yet a health problem as the noise levels are normally low. Most of the urban area problems are due to movement of various types of vehicles. Every moving vehicle produces vibrations because of the construction of that vehicle, age of the vehicle and the way it is handled. Automobile, the result of the invention of the internal combustion engine, is prominent in producing noise.

The noise production from an automobile can be examined starting from the tires onwards. If the tread of the tire is broad, the area touching the ground will be more and there will be more noise production. If there is not enough tire pressure, then also there will be more noise due to higher friction. So one of the factors to be kept in mind is the tire pressure. Less pressure increases petrol consumption due to higher frictional drag. The noise from the tires also increases with the speed of the car [4] as depicted in the graph (Fig. 7.4).

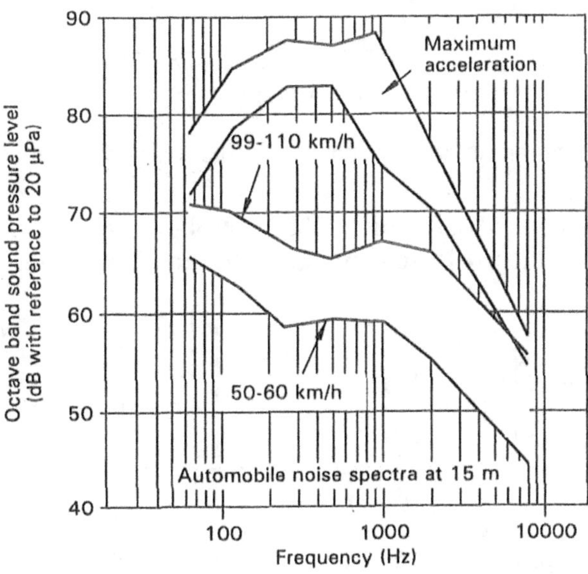

Fig. 7.4 Noise of tyres increasing with speed.

More prominent is the exhaust pipe noise. The compressed exhaust gases from the cylinders enter the exhaust manifold and enter the 'silencer' where the gases expand and noise level also drops down considerably. Depending on the design of the silencer, the gases and noise also have to take a devious route, thus reducing the noise. Much of the noise produced is curtailed at the silencer level.

The connecting cod from the gear box to the rear wheels, the transmission mechanism produces noise due to the coupling mechanisms which have to be free-moving for efficiency, but produce noise. By proper maintenance, servicing with oil and grease this noise can be reduced.

7.2.2 Highway Vehicle Noise

Exhaust noise is the predominant source for normal operation below about 55 km/h (Fig.7.5) for most automobiles while the tire noise becomes the dominant noise source at speeds above 80 km/h. The contribution of automobiles to the noise environment is significant because of the large number of vehicles in operation.

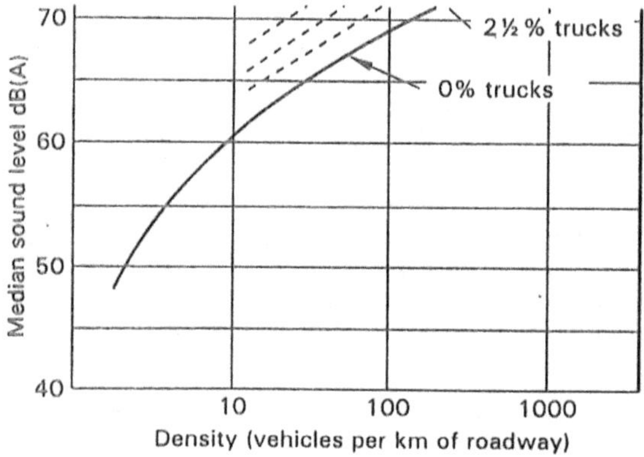

Fig. 7.5 Median noise level estimates of mixed traffic at 80 kmph.

Diesel trucks are 8 to 10 dB are noisier than petrol-powered ones. At speeds above 80 km/h, tire noise source of the truck predominates. The crossbar tread is the noisiest.

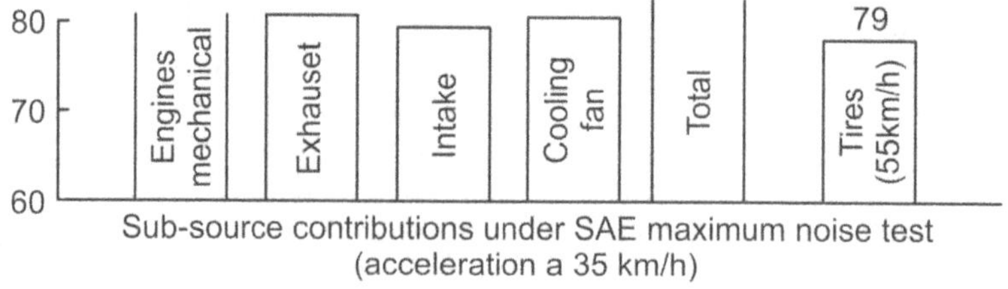

Fig. 7.6 Typical example of diesel truck noise

Table 7.2 Summary of noise characteristics of internal combustion engines

Source	A weighted noise energy (kWh/d)a	Typical A weighted noise level at 15.2m [dB(A)]	8 h exposure level [dB(A)]b Average Maximum		Typical exposure time (h)
Lawn mowers	63	74	74	82	1.5
Garden tractors	63	78	N/A	N/A	N/A
Chain saws	40	82	83	95	1
Snow blowers	40	84	61	75	1
Lawn mowers	16	78	67	75	½
Model aircraft	12	78	70c	79c	¼
Leaf blowers	3.2	76	67	75	¼
Generators	0.8	71	-	-	-
Tillers	0.4	70	72	80	1

a--Based an estimates of the total number of units in operation per day; b--Equivalent level for evaluation of relative hearing damage risk; c--During engine trimming operation.

Internal combustion engines due to their wide usage are important sources of noise (Table 7.2).

7.3 Noise Measurement

Noise measurement specifications require definition of the period of measurement, the noise parameter to be recorded, and the position of the recording instrument relative to the road and adjacent properties.

7.3.1 Measurement units

The indicator used to measure sound levels is a logarithmic function of acoustic pressure, expressed in decibels (dB). The audible range of acoustic pressures is expressed in dB (A). The human ear perceives a constant increase in sound level whenever the acoustic pressure is multiplied by a constant quantity. The scale of sound levels shows that calm environments correspond to a level of 30 to 50 dB (A), and that beyond 70 dB (A) sound becomes very disruptive Fig.7.6. Since noise is variable over time, measurements and forecasts are expressed as mean values or other indicators over a given period of time.

7.3.2 Measuring Instruments

Existing noise levels can be measured using devices called sonometers, which convert sound wave energy into an electrical signal, the magnitude of which is displayed or recorded.

Measurements obtained using these instruments can become valuable baseline data, but their further usefulness is somewhat limited, both in terms of sampling period and as a result of their inability to distinguish separate sources of noise.

Table 7.3 summarizes the effects of noise on humans in residential areas. Several factors other than the magnitude of exposure have been found to influence community reaction to noise. These factors include (5):

- Duration of intruding noises and frequency of occurrence.
- Time of year (windows open or closed).
- Time of day of noise exposure.
- Outdoor noise level is community when intruding noises are not present.
- History of prior exposure to the noise source.
- Attitude toward the noise source, and
- Presence of pure tones or impulses.

7.4 Effects of Noise on People

- Noise leads to emotional and behavioral stress. A person may feel disturbed in the presence of loud noise such as produced by beating of drums.

- Noise may permanently damage hearing. A sudden loud noise can cause severe damage to the eardrum.

- Noise increases the chances of occurrence of diseases such as headache, blood pressure, heart failure, increased heart beat, constriction of blood vessels and dilation of pupil, damage to liver, brain and heart.

- Frequent and long use of headphones at a loud volume for a prolonged period of time could affect one's hearing.

7.4.1   Normal Hearing

Frequency range and sensitivity

The ear of the healthy adult male responds to sound waves in the frequency range of 20 to 16,000 Hz. Young children and women often have the capacity to respond to frequencies up to 20,000 Hz. The speech zone lies in the frequency range of 500 to 2,000 Hz. The ear is most sensitive in the frequency range from 2,000 to 5,000 Hz.

Table 7.3 Effects of noise on people (Residential land use only).(6.)

			Speech Interference			
Effects[a]	Hearing Loss	Indoor	Outdoor	Annoyance[b]		
Day-night average sound level in decibels	Qualitative description	% sentence intelligibility	Distance in meters for 95% sentence intelligibility	% of population highly annoyed[c]	Average community reaction[d]	General community's attitude towards area
75 and above	May begin to occur	98	0.5	37	Very Severe	Noise is likely to be the most important of all adverse aspects of community environment
70	not likely to occur	99	0.9	25	Severe	Noise is one of the most important adverse aspects of the community environment.
65	Will not occur	100	1.5	15	Significant	Noise is one of the important adverse aspects of the community environment
60	Will not occur	100	2.0	9	Moderate to Slight	Noise may be considered an adverse aspect of the community environment
55 and below	Will not occur	100	3.5	4		Noise considered no more important than various other environmental factors.

Note : Research implicates noise as a factor producing stress-related health effects such as heart disease, high-blood pressure and stroke, ulcers and other digestive disorders. The relationships between noise and these effects, however, have not as yet been quantified.

[a] "Speech Interference" data are drawn from other U. S. Environmental Protection Agency Studies.

[b] Depends on attitudes and other factors.

[c] The percentages of people reporting annoyance to lesser extents are higher in each case. An unknown small percentage of people will report being "highly annoyed" even in the quietest surroundings. One reason is the difficulty all people have in integrating annoyance over a very long time.

[d] Attitudes or other non-acoustic factors can modify this. Noise at low levels can still be an important problem, particularly when it intrudes into a quiet environment.

Source : Federal Interagency Committee on Urban Noise, (5).

The smallest perception of sound pressure in this frequency range is 20 mPa. A sound pressure of 20 mPa at 1,000 Hz in air corresponds to a 1.0 nm displacement of the air molecules. The thermal motion of the air molecules corresponds to a sound pressure of about 1mPa. If the ear were much more sensitive, you would hear the air molecules crashing against your ear like waves on the beach.

7.4.2   Repeated Interference with Sleep

Noise can awaken people from sleep and it can keep them awake, frequent awakening or awakening for long periods can be very disruptive. Even if not awakened by noise, a person's sleep pattern can be significantly disturbed, and a reduced feeling of well-being can result next day. Frequent and prolonged sleep disturbances can result in physical, mental or emotional illness, Fig. 7.6(a) and (b).

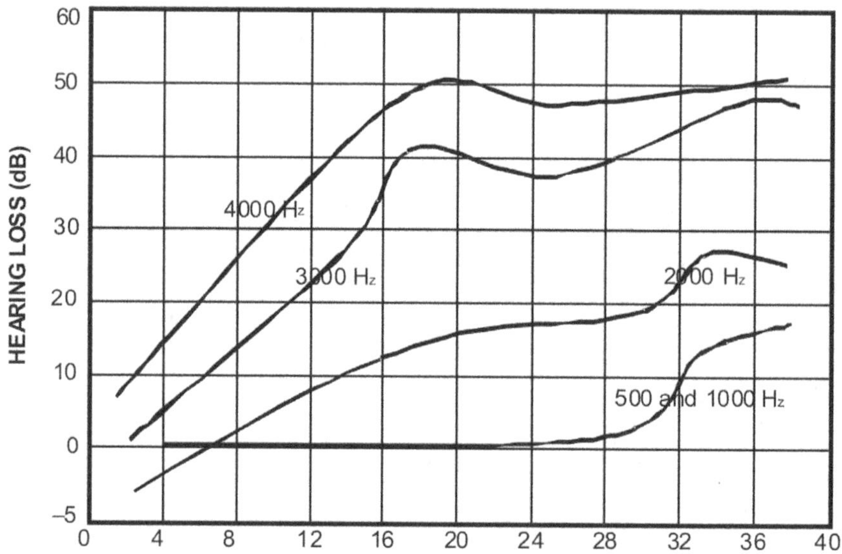

Fig. 7.6(a) Hearing impairment with years of exposure.

7.4.3 Effects on Communication

External sounds are able to interfere with conversation's and use of the telephone as well as the enjoyment of radios and television programs Fig.7.6 (c). It can thus affect the efficiency of the offices, schools and other places, where communication is of vital importance. The maximum acceptable level of noise under such conditions has been 55 dB. 70 dB is considered very noisy and serious interference with verbal communication is inevitable. Fig. 7.6 (d) presents how the quality of speech communication will be affected by different background sounds and speaker to listener distance.

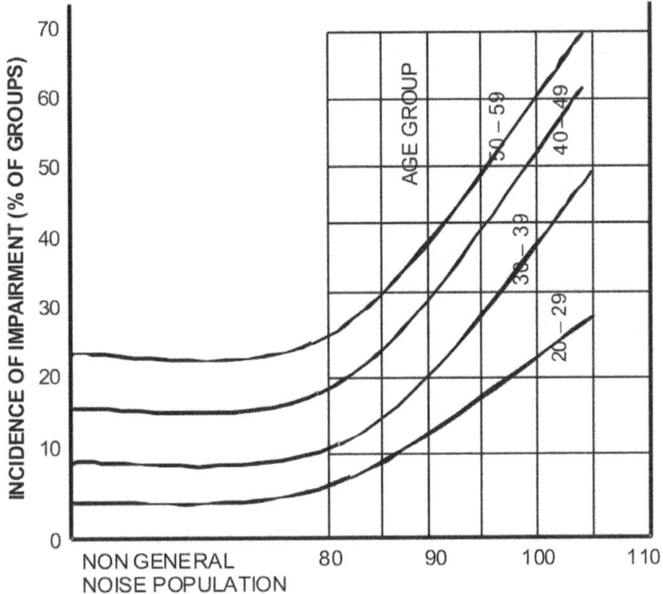

Fig. 7.6(b) Incidence of hearing impairment (average hearing threshold level in excess of 15db ASA at 500, 10000 and 2,000 cps).

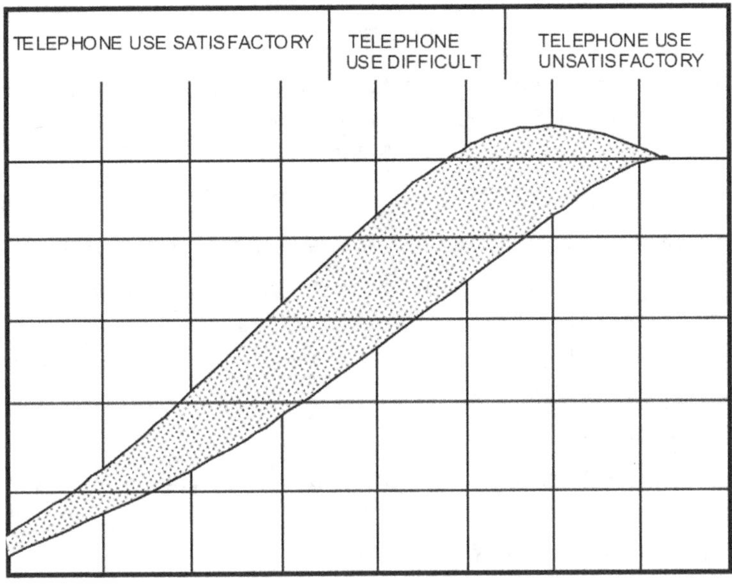

Fig. 7.6(c) Rating chart for office noises.

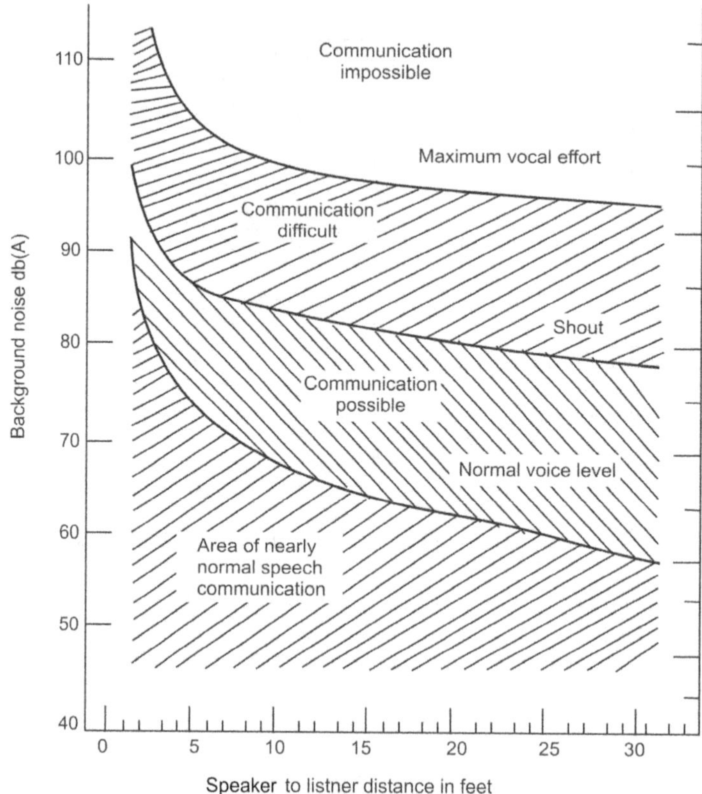

Fig. 7.6(d) Quality of speech communication with distance (in relation to the a-weighted sound level of noise, db (A) and the distance between the speaker and the listener).

7.5 Physiological Responses

Physiological responses accompanying response and other noise exposures include:

1. A vascular response characteristic by peripheral vaso-constriction, changes in heart beat rate and blood pressure.

2. Various glandular changes such as increased output of adrenaline evidenced by chemical changes in blood.

3. Slow, deep breathing.

4. A change in the electrical resistance of skin with changes in activity of the sweat glands.

5. Brief changes in skeleton muscle tension.

Constant noise may cause blood vessels to contract, skin to become pale, muscles to contract and adrenaline to be shot into blood stream. The adrenaline is responsible for both excitery and inhibitory responses in living beings. This is the reason why factory workers develop abnormal heartbeat rates and suffer from insomnia, nervousness and impaired motor co-ordination. Considering these points the U.S. Government has kept 90 decibels as a health hazard limit for an 8 hrs a day working environment. It has been proved that high noise is harmful particularly to those suffering from hyper tension and diabetics. Noise also produces startling effects on babies and they may even develop a fear psychosis as a result of sharp and sudden noises.

7.5.1 Epidemiological Studies

High levels of occupational noise remain a problem across the world and there is evidence of its increasing prevalence in the work place (Muigui K, 2013) (7). In the United States of America (USA), for example, more than 30 million workers are exposed to hazardous noise exceeding 85 dB (A). In Germany, 4-5 million people (12- 15% of the workforce) are exposed to noise levels defined as hazardous by WHO, (2004) (8). Data for developing countries is scarce, but available evidence suggests that average noise levels are well above the recommended standards in most developing nations (Melamed, S., Yitzhak F., Froom P. (2001) (9).

Several researchers have conducted field studies testing industrial workers and or collating their health records in an attempt to overcome the limitations of duration and realism in laboratory studies. The physiological problems amongst two groups of workers were classified as from "very noisy industries" to "less noisy industries" respectively. The difference between the two groups indicates a higher incidence of problems amongst the high- noise group than in the low- noise group. Noise level, however, is not isolated in such a study as the cause or time factor. The two groups are based on two separate industries. The high noise group came from the light industries, such as, textile industries. Between these industries there were numerous other differences which could have effects on health, such as, heat, physical workload, anxiety, and the type of people.

7.6 Systematic Methodology for Assessing Environmental Impacts of Noise

7.6.1 EIA Methodology for Noise Impacts

Sound and noise may be emitted to the environment from stationary sources (industry, equipment), road traffic and railways, aircraft operations, and blasting. The emissions can result in changes in ambient sound and vibration levels as well as noise levels which may adversely affect health and well being of people living and working in the vicinity of the noise source.

To provide a basis for addressing noise-environment impacts, a seven-step or six-activity model is suggested for the planning and conducting impact studies.

The seven main steps associated with noise environment impacts are:

- Identification of levels of noise emissions and impact concerns related to the construction and operation of the development project.
- Description of the environmental setting in terms of existing noise levels and noise sources, along with land-use information and unique receptors in the project area.
- Procurement of relevant laws, regulations or criteria related to noise levels, land-use compatibility, and noise emission standards.
- Conducting impact prediction activities, including the use of simple noise-attenuation

models, simple noise-source-specific models, comprehensive mathematical models, and/or qualitative-prediction techniques based on the examination of case studies and the exercise of professional judgement.

- use of pertinent information from step 3, along with professional judgement and public input, to assess the significance of anticipated beneficial and detrimental impacts.

- identification, development, and incorporation of appropriate mitigation measures for the adverse impacts, and

- Preparation of final environmental impact statement. Fig. 7.7 delineates the relationship between the seven steps or activities in the suggested conceptual approach.

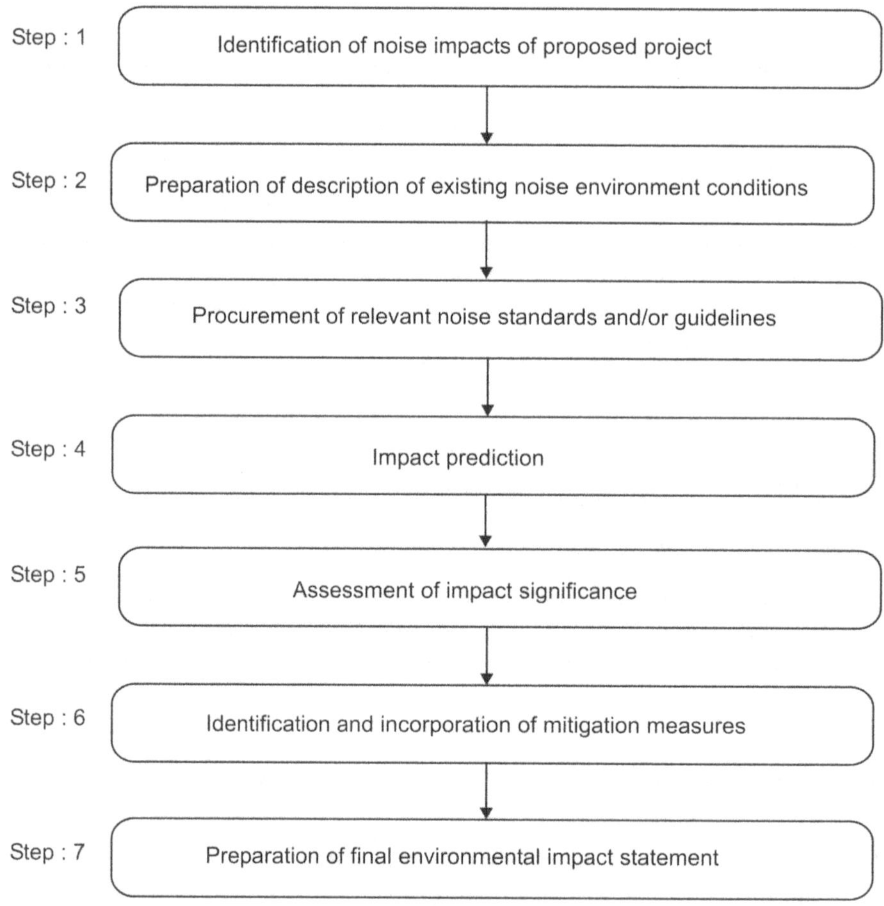

Fig. 7.7 Conceptual approach for study focused on noise-environment impacts.

7.6.2 Step 1 : Identification of Noise Impacts

The first step in the methodology is to determine the potential impacts of the proposed project (or activity) on the noise environment. This requires the identification of the noise levels associated with the project. A considerable body of information exists on noise levels associated with a variety of projects and related activities.

The sources of noise are numerous, so they may be broadly classified into two classes, namely, industrial and non-industrial. The industrial category may include noises from various industries like transportation, vehicular movements, rockets, defense equipment and explosions. Among the non-industrial category, the notable sources of noise are loudspeakers, traffic, aircrafts, trains, construction works, radios, TVs, vacuum cleaners, mixers, power lawn movers and air conditioners in the domestic or commercial use.

Noise from Industry

No environment factor has caused so much confusion regarding its effect on workers' efficiency and health as industrial noise has. Noise in industry originates from processes causing impact, vibration or reciprocation movements, friction and turbulence in air or gas streams.

Construction activities generally generate noise levels in excess of those typically found in the project environs. Construction sites can be categorized into four major types: domestic housing including residences for from one to several families, nonresidential buildings including offices, public buildings, hotels, hospitals, and schools, industrial buildings including religious and recreational centers, stores, and repair facilities, and public works including roads, streets, water mains, and sewers (10). Noise from construction of major civil works, such as dams, generally affects relatively few people other than those employed at or near construction sites, so these sites are not included in these categories.

Noise from project operations includes sound emissions from highway vehicles, aircraft, recreation vehicles, internal-combustion engines, and industrial machinery. Noise produced by highway vehicles can be attributed to three major generating systems: rolling stock, such as, tires and gearing, propulsion systems related to engine and other accessories, aerodynamic and body systems.

7.6.3 Step 2 : Description of Existing Noise-Environment Conditions

In analyzing the potential noise impacts of a proposed project (or activity), it is necessary to consider the study as area (potential project area or region of influence) associated with the noise emissions. The delineation of a study area can be made based upon the boundaries of the land associated with the project, or the delineation can include a larger area by considering the area of noise influence within the vicinity of the proposed project.

The primary information which should be accumulated in step 2 is data on existing noise levels and noise sources within the study area. Land-use and human-population-distribution maps in relation to the proposed project would also be needed.

If no specific data on existing noise levels is available for the study area, it might be possible to use published noise-level information developed for project involving similar land use. For example, Table 7.2 shows typical day-night noise levels in urban areas in the United States. Quiet suburban residential areas have an average L_{dn} of 50 dBA, while very noisy urban residential areas exhibit L values of 70 dBA. Typical noise levels in rural settings are 30 to 35 dBA, and in wilderness locations they are in the order of 20 dBA. Seasonal and daily variations in noise levels may occur, particularly at national and state parks and recreational areas (11).

Sampling methods: For highways and other roadways project the most commonly used method of sampling is the grid method. In the grid method, a grid is superimposed over a study area map and the measurement points are located at the nodes of the square or at the nearest location when the nodes are inaccessible. The selection points of the grid is important for accuracy of the results.

The A-weighted equivalent sound level (L_{Aeq}) is normally used to analyse and interpret the results at different time-intervals of the day. The L_{Aeq} registered in the diurnal period (from 07:00 to 19.00) and evening period (from 19.00 to 23.00) is normally compared and correlated with other relevant information of traffic flow, types of vehicles, meteorological conditions, urban variables, *etc.*

7.6.4   Step 3 : Procurement of Relevant Noise Standards and/or Guidelines

The primary sources of information on noise standards, criteria, and policies will be the relevant local, state, and federal agencies, which have a mandate for overseeing the noise environment of the study area. Additional information may be available from international agencies such as the World Health Organization (WHO) or the United Nations' Environment Program. This information can be used to determine the baseline quality and the significance of noise impacts incurred during projects (or activities); it could also aid in deciding between alternative actions or in assessing the need for mitigation measures for a given alternative.

General Noise Criteria

Table 7.4 summarizes noise criteria developed by the EPA for the protection of public health and welfare with an adequate margin of safety. The phrase "public health and welfare" is defined as complete physical, mental, and social well-being, and not merely the absence of disease and infirmity. Table.7.4 is useful for noise-impact assessment in the absence of specific noise standards for a given area. The two key terms in Table.7.4 are L_n (or DNL) and L.

Table 7.4 Yearly Average[a] equivalent sound levels identified as requisite to protect public health and welfare with an adequate margin of safety.

Indoor	To perfect	Outdoor		To protectagainst		against both effects[c]	
Land use	Measure	Activity interference	against both effects[c]	Activity Interference	Hearing loss consideration[b]		
Residential with outside space and farm residences	L_{dn}	45		45	55		55
	$L_{eq(24)}$		70			70	
Residential with no outside space	L_{dn}	45		45			
	$L_{eq(24)}$		70				
Commercial	$L_{eq(24)}$	d	70	70[f]	d	70	70[f]
Inside transportation	$L_{eq(24)}$	d	70	d			
Industrial	$L_{eq(24)}$	d	70	70[f]	d	70	70[f]
Hospitals	L_{dn}	45		45	55		55
	$L_{eq(24)}$		70			70	
Educational	$L_{eq(24)}$	45		45	55		55
	$L_{eq(24)}$		70			70	
Recreational areas	$L_{eq(24)}$	d	70	70[f]	d	70	70[f]
Farm land and general unpopulated land	$L_{eq(24)}$		70	d		70	70[f]

[a] Refers to energy rather than arithmetic averages.
[b] The exposure period that results in hearing loss at the identified levels in 40 yr.
[c] Based on lowest level
[d] Since different types of activities appear to be associated with different levels, identification of a maximum level for activity interference may be difficult except in those circumstances where speech communication is a critical activity.
[f] An $L_{eq(a)}$ of 75 dB may be identified in these situations so long as the exposure over the remaining 16 hr/day is low enough to result in a negligible contribution to the 24-hr average.
Source: U. S. Environmental Protection Agency, (8)

Noise Emissions Standards

Standards for noise emissions from various sources have been established by the EPA, Table 7.5.

U.S EPA has given guidelines for classification and land-use-compatibility guidelines. Table 7.6.

In Table.7.6 noise zones are identified in order of increasing noise levels by the letters. A through D. The day-night average sound level (DNL) descriptor L_{dn} can be used for all noise sources.

Noise emission standards have only an indirect control over the noise radiated by a machine. They state maximum permissible sound levels in work places, acceptable levels for day- time and night- time in residential, commercial and industrial areas and maximum permissible noise crossing industrial and construction site boundaries. The measured noise in such cases may be produced by a single machine or by a combination of many kinds of machinery.

The maximum permissible sound level at a worker's ears and the time of exposure are not related directly to the noise produced by any one machine but depend upon the total noise in the area where the workers are located with respect to the machine and other factors. For this reason, noise emission standards or their intent must be confined to product noise emission regulations.

The maximum permissible sound levels crossing industrial site into residential and commercial zones may be stated in terms of overall A-weighted sound pressure levels. It has been reported that high intensities, high frequencies and intermittent nature of noise are the factors of annoyance for the workers. Such a situation not only causes physical and physiological damages but also impair workers' efficiency resulting in low production and ultimate dissatisfaction. Community response to industrial noise is seen in setting up acceptable limits for community areas but it is difficult to establish this precisely because of the variety and complexity of the different factors involved.

Basic noise levels for industrial zone should not exceed 55 dB at night and 65 dB during the day time. Noise results in cardiovascular problems like heart diseases and high blood pressure. Workers exposed to high noise levels have acute circulatory problems, cardiac disturbances, neuro-sensory and motor impairment and even social conflicts at home and work.

However, it is not a regulatory goal. It is a level defined by a negotiated scientific consensus without concern for economic and technological feasibility or the needs and desires of any particular community. The Federal Highway Administration (FHWA) noise policy used this descriptor as an alternative to L_{10} (noise level exceeded ten percent of the time) in connection with its policy for highway noise mitigation. The L (design hour) is equivalent to DNL for planning purposes under the following conditions: 1. heavy trucks equal ten percent to total traffic flow in vehicles per 24 hours. 2. traffic between 10 p.m. and 7 a.m. does not exceed fifteen percent of the average daily traffic flow in vehicles per 24 hours. Under these condition, DNL equals L_{eq} − 3 decibels.

Table 7.5 Osha noise exposure limits for the work environment (noise exposures in dBA).

Noise	Permissible exposure Noise (hours and minutes)
85	16 hrs
87	12 hrs 6 min
90	8 hrs
93	5 hrs 18 min
96	3 hrs 30 min
99	2 hrs 18 min
102	1 hr 30 min
105	1 hr
108	40 min
111	26 min
114	17 min
115	15 min
118	10 min
121	6.6 min
124	4 min
127	3 min
130	1 min

Note: Exposures above or below the 90 dB limit have been "time weighted" to give what OSHA believes are equivalent risks to a 90 dB eight-hour exposure.

Source : Marsh (12)

Table 7.6 Noise-Zone Classification

Noise zone	Noise exposure class	Noise descriptor			HUD noise standards
		DNL[a] day-night average sound level	L_{eq} (hour)[a] equivalent sound level	Nef[d] noise exposure forecast	
A	Minimal Exposure	Not Exceeding 55	Not Exceeding 55	Not Exceeding 20	"Acceptable"
B	Moderate Exposure	Above 55[b] But Not Exceeding 65	Above 55 But Not Exceeding 65	Above 25 But Not Exceeding 30	
C-1	Significant Exposure	Above 65 Not Exceeding 70	Above 65 Not Exceeding 70	Above 30 Not Exceeding 35	"Normally unacceptable"
C-2		Above 70 Not Exceeding 75	Above 70 Not Exceeding 75	Above 35 Not Exceeding 40	
D-1	Severe Exposure	Above 75 Not Exceeding 80	Above 75 Not Exceeding 80	75 Not Exceeding 45	Above 40
D-2		Above 80 Not Exceeding 85	Above 80 Not Exceeding 85	Above 45 Not Exceeding 50	"Unacceptable"
D-3		Above 85	Above 85	Above 50	

[a]CNEL – Community Noise Equivalent level (California only) uses the same values.

BHUD, DOT and EPA recognize L_{dn} – 55 dB as a goal for outdoors in residential areas in protecting the public health and welfare with an adequate margin of safety.

7.6.5   Step 4 : Impact Prediction

Step 4 involves predicting the propagation of noise from a source and determining the type of affected land- use. Several approaches for predicting noise contours are outlined in the discussion of this step.

One method of expressing both existing noise and predicted noise levels is by using a level-weighted population value (13). A sound-level-weighted population is a single-number representation of the significance of a noise environment to the exposed population. The assumptions are that the intensity of human response is one of several consequences of average sound level, depending upon the response mode of interest (annoyance, speech interference and hearing loss) and that the impact of high noise levels on a small number of people is equivalent to the impact of lower noise levels on a larger number of people in an overall evaluation. Based on these assumptions, the "fractional impact" can be determined as the product of a sound-level-weighting value and the number of persons exposed to a specified sound level. Summing the fractional impacts over the entire population provides the sound-level-weighted population (LWP). The calculation is as follows (13):

$$LWP = P(L_{dn}).W(L_{dn})d(L_{dn})$$

where $P(L_{dn})$ is the population distribution function, $W(L_{dn})$ is the day-night average sound-level-weighting function characterizing the severity of the impact as a function of sound level (its derivation is described below), and $d(L_{dn})$ is the differential change in day-night average sound level. Sufficient accuracy can be obtained by taking average values of the weighting function between equal decibel increments – say, up to 5 dB – and replacing the integrals by summations of successive increments in average sound level (13).

The weighting function $W(L_{dn})$ is based on the reaction of populations to living in noise impacted environments and other social survey data relating the fraction of sampled population expressing a high degree of annoyance to various L_{dn} values. The weighting function is normalized to unity at 75 dB; value of $W(L_{dn})$ is listed in Table.7.7(a).

A Noise Impact Index (NII) can then be used for comparing the relative impact of one noise environment with that of another. It is defined as the sound-level-weighted population LWP divided by the total population P_{total} under consideration:

$$NII = LWP/P_{total}$$

An example calculation for this index is in shown Table 7.7(b).

Forecasting Noise Levels

Forecasting methods include equations, computer models, and physical models. The simplest are equations, which estimate noise from information on traffic flow, composition, and speed. Computer models are perhaps more widely employed and can be used to forecast future changes in baseline condition and the likely actually autonomous and difficult to integrate into the surrounding social environment.

Table 7.7(a) Sound-level-weighting function for Overall Impact Analysis

Ldn, in dB	$W(L_{dn})$	$W(L_{dn}) + W(L_{dn} + 5) / 2$
35	0.006	0.010
40	0.013	0.021
45	0.029	0.045
50	0.061	0.093
55	0.124	0.180
60	0.235	0.324
65	0.412	0.538
70	0.664	0.832
75	1.000	1.214
80	1.428	1.697
85	1.966	2.307
90	2.647	

Note: The right-hand column is included for conventional for finding the weighting of certain 5 dB increments.

Source: Von Gierke H. E (14)

Table 7.7(b) Example of level-weighted population LWP and noise-impact-index. computation.

Ldn in dB	Cumulative population[a]	Incremental Population[a]	Weighting Function[b]	Level-weighted population[a]
80	0.1	0.1	1.697	0.17
75	1.3	1.2	1.214	1.46
70	6.9	5.6	0.832	4.66
65	24.3	17.4	0.538	9.36
60	59.6	35.3	0.324	11.44
55	97.5	37.9	0.180	6.82
	Total: 97.5	NII = 33.91/97.5 = 0.35		Total: 33.91

Prediction of Traffic Noise Levels

The methodology developed under National Cooperative Highway research program of USA (15) is a widely adaptable methodology for the prediction of highway noise as it is very simple with a high success rate. This methodology involves a four step prediction procedure.

The first step is known as a short method in which a gross quick and gross prediction of the noise levels will be made, the potential problem areas will be identified and no problem areas in terms of noise levels will be eliminated to make the assessment process simple. Because of the complexity involved in the assessment of true highway noise levels in the short method, the prediction will be always on the higher side.

In the short method, prediction can be carried out using two nomographs (a nomograph provides the solution to an equation or series of equations containing three or more variables) of traffic and road way parametric data (16)ERL 1984). As the short method is based on a number of assumptions and approximations, it should be further refined for arriving at final conclusions.

The flow diagram presenting the methodology of short methods is given in Fig.7.8. In this method the roadway will be approximated by one infinite element with constant traffic parameters and roadway characteristics.

The first step in this method involves defining an infinite straight-line approximation to the real highway configuration omitting on-ramps and interchange ramps. After selecting approximate roadway, the following parameters must be estimated or computed (a) the traffic parameters which include the speed and volume of each class of vehicles (b) the propagation characteristics which describe the location of the receiver relative to the roadway and (c) the roadway shielding parameters which describe the shielding provided by the roadway, if any.

Then the traffic and propagation parameters have to be combined in the L_{10} nomograph to determine for each type of source, the observer unshielded L_{10} level at the observer Fig. 7.8. The final result should then be compared to the criteria level, L_c, at the observer to define a "no problem" or "potential problem" condition. After identification of potential problem area, the observer location in question should be evaluated using the complete method.

The second step of methodology is termed as NCHRP 174 "complete method" which utilizes-a fairly large computer program to refine the predictions made in the first step. The third step in this procedure is termed as "selection of a noise control design". The fourth step is termed as "check the design operation" in which the second step will be repeated and refined to arrive at the final conclusions.

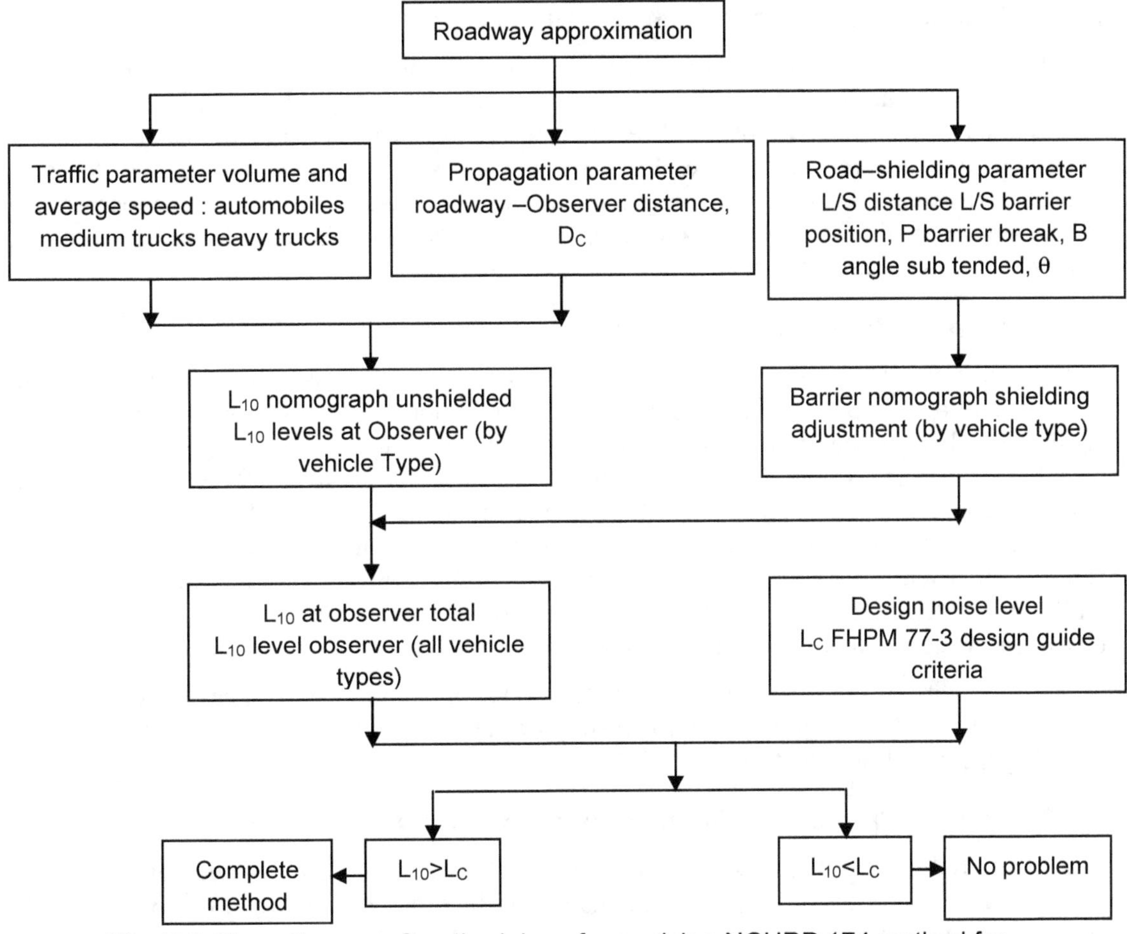

Fig. 7.8 Flow diagram of methodology for applying NCHRP 174 method for estimation from traffic
(**Source :** 15).

Expert systems technology such as artificial neural networks (ANNs) has been thought of as a viable alternative method to model noise levels.

Details of Various Computer Models

STAMINA: STAMINA was developed for the Thonburi Road Extension Project in Thailand to forecast traffic conditions

NOISECALC: The New York State Department of Public Service's NOISECALC computer program was used to predict noise impacts associated with the Mangalore Thermal Power Station project.

JICA Noise Model: The Masinloc Coal Fired Thermal Power project analysis used a noise model developed by JICA based on a report undertaken for the Coal-Fired Thermal Electric Development Project in Luzon Island in1990. The model considered the attenuation by distance, the effects of the barrier, and adsorption in the air. All noise sources are assumed to be point noise sources with no directivity

7.6.6　Step 5 : Assessment of Impact Significance

One basis for evaluation of significant impact is public input. This input could be received through a continued scoping process of the conduction of public meetings or public participation programs or both. The general public can often delineate important environmental resources and values for particular areas, and this should be considered in impact assessment. Professional judgment can also be useful to assess the percentage changes from baseline conditions in terms of noise levels and/or exposed human population, or a noise index as discussed in step.4.

7.6.7　Step 6 : Mitigation Measures

Mitigation measures refer to steps that can be taken to minimize the magnitude of the detrimental noise impacts. The key approach to mitigation is to reduce or control the noise expected to be emitted from the project (or activity). Mitigation can proceed along three possible courses of action, either by changing 1. the source of noise, 2. the path of noise from the source to the receiver, or 3. the receiver of noise. Some additional principles of noise control include the reduction of the number of vibrating sources, enclosure of the source, and attenuation of noise by absorbing barriers methods. Simple hand calculations and/or computer models described earlier can assist in forecasting the relative effectiveness of various designed and/or operational phase mitigation techniques.

Further, various designs can be used to reduce the noise from specific sources, for example, the mechanical noise from the gearbox of large wind turbines can be minimized by adapting specific design features (17-22).

7.6.8　Step 7 : Prediction of Final Impact Statement

The final environment impact statement should include summary tables and conclusions of discussion with industries. The results of analysis will help in decision making. Maps should show location of surface receptors and measurement rates and size balance.

Table 7.8　Transmission loss values for common barrier materials.

Material	Thickness, (inches)	Transmission loss, dBAa
Woods	½	17
Fire	1	20
	2	24
Pine	½	16
	1	19
	2	23
Redwood	½	16
	1	19
	2	23

Table 7.8 *Contd...*

Material	Thickness, (inches)	Transmission loss, dBAa
Cedar	½	15
	1	18
	2	22
Plywood	½	20
	2	23
Particle Board	½	20
Metals		
Aluminum	1/16	23
	1/8	25
	¼	27
Steel	24 ga	18
	20 ga	22
	16 ga	15
Lead	1/16	28
Concrete, Masonary, etc.		
Light Concrete	4	38
	6	39
Dense Concrete	4	40
Concrete Block	4	32
Cinder Block (Hollow Core)	6	28
Brick	4	33
Granite	4	40
Composites		
Aluminum Faced Plywood	¾	21 – 23
Aluminum Faced Particle Board	¾	21 – 23
Plastic Lamination Plywood	¾	21 – 23
Plastic Lamina on Particle Board	¾	21 – 23
Miscellaneous		
Glass (Safety Glass)	1/8	22
	1/4	22

Table 7.8 Contd…

Material	Thickness, (inches)	Transmission loss, dBAa
Plexiglass (Shatterproof)	-	22 – 25
Masonite	1/2	20
Fiberglass/Resin	1/8	20
Stucco on Metal Lath	1	32
Polyester with Aggregate Surface	3	20 – 30

A Weighted TL based on generalized truck spectrum.

Source: (3)

Summary

Noise as a pollutant produces contamination in environment becoming a nuisance and a cause for annoyance, and affects the health of a person, his activities and mental abilities. The extent to which noise contributes to the deterioration of our environment could not be easily determined like that of pollution from other sources. It has been possible to assess impurities in the air or in water quantitatively but all persons are not affected to the same degree by the same noise. There occurs a vast variation in individual sensitivity to noise and people are affected differently when they are at home or outside or at work. The intensity of noise can have a direct effect on biological and human communities. Constant noise may cause blood vessels to contract, skin to become pale, muscles to contract and adrenaline to increase into blood stream. The adrenaline is responsible for both excitery and inhibitory responses in living beings. To provide a basis for addressing noise-environment impacts, a seven-step or six-activity model is suggested for the planning and conducting impact studies. The seven main steps associated with noise environment impacts assessment are 1. identification of levels of noise emissions and impact concerns related to the construction and operation of the development project. 2. description of the environmental setting in terms of existing noise levels and sources, along with land-use information and unique receptors in the project area. 3. procurement of relevant laws, regulations or criteria related to noise levels, land-use compatibility, and noise emission standards. 4. conducting impact prediction activities, including the use of simple noise-attenuation models, simple noise-source-specific models, comprehensive mathematical models, and/or qualitative-prediction techniques based on the examination of case studies and the exercise of professional judgement. 5. use of pertinent information from step 3, along with professional judgement and public input, to assess the significance of anticipated beneficial and detrimental impacts. 6. identification, development, and incorporation of appropriate mitigation measures for the adverse impacts, and 7. Preparation of final environmental impact statement which are discussed with examples in this program.

Case Studies

Case study 1 Noise pollution in Mining industry (CPCB report 2017)

Limestone mining in Rajasthan, in India is common since it has large deposits of cement grade limestone as well as splittable limestone. The Tilakhera limestone lease area is located at a distance of about 2 Kilometers from Gambhiri Road Railway Station in the eastern direction near Mangrol Village. This mining area is situated in the Nimbahera Tehsil of Chittorgarh District of Rajasthan. Noise level data was collected from the mining operations and the limestone crusher was be the main sources of noise pollution.

The major sources of noise pollution include drilling; blasting; operation of HEMM; crusher and workshop and belt conveyor. Data was collected in these regions and results are presented below:

Equipment	Measurement Location	Noise Level dB (A)
Drill Operating	Operator's position	92-93
Dumper	Operating 10 m away	86-100
Dozer & Dumper both,	Operating 05 m away	90-111
Diesel Excavator	Near location area	83-87
Vehicular Movement	Road side 10 m away	60-66

The Proposed Environmental Mitigation Measures include:

1. Providing of ear plugs and ear muffs to reduce noise level exposure.

2. Use of noise abatement padding's in fixed plant installations.

3. Use of closed and advanced blasting technology.

4. Providing silencers or enclosures for noise generating machines such as compressors etc.

5. Creating a green belt around potential noise prone area.

Case Study 2 Analysis of ACC Cement Plant (CPCB Report 2016)

Dust emission sources are kiln, crusher, grinders, clinker, cookers and material handling equipments are common in cement plants. Crusher department is one of the major sources of environmental pollution. In this study aspects and associated impacts are first identified for analysis. Fugitive dust emission, stack emission and noise have been identified during the EIA study as significant aspects during activities like receipt of limestone, primary crushing, screening. These significant aspects impact on the human health like respiratory disorders, hearing impairment, etc.

Table 7.9 Aspect/Impact analysis form for crusher unit

Activity	Aspect	Impact
1. Receipt of limestone	Fugitive dust emission	Health effects
	Noise	Hearing impairment
	Loss of resource	Wastage of resources
2. Primary crushing	Fugitive dust emission	Health effects
	Stack emission	Air pollution
	Noise	Hearing impairment
	Oil leakage	Land contamination
	Use of water for cooling	Leakage of recirculation water

Table 7.9 *contd….*

Activity	Aspect	Impact
3. Conveying of crushed limestone after primary crushing	Fugitive dust emission	Health effect
4. Secondary crushing	Fugitive dust emission	Health effects
	Stack emission	Air pollution
	Noise	Hearing impairment
	Oil spillage	Land contamination
5. Screening	Fugitive dust emission	Health effects
	Noise	Hearing impairment
6. Conveying of crushed limestone from secondary crusher	Fugitive dust emission	Health effects
	Noise	Hearing impairment
	Oil leakage	Land contamination
7. Replacement of gear box of belt conveyors	Leakage of lube oil during transfer	Resource consumption

References

1. Council of Environmental Quality. USA Report, 1997.

2. Transportation Noise Pollution Control and Abatement. 1970. Langley Research Center.

3. Chanlett E.T. (1973) Environmental Protection. McGrawhill Book New York pp 523.

4. USEPA, (1971). Transportation noise. US Govt. Printing Office, Washington DC.

5. U. S. Environmental Protection agency (EPA), (1973) "Public Health and Welfare Criteria for Noise," EPA 550/9-73-002, U. S. Environmental Protection Agency, Office of Noise Abatement and Control, Washington, D. C.

6. Federal Interagency Committee on Urban Noise, (1980) "Guidelines for considering Noise in Land Use Planning and Control," Federal Interagency committee on Urban Noise, Washington, D. C.

7. Muigua K., (2013). Realising Occupational Safety and Health as a fundamental Human right in Kenya. http://www.kmco.co.ke/attachments/article/85/. Accessed on 21/04/2015.

8. WHO. (2004). Protection of the Human Environment: Occupational noise - Assessing the Burden of Disease from Work-related Hearing Impairment at National and Local Levels.

9. Melamed, S., Yitzhak F., Froom P. (2001). The Interactive Effect of Chronic Exposure to Noise and Job Complexity on Changes in Blood Pressure and Job Satisfaction: A Longitudinal Study of Industrial Employees. Journal of Occupational Health Psychology 2001, Vol. 6, No. 3, 182-195.

10. U. S. Department of Housing and Urban Development, (1985) The Noise Guidebook, Washington, D. C.

11. Marsh W. M. (1991) Landscape Planning. Environmental Applications 2nd ed., John Wiley and Sons, p322. Von Gierke H. E. (1977) Guide lines for preparing Environmental Impact Statement on Noise. National Research Search Council, Washington D.C.

12. National Cooperation Highway Research Program of USA, 174, (1976)

13. ERL (Environmental Resources Limited). (1984). Prediction in Environmental Impact Assessment, a summary report of a research project to identify methods of prediction for use in EIA. Prepared for the Ministry of Public Housing, Physical Planning and Environmental Affairs and the Ministry of Agriculture and Fisheries of the Government of Netherlands.

14. L. Junggren S. and Johanson M. (1991) Measures against Mechanical Noise from large Wind Turbines. A Design Guide FFA-TN-1991-26 Aeronautical Research Institute of Sweden, Stockhom.

15. U. S. Environmental Protection Agency (EPA), (1978) Protective Noise Levels: Condensed version of EPA levels Document, EPA 550/9-79-110, U.S. Environmental Protection Agency, Office of Noise Abatement and Control, Washington, D. C., Nov. 1978.

16. U. S. Environmental Protection Agency (EPA), (1972) "Report to the President and Congress on Noise, 92nd congress, 2nd session, DOC., 92-63, Washington, D.C.

17. Bowlby W. Harris R. A and Cohn L.F. (1990) Seasonal Measurements of Aircraft Noise in a National Park, Journal of the Air and Waste Management Association Vol. 40 pp 68-76.

18. Landscape planning. Environmental applications 2nd ed., John Wiley and Sons, p322.

Questions

1. Define noise. How it is measured? Explain how it can create impacts on environment?
2. Discuss the different effects of noise on people.
3. Discuss the systematic methodology for assessing Environmental Impacts due to noise of any project activity.
4. Discuss level weighted population value method for assessing significance of any noise impact.
5. Discuss various methodologies used for assessment of traffic noise.
6. Discuss the criteria used for mitigation of noise impacts.

CHAPTER 8

Prediction and Assessment of Socio-Economic and Human Health Impacts

8.A PREDICTION AND ASSESSMENT OF IMPACTS ON THE SOCIO-ECONOMIC ENVIRONMENT

8.A.1 Introduction

Many times development programs in addition to community benefits also bring deleterious effects on ecosystem resources resulting disruption of social and communal harmony, loss of human livelihood and life, introduction of new diseases, and destruction of renewable resources. These impacts known as social impacts are the result of developmental interventions on human environment. As in certain cases these negative effects neutralize the benefits of economic development it is necessary to evaluate the consequential negative impacts of developmental activities on human environment. Further the negative impacts not only to be identified and measured, they need to be managed in such a way to maximize positive impacts and minimize negative impacts. Thus a judicious and balanced need to be adopted in planning by integrating environmental, social and biodiversity impacts with overall economic development.

Social impact assessment (SIA) in association with Environmental impact (EIA) facilitate to provide a overview of positive and negative impacts of proposed policy actions, tradeoffs, and synergies to arrive at informed decision making. Thus SIA and EIA, help in enhancing positive and sustainable outcomes from implementation of various developmental projects.

8.A.2 Social Impact Assessment

8.A.2.1 Basic Aspects of Social Impact Assessment

Social Impacts Assessment (SIA) is a systematic process for identifying and mitigating impacts on individuals or society. SIA need to be carried-out in consultation with the stake holders/actors in the society who are exposed to the impacts of any new development. (SIA) ensures that development activities are implemented after the following concerns are addressed: (i) informed by and take into account the key relevant social issues and formulate mitigative measures, and (ii) incorporate a strategy for participation of wide range of stakeholders. SIA facilitates identifying types and extent of impacts, identifying impacts that can be minimized by good engineering practices and planning mitigation measures for impacts which cannot be minimized during implementation.

Environmental Impact Assessment (EIA) and Social impact Assessment (SIA) tools have been applied internationally to ensure that proposed actions are economically viable, socially equitable and environmentally sustainable.

8.A.2.2 Social Assessment Process

SIA has to be organized in a phased manner in several stages as it is a iterative process. Fig. 8.1 depicts various phases of actions of the social assessment process.

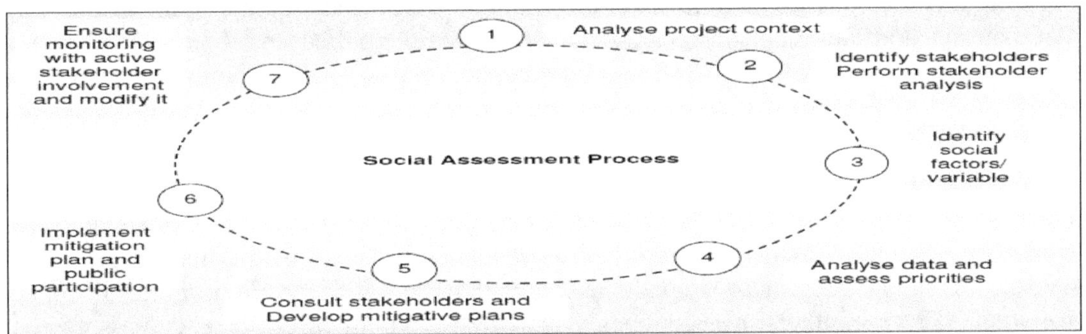

Fig. 8.1 Social Assessment Process Cycle

(**Source:** Rietberg-McCracken and Narayan 1998 (1)

Social Impact Assessment (SIA) which is a processes of analyzing, monitoring and managing the intended and unintended social consequences, both positive and negative, of planned interventions and any social change .SIA is most frequently used to communicate the process of understanding the way human communities may change as a result of an intended action.

Identifying Social Impact Assessment Variables

The following are suggested variables to be measured for SIA on any development project

1. Population Characteristics
2. Community and Institutional Structures
3. Political and Social Resources
4. Individual and Family Changes
5. Community Resources

1. **Population Characteristics** refers to present population and expected change, ethnic and racial diversity, and influxes and outflows of temporary residents as well as the arrival of seasonal or leisure residents.

2. **Community and Institutional Structures** mean the size, structure, and level of organization of local government including linkages to the larger political systems. They also include historical and present patterns of employment and industrial diversification, the size and level of activity of voluntary associations, religious organizations and interests groups, and finally, how these institutions relate to each other.

3. **Political and Social Resources** point to the distribution of power authority, the interested and affected publics, and the leadership capability and capacity within the community or region.

4. **Individual and Family Changes** are factors which influence the daily life of the individuals and families, including attitudes, perceptions, family characteristics and friendship networks. These changes range from attitudes toward the policy to an alteration in family and friendship networks to perceptions of risk, health, and safety.

5. **Community Resources** include patterns of natural resource and land use; the availability of housing and community services to include health, police and fire protection and sanitation facilities. A key to the continuity and survival of human communities are their historical and cultural resources. Under this collection of variables we also consider possible changes for indigenous people and religious sub-cultures.

Social assessments must go well beyond determining a project's adverse impacts. As a methodology, social assessment refers to a broad range of processes and procedures for incorporating social dimensions into development projects. In some jurisdictions and agencies, the social assessment is conducted in conjunction with the Environmental Impact Assessment (EIA); in others, it is conducted separately. In both cases, the social assessment influences project design and the overall approval of the project. In socio economic assessment people are to be considered vulnerable groups. Since some of them will benefit and some groups may be harmed by the project activity. The social assessment aims to determine the social costs of the project and the degree to which the benefits of a project will be distributed in an equitable manner. Social assessments are necessary to help ensure the project will accomplish its development goals (for example, poverty reduction, enhancement of the role of women in development, human resources development, including population planning, and avoiding or mitigating negative effects on vulnerable groups, and protecting these groups).

By addressing the specific development goals in the assessment of development projects, developers, lenders and governments can help ensure that project benefits are realized and negative social impacts are minimized. Various methods and approaches have been developed to consider social dimensions, including:

- social analysis.
- gender analysis.
- indigenous peoples plans.
- involuntary resettlement plans.
- cooperation with non-governmental organizations.
- use of participatory development processes and
- benefits monitoring and evaluation.

8.A.2.3 Social Impacts of Multi-Dimensional Developmental Projects

Major multi - dimensional development projects can include significant requirements for associated infrastructure, such as, streets, highways, or railroads, water supply, sanitary sewers, storm-water drainage, erosion control, sediment control and grading, electrical systems; gas systems, and telephone communication systems. The provision of such needed infrastructures can also generate environmental impacts. Proposed projects involving the decommissioning and closure of major governmental installations or industrial sector developments can also have significant socio-economic consequences in terms of local and/or area wide decreases in jobs and revenue, decline in human population, and leftover societal debts for local infrastructure and educational facilities (2).

Many government or private programs, policies and projects can cause potentially significant changes in many features of the socio-economic environment. In some cases the changes may be beneficial while in others they may be detrimental. Accordingly, environmental impact studies must systematically identify and quantify where possible, and appropriately interpret the significance of these anticipated changes. The multidimensional nature of development interventions call for identification of not only potential economic impacts but also potential social and environmental impacts (Figure 8. 2).

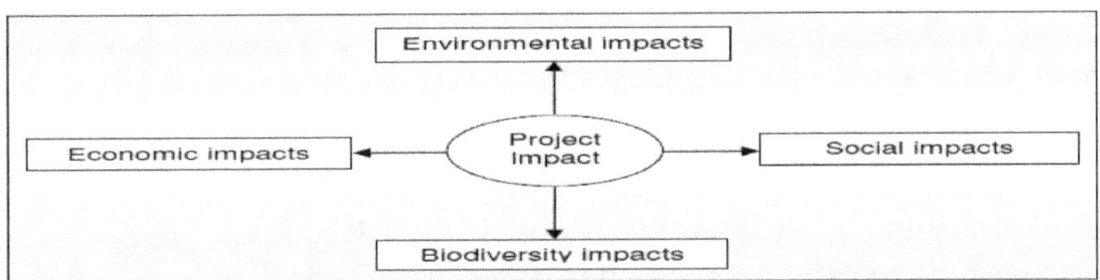

Fig. 8.2 Multi-dimensional impacts of projects.

Increased pace in urbanization, population growth and globalization bring in addition to environmental degradation (increase in air and noise pollution, water pollution, land degradation etc.), adverse social impacts like poverty increase, disturbance/dislocation of vulnerable sections of the society, loss of livelihood etc. While these impacts are properly taken care in project/policy implementation, biodiversity considerations are often inadequately addressed.

The following are salient features and specific advantages of Social Impact Assessments (SIA) of major multi dimensional development projects

1. Processes through which the government departments/agencies can better understand how the socio-cultural, institutional, historical and political contexts influence the social development outcomes of specific investment projects and sector policies.

2. The means to enhance equity, strengthen social inclusion and cohesion, promote transparency and empower the poor and the vulnerable in the design and/or implementation of the project.

3. The mechanisms to identify the opportunities, constraints, impacts and social risks associated with policy and project design A framework for dialogue on development priorities among social groups, civil society, grassroots organizations, different levels of government and other stakeholders.

4. Approaches to identify and mitigate the potential social risks, including adverse social impacts, of investment projects.

8.A.2.4 Role of Social Impacts in Framing Sustainable Development Policies

Sustainable development which includes economic, environmental and social dimensions of the development process has been accepted as main policy goal in any development activity and stimulated interest in Social Impact assessment. Assessing impacts of human interventions on sustainable development at aggregate, sectoral or project levels are essential for a good environmental and social management of any developmental project.

SIA which mainly includes analysis, monitoring and managing the intended and unintended social consequences, both positive and negative, of planned interventions (policies, programs, plans, projects) and any social change processes invoked by those interventions, enables the project implementing authorities to identify social and environmental impacts. Further it also to put in place suitable institutional, organizational and project-specific mechanisms to mitigate the adverse effects which helps in bringing about greater social inclusion and participation in the design and implementation stages of the project.

Social Impact Assessments for any project must:
- enhance positive and sustainable outcomes associated with project implementation;
- support the integration of social and environmental aspects associated with the numerous subprojects into the decision making process.
- enhance positive social and environmental outcomes.
- minimize social and environmental impacts as a result of either individual subprojects or their cumulative effects.
- They protect human health and minimize impacts on cultural property.

For framing a common process flowsheet the following nine principles to be included in any SIA as per the guidelines provided by the Inter-organizational Committee on Guidelines and Principles for Social Impact Assessment consisting of U.S. Department of Commerce, Oceanic and Atmospheric Administration and National Marine Fisheries Service are considered.

8.A.3 General Aims and basic Principles to be followed for any Social Impact Assessment (SIA) (4)

8.A.3.1 General Aims for Conducting any Social Impacts Assessment
- to provide a satisfactory and impartial description for society about the local communities affected and individuals who will gain from the project.
- to inform and involve relevant and affected individuals and stakeholders early on in the process via ongoing dialogue and specific procedures.
- to provide a detailed description of the social pre-project baseline situation, on the basis of which planning, mitigation initiatives and future monitoring will be undertaken.
- to provide an assessment based on collected baseline data to identify both positive and negative social impacts at local level.

- to optimize positive impacts and mitigate negative impacts throughout the project lifetime and through this ensure sustainable development to involve in a systematic manner settlements and respecting local knowledge, experience, culture and values.

8.A.3.2 Basic Principles to be followed for any SIA

In any SIA as per Inter-organizational Committee on Guidelines and Principles for Social Impact Assessment consisting of U.S. Department of Commerce, Oceanic and Atmospheric Administration and National Marine Fisheries Service (3) the following principles need to be followed.

1. **Involve the diverse public**

   The first step is to identify and involve all potentially affected groups particularly those which do not routinely participate in government decision making because of cultural, linguistic, and economic barriers, by an active and interactive process in SIA preparation.

2. **Analyze impact equity**

   Impact equity (Identification of all groups likely to be affected by a developmental activity both positive and negative) must be considered in close and sympathetic consultation with affected communities, neighborhoods, and groups, especially low-income and minority groups even from fixing scoping of the project so that important issues are not left out. Trade-off need to be examined before taking a decision for example to construct a dam, build a highway or close an area to timber harvesting.

   SIA need to take care that the cost of adverse social impacts should not be borne by one single category of persons or sections of the society that are considered as vulnerable due to age, gender, ethnicity, race, occupation or other factors. Further SIA has a special duty to identify those whose adverse impacts might get lost in the aggregate benefits.

3. **Focus the assessment**

   In order of priority significant impacts need to be focused first and all impacted groups for significant impacts need to be identified early adopting rapid appraisal or investigative techniques.

4. **Identify methods and assumptions and define significance**

   The detailed process informing how SIA is conducted and what assumptions are made and how significance of an impact is determined need to be published prior to a decision in order to allow decision makers as well the public to evaluate the assessment of impacts. Significance should be based on considerations such as society as a whole, affected regions, affected interests and locality (e.g., when considering site-specific projects, local impacts assume greater importance than those of a regional nature).

5.   Provide feedback on social impacts to project planners

Project planners   need to design SIA as a dynamic process with cycles of project design, assessment, redesign and reassessment so that feedback from SIA can be given as input in the redesign with alternative process of the project to mitigate negative impacts while enhancing positive ones. Public comments on a draft EIS can contribute importantly to this process of feedback and modification.

6.   Use SIA practitioners

SIA practitioners who are conversant with the technical and biological perspectives will provide the best results as they will be familiar and conversant with existing social science evidence pertaining to impacts that have occurred elsewhere, which may be relevant to the impact area in question. An expert social scientist with breadth of knowledge and experience can prove invaluable in identifying important impacts that may not surface as public concerns, will be able to identify the full range of important impacts and to select the appropriate measurement procedures. Having social scientist as part of the interdisciplinary EIS team will also reduce the probability that an important social impact could go unrecognized. In assessing social impacts, if the evidence for a potential type of impact is not definitive in either direction, then the appropriate conservative conclusion is that it cannot be ruled out with confidence.

7.   Establish monitoring and mitigation programs

Monitoring and mitigation programs should be taken up on iterative basis through-out the project cycle jointly by project agency and impacted local communities with trust and expertise as key factors. Local communities should be provided resources to assume a portion of the monitoring and mitigation responsibilities as only few agencies can continue monitoring for extended periods

8.   Identify data sources

SIA data sources include published scientific literature from similar projects, secondary data and primary data from the affected area. Census, vital statistics, geographical data, relevant agency publications, and routine data collected by state and federal agencies are considered as secondary data while primary Data from the Affected Area includes surveys, oral histories and informant interviews.

9.   Plan for gaps in data

In the absence of relevant/necessary data, the SIA practitioners need to plan for assessment adopting statistical strategies for evaluation of missing information

These principles are suggested based on the expert judgment of widely varied professionals like sociologists, anthropologists, social psychologists, geographers, land-use planners, economists, natural resource social scientists and landscape architects which are meant to ensure sound scientific inquiry.

8.A.4   Conceptual Frame Work for Socio Economic Assessment (SIA) Process

The conceptual general flow chart for carrying out SIA for any project is shown in Fig 8.3.

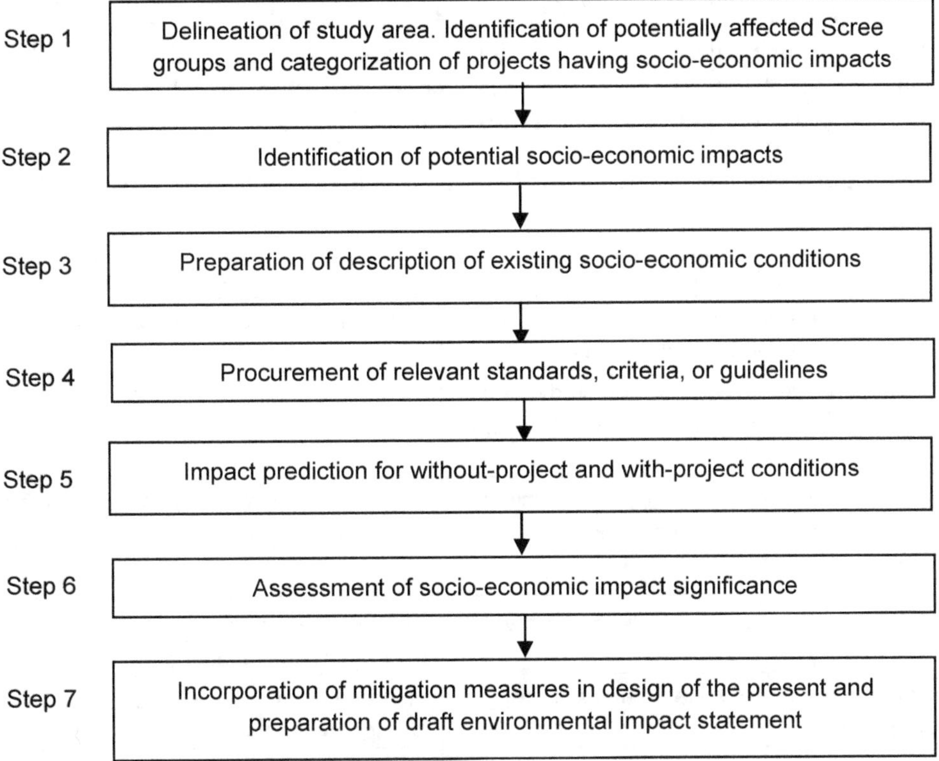

Fig. 8.3 Conceptual flow chart for prediction and assessment of socio-economic Impacts.

8.A.4.1   Step 1 : Study Area Delineation, Categorization of Present Activities

The delineation of study area for the analysis of social and community effects requires a sufficient knowledge of the characteristics of the proposed project or activity. The proposed project alternatives should give sufficient detail to enable understanding of both the construction process and the long-term or operational, characteristics.

Definition of the boundaries of the study area may also be influenced by the availability of population and employment data. Census tract data is most often used to define population characteristics. The study area boundaries are, therefore, in some cases defined by the boundaries of census tracts or block data.

The basic impact area associated with predicting and assessing impacts on the socio-economic environment is called the "Regions of Influence" (ROI). This represents the geographical area, or region, wherein the project-induced changes to the socio-economic environment will occur. In an analogous sense, the ROI for addressing socio-economic impacts would be comparable to 1. an air quality control region for addressing air quality impacts, 2. a watershed for addressing surface-water quantity and quality impacts, and 3. an ecoregion or habitat type for addressing biological impacts.

8.A.4.2 Step 2 : Identification of Socio-Economic Impacts and Screening criteria

Potential socio-economic impacts can be identified through the use of interaction matrices, networks, simple checklists, and/or description checklists. Case studies of similar project types can also be helpful. Some of the socio-economic factors and their potential changes resulting from project implementation are given in Table 8.1.

Table 8.1 Samples of socio-economic factors and their potential changes resulting from project Implementation.

Factor	Potential change
General characteristics and trends in population for state, substate region, country, and city	Increase or decrease in population
Migrational trends in study area (The study area is a function of the alternatives being considered and the available database.)	Increase or decrease in migrational trends
Population characteristics in study area, including age, sex, ethnic group, educational level and family size	Increase or decrease in various distributions: distributions people relocations
Distinct settlements of ethnic groups or deprived groups in study area	Disruption of settlement patterns; economic or minority people relocations
Economic history for state, substate region, country and city	Increase or decrease in economic activities; change in economic patterns
Employment and unemployment patterns in study area, including occupational distribution and location and availability of workforce.	Increase or decrease in overall employment or unemployment levels; change in occupational distribution
Income levels and trends for study area	Increase or decrease in income Levels
Land-use patterns and controls for study area	Change in land usage; project may or may not be in compliance with existing land-use plans
Land values in study area	Increase or decrease in land values
Tax levels and patterns in study area, including land taxes, sales taxes, and income taxes	resulting from changes in land usage and income levels
Housing characteristics in study area, including types of housing occupancy levels, age and condition of housing	Changes in types of housing and occupancy levels
Health and social services in study area, including health manpower, law enforcement, fire protection water supply, waste water-treatment facilities, solid-waste collection and disposal, and utilities	Changes in demand for health and social services

Table 8.1 Contd...

Factor	Potential change
Public and private educational resources in study area, including grades K-12 schools, junior colleges and universities	Changes in demand for educational resources
Transportation systems in study area, including highway, rail, air, and waterway systems	Changes in demand for transportation systems; relocations of highways and railroads
Community attitudes and lifestyles, including history of area voting patterns	Changes in attitudes and lifestyles
Community cohesion, including organized community groups	Disruption of cohesion
Tourism and recreational opportunities in study area	Increase or decrease in tourism and recreational potential
Religious patterns and characteristics in study area	Disruption of religious patterns; change in characteristics
Areas of unique significance, such as cemeteries or religious camps	Disruption of activities in or changes to unique areas.

8.A.4.2.1 Screening Criteria

The significance of expected socio-economic impacts due to any of the present activities can be considered with reference to the screening criteria given in Table 8.2 for assessing the impact significance with respect to 1. nature of the impact, 2. the reverse of the impact both absolute and perceived, and 3. the potential for mitigation.

Table 8.2 Descriptive-checklist methodology for addressing socio-economic impacts.

Category	Questions to identify	Necessary information	Methodology
Educational facilities	Can projected enrollments be properly handled in existing or proposed facilities with proper spacing for all activities (including classrooms, recreational areas, and staffing needs)? Will the project impact the pupil/teacher ratio as to impede the learning process? Is the school located such that presents a hardship for students' enrollment in terms of too great a travel time or distance or through the existence of safety hazards?	Number and type of housing units; existing pupil/teacher ratio, in local schools; size (number of pupils) of existing school facilities; miles to nearest schools with available capacity; average local bus speed	Estimate numbers, by school age, of students generated by project; compare number of students generated with existing pupil/teacher ratios; compare number of pupils in and size of present schools with number of pupils generated by project; measure distance; calculate travel time, as follows; Distance from Project to school ————————— Average local bus speed

Category	Questions to identify	Necessary information	Methodology
Commercial facilities	Will there be an adequate supply of and access to commercial facilities (existing or proposed) for the project?	Location, type, and size of commercial facilities; income of future residents; transit availability and other transportation alternatives; car ownership profiles of future residents	Compare retail market demand of future residents to available type and size of commercial facilities; calculate travel time, as follows : Distance to facilities from project / Average speed by mode.
Health care and social services	Are provisions for and access to quality health care and social services adequate to meet the needs of the residents of the project area?	Local type and size of existing facilities; beds or population; socioeconomic characteristics of future population in project area ; transportation availability; waiting time for existing services	Project health, hospital-bed and day-care needs of future project population to compare with those existing facilities; calculate travel time, as follows: Distance between project and stores / Average speed by mode
Liquid-waste disposal	Is provision for sewage capacity adequate to meet the needs of the project without exceeding water quality standards? Will the project be exposed to nuisances and odors associated with wastewater treatment plants?	Number of people or dwelling units in project; type and size of other facilities, e.g., commercial sites; offices in project; lot size and soil conditions; location of treatment facilities; Section 208 water-quality Management plan	Compare future sewage needs with existing availability of public services or lot and soil suitability (percolation test) for septic tanks.
Solid-waste disposal	Is there provision for environmentally sound disposal of solid wastes generated by the project?	Number of people in project; type and size of other facilities, e.g., commercial office, health care, etc.; existing disposal methods and capacities	Compare quantity (one kg per day) and type of solid waste generated with available capacity for disposal.

Table 8.2 *Contd…*

Category	Questions to identify	Necessary information	Methodology
Water supply	Are there provisions for adequate quantity and quality of water supply to meet the needs of the project?	Number of people or residential units in project; type and size of other facilities (commercial, office, etc.); capacity of local public water system; existing water quality	Compare water supply needs of project (gal. per day) to available excess capacity of local systems.
Police	Will the present or proposed police system adequately protect the additional population and facilities generated by the project?	Number and type of people in project; existing police/population ratios; existing police response time (min)	Compare police needs for new population to existing resources.
Fire protection	Will the project be provided with adequate fire protection services?	Location of nearest fire station; type of equipment; staff; response time; type, number, and density of buildings in project; existing available fire flow (gal. per min)	Determine distance from project to fire station; divide by average speed to obtain required fire flow (estimated).
Recreation	Will project have access to adequate facilities to meet the recreational needs of residents?	Location, size, type, and capacity of recreational facilities; number and socioeconomic characteristics of future population in project	Determine distance to facilities with necessary capacity; compare available types and capacity of recreation facilities with needs of population.
Transportation	Are the transportation facilities which serve the project part of a well-integrated multi model system and are they adequate to accommodate the project's travel demands?	Location of travel way; width and type of travel way; frequency of bus service; socioeconomic characteristics of resident population	Locate transportation facilities in relation to project; document size – maps or field measurements, e.g., size, number of lanes; estimate optimal peak-hour flows; estimate trips by mode, based on type and density of land use and on socioeconomic characteristics of resident population; compare needs to existing capacity for all modes, compare needs to existing levels of service.
Cultural facilities	Are the transportation facilities available to the project residents?	Location, size, type, and capacity of facilities; number socioeconomic characteristics of project population	Determine distance to facilities with capacity; compare available types and capacity with needs of population.

Source : Voorhees and Associates, 1975.(5)

8.A.4.3 Step 3 : Description of existing Socio-Economic condition of the Study Area-Base Line Data

8.A.4.3.1 Identification of Region of Influence

After identifying the area of influence and potential impacts likely to occur on various social factors due to project activities, the existing socio-economic conditions and physical parameters of the Region of Influence (ROI) have to be examined on the basis of the following points:

(a) Areas separated from the surrounding areas by physical boundary obstruction like rails, roads, high ways, and rivers.

(b) Area of residential land-use surrounded by other uses such as commercial and industrial.

Area with a concentration of special population groups, such as, elderly, low-income or a specific ethnicity.

(c) Area with like housing types, such as, mobile homes, single family homes, or high-density condominiums or apartments.

(d) Area of distinct housing value, compared with surrounding areas.

(e) Area of predominantly one type of population employment, such as, professional.

(f) Area with an established community group or organization.

(g) Area where the average length of residence in the same housing unit is more than five years, as opposed to areas of more-transient residents.

Any special characteristics of each defined neighborhood should be identified, as high percentage of ethnic minority population or elderly persons or a high degree of community cohesion. Community cohesion can be estimated through examination of many of the same factors listed above for identifying neighborhoods. Generally, communities with long-term residents or with established neighborhood organizations are considered highly cohesive.

The environmental document should include a map of study area neighborhoods with tables identifying special demographic characteristics, if any. This map may be combined with the land-use map, if appropriate.

8.A.4.3.2 Baseline Data

Baseline data means a geographical and time line data to start the assessment.

Some of the baseline data which need to be collected as part of the social impact assessment includes:

(a) ***Demographic factors:*** number of people, location, population density, age etc.

(b) ***Socio-economic determinants:*** factors affecting income and productivity, such as risk aversion of the poorest groups, land tenure, access to productive inputs and markets, family composition, kinship reciprocity, and access to labour opportunities and migration.

(c) ***Social organization:*** organization and capacity at the household and community levels affecting participation in local level institutions as well as access to services and information.

(d) ***Socio-political context:*** implementing agencies' development goals, priorities, commitment to project objectives, control over resources, experience, and relationship with other stakeholder groups.

(e) ***Needs and values:*** stakeholder attitudes and values determining whether development interventions are needed and wanted, appropriate incentives for change and capacity of stakeholders to manage the process of change.

(f) Nature and land use.

(g) Industry and business structure.

(h) Labour market structure and education structure.

(i) Infrastructure.

(j) Health.

(k) Values, heritage, knowledge and social/cultural well-being.

(*l*) Listed/conservation areas.

It is essential to describe the data collection methodology used in the SIA, including the data collection methods and analytical tools used like qualitative versus quantitative data, mix of data from different units of analysis for triangulation of results etc; need to be provided in detail. The representative sampling (rather than subjective sampling) needs to be employed wherever possible. Further in designing the research methodology, it is important to keep in mind how much time and resources will be required from the communities.

8.A.4.3.3   Step 4 : Procurement of Relevant Standard/Criteria/Guidelines/ and Scoping

Socio-economic impact assessment involves relative comparison of the effects with standard/criteria published by professional stamps/organization with public guidelines or standard for various social activities which need baseline data which includes identification and prioritization of the range of likely social impacts through a variety of means. This covers discussions or interviews with members of all potentially affected, reviews of the existing social science literature by experts, public scoping, public surveys and public participation techniques.

Figure 8.4 gives an illustration of **the scoping process**. The methods for social analysis and participation include (Rietbergen-McCracken and Narayan 1998) (1).

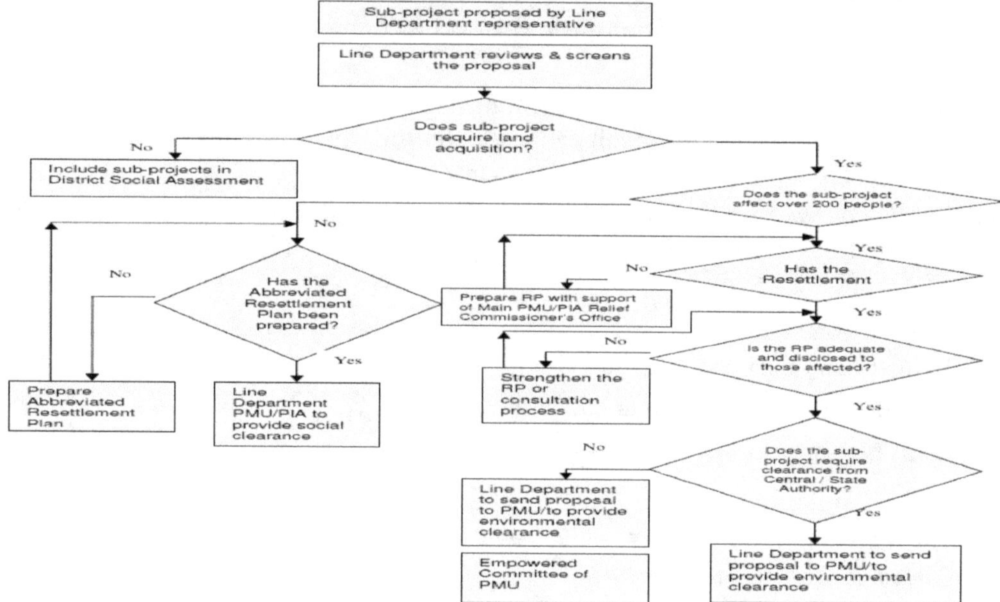

Fig. 8.4 The Scoping process for social Impacts.

8.A.4.3.3.1 Methods for Scoping the Impacts

(a) *Workshop based methods:* Collaborative decision making often takes place in the context of stakeholder workshops, which bring stakeholders together to assess issues and design development projects collaboratively. A trained facilitator guides stakeholders through a series of activities to promote learning and problem solving.

(b) *Participatory assessment methods:* Social assessments can also be informed by field visits to communities and other local-level stakeholders to learn about their perspectives and priorities. The consultations make use of participatory assessment methodologies such as participatory Rural Appraisal (PRA), or beneficiary assessment.

These methodologies provide tools for collaborating with local people in analysis planning, and can contribute to the development of action plans and participation strategies.

8.A.4.4 Step 5 : Impact Prediction with and without Project

The following are important impacts need to be considered for prediction

> Direct or indirect impacts occurring at the project site or within the project's area of influence.

> Indirect impacts include side-effects of the project given the complexity of social processes and the human-environment interface.

> Impacts within the project's like transboundary impacts.

> Negative impacts both immediately as well as longer term.

> Cumulative effects that materialize through interaction with other developments at the project site.

Fig. 8.5 presents the general information flow relative to prediction and assessment of socio-economic impacts. Modeling of economic demographic impacts provides basic information for addressing public service impacts (education, health services, police and fire protection, utilities, and solid-waste management), social impacts (housing, transportation, urban land use, and land ownership), and fiscal impacts. Fiscal impacts are themselves dependent upon many public services and social impacts. The Quality Of Life (QOL) represent, a composite indication of economic, demographic, public service, social, and fiscal impacts along with impacts caused by still other factors related to a sense of well being at a given time and location.

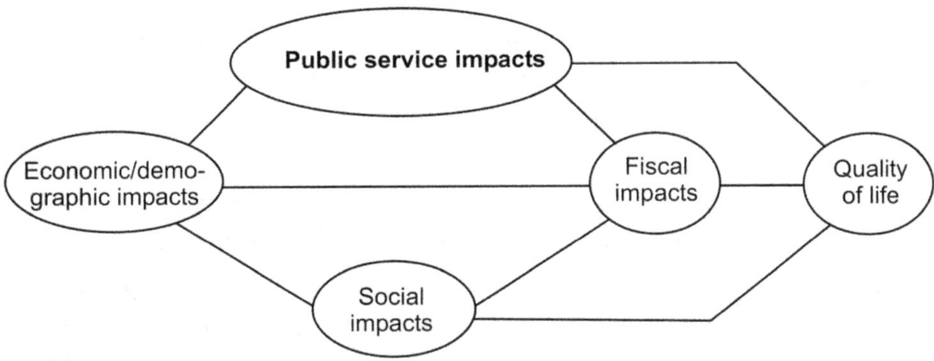

Fig. 8.5 Example of information flow in prediction and assessment of socio-economic impacts.

8.A.4.4.1 Analyzing and predicting probable impacts of the project proposal and the alternatives against baseline conditions (with Vs without the action).

8.A.4.4.1.1 Analysis and Prediction of Impacts

The following information is needed for analysis and prediction of impacts:

(i)   predicted conditions without the actions (baseline condition) and

(ii) predicted conditions with the actions and the predicted impacts.

Investigation of the probable impacts involves five major sources of information:

(a) detailed data from the sponsoring agency on the proposed action;

(b) record of previous experience with similar actions as represented in reference literature to include other SIAs.

(c) census and vital statistics, documents and secondary sources.

(d) field research, including informant interviews, hearings, group meetings and, if funds are available.

(e) surveys of the general population.

(f) Methods of predicting the future impacts are at the heart of the SIA process. Care must be taken to ensure the quality and transparency of methods sond data, and to provide for critical review.

8.A.4.4.1.2 Methods for Analyzing and Predicting Impacts

The following are some of the methods for analyzing and predicting social impacts [adapted from Taylor et al., 1998 and Impact Assessment and Project Appraisal, 2003] (6)

Comparative method: In this method the present impacts are compared to the future with the proposed action. For impact analysis one has to examine how an affected community has responded to change in the past, or the impact on other communities that have undergone a similar action. Based on past research and experiences in similar cases, determination of significance is made based on the comparative data presented.

Straight-line trend projection: *By assuming what happened in the past likely to happen in the future also a*n existing trend will be simply projected the same rate of change into the future. For example, visitations for recreation increase each year at about the same rate they did in the past.

Population multiplier methods: Each specified increase in population implies designated multiples of other variables, such as jobs, housing units and other infrastructure needs.

Statistical significance means: Calculations to determine probabilistic differences between with and without the proposed action. A social assessor could employ comparative statistical methods to determine statistical significance for appropriate SIA variables.

Scenarios: These refer to logical-imaginations based on construction of hypothetical futures through a process of mentally modeling the assumptions about the SIA variables in question. Scenarios include exercises to develop the likely, alternative or preferred future of a community or society. Scenarios can be used to compare different outcomes (e.g., best versus worst case).

Consulting experts: Use of expert knowledge such as researchers, professional consultants, local authorities, or knowledgeable citizens. Such persons familiar with the study area could be asked to present scenarios and assess the significant implications for the proposed action.

Calculation of 'futures forgone': a number of methods have been formulated to determine what options would be given up irrevocably as a result of a plan or project, for instance, river recreation and agricultural land use after the building of a dam. The wetlands mitigation strategy is such an example.

8.A.4.4.3 Evaluation of alternatives and Impact Mitigation

This involves evaluating alternatives in terms of projection of their consequences for affected and interested stakeholders. Each alternative or modification to the proposed action should be assessed separately. Mitigation plans first to avoid, secondly to minimize and thirdly to compensate the adverse impacts need to be developed and executed. Mitigation plans may be in the form of modification of the specific event in the project, operation and redesign of the project or policy or compensation for the impact by providing substitute facilities, resources and opportunities.

8.A.4.4.4   The Indirect and Cumulative Impacts

These are estimated to identify the subsequent flow-on effects of the proposal, including the second/third order impacts and their incremental impacts when added to other past, present and foreseeable current activities. Secondary or indirect impacts are those caused by the primary or direct impacts; they often occur much later, both in time and geographic distance, than primary impacts. Cumulative impacts are those resulting from the incremental impacts of an action added to other past, present, and reasonably foreseeable future actions regardless of which agency or person undertakes them.

8.A.4.5 Step 5 Assessment of Significance of Socio-Economic Impacts

8.A.4.5.1 Procedure for Assessing Socio-Economic Impact Significance

Assessment of the significance of predicted changes in the socio-economic environment requires considerable exercise of professional judgment. Every attempt should be made to use systematic and scientific rationale for significant assessment.

Figure 8.6 presents the flow chart for the procedure for assessing socio-economic impact significance.

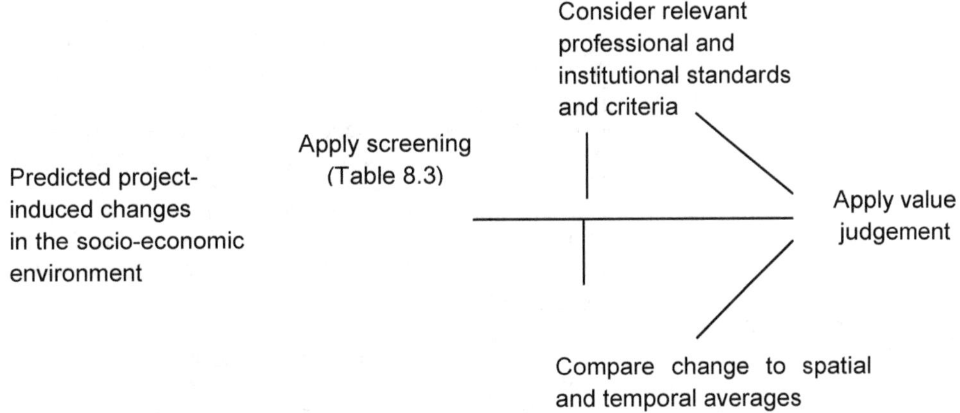

Fig. 8.6 Procedure for assessing socio-economic impact significance

The criteria for assessing impact significance are presented in Table 8.3.

Table 8.3 Criteria for assessing impact significance.

Criteria	Definition, Measurement
1. Nature of the impact	
A. Probability of occurrence	Likelihood that a given impact will occur as a policy, and/or project For many socioeconomic impacts, qualitative assessments would be appropriate (high, medium, low).
B. People affected	How pervasive will the impact be across the population? This criteria should be used to assess both the percentage of the population affected and the extent to which it will affect different demographic groups.
C. Geographic pervasiveness	The extent to which the impact is experienced across a widespread area. Can it frequently be addressed by mapping or use of data sources which are geographically specific (census data)?
D. Duration	How long the impact will last, assuming no direct public or private sector attempts to mitigate. Can it be addressed by identifying short-term, long-term, and permanent impacts?
2. Severity	
A. Local sensitivity	To what extent is the local population aware of the impact? Is it perceived to be significant? Has it been a source of previous concern in the community? Are there any organized interest groups likely to be mobilized by the impact.
B. Magnitude	How serious is the impact? Does it cause a large change over baseline conditions (e.g., will crime rates double)? Does it cause a rapid rate of change (large changes over a short time period)? Will these changes exceed local capacity to address or incorporate the change? Does it create a change which is unacceptable? Does it exceed a recognized threshold value?
3. Potential for mitigation	
A. Reversibility	How long will it take to mitigate the impact by natural or man-induced means? Is it reversible; if so, can it be reversed in the short term or the long term?
B. Economic costs	How much will it cost to mitigate this impact? How soon will finances be needed to address this impact?
C. Institutional capacity	What is the current institutional capacity for addressing the impact? Is there an existing legal, regulatory, or service structure? Is there excess capacity, or is the capacity already overloaded? Can the primary level of government (e.g., local government) deal with impact or does it require other levels or the private sector?

8.A.4.5.2 Evaluation of Responses of Client Groups to Impacts

This is made to determine the significance of the identified social impacts to those who will be affected (client groups). This involves identifying Client Groups, Client Needs, Client Demands, Absorptive Capacity and Gender Issues.

Identifying Client Groups

Client groups are those groups that will either benefit or be adversely affected by the project. The first step in any social analysis is to gather baseline information on client groups. As not all project beneficiaries will have the same needs and demands, the population may have to be divided into sub-groups. A basic profile for each sub-group should be developed. This profile should include:

- the number in each sub-group,
- differentiation by gender,
- number of single-headed households,
- household size,
- occupations,
- income and asset levels,
- levels of education and access to education,
- health problems and access to health services,
- social organization and group formation, and
- ethnic or cultural distinctions.

This information should describe the socioeconomic traditions of the client group which affect life-styles, beliefs and patterns of use of facilities to be affected by the project. It is not always necessary to conduct detailed socioeconomic surveys to gather this information. In fact, current practice is usually to conduct participatory rural appraisals as part of the participatory development activities associated with the project.

Client Needs

Once the baseline information on the client group has been identified, the social assessment team should assess the expressed need of client groups in relation to the benefits to be provided by the project. In assessing client needs, the social analysis team should:

- describe the quantity and quality of related facilities available to each of the sub-groups, including any problems of access, cost, quality, etc., and the level of service to be provided to each subgroup under the project.

- assess the priority given by the expected clients to acquiring the facilities to be established by the project in relation to their willingness to allocate their resources (for example, time, capital, effort) for the acquisition of such facilities if the facilities are a priority, determine the clients' preference with respect to type, quality, and cost of project. and

- determine the potential to maximize the project's benefits through the addition of project components designed specifically to ensure benefits flow to affected people.

Client Demands

The SIA team after identifying client needs have to

- assess the client group's demand for the project, for example, by assessing present expenditures and efforts by clients to access such facilities through formal, informal, or traditional means.

- for each client group, assess the client's ability and willingness to pay for access to the project. and

- assess the project's potential to change the demand for the project (for example, through better client-provider relations).

As an example of client needs and demands, the inventory of Natural Resources and Access Restrictions – template which need to be collected during the study is given in table 8.4.

Table 8.4 Client needs and demands, the inventory of Natural Resources and Access Restrictions

Resource	Use/ rights/ restrictions	Core zone	Buffer zone	Other community land
Timber	Current use/ mportance of livelihoods			
	Use value			
	Restrictions			
Grazing livestock	Current use/ importance of livelihoods			
	Use value			
	Restrictions			
Fuel wood	Current use/ importance of livelihoods			
	Use value			
	Restrictions			
Fodder collection	Current use/ importance of livelihoods			
	Use value			
	Restrictions			
Medicinal plants	Current use/ importance of livelihoods			
	Use value			
	Restrictions			

Absorptive Capacity

Absorptive capacity is the capacity of the client group to reap the benefits from the project and/or adapt to the adverse impacts associated with the project.

The social analysis should:

- examine the variations in existing knowledge, attitudes, and practice which may influence the extent and manner in which the project may be used;

- describe the behavioral changes which may be required for clients to use and sustain the benefits which may be provided through the project; and

- assess their ability and willingness to make these changes in terms of their motivation to change, including aspirations, level of knowledge, skills and experience, social cohesion of the client groups, and constraints. Many projects concerned with forestry enhancement and watershed rehabilitation have to be concerned with increasing the absorptive capacity of target beneficiaries. Potential beneficiaries are often poor with low cash income, little education, and are in poor health. They are likely dependent on a natural resource base which has been degraded through unsustainable practices.

Gender Issues

The social analysis must include an examination of gender issues, including:

- an assessment of differences in values, roles, and needs of men and women in terms of the impact of these factors on decisions to use the project; and

- an assessment of the access of men and women to the project, and to related training and employment opportunities including identification of constraints (for example, time, finances, transportation, literacy, health, social, cultural, legal or religious constraints) faced by women or men in gaining access to the project.

To assess the significance of impacts, the potential of adverse impacts for specific client groups need to be evaluated

Potential Adverse Impacts

The assessment of potential adverse impacts should include:

- identification of those groups which may be adversely affected by the project, including groups who may be required to relocate, or groups adversely affected by loss of income, loss of traditional lands and cultural property and possible exposure to health hazards (for example, noise or air pollution, traffic hazards, etc.);

- determination of the possibility of conflict over rights to key resources such as water;

- determination of how pricing policies would affect the distribution of and access to project benefits by poor clients, including an assessment of the ability of client groups who are defined as too poor to afford a basic level of service to access the project, and identification of financing measures which are affordable for the poor groups on a sustainable basis;

- determination of any significant changes in affected groups' life-styles;

- identify and assess options for avoiding, mitigating, or compensating groups which may be adversely affected; and

- consult with affected groups to obtain feedback through such means as community dialogues, public hearings, referendum, formation of multipartite negotiating, or monitoring teams concerning the proposed solution.

8.A.4.6 Step 7: Incorporation of Monitoring and Mitigation Measures in the Project and Preparation of Draft Environmental Impact Statement

Mitigation of social and neighborhood effects is often very site-specific. Techniques to avoid impacts may include providing new or revised access to communities; redesigning of particular features of the proposed project or action to avoid relocations; constructing noise control walls or security fencing and adding parking areas.

Other mitigation measures may be intended to offset any secondary impacts of increased population growth, such as, supplying accessory fire protection, security, or water and sewage capability.

8.A.4.6.1 Establish Monitoring and Mitigation Programs

Monitoring significant social impact variables and any programs that have to put into place to mitigate them are crucial to the social impact assessment process.

Monitoring Plan: This involves developing and implementing a monitoring programme to identify deviations from the proposed action and any important unanticipated impacts. This should track project and program development and compare real impacts with projected ones. It should spell out (to the degree possible) the nature and extent of additional steps that should take place when unanticipated impacts or those larger than the projections occur.

Identifying a monitoring infrastructure needs a key element of the local planning process. Monitoring and mitigation should be a joint agency and community responsibility and both activities should occur on an iterative basis throughout the project life cycle. Trust and expertise are key factors in balancing agency and community monitoring participation. Few agencies have the resources to continue these activities for an extended period, but local communities should be provided resources to assume a portion of the monitoring and mitigation responsibilities.

Provide feedback on social impacts to project planners

Identify problems that could be solved with changes to the proposed action or alternatives. Findings from the SIA should feedback into project design to mitigate adverse impacts and enhance positive ones. The impact assessment, therefore, should be designed as a dynamic process involving cycles of project design, assessment, redesign, and reassessment. This process is often carried out informally with project designers prior to publication of the draft assessment for public comment; public comments on a draft EIS can contribute importantly to this process of feedback and modification.

8.A.4.6.2 Preparation of Draft Environmental Impact Statement

The Environmental Assessment (EA) or Draft Environmental Impact Statement (DEIS) should summarize social and community effects relevant to the comparison of proposed alternatives. If large amounts of data and analysis material have been generated, they should be contained within a separate technical report supporting the environmental document.

Factors for Overall Success of SIA

Some important factors which improve quality of Socioeconomic Analysis

The following factors will improve the quality of socioeconomic analysis

> Hire qualified social impact specialist with a solid background in social sciences.

> Participatory development process need to be implemented hiring local experts.

> Instead of using ideal or worst case scenario data, use of local knowledge as well as scientific data; and use realistic assumptions for development practices such as construction practices will improve SIA.

> Including analyses of cultural and historical relationship to the resource being restricted/used (natural resources) will improve quality.

> current rights of communities to the natural resources – legal rights; customary and non-legal and the degree of their dependency on these resources for livelihoods to be considered.

> Evaluating the relationship between the use of resources and conservation objectives and the extent of their positive and negative impacts on resource sustainability will improve the quality of SIA.

8.A.5 Case Studies

The importance of social impact assessment and its application in specific projects can be understood clearly by some case studies

Case Study 1. Social impact of the Sardar Sarovar Scheme, India: Key findings and conclusions from SIA (Berger, 1994) (7)

Sardar Sarovar became the focus of the debate, in India and internationally, on how to balance economic development on the one hand, and human rights and environmental protection on the other.

The environmental and social impact of the project components is immense and extends over a wide area.

At least 100,000 people, in 245 villages, live in the area affected by submergence.

In Gujarat and Maharashtra almost all of those affected are tribal people.

In addition, there are 140,000 families who will be disrupted by the construction of the canal and irrigation system.

The issues in Sardar Sarovar were complicated because the majority of those displaced were tribal people who usually have no formal title to the land they occupy and were considered by two state governments of Gujarat and Maharashtra to be encroachers and not entitled to resettlement.

The review found this position to be non-compliant with recognized norms of Human Rights.

In addition, it concluded that a number of issues of related to the environmental impact of the scheme were unresolved and questioned the assumptions used in project design and mitigation.

Case Study 2.: Environmental and social impact assessment of the Vanimo Timber Area, Malaysia: Key findings

The WTK Reality Group of Malaysia through their operation of Vanimo Forest Products (VFP) has violated at least 13 of the key standards from the PNG Logging Code of Practice.

➢ VFP failed to fully comply with the project agreement concerning the Vanimo Timber Area.

➢ The East-West Highway has never been completed by VFP.

➢ Poor quality road and bridge construction has severely limited development and transport options for people in the Vanimo region.

➢ Culturally significant areas, including gravesites, have been negligently damaged by VFP.

➢ A fledgling palm seed export business has been threatened by VFP.

➢ Sediment and nutrient runoff from negligent logging operations damaged stream ecosystems.

➢ Clean water sources near villages for drinking/processing sago have been damaged by VFP.

➢ Undersized logs are regularly cut from the forests.

➢ Extensive damage from logging has retarded the regeneration capacity of the forest.

➢ Sago palms have been damaged by logging operations.

➢ The time taken for villagers to find food sources in the forests have increased.

➢ Malnutrition, low birth weight babies, malaria and sexually transmitted diseases have become more common in the region.

➢ Limited numbers of local men and no local women are employed by VFP.

➢ Birds of Paradise and New Guinea Pigeons are reportedly smuggled out on the logging boats.

➢ Police have been used to protect the interests of VFP.

➢ The Forestry Authority has not enforced the PNG Logging Code of Practice.

8.B. PREDICTION AND ASSESSMENT OF IMPACTS OF PROJECT ACTIVITIES ON HUMAN HEALTH

8.B.1 What is Health Impact Assessment?

For assessing the impact of any new policy/program on an environmental health issue in any ecosystem with fixed population, health impact assessment (HIA) will be useful as it provides a systematic frame work and procedures for carrying out HIA. HIA can be made as a part of Environmental Impact assessment EIA, Strategic Environmental Impact Assessment (SEA). Social Impact Assessment (SIA), or Integrated Impact Assessment (IIA)

Objectives

(i) To act as a prospective assessment of program or intervention before implementation,

(ii) To carry- out concurrently or retrospectively.

(iii) To gather opinions and concerns regarding the proposed policy

(iv) To use knowledge of health determinants to understand the expected impacts of the proposed policy or intervention,

(v) To describe the expected health impacts using both quantitative and qualitative methods as appropriate

(vi) To improve the quality of policy decisions by evaluating the likely positive and negative health impacts from proposed programs or policies,

(vii) To make recommendations to improve positive health impacts and mitigate negative ones due to implementation of any new program/ policy

(viii) To provide a social model with the participation of public/stakeholders for the health and wellbeing of a society with special focus on equity, sustainability, social justice, commitment to openness and public scrutiny

8.B.2 Tasks of Health Impact Assessment

The importance of health impact considerations in project planning have been stressed by the World Health Organization [8, 9]. The main tasks of a Health Impact Assessment process are:

1. *Definition of Project Type and Location*: Project title, location, department, executing agency and major project components are to be defined as part of the screening process and project classification.

2. *Health Hazard Identification*: It is based on existing experience and the screening tools provided are for a list of health hazards.

3. *Initial Health Examination (IHE)*: This is the secondary screen. It uses rapid appraisal, secondary data and a fact finding mission (if necessary). It is a part of the initial environmental evaluation (IEE) and should normally be undertaken at the prefeasibility stage. The outputs are: a short list of the health hazards which may carry the most significant health risks. The identification of a hazard short is part of the scoping process.

4. *Requirement for Health Impact Assessment* (HIA): A decision is made based on the experience of previous projects and the need to obtain further experience.

5. *Terms of Reference (TOR) Definition for HIA*: A TOR is prepared for an HIA which specifies the scope of the assessment. It includes, but is not limited by, the short list of health hazards identified by the IHE.

6. *Health Impact Assessment*: The assessment is undertaken by a specialist consultant. The output is a Health Impact Statement. The HIA may be a stand – alone study, but more typically, it will be a part of an EIA.

7. *Health Risk Management*: The Health Impact Statement recommends health risk management actions including environmental management and health monitoring. Health monitoring data is an output.

8. *Benefit Monitoring and Evaluation*: The project may be evaluated by an appropriate agency and the output should include a health impact evaluation report which can be used in future projects.

8.B.3 Methodology

A systematic approach for health-impact-prediction and assessment methodology involves the principles of Risk Assessment (RA) methods and of traditional approaches which should be integrated into the unified analytical process in an EIA study. The Health Risk Assessment methodology is discussed in Chapter 9. Fig. 8.7 depicts a flow diagram of the methodology.

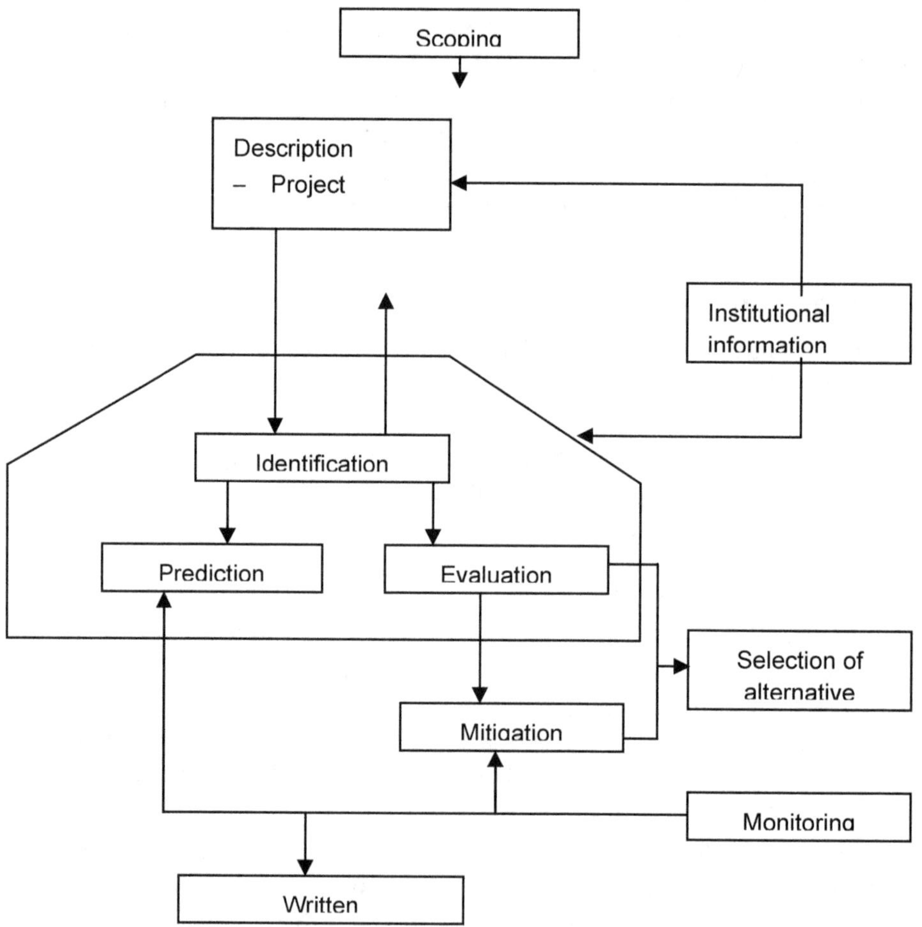

Fig. 8.7 Flow diagram of generic health impact prediction and assessment

It consists of a sequence of 10 operational activities, or components; the information derived in each component serves as the input for the next. With the exception of scoping and written documentation, the activities revolve around the three key activities and, therefore, are enclosed in the larger box. The scoping and written – documentation activities are shown outside the larger box because they represent the initial and final stages, respectively, in the methodological process and are not directly connected to the fundamental functions.

8.B.3.1   Screening

Screening determines the Potential Health Implications of the Policy or project under consideration to determine if an HIA is, in fact, required. The need for a HIA should be determined based on input from regulatory agencies, other pertinent organizations, and the general public and on the professional knowledge and judgement of the EIA study preparers, as a part of the EIA.

8.B.3.2 Scoping

In scoping the potential health gains or losses are estimated using available information on who will be benefitted/affected by he new program/policy intervention or a cluster of interventions. A review of baseline data on present health status of the population in area of study will useful to identify determinants.

In general a health impact focus should be included in any EIA study if the answer to any of the following questions is in the affirmative:

1. Does the nature of the proposed project (of activity) involve the handling of or emissions to the environment of materials such that their physical, chemical, radiological or biological nature may be harmful to human health?

2. Is the location of the proposed project, together with its nature, likely to give rise to conditions that would alter the occurrence of natural hazards in the study area?

3. Could the implementation of the proposed action eventually give rise to conditions that would reduce or increase the number of adverse health-impact-causing factors?

Review and Analysis of Pertinent Institutional Information

The institutional information pertaining to the project should be reviewed and analyzed prior to any intensive effort in the EIA process. For the purpose of assessing human health impacts any specific law, regulation, executive order, or guideline which directly or indirectly relates to human health should be identified for each pertinent administrative grouping (federal, state, regional, and/or local). In general, the institutional information will probably pertain to specific levels (or concentrations) of given health-impact causing agents. Thus the institutionally set level guidelines may be used to 1. determining reference doses without having to go through a dose-response assessment process and/or 2. interpreting health impacts by comparing the institutionally set levels against predicted exposure levels. In both cases, it is important to analyze the institutional information in order to identify the specific health effects and conditions for which the levels were established and to determine whether they should be applied to the conditions of the project.

Description of Project and Affected Environment

The description of the project and the effected environment is also needed for the assessment of other impacts; therefore, the collection effort can be minimized by coordinating the informational needs of all potential impacts to be addressed. The procedure for the procurement of health-related information, in particular, needed for the description of the project and that of the environment should consist of (1) structuring the description process in organizational units according to the health-related characteristics of the project and (2) collecting appropriate information for each organization unit.

Correctly organizing the description process in an EIA study in order to ensure thoroughness is important because of the great number of alternatives, phases, activities, processes, and environments that may be pertinent. Organizational units for the project and the environment may be delineated as follows:

(a) Project

- Identify the main components or activities of the project and its alternatives. Do not repeat components or activities common to two or more alternatives.

- For each component or activity, identify phases (for example, construction, operation and closure).

- For each phase of each main component or activity, identify sub-components or sub-activities.

(b) Environment

- Identify sub-components or sub-activities that may affect one or more environments; if this is not applicable, proceed with the following identification step

- Identify each environment affected by each main component or activity of the project and its alternatives.

In the context of a HIA, project information needed for each organizational unit includes the size and characteristics of the area affected, time schedules, labor – force characteristics, methods and equipment used, sources and levels of physical phenomena that may have the potential to harm human health, and sources and characteristics of any hazardous materials (chemical, biological, or radiological).

The description of the affected environment provides the basis for determining the ways in which humans may become exposed to health-impact-causing agents. It may also lead to the identification of naturally occurring health-impact-causing agents, that, if expected to be affected by the project, should be addressed in the HIA. Each environment affected by the components or activities of the project should be described in terms of three components:

1. The physical–chemical environment, including information pertaining to the meteorology and the geologic and hydrologic settings of the affected area;

2. The biological environment, including information on pertinent food chains, pathogenic organisms, and disease vectors; and

3. The human environment, including information on population, disease, land usage, health care systems, and pollution.

The risk of injury to people is an important consideration. Most industries and development projects have inherent risk of injury but here what we should assess is whether the level of risk is acceptable. The two factors which will affect the acceptable level of risk are: Whether the people at risk e.g., employees in a chemical factory and, the measures that the population placed at risk by the project, would have to take to protect themselves. We should also assess what action the project initiator should reasonably be expected to take to minimise the risk to unwitting public.

A feeling of well-being is closely related to the quality of one's environment; a perceived threat whether the real or imaginary, can have a real psychological impact on people.

Diseases such as malaria, dengue and schistosomiasis which are caused by parasitic organisms, are transmitted to people by vectors e.g., mosquitoes, snails. Control or eradication of the disease is affected through vector control. Any project development which provides a habitat for, or otherwise assists the proliferation of the vector, can cause increased incidence of the disease.

Communicable diseases can be spread through project development. Immigrant workers or their families may introduce communicable diseases such as typhoid, cholera etc. The indigenous population may be exposed to diseases with which they have not had previous contact and against which they have no immunological defense.

Many physiological diseases particularly diseases of the circulatory, respiratory and alimentary systems are provoked or aggravated by environmental conditions. Dietary imbalance, mental stress, sedentary lifestyles, air and water quality and socio-behavioral trends appear to be directly related to the rate of occurrence of one or more of these diseases. Included in this group are well known environmental diseases such as silicosis, heavy metal poisoning, radiation sickness etc.

Safety comes from Man's mastery of his environment and of himself. It is won by individual and group cooperation. It can be achieved only by informed, alert, skillful people who respect themselves, and have a regard for the welfare of others.

8.B.3.3   Identification of Potential Health Impacts

The identification of health impacts consists of three steps: 1. identifying the sources of potential health effects within each organizational unit defined during the description process, 2. defining those circumstances or scenarios under which health impacts from those sources may occur, and 3. identifying health-impact-causing agents associated with the sources and their health effects. The identification of health-impact sources requires professional knowledge, experience, and the exercise of professional judgement. In general, however, the following categories can be used to delineate potential sources:

- Processes or activities that involve the use, production, and/or handling of materials that may contain radioactive, chemical, and/or biological hazards.

- Processes or activities that involve the generation of conditions that increase (or decrease) the occurrence in the environment of natural hazards.

The scenarios under which health impacts from the identified sources occur can include 1. a "routine scenario," where emission into the environment of potential health-impact – causing agents occur on a regular basis. 2. an "extraordinary scenario," which considers health impacts resulting from low-occurrence but foreseeable events. and 3. a "maximum scenario," often used to describe a more realistic scenario in which routine emissions are combined with maximum levels of emissions that may occur periodically.

In the extraordinary scenario, it is necessary to estimate the probability of occurrence of the extraordinary events that will be considered in the assessment. Methods to identify extraordinary events include hazard and operability studies, technical audits, and examination of historical records (10). The probability of occurrence can be determined either through analysis of existing statistical data for the same or similar types of project components or activities, or through the systematic usage of events and fault trees (5).

The task of identifying health-impact-causing agents and their associated health effects is equivalent to the hazard-identification step in conventional Rig Analysis studies. In general, and for any type of health-impact-causing agent, the process involves an extensive review of studies and statistical records to determine whether exposure to the elements in or from the sources previously identified, and which are known or suspected to have some effect on human health, is likely to cause a change in the incidence of a health condition. The outcome of this process should be one or more lists of health-impact-causing agents, indicating their sources, levels (and the degree of uncertainty regarding these levels), associated health effects, and the nature and extent of evidence of health effects in humans that the agent causes.

8.B.3.4 Prediction of Health Impacts

The prediction methods included in this generic methodology are based on existing Rig Analysis techniques; these techniques include exposure assessment, dose-response assessment, and health impact characterization.

EIA studies often involve the analysis of several alternatives that may encompass multiple components, activities, and/or environmental settings. Therefore, the use of structured approach is needed for conducting the exposure assessment and subsequent prediction steps. For each combination of project components, the exposure assessment should include several levels of analysis the highest level of analysis is represented by the routine, extraordinary, and no-action scenarios, and a more detailed level of analysis corresponds to each sub-component (or sub-activity). Each sub-component should be analysed with regard to the worker's population and the general population.

The purpose of the no-action scenario is to delineate the baseline conditions. The exposure analysis for the no-action scenario is important because it can serve a predictive function by providing background exposure levels and an evaluation function by providing the basis against which the health effects resulting from the project action can be compared.

An "exposure assessment" involves characterizing exposure pathways and quantifying exposure levels. The characterization of exposure pathways should be conducted for each project component and its associated environment. The aim is to identify and quantify individual exposure pathways so that, when the project (or any alternative) is considered as a whole, they can be aggregated, where appropriate, into exposure scenarios. "Exposure scenario" refer to situations in which the same people are likely to be exposed to health – impact-causing agents through several exposure pathways. A key objective of the exposure assessment as conducted for the EIA process is to coordinate efforts in the prediction of impacts. That is, the exposure assessment should make all possible use of the predictive techniques used in the overall EIA process. This approach not only minimizes work, but it also fosters consistency in the prediction of impacts, since the same assumptions and criteria are used for all comparable impact calculations.

"Dose-response assessment" consists of describing the relationship between the dose of the health-impact-causing agent and the predicted occurrence of a health effect in an exposed population. Quantitative methods that have been developed or used for each type of health-impact-causing agent should be used in this methodology (5). The results from the dose-response and exposure assessments should then be integrated into quantitative expressions (if possible) of anticipated health-impact occurrences.

The characterization of anticipated health impacts involves 1. matching the estimated exposure doses with the appropriate dose-response values, 2. quantifying the health – effect incidence, 3. assessing the degree of uncertainty, and 4. summarizing the results. Quantitative characterization of measurable health effects involves a two-stage process.

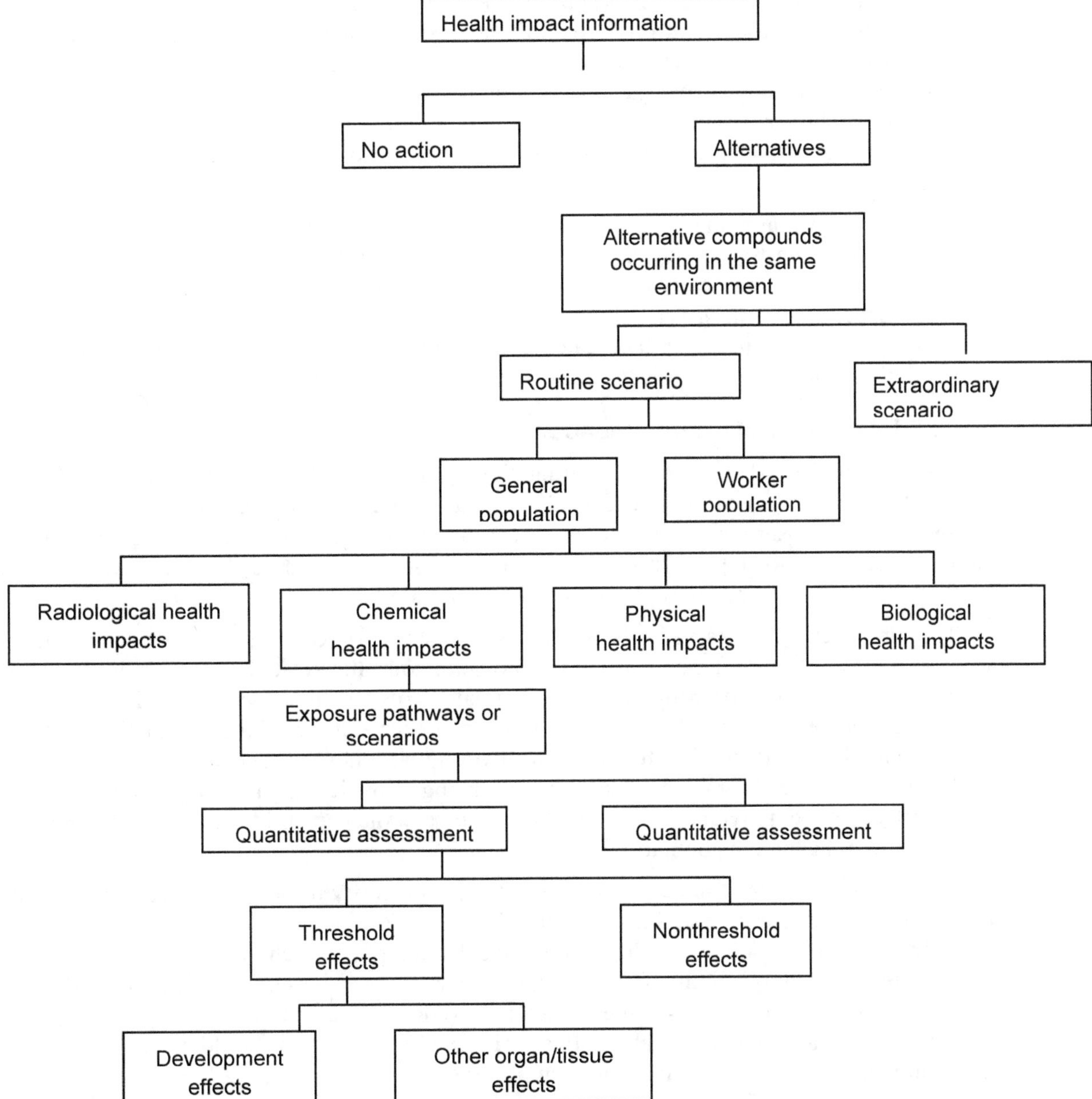

Fig. 8.8 Flow diagram of qualitative and quantitative results of the characterized health impacts.

Source: Vorhees and Associates: (1975) Interim guide for Environmental Assessment prepared for U.S department of housing and urban development Washington D.C.

In the context of the EIA process, the evaluation of uncertainties associated with the prediction of health impacts is of utmost importance for the rational interaction of these impacts. The most sophisticated methods for estimating uncertainty include numerical methods such as the Monte Carlo simulation method, and series approximation (10). Use of these methods, however, requires a substantial input of data are sufficient to describe the potential range the parameters might exhibit, a sensibility analysis can be used to identify influential variables and to develop bounds on the distribution of health effects. The most practical approach to characterize uncertainty in most EIA studies will be a qualitative description of the uncertainty for each parameter developed, followed by a qualitative indication of the possible influence of these uncertainties on the numerical estimate of the health effects.

The qualitative and quantitative results of the characterized health impacts can be organized according to the flowchart shown in Fig. 8.8. In this figure, only the chemical health impacts associated with a given project component, on the general population, for a routine scenario and a given alternative, are fully depicted. The same method of organization could be used for each category of health impact, project component, type of exposed population, type of project scenario, and project alternative.

8.B.3.5 Evaluation of Health Impacts

The method discussed here for evaluating the significance of anticipated health impacts should be used for assessing those health effects associated with each of the exposure pathways. On exposure, scenarios delineated in the evolution method consists of two main steps and several sub steps such that when the significance could not be established in the previous step or sub step, a new step or sub step is undertaken.

The first step involves establishing the significance of each impact by determining whether any health-impact-causing agent is projected to health-related regulatory limits. The second step involves determining the significance of health effects, first in terms of magnitude and regulatory criteria, then in terms of uncertainty and other evaluation factors such as cumulative effects, institutional information that might be especially relevant to the health effects or health-impact-causing agents being considered, the repercussions of the health effects at the individual and population levels and the perceptions related to the health effects among the human population.

One of the purposes of the evaluation process is to identify those health effects that need to be considered during the impact-mitigation planning process. These health effects, therefore, need to be included in the alternative selection process, and included in a way that is consistent, not only within each alternative, but across alternatives. Thus in order to ensure this consistency of use, these factors should not be influenced in any way by variations that may exist among alternatives. The evaluation factors should be considered by decision-makers in the alternative-selection process.

8.B.3.6 Identification and Evaluation of Mitigation Measures

In most EIA studies, mitigation measures for undesirable health effects can fall into one or more of the following three categories (10): 1. mitigation through control of source, 2. mitigation through control of exposure, and 3. mitigation through health-services development. Mitigation measures to control sources act by preventing or limiting the introduction into the environment of the health-impact-causing agents. The control of sources can be achieved through engineering techniques such as modifications in the design of the project (e.g., introducing water or product recycling) or through management methods which involve adjustments of project components according to changes in environmental conditions (e.g., reducing the activity of processing plant during thermal inversions). Control of exposure is usually achieved by preventing or limiting the access of individuals to the contaminated or potentially contaminated medium, by preventing the contaminant from reaching individuals, by stopping or reducing contact with the health-impact-causing agents, by warning individuals of potential hazards, or by devising contingent plans for both workers and the general public.

Mitigation through health-services development may involve the implementation of health education programs or the development of health-impact-prevention and health care systems. Preventive-health-oriented measures (such as vaccination programs) are aimed at protecting individuals from acquiring a disease. Health care approaches might include planning systems for the treatment of diseases or other effects of physical, chemical, or radiological exposures should such symptoms develop in the affected individuals.

Two major considerations in identifying health-effect mitigation measures are that not all the identified measures are necessarily technically or economically feasible, and that public participation is important for selecting acceptable measures. In identifying any quantifiable mitigation measure, sufficient information should be aggregated to allow quantification of the reduction of the impact.

For mitigation measures which do not allow quantification of the mitigation effect or for health effects that were characterized on a qualitative basis, a qualitative estimation of the mitigation effect should be developed using professional knowledge and judgement.

Selection of Proposed Action Alternative

Selection of Proposed action alternative to assist in the process by organizing and presenting the information on health impacts is a way most useful to the decision. The suggested approach for organizing and presenting health-impact information consists of the following steps:

Table 8.4 Potential scale for rating health effects.

Sl. No.	Signification	Rating
1.	No significant impacts	0
2.	Nondisabling, reversible adverse health effects affecting a limited number of people	– 1
3.	Nondisabling, reversible adverse health effects affecting a large number of people	– 2
4.	Disabling (not leading to death), reversible adverse health effects affecting a limited number of people	– 3

Table 8.4 *contd...*

Sl. No.	Signification	Rating
5.	Disabling (not leading to death), reversible adverse health effects affecting a large number of people	− 4
6.	Disabling (leading to death), reversible adverse health effects, either short-term or long-term, affecting a limited number of people	− 5
7.	Irreversible, long-term adverse health effects affecting a limited number of people	− 6
8.	Disabling (leading to death), reversible adverse health effects, either short-term or long-term, affecting a large number of people	− 7
9.	Irreversible, long-term adverse health effects affecting a large number of people	− 8

[**Note :** Use same system for beneficial effects, but change − sign to + sign].

1. Classify the health effects that were found to be significant, using the following categories – carcinogenic effects, hereditary effects, teratogenic effects, organ-tissue effects (including traumatic effects), and infections from biological health-impact-causing agents.

2. For each alternative and for each project scenario, estimate the number of people that would be affected by each category of health effects given above.

3. Assign a rating value to each category of health effects for which population numbers were estimated.

4. Present the results obtained in steps 1, 2 and 3. These results should be accompanied by evaluation factors that are affected differently for each alternative.

Monitoring of Health Impacts

Normally, and in contrast with other environmental impacts, health impacts are not easy to detect, and, if they are detected, it may be difficult to establish a clear relationship between the health effect and the project. This is especially true for carcinogenic and hereditary effects, which may take a long time to develop, to be influenced by several confounding factors and be difficult to discern from background level effects. Thus, while one of the purposes of environmental – monitoring programs is to provide an early warning of unanticipated adverse impacts, sudden changes in impact trends, or approaches to pre-selected critical levels, it has to be recognized that as of today, with the techniques available to detect environmental health effects, health-impact monitoring is to provide information which can be used to 1. document health impacts that might result from a proposed action, 2. review and validate impact prediction techniques, 3. evaluate the effectiveness of implemented mitigation measures, and 4. enable a more accurate prediction, in the future, of health impacts associated with similar actions. For short-term health effects, however, monitoring could possibly also serve as a warning system.

Because the ultimate purpose of health monitoring is to determine whether a cause-response relationship exists, and to quantify the relationship, if possible, the monitoring activities should be established at two levels. One level would account for the potential causes of the predicted health effects and, therefore, consist of monitoring the levels of the health-impact-causing agents at their sources, as well as predicted exposure points. The other monitoring level would involve the detection and recording of the health effects that develop in the non-exposed and potentially exposed populations. This monitoring level necessarily involves the health-care community of the affected area.

Finally, in terms of data collection, the monitoring of health-impact-causing agents would probably require the use of monitoring stations and systems developed specifically for the project. These monitoring stations and systems should be integrated, whenever possible, into the monitoring systems for other environmental impacts so as to minimize expenditure of resources and avoid duplication of efforts.

8.B.3.7 Preparation of Environmental Impact Statement

The findings of the environmental impact study, and in particular the information and findings corresponding to the selected alternative, need to be summarized and organized in a written EIS. The information generated and used in predicting and assessing health impacts is normally extensive. Therefore, it is recommended, as part of this methodology, to document each of the previous activities. This information can then be incorporated, so that the information pertinent to the prediction of health impacts is contained in the appropriate sections of the EIS; the most-extensive material should be presented in appendices, and the external supporting data and literature should be adequately referenced.

8.B.3.8 HIA for Policy Making

Either HIA alone or as integrated with any broader impact assessment play important role in placing health on agenda for all development programs, bringing public participation and awareness. Further it will be useful to develop procedures for deciding on policy alternatives. HIA are now made as part of SEA and EIA.

Summary

Major Societal Developmental projects can cause potentially significant changes in many features of the socio-economic environment. Large scale development projects also involve significant requirements for associated infrastructure, such as, streets, highways, or railroads, water supply, sanitary sewers, storm-water drainage, electrical systems, gas systems, and telephone communication systems which can also generate environmental impacts. In some cases the changes may be beneficial while in others they may be detrimental. Accordingly, environmental impact studies must systematically identify and quantify where possible, and appropriately interpret the significance of these anticipated changes. The methodologies to be followed for Prediction and Assessment of Impacts of major development Projects on Socio-Economic and Human Health and the technical details for collecting data/information for each criteria of impact assessment are discussed in this chapter.

References

1. Rietbergen-McCracken, J. and Narayan, D. (1998): *Participation and Social Assessment: Tools and Techniques*, The World Bank, Washington DC.

2. Grady S. Braid .R Bradbury J. and Kerley. C. (1987) Socio-economic assessment of plant closcese : Three case studies of Large manufacturing facilities. Environmental Impact Assessment Review. Vol.7 pp 151-165.

3. Guidelines and Principles for Social Impact Assessment, 1994. Prepared by the Inter-organizational Committee on Guidelines and Principles for Social Impact Assessment.

4. Impact Assessment and Project Appraisal, 2003 Volume 21, number 3.

5. Vorhees and Associates: (1975) Interim guide for Environmental Assessment prepared for U.S department of Housing and Urban Development Washington DC.

6. Taylor N, Goodrich C and Bryan H, 1998. Social Assessment. In Porter A and Fittipaldi J (eds) Environmental Methods Review: Retooling Impact Assessment for the New Century (pp.210-218). The Press Club, Fargo, USA.

7. Berger T (1994) The Independent Review of the Sardor Sarovar Projects, 1991-1992. *Impact Assessment* 12:1, 3-20.

8. World Health Organisation (WHO), (1985) "Environmental Health Impact Assessment of Urban Development projects," WHO Regional office for Europe, Copenhagen.

9. World Health Organisation (WHO), (1986) "Health and Safety Component of Environmental Impact Assessment," report on a WHO Meeting, Regional Office for Europe, Copenhagen.

10. Arquiaga M.C. (1991) Generic Health Impact Prediction and Assessment Methodology for Environmental Impact studies, Ph.D thesis, University of Oklahoma, Norman.

Further Reading

1. Asian Development Bank. 1994. Handbook for Incorporation of Social Dimensions in Projects. Social Dimensions Unit, Asian Development Bank. Manila, Philippines. 105 pp

2. Glasson, J., 2000. Socio-economic Impacts 1: Overview and Economic Impacts, In: Morris, P. and Therivel, R. (2000) (ed), *Methods of Environmental Impact Assessment*, Spon Press, London and New York.

3. *Impact Assessment and Project Appraisal*, volume 21, number 3, September 2003, pages 231–250, Beech Tree Publishing, 10 Watford Close, Guildford, Surrey GU1 2EP, UK.

4. International Association for Impact Assessment. Social Impact Assessment, 2003.

5. International Principles, Special Publication Series No. 2, May. National Environmental Policy Act (NEPA) , 1998. Fact sheet: Social Impact Assessment. http://www.gsa.ene.com/factsheet/1098b/10_98b_1.htm

6. World Bank, 2003. Social Analysis Sourcebook www.worldbank.org/

7. Social analysis http://www.forests monitor.org/reports/vanimo/summ.htm

8. http://www.nmfs.noaa.gov/sfa/social_impact_guide.htm#sectV

Questions

1. Explain what is meant by social assessment? Why Socioeconomic impacts are important in EIA of a major development project?

2. Explain the essential features of any systematic approach for Socioeconomic impacts of any major project activity.

3. How can the Socio-economic impacts of a Project be identified. Discuss the significance of social factors in the implementation of project activities.

4. What are the basic steps to be followed for incorporating social dimension in a project?

5. Discuss the factors to be considered for fixing Region of Influence (ROI) for assessing Socioeconomic Impacts.

6. Discuss various criteria to be followed and general methodology to be adopted with a flow chart for predicting the Socio Economic Impacts of any project activity.

7. Discuss the procedure to be adopted for assessing in Socio Economic Impact.

8. Discuss the main tasks to be addressed as given by WHO in any Health Impact Assessment of any major project activity.

9. Discuss the systematic approach for assessment and prediction of health impacts of any major project activity.

10. Discuss the basic aspects of Social Impact Assessment and Social Assessment Process.

11. Explain Social Impact Assessment Variables. Discuss salient features and specific advantages of Social Impact Assessments (SIA) of major multi dimensional development projects.

12. Discuss the role of Social Impacts in framing Sustainable Development Policies.

13. Give the general aims and basic principles to be followed for any Social Impact Assessment.

14. Discuss the various methods for scoping the Social Impacts.

15. How do you analyze and predict probable impacts of the project proposal and the alternatives against baseline conditions (with versus without the action)? Discuss with examples.

16. How do you assess the significance of Socio-Economic Impacts?

17. Describe the process for evaluation of alternatives and impact mitigation.

18. Discuss the procedure for assessing Socio-Economic Impact significance.

19. What are the criteria for assessing impact significance?

20. Evaluation of responses of client groups to impacts will help to assess impact significance. Explain.

21. Explain the salient features of monitoring and Mitigation Measures in the SIA of any Project.

22. What are various general factors that help for Overall Success of SIA?

23. What is Health impact assessment? Explain its objectives.

24. Explain different Tasks of Health Impact Assessment.

25. Describe a systematic approach for health-impact – prediction and assessment.

26. How will Health Impact Assessment (HIA) be useful for policy making? Explain.

CHAPTER 9

Environmental Risk Assessment (ERA) and Risk Management in EIA

9.0 Introduction

Environmental risk is a probability value of an undesirable event and its consequences that arise from a spontaneous natural origin or from a human action that is transmitted through the environment, and cause harmful, damaging and even destructive effects on human society and natural environment (1, 2). It commonly exists in all kinds of human activities and varies in characteristics and modes.

A variety of frameworks of environmental assessment and risk assessments such as Environmental Impact Assessment (EIA), Human Health Risk Assessment (HHRA), Ecological Risk Assessment (ERA), Strategic Environmental Assessment (SEA), the precautionary principle, and Sustainability Assessment (SA) etc are widely used in different scenarios of environmental management. These frameworks are designed for different purposes and are used in different ways.

Since the concept of Risk has been applied over a broad range of disciplines and activities to avoid confusion Royal Society (3) defined some of the terms used in Risk Assessment as given below:

Hazard: a property or a situation with the potential to cause harm.

Risk: a combination of the probability or frequency of the occurrence of a particular hazard and the magnitude of the adverse effects or harm arising to the quality of human health or the environment.

Probability: the occurrence of a particular event in a given period of time or as one amongst a number of possible events.

Risk Management: the process of implementing decisions about accepting or altering Risks. Stressors can be chemical, physical, biological, or radiological in nature. A given ecological risk assessment study could be simple or comprehensive (multiple ecological components, stressors, and activities).

9.1 Environmental Impact Assessment (EIA)

EIA involves prediction of Risks based on quantification of cause-effect relationships. However uncertainty is an inherent character of all natural systems (4) and any predictions or assessments made on the impacts of various activities on these natural systems are bound to suffer with these uncertainties. Environmental Impact Assessment provides the scope and data for dealing with these uncertainties. As large opportunities exist for variability of data, and partial understanding of development activities and their consequences, impact predictions are bound to suffer from these uncertainties. The EIA may use the average (mean) or expected value, or alternatively, a worst case value. The implied choice may be conservative or optimistic, and is usually internally inconsistent. The use of mean values for measurements of effects, when the actual data are widely scattered and/or skewed from a symmetrical bell shaped pattern, is also misleading. The correct and appropriate way to characterize data is to describe the statistical distribution of a range of values and the confidence with which that range is held to be true. EIA studies can identify linkages between injurious events and/or activities and their damaging consequences which will help in proper risk management decisions. The relationships amongst ecology, society, and economic concerns which are three pillars of EIA, the investigation of the correlations amongst different types of environmental assessments and their merits and demerits is necessary using integrated environmental assessment. This facilitates decision makers in selecting proper assessment framework and/or some combinations of them.

9.2 Environmental Risk Assessment (ERA)

It is an emerging technique which involves structured gathering of available information about the environmental risks and then formation of judgment about them (5,6). The communication of ERA results should take the form of decision analysis, that is, what options are available and, for each option, what are the risks, costs, and benefits, and how are these distributed within society. Proper comparison and communication can change lay people's misperceptions of risks so participatory decision making may proceed on a more rational, less emotional basis. Environmental Risk Assessment and Management with reference to issues relating to human health and ecological issues is a rapidly growing field which can provide information to decision makers about the frequency and magnitude of adverse environmental consequences arising from human activities or planned interventions.

9.3 Ecological Risk Assessment (EcoRA)

EcoRA deals with the condition of ecosystems rather than human health or individual organisms. Ecological risk assessment is defined as a process that evaluate the likelihood of adverse ecological effects that may occur or are occurring due by exposure to multiple agents (e.g., physical disturbances, and toxical chemicals [3], (7) which must be based on sound ecological theory.

EcoRA has a host of special problems (for example, choosing assessment endpoints) that are well recognized and that add to the uncertainty of the results.

In Ecological risk assessment as there are no definable harm like that of premature death used in human health risk assessment, appropriate criteria have to be chosen based on scientific information and social judgments. Ecological risk assessment has been defined as a process that evaluates the probability or likelihood that adverse ecological effects will occur or are occurring as a result of exposure to stressors from various human activities. Such effects can occur on non-human ecological components ranging from organisms, to populations and communities, to ecosystems. Both human health risk assessment and ecological risk assessment within the EIA process can be accomplished using one to several of the following approaches:

- addressing actual or perceived risks using a descriptive or qualitative approach.
- calculation of a relative risk index based on information on several selected factors.
- comparisons of the perceived risks of the alternatives being evaluated.
- a quantitative, probabilistic approach focused on actual risks of the alternatives being evaluated.

9.4 Comparison between EIA and ERA

Both EIA and ERA are structured tools which give recommendations concerning environment to the decision makers   for project implementation which can be extended to strategic levels of decision making. Though they have similarities they have some fundamental differences. While EIA involves consideration of development alternatives ERA does not. Both are iterative processes and in final stage after implementation of a project or proposal monitoring and auditing are needed. After learning from experiences and mistakes committed, decisions are to be improved next time. While EIA gives thrust to public participation and consultation, risk perception and risk communication are advocated in ERA. The salient features of these two processes are compared in Table 9.1.

Table 9.1 Salient features of ERA & EIA

Framework for ERA	Frame work For EIA
Screeningto determine the range of risks and the factors that control and whether they are likely to result in damage to the environment. When all risks have been identified prioritization or ranking is conducted to ensure that resources for further work are targeted at the highest priority risks. Defining the problem is also known as **hazard identification**	**Screening** of the project or proposal and preliminary assessment of the existing environment to decide whether to carry out a detailed EIA followed by Scoping of the key environmental issues likely to be affected by the project or proposal.
Hazard Analysis involves identification of the routes by which hazardous events could occur and estimation of the probability or chance of occurrence. **Consequence analysis** involves determining the potential consequence of hazard. **Risk determination** combines the results of hazard and cosequence analysis.	**Baseline Studies-** collection of existing information **Impact Prediction – determining the magnitude spatial extent and probability of impacts including direct and indirect effects.**
Risk evaluation i.e., whether the environment is likely to withstand the effects. It may be partly correct to take decisions in response to pressures generated by risk perception. Risk management options include tolerating or altering risks.	**Assessment**: of the relative importance of the predicted effects taking into account the present condition and future condition that would result as well as mitigation measures. **Evaluation** of the overall acceptability of the proposal or project and each of its alternatives leading to selection of one or more preferred options.
Monitoring and Audit e.g., leading to confirmation or rejection of predicted effects.	**Monitoring and Audit** Confirmation or rejection of predicted effects.

9.5 Human Health Risk Assessment and Ecological Risk Assessment

Human health risk assessment (HHRA) is evaluation of potential adverse health effects on humans exposed to environmental hazards together with uncertainties (4).

Four major components are included in a risk assessment process (8):

1. Hazard identification: This step identifies whether the contaminants present lead to adverse effects on human health;

2. Exposure assessment: This step estimates the intensity, frequency, and duration of exposure;

3. Dose-response assessment; This step determines the relationship between dose and the adverse effects based on animal experiments and epidemiological studies; and finally,

4. Risk characterization.

This four-step process is used to describe the nature and magnitude of risk for each exposure pathway and derive the total risk.

Assuming protection of human health then will protect non-human receptors is not sufficient as some chemicals like chlorines, ammonia and aluminium will cause serious effects for aquatic organisms, but pose no risk or negligible risk to humans (9).

Suter II (9) presented that human health risk assessment is a complement of ecological risk assessment. Thus integration of human health risk assessment and ecological risk assessment is necessary for environmental management in many cases.

There are four significant differences between human health risk assessment and ecological risk assessments:

1. The human health risk assessment has determined receptors, i.e., humans, while there are no universal ecological risk assessment endpoints;

2. Ecological risk assessment considers the effects on population, communities, or system level while the human health risk assessment most often focuses on a hypothetical adult (e.g., an adult with average weight is 70 kg and life time is 70 years for cancer risk assessment etc.);

3. Usually ecological risk assessment hasto capture the effects caused by the interactions among multiple agents such as physical disturbances and toxical chemicals;

4. It is easy to determine the exposure pathways for human health risk assessment, e.g., ingestion, inhalation, and dermal contact, while exposure analysis in more complicated in ecological risk assessment as an agent will act periodically and has significantly variation in exposure sites, or take different alternatives in the future. It is not easy to track the fate and transport of contaminants in ecosystems (10).

9.6 Risk Assessment and Treatment of Uncertainty

Risk assessment (RA) is the scientific method of confronting and expressing uncertainty in predicting the future. Risk is (has the dimensions of) the chance of some degree of damage in some unit of time, or the probability of frequency of occurrence of an event with a certain range of adverse consequences.

These probabilistic expressions, as opposed to a single (mean or expected) value, are what distinguish RA from mere impact assessment. The irreducible uncertainties leave uncertain the calculated absolute value of a risk, often by as much as an order of magnitude or more. Risk itself, therefore, should be expressed and communicated as a mean plus its standard deviation, in addition to any upper bound or worst case calculation that may be used for purposes of conservative policy. Environmental risk assessment is the process of evaluating the likelihood of adverse effects in, or transmitted by, the natural environment from hazards that accompany human activities.

9.6.1 What is Uncertainty

Science is the activity of understanding the regularities of the universe and revealing the simple laws that produce them. Prediction (for example, EIA) to guide human actions is also a primary goal of science; however uncertainties interfere, causing what occurs to differ from what was expected. Scientific truth is always somewhat uncertain, and information is characterized by kinds and degrees of doubt, change, and availability.

Ecological uncertainty is of two basic kinds: what is not known at all, and errors in what is known. The latter type is a quantitative departure from the truth, and can sometimes be expressed statistically as a distribution of a number of repeated measurements around a mean value. Other forms of quantitative uncertainty are incomplete data, "anecdotal" data which is not gathered with a statistical design, inappropriate extrapolation, and temporal and spatial variability of the measured parameter. Most important environmental problems, however, suffer from true uncertaintythat is, indeterminacy, or events with an unknown probability. Totally unanticipated rapid and adverse changes in ecosystems may arise from apparently unrelated policy or social events. Long-term effects sometimes become evident only much after the original cause, and explanation may be confounded. For example, surprising consequences in ecosystem behavior are likely as a result of rapid climate change, carbondioxide fertilization, and enhanced UV radiation. Ocean warming may bring larger and more frequent typhoons, previously not experienced by tropical coastal zones. Multiple causes and nonlinear response are also sources of "unknowable" outcomes.

Uncertainties have relative importance, depending on their size. Stochasticity is the variation in response of an ecosystem due to random, uncontrollable factors such as weather, and not to the stressor being studied. If stochasticity is large, a smaller degree of error in measurement is not of much help, and the predictive power of the EIA/ERA is limited.

Another source of uncertainty is bias resulting from measurement and sampling errors of the parameters in a model. These may be reduced at some cost.

Natural Variation

A familiar source of uncertainty is natural variation. Any "signal" that a change in condition is due to human action is often hidden in the "noise" of natural changes in the value measured. For example, no statistically significant change in the volume of water discharged by the Amazon River, or the amount of sediment delivered from the deforested Rondonia region, has yet been detected. In this case, a signal that deforestation has altered the hydrologic cycle or soil erosion rate in that basin is obscured by the high natural variability of rainfall.

Human and ecological risk assessments use quantitative methods to estimate the human and ecosystem risks posed by agents or stressors. Thus both of them can be used in each assessment framework to enhance the evaluation of the effects caused by a variety of stressors on human and ecosystem health. Uncertainties involved in each step and their cumulative on final results are represented and transparently propagate through the whole risk assessment paradigms. The methodologies developed for handling uncertainty in risk assessment can also be used inother environmental assessments.

9.7 Key Steps in Performing an Environmental Risk Assessment (ERA)

9.7.1   ERA Addresses Four Questions

- What can go wrong to cause adverse consequences?
- What is the probability of frequency of occurrence of adverse consequences?
- What are the range and distribution of the severity of adverse consequences?
- What can be done, at what cost, to manage and reduce unacceptable risks and damage?

Environmental hazards one should consider:

- **environmental impact of raw materials**, e.g., potentially toxic metals or other materials
- **environmental impact of packaging**
- **waste storage and disposal**, e.g., making sure that proper containers are used, and are located away from drains and watercourses
- **emissions**, e.g., dust and other substances to the air
- **hazardous substance** storage, use and disposal
- **liquid waste** drainage and disposal

EIA should answer the above question, and give at least a qualitative expression of the magnitude of the impacts. The major additional consideration in ERA is the frequency of adverse events. Risk management is integrated into ERA because it is the attitudes and concerns of decision makers that set the scope and depth of the study. ERA attempts to quantify the risks to human health, economic welfare, and ecosystems from those human activities and natural phenomena that perturb the natural environment. Fig. 9.1 gives the frame work of basic elements within which Environmental Risk assessment can be carried out along with options of generic and tailored Quantitative Risk Assessment (QRA).

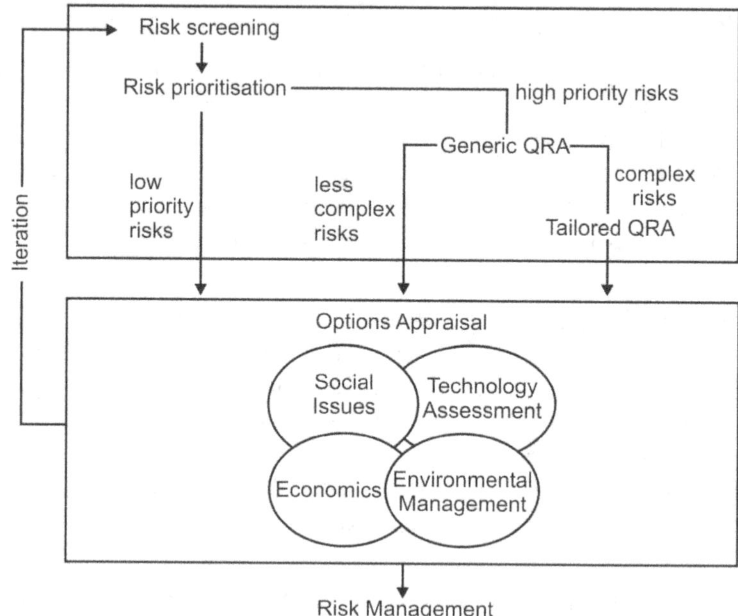

Fig. 9.1 A framework for environmental risk assessment
(**Source** :DOE1995).(11)

Environmental Risk assessment with reference to human health and ecological issues from human activities involve analyzing uncertainities which require the following basic steps which are iterative.

The five step sequence in performing ERA is:

1. hazard identification - sources of adverse impacts;
2. hazard accounting - scoping, setting the boundaries of the ERA;
3. scenarios of exposure - how the hazard might be encountered;
4. risk characterization - likelihood and severity of impact damage; and
5. risk management - mitigation or reduction of unacceptable risk.

9.7.2 Step 1 - Hazard Identification

This step, which is akin to the qualitative prediction of impacts in EIA, begins to answer the question "What can go wrong?" It lists the possible sources of harm, usually identified by experience elsewhere with similar technologies, materials, or conditions. This is, in fact, a preliminary risk assessment, immediately useful to managers in appraising the project or activity upon which they are embarking. Hazardous chemicals are a major topic for ERA. Elaborate screening procedures have been devised to judge when a chemical merits full investigation (Carpenter et. al., 1990) (10). The U.S. Environmental Protection Agency (EPA) and the World Bank issue threshold guidelines based on frequently revised lists of highly toxic chemicals. These thresholds indicate the amounts of each chemical, if present at any one location, that trigger risk assessment and emergency planning. Similar quantity-related guidelines are issued for highly reactive and flammable materials. Some of the major hazards associated with some important development projects are listed in Table 9.2.

Table 9.2 Major hazards associated with development projects.

Type of Project	Toxic Chemical	Flammable or Explosive Material	Highly Reactive or Corrosive	Extreme Conditions of Temperature or Metal	Large Mechanical Equipment Pressure	Collisior
Refinery/Petrochemical	×	×		×		
Pesticide Manufacturing			×	×		
Fertilizer (N &P)	×			×	×	
Pulp and Paper				×		
Thermal Electric	×		×	×		
Oil and Gas Transport	×	×	×			
Heavy Chemicals	×	×				
Light Chemicals	×		×			
Smelting	×		×	×	×	
Cement	×			×		
Railroad	×	×	×	×		
Highway	×		×		×	
Hazardous Waste	×		×		×	
Ports and Harbors	×		×	×	×	×
Dam and Reservoir				×	×	

(**Source** :DOE19995)(11).

9.7.3   Step 2 - Hazard Analysis

In the second step, ERA (a) considers the total system of which the particular problem is a part, (b) begins to answer questions about the frequency and severity of adverse impacts, and c) sets the practical boundaries for the assessment. Much of the hazard accounting will be covered during the scoping of the EIA. For example, a hazardous chemical may pose a risk in any stage of its life cycle (Fig. 9.2) that is, from mining/refining or synthesis through manufacturing, processing and compounding, to storage and transportation, to use and misuse, and finally, to post-use waste disposal or recycling. The scope should include the social and natural systems around a project, not just a single pollutant path. For example, it would be wrong to assess the risk posed by small concentrations of halomethanes produced incidentally to the chlorination of drinking water without comparison of the risks to the same public from not killing the pathogenic organisms with chlorine. During the hazard analysis step, it is to be decided which parts of the flow cycle are appropriate to be included in the ERA (dependant on the management questions being asked).

Risk managers must state their concerns and indicate possible linkages of operations to mitigation measures. Some of the scoping choices to be made are:

- geographic boundaries;
- time scale of impacts;
- stages of the causal chain of events;
- phase or phases of the technological activity;
- whether to include routine releases or just accidents;
- whether to include workers or just the general population;
- definitive end points for health or ecosystem effects; and
- cumulative effects and interactive risks that result from other projects.

The time covered should include all phases of an activity where risk is important, not just the operational period. Construction, maintenance and dismantling may pose special hazards. For example, it is well known that the Chernobyl nuclear reactor was being tested, and normal safety systems were disabled, at the time of the disaster. Toxic effluents such as heavy metals may circulate for a long time and nuclear wastes may have half lives of thousands of years. It is common practice to look at least one lifetime (about 75 years) into the future.

The important point is that the time horizon should be consciously chosen and recorded as one of the assumptions of the ERA.

A causal chain for a risk may stretch from an original decision to satisfy some wants and needs, through the choice of technologies, to adverse events, to exposure conditions, and finally to health impacts. In a sense, the Bhopal accident originated with India's desire to be self sufficient and to invite the local manufacture (incidentally by a multinational concern) of pesticides necessary for the protection of food crops. Such a comprehensive analysis of all related human activities is difficult and infrequently attempted.

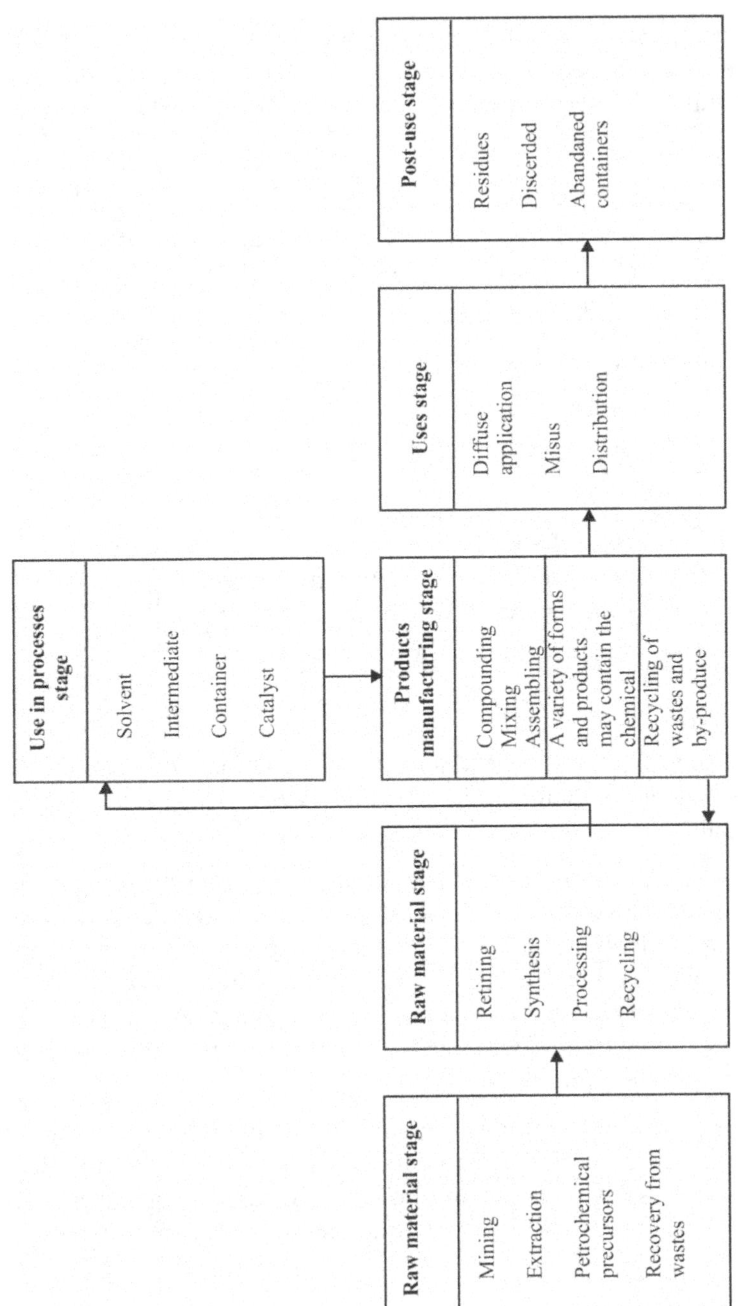

Fig. 9.2 Generic flow cycle for hazardous chemicals. Any of the stages may involve suboperations packaging, storage and transportation during the hazard accounting step, it is decided which parts of the flow cycle are appropriate to be included in the ERA (dependent on the management questions being asked).

(**Source** :*Smith et.al., 1988*)(12)

Environment consists of biotic and abiotic things. If for example one needs to calculate the risk in a flowing river, it is required to develop risk ratio for all receptor types and then use them together to come up with overall environmental risk ratio estimate.

Receptor type	Total Risk	Acceptable Risk	Risk Ratio
Humans	$Risk_{total,H}$	$Risk_{acceptable,H}$	r_H
Aquatic life	$Risk_{total,aq}$	$Risk_{acceptable,aq}$	r_{aq}
Physical environment	$Risk_{total,river}$	$Risk_{acceptable,river}$	R_{river}

Overall risk ratio for environment is given by:

$$r_{envt} = \sum_{i=1}^{N_r} w_i r_i = w_1 r_1 + w_2 r_2 + ... \qquad .....(9.1)$$

R_{envt}	Risk ratio for environment
w_i	Important weights for different receptors (say W_H for humans, w_{aq} for aquatic species)
N_r	Total number of receptors

9.7.3.1 Fault Tree Analysis

During hazard analysis the sequence of events which could lead to hazardous incidents is set out. The likelihood of the incident is then quantified. Fault tree analysis plays a key role in this part of the risk assessment.

Fault tree analysis is normally used to evaluate failures in engineering systems. The analysis provides a graphical representation of the relationships between specific events and the ultimate undesired event (sometimes referred to as the "top event"). For example, the ultimate undesired event might be a large fire for which the preceding events might be both spilling a large quantity of flammable liquid and introducing a source of ignition. Fault tree analysis allows systematic examination of various materials, personnel, and environmental factors influencing the rate of system failure.

The method also allows for the recognition of combinations of failures, which may not otherwise be easily discovered. The fault tree analysis is sufficiently general to allow both qualitative and quantitative estimates of failure probabilities within the analysis. A typical fault tree is given in Fig. 9.3, Fig. 9.4 and Fig. 9.5 is an example of a fault tree applied to biotic systems.

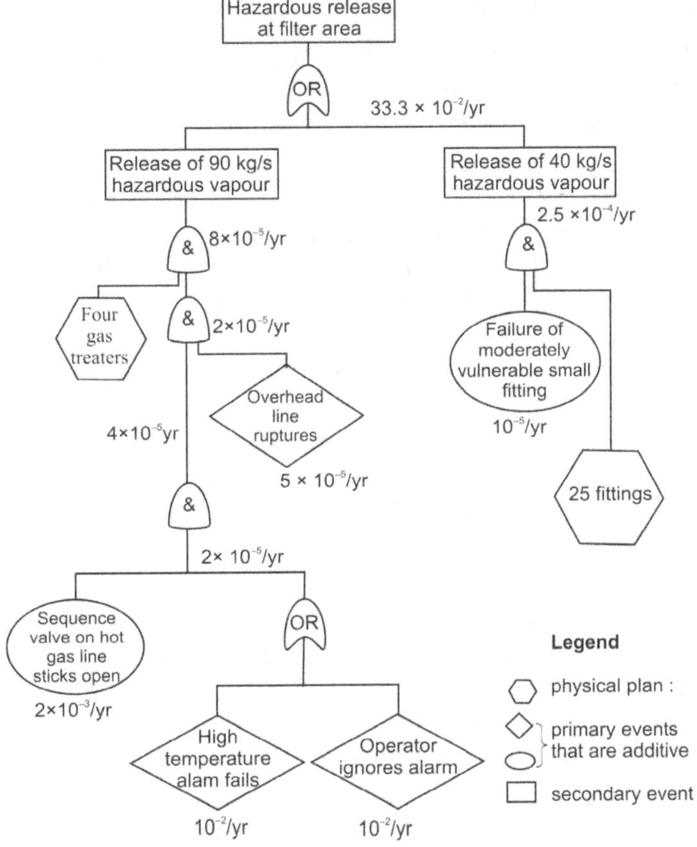

Fig. 9.3 A fault tree.

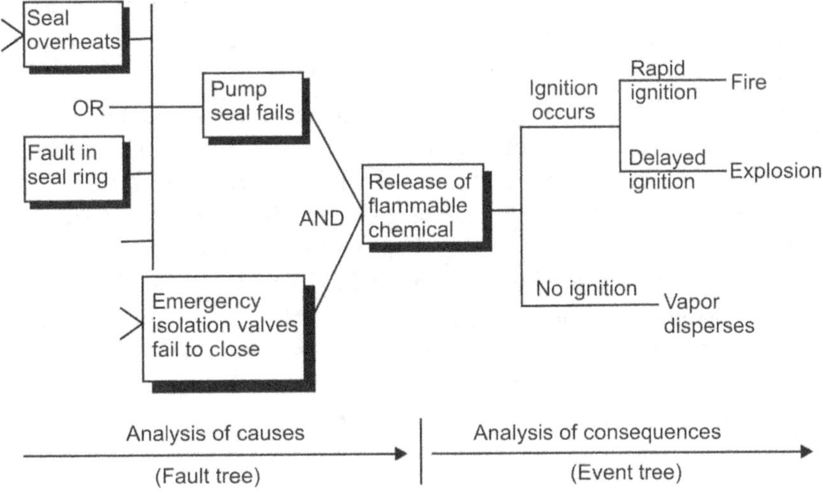

Fig. 9.4 A fault tree.

9.7.3.2 Uncertainty in Hazard Analysis

The major uncertainties in hazard analysis are the ability of the analyst to include all important initiating events and the reliability of the figures used to quantify likelihood. Several points should be noted concerning these uncertainties, including:

1. The *magnitude of the final likelihood figures* for hazardous events provides an immediate indication of whether all initiating events have been considered. As a broad generalization, if the predicted likelihood of a hazardous event, to which many different initiating events may contribute, is much below 10^{-6} yr-1, the chances are some initiating events have been missed. This is based on practical observation in Western Europe of the likelihoods of failures of well-engineered structures and of major natural disasters. This is not to say, however, that events with lower predicted likelihoods should be excluded from analysis.

2. The *probability distribution* associated with particular events is a second concern. Say a bank of data on equipment failures gives a statistical confidence that a particular failure rate is $2 \pm 0.5 \times 10^{-4}$ yr-1. One could therefore use a figure of 2×10^{-4} yr-1 with confidence in a "best estimate" approach and 2.5×10^{-4} yr-1 in a conservative approach. Clearly the results of both approaches would be similar.

3. *Precision and accuracy* is the third concern. It should be clear by now that results of the form 2.56×10^{-3} yr-1 immediately convey a lack of appreciation of the uncertainties inherent in quantification. In short, precision of better than a few percent is worthless. In terms of overall accuracy, one should be wary of claims that an accuracy of much better than an order of magnitude has been achieved.

4. There is the issue of *"operator error,"* for which data do exist. A routine error rate of one in 1000 is often used for situations where it does not apply. For example, an operator and a supervisor are claimed to have a combined probability of failure of 10^{-6} (1/1000 × 1/1000). This is spurious since the two are not independent; the supervisor will tend to assume that the operator is competent and so will not be expecting errors. A combined failure probability of 10^{-4} (1/1000 × 1/10) is more realistic. Also, operator error rates under abnormal conditions can be much higher than one in a 1000.

5. Finally, one must consider the operability of protective instrumentation and equipment. A proper design philosophy is that control and protective items should be independent. Thus, instruments which control the temperature of a reaction should not be used to sound the alarm for "high" temperature since one initiating event, temperature control failure, will also fail the warning system. It is important when quantifying the hazard to ensure that the events which are being awarded independent probability/likelihood values are in fact independent. Another fault encountered is the evaluation of complex protective instrumentation, which incorporates several redundant systems to give a high reliability, in which no allowance is made for "common mode failure." Such failures include a loss of main power or pneumatic pressure which could disable the entire protective system.

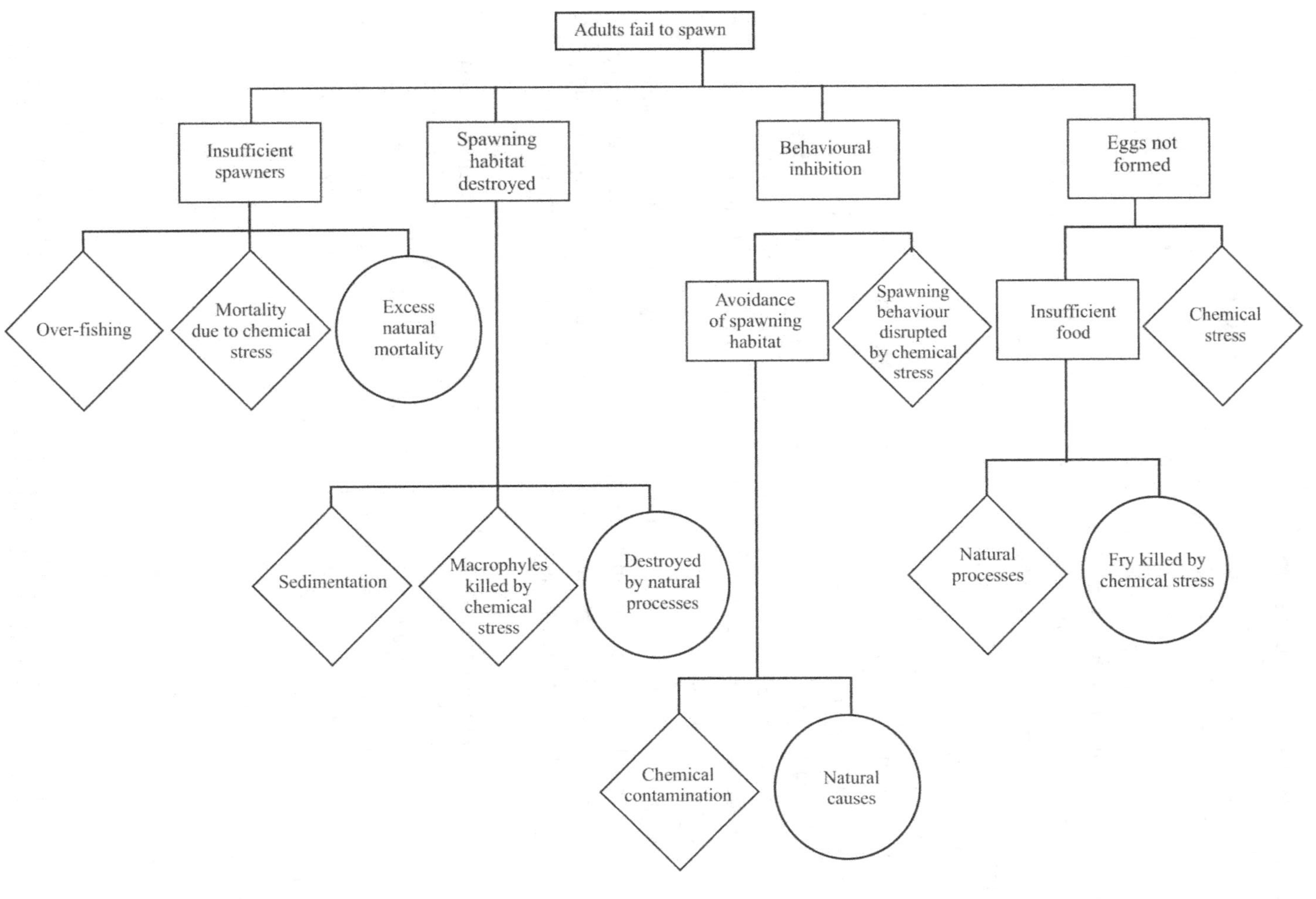

Fig. 9.5 A fault tree.

9.7.4   Step 3 – Exposure Assessment

No exposure means no risk, so imaginative constructions (models) are made of how the hazard might be encountered. For the environmental pathway, the bodily dose/response calculation is only one step. Knowledge of earlier parts of the exposure sequence can reveal chances to reduce risk. For example, a toxic chemical may ultimately poison people when inhaled, but ERA seeks information (Fig. 9.6) on:

- the type, amount, location, and storage conditions of the chemical (an inventory).
- releases to the environment, whether deliberate or accidental.
- type and duration of exposure.
- ambient concentrations.
- the actual bodily dose; and then
- the physical condition of specific victims that might affect how they respond.

At each step, different units and techniques of measurement are used with differing degrees of reliability and specificity. A complete understanding of the risk represented by a pollutant and the potential ways to manage it would entail exploring all the links shown. In practice, lack of data, time or money substantially limit the direct relationship with human health (*source:* Carpenter et. al., 1990 (10).

Reasonable sequences of events and environmental pathways by which the source of harm could impact health and welfare, including the condition of ecosystems, are devised. For example, a toxic chemical might move from any point in its life cycle through air, water, plants, animals, or soil to cause an exposure by skin contact, inhalation or ingestion Fig. 9.6.

Event and fault trees are approaches to schematically breaking down complex systems into manageable parts for which failure rates or other risk-related data can be found. It is thus possible to construct some idea of the failure rate and resultant risk of a large, complex, and new entity, such as a chemical plant, even if no data about its performance exists. This example shows the differences between the two approaches and also how they can be used in a complementary manner (*source:* Smith et. al., 1988(12).

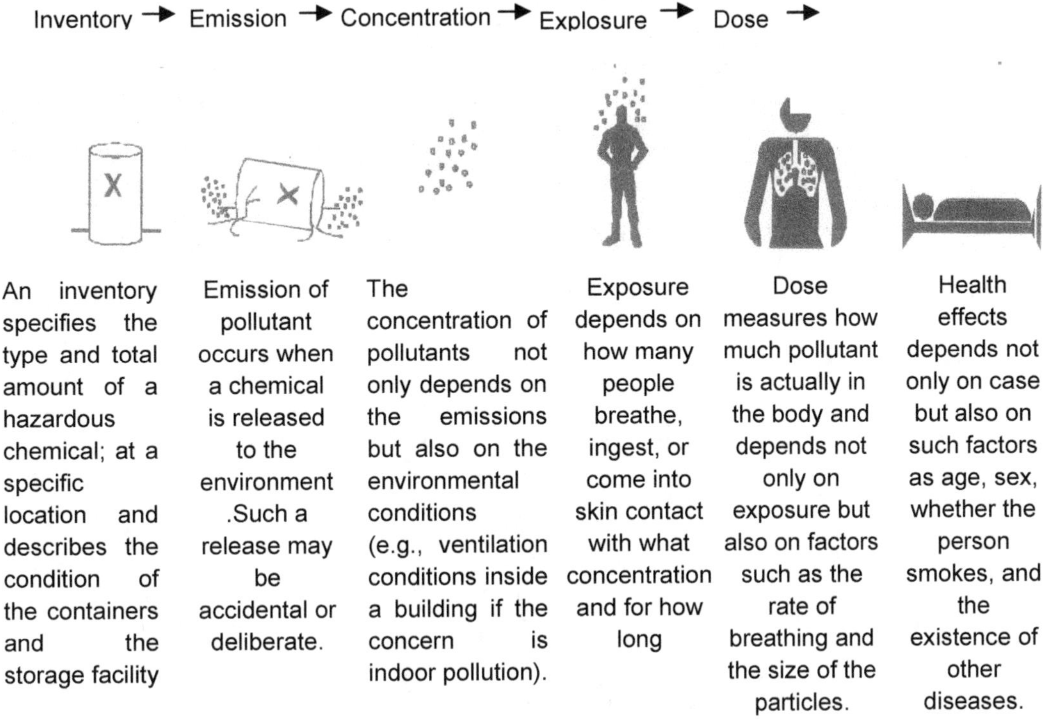

An inventory specifies the type and total amount of a hazardous chemical; at a specific location and describes the condition of the containers and the storage facility	Emission of pollutant occurs when a chemical is released to the environment .Such a release may be accidental or deliberate.	The concentration of pollutants not only depends on the emissions but also on the environmental conditions (e.g., ventilation conditions inside a building if the concern is indoor pollution).	Exposure depends on how many people breathe, ingest, or come into skin contact with what concentration and for how long	Dose measures how much pollutant is actually in the body and depends not only on exposure but also on factors such as the rate of breathing and the size of the particles.	Health effects depends not only on case but also on such factors as age, sex, whether the person smokes, and the existence of other diseases.

Fig. 9.6 The relationship between quantity, emissions, environmental considerations, human exposure doses, and health effects. At each step, different units and techniques of measurement are with differing degree of reliability and specifity. A complete understanding of the represented by a pollutant and the potential ways to manage it would entail exploring all the shown. In practice, lack of data, time or money substantially limit the direct relationship human health.

(**Source :** Carpenter et. al., 1990).(10)

9.7.5   Step 4 - Risk Characterization

Methodically observing or estimating the likelihood of occurrence and the severity of impacts for each scenario can produce curves as illustrated in Fig. 9.8 plotting the probability of frequency of adverse events of a given severity vs. the severity per event (for example, the number of fatalities). Known as F/N curves, they present the "how often" and "how bad" aspects of risk. As shown in Fig. 9.8 the integral under the curve is not the whole story. Hypothetical project (or technology) A has a lower mean risk than B (for example, spill and fire from a tank truck in transit), but A also has a larger probability of a catastrophic accident (for example, explosive dispersion of a toxic material in a highly populated area). There is no objective way to combine these two criteria and different societies or individuals will make different choices between the two.

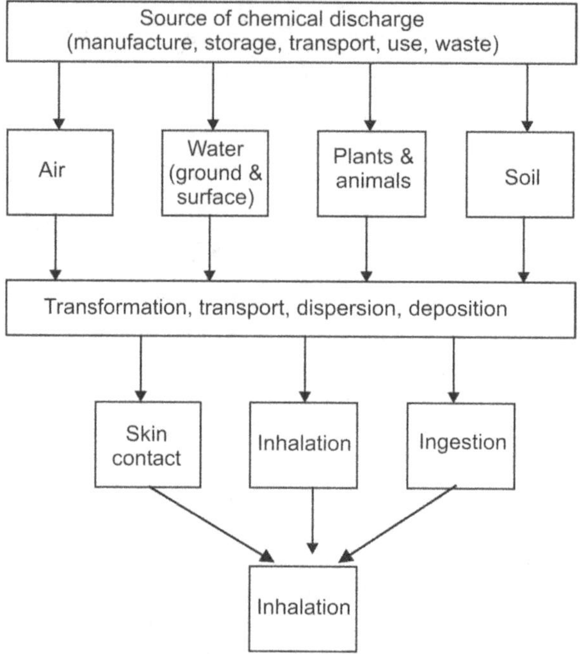

Fig. 9.7 Exposure pathways.

(**Source:** Carpenter, R., C. Claudio, L. Habegger, and K. Smith ADB Environmental Paper No. 7 1990. Asian Development Bank, Manila)

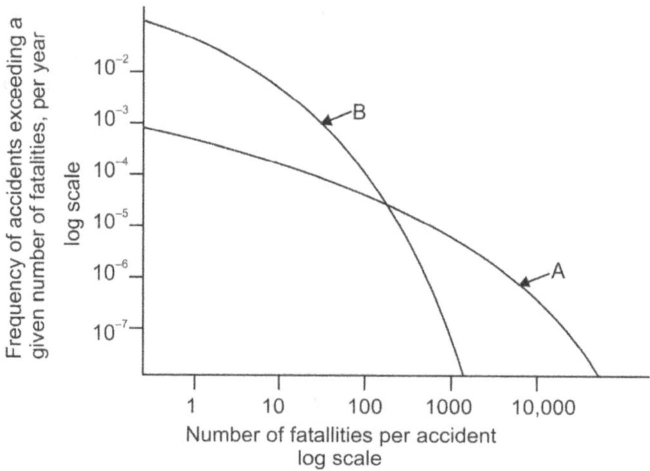

Fig. 9.8 Risk distribution for two hypothetical alternative industrial facilities. Plant A has lower means (expected value of damage) than Plant B. A, on the other hand, has considerly probability (although still small) of causing a large accident that kills many people. There objective way to combine these two criteria (expected value and distribution), and different groups of people will make different choices.

(**Source:** Carpenter, R., C. Claudio, L. Habegger, and K. Smith ADB Environmental Paper No. 7 1990. Asian Development Bank, Manila).

However, the explicit depiction of risk is valuable information. :Risk is a function of frequency of occurrence of adverse events and the magnitude of their consequence. Note that while risk deals with uncertainty, there is also uncertainty in the expression of risk due to the variability of data used to estimate frequency and severity.

Risk may also be indicated by the breadth and shape of the distributions or probability densities of the severity values (Fig. 9.9).

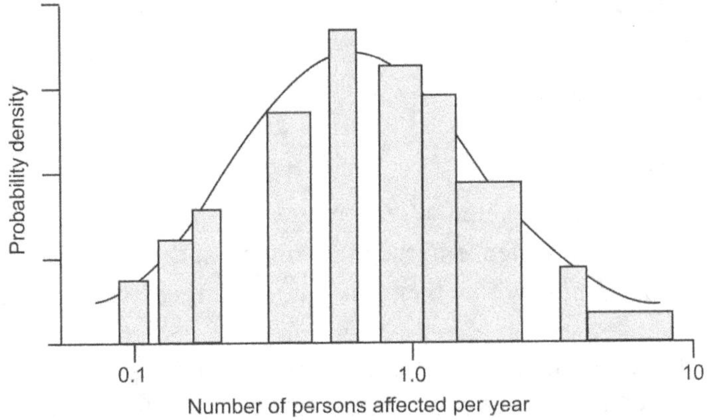

Fig. 9.9 Plots of probability density functions (PDF) The probability that a variable will have a value within a small interval around x approximated by multiplied f(x) (that is, the value of y at x in a PDF lot) by the width of interval.

If the standard deviation is small and the distribution approximately log-normal (bell-shaped), the mean can adequately represent the impact. If the standard deviation is large and there is a pronounced positive skew (tail) with low frequency but high severity outcomes, an expression of this risk and a more thorough investigation are warranted. Even a qualitative presentation of risk is useful (Fig. 9.10).

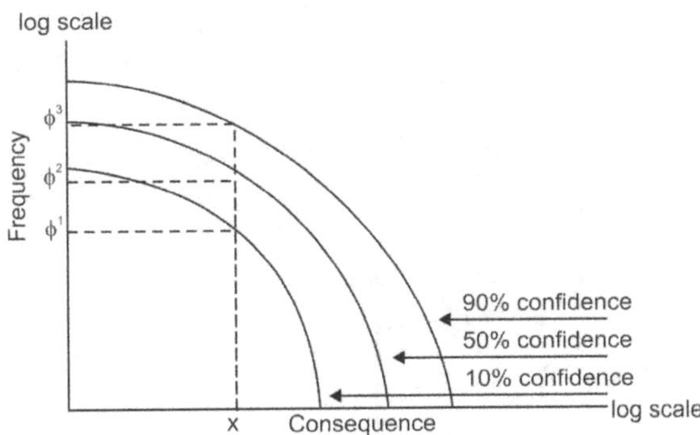

Fig. 9.10 Risk is a function of frequency of occurrence of adverse events and the magnitude of the consequence. Note that while risk deals with uncertainty, there is also uncertainty in expression of risk due to the variability of data used to estimate frequency and severity.

It is obvious that whenever frequent occurrence is combined with catastrophic or critical severity, the risk must be reduced if the project is to proceed. Occasional or infrequent adverse events that have only negligible or marginal consequences may be acceptable because of the benefits of the project or activity.

9.7.5.1 Risk acceptability

Risk characterization facilitates the judgment of risk acceptability. Risks to health are typically characterized in terms of:

- exposure period;
- potency of a toxic material;
- number of persons involved;
- quality of models;
- quality of data, assumptions, and alternatives;
- the uncertainties and confidence in the assessment; and
- appropriate comparisons with other risks. Useful risk characterization expressions include:
- probability of the frequency of events causing some specified number of prompt fatalities (for example, equipment failure releasing toxic gas that kills ten or more people is estimated to occur every fifty years);
- annual additional risk of death for an individual in a specified population (for example, one in a million);
- number of excess deaths per million people from a lifetime exposure (for example, 250 people in the exposed population);
- annual number of excess deaths in a specific population (for example, living within a certain distance from a hazard); and
- reduction in life expectancy due to chronic exposure, or chance of an accident.

Figures 9.11 through 9.13 are examples of different means of characterizing.

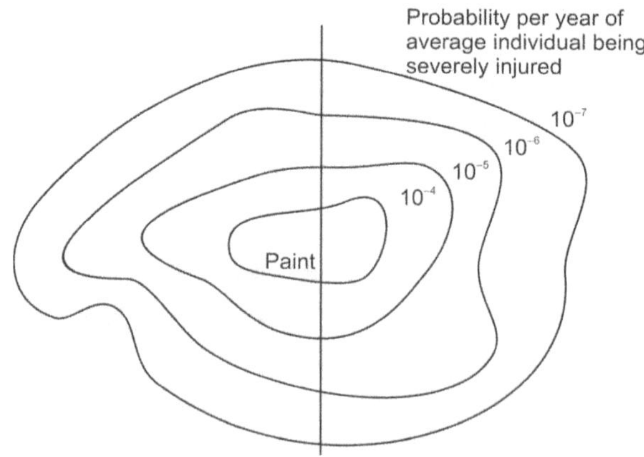

Fig. 9.11 A simple example of a stressor-response relation.

(**Source** :U.S. environmental protect) (13)

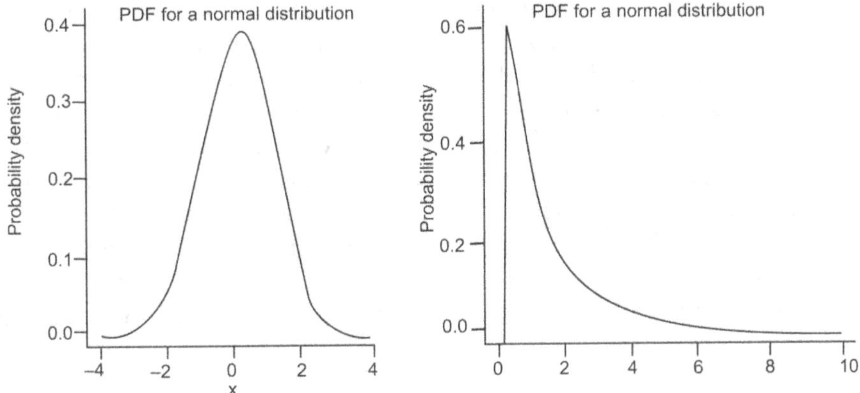

Fig. 9.12 Plots of probability density functions (PDF).

(**Source:** U.S. Environmental protection agency 1997).(13). The probability that a variable will have a value within a small interval around x can be approximated by multiplying f(x) (that is, the value of y at x in a PDF plot) by the width of the interval.

9.7.6   Step 5 - Risk Management

The communication of ERA results should take the form of decision analysis that is, what options are available, and for each option the risks, costs, and benefits involved, and the way these distributed within society.

Fig. 9.13 presents risk management strategies

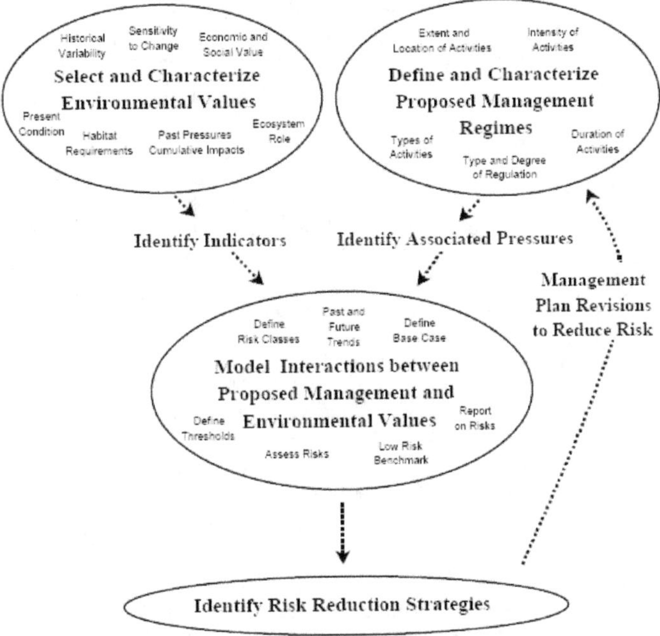

Fig.9.13 Risk Management Strategies.

Proper comparison and communication can actually change lay people's misperceptions of risks so participatory decision making may proceed on a more rational, less emotional basis. Risk management is the use of ERA results to mitigate or eliminate unacceptable risks. It is the search for alternative risk reduction actions and the implementation of those that appear to be most cost-effective. Most human activities are undertaken for obvious and direct benefits and risks are intuitively compared with these benefits. Avoiding one risk may create another (risk transference); net risk is a consideration facilitated by ERA. There are strong reiteration and feedback between risk management and hazard analysis because a changes in the scope of the ERA may be necessary to fully answer the questions of management, and b) relatively simple changes in the project may alter the hazard and reduce risk (for example, different siting).

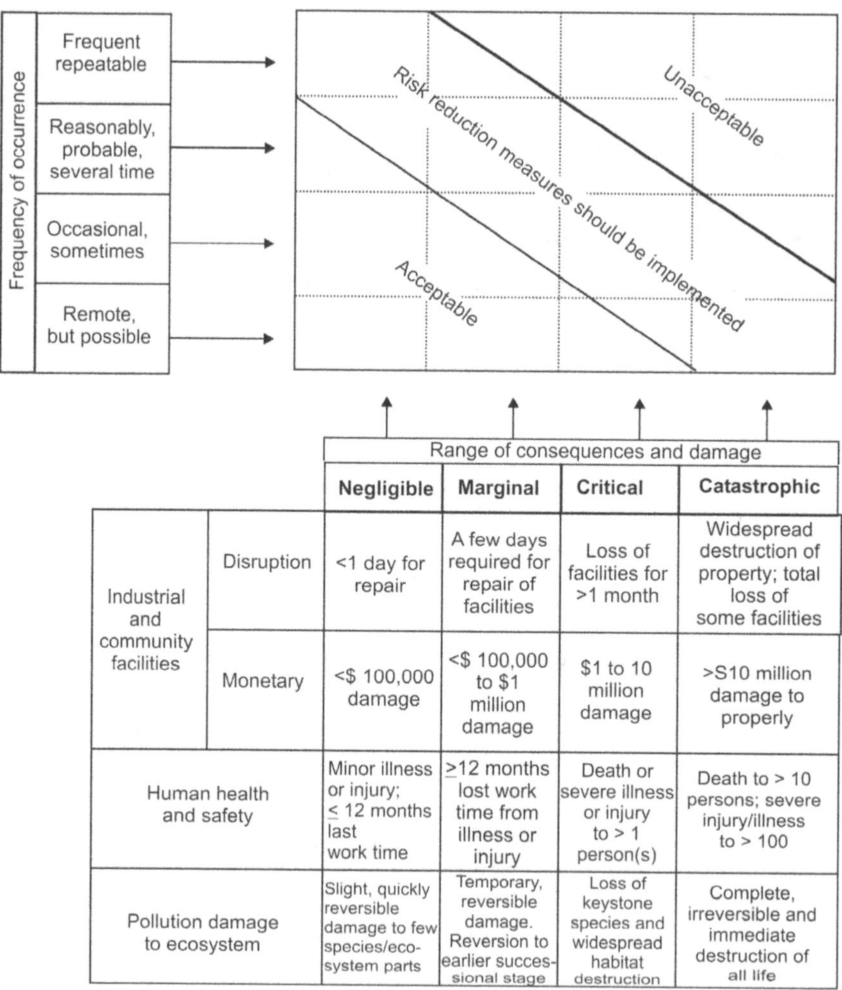

		Negligible	Marginal	Critical	Catastrophic
Industrial and community facilities	Disruption	<1 day for repair	A few days required for repair of facilities	Loss of facilities for >1 month	Widespread destruction of property; total loss of some facilities
	Monetary	<$ 100,000 damage	<$ 100,000 to $1 million damage	$1 to 10 million damage	>S10 million damage to properly
Human health and safety		Minor illness or injury; ≤ 12 months last work time	≥12 months lost work time from illness or injury	Death or severe illness or injury to > 1 person(s)	Death to > 10 persons; severe injury/illness to > 100
Pollution damage to ecosystem		Slight, quickly reversible damage to few species/eco-system parts	Temporary, reversible damage. Reversion to earlier successional stage	Loss of keystone species and widespread habitat destruction	Complete, irreversible and immediate destruction of all life

Fig. 9.14 Risks may be categorized on the basis of their frequency of occurrence and severity, consequences or damage.
(**Source:** Carpenter, R., C. Claudio, L. Habegger, and K. Smith; ADB Environmental Paper No. 7 1990. Asian Development Bank, Manila).

9.7.6.1 Risk - Cost - Benefit

The effectiveness and efficiency of risk management depend on deploying limited resources where they are most needed. Comparing risks and the costs of their reduction is a valuable decision tool. For example, hazardous waste sites are perceived by many citizens as posing a high health risk, and large expenditures are made to clean them up. Yet, when quantitative probabilistic risk assessment is performed on these sites they usually turn out to be relatively low threats. This is because, in most cases, the chance of exposure is slightly due to isolation from drinking water supplies and prevention of access. In contrast, the risk from indoor air pollutants is found to be relatively high and worthy of greater reduction efforts than the public might demand. People spend most of their time indoors, often in poorly ventilated areas, exposed to vapors of hazardous household products, to second hand tobacco smoke, and, in some locations, to radon.

Finding a small residual risk does not mean the management activities that have brought the risk down should be decreased, although they should be reviewed for cost effectiveness. It is the further expense of reducing the small residual risk that is subject to question. For example, in the case of public water supply in most Western countries, the low risk is testimony to good sanitation and water treatment practices. But, often proposed drastic and expensive measures necessary to remove trace amounts of pesticides that may pose only a small residual risk should be judged against other opportunities for protecting public health. So there is a need to iterate between risk management and hazard analysis.

There will be various levels of sophistication in ERA and it is necessary to recognize the value of different stages. In many circumstances there will not be any justification to go beyond initial stages that may be of low cost. The degree of sophistication should be determined by the magnitude and significance of the risks being studied, the sensitiveness of the receptor, the quality of available data and the means by which risks are being communicated and outputs utilized (13). Fig. 9.15 shows the different levels of sophistication that might be used with increasing risks and costs. It is important to adopt the most appropriate technique.

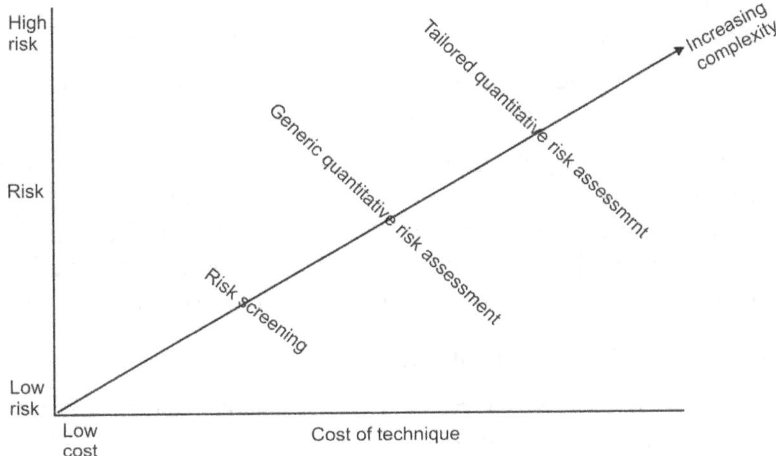

Fig. 9.15 Levels of sophistication with increasing costs.

Global problems like green house gases or ozone depletion require different approach from that of any local industrial pollution.

9.8 Advantages and Limitations of Environmental Risk Assessment

ERA should be considered as tool for assisting decision makers to get information for the question what if in a given situation based on available data in a structured way, with precision and quantitative output. However it should be transparent recording the assumptions made and uncertainties in the estimates. It should be considered as an iterative process leading to future refinement.

In dealing with environmental health risks the human exposure plays dominant role. In Ecological Risk Assessment the uncertainties of the complex and dynamic natural systems require different treatment.

9.9 Risk Assessment Methodology

9.9.1 Human Health Risk Assessment Methods

Exposure and Dose

From the scenarios of exposure it is possible to estimate the amount of a toxin that affects any one person or a population. The procedure varies with the mechanism of exposure; that is, ingestion in food or water, inhalation, or through the skin. For example, the concentration of a toxin in air is calculated as mg/m^3, or it may be converted to an inhaled dose, mg/kg/day, by dividing by 70 kg (an assumed body weight) and multiplying by 20 m^3/day (an assumed human inhalation rate). Published reference concentrations or daily doses are then used for comparison with the measured values to estimate risk. Oral exposure for drinking water is measured in mg/kg/day of intake, and the risk is compared with animal-derived data for a safe concentration (or reference intake that is safe). The concentration in the drinking water supply is measured in mg/l, and then compared with a safe concentration which is calculated as follows:

safe concentration (mg/l) = reference intake dose (mg/kg/day) × 70 kg / 2 l/day

2l/day is assumed to be the amount of water a person uses. All of these individual assumptions need to be adjusted for specific groups and life styles.

Similarly, diet patterns can yield estimates of a dose by studying the amount of a certain food that is eaten and the concentration of the contaminant in that food. Reference safe concentrations (RfC) and doses (RfD) are available from the World Health Organization. The risk assessor must carefully link exposure pathways and personal habits to estimate a dose and consequently a risk of daily continuous or one time exposure.

9.9.2 Ecological Risk Assessment (EcoRA)

The objective of ERA as applied to ecosystems is usually comparative and qualitative because of the lack of data on stress/response. It is useful to decision makers to (a) rank a comprehensive set of environmental problems (stressors on VECs at specific sites) relative to one another, using broad levels of risk and (b) target risk reduction actions toward those geographical areas or ecosystem sites that are of greatest value and at greatest risk. There is not yet any widely applicable, established procedure for EcoRA.

In general, information is gathered about (a) hazards or sources of harm (b) stressors and their pathways to target organisms (c) adverse responses of species and communities; and d) measurable changes in the condition (integrity, resilience, productivity, health, sustainability) of the ecosystem. (d) is termed an endpoint attribute similar to mortality or morbidity in humans.

The following ideas on ecosystem integrity are adapted from Regier et. al., 1994 (14) An ecosystem with integrity:

- is an ever changing set of organisms, within adapting populations of evolving species, and with a capability for creativity.

- contains organisms that purposely modify their surroundings, but not so as to impair self organizing capabilities.

- contains some larger and longer lived organisms that cumulate, integrate, and regulate many features of the system.

- processes energy and information from outside the system in a trophic network so as to increase energy and information per unit of biomass.

- exhibits interactions through organization within a complex spatio-temporal domain so that relatively persistent structures are overlain with transient, perhaps cyclical processes.

- achieves organizational flexibility and redundancy to cope with inevitable surprises that are simplifying in the short term but complicating in the long term.

- interrelates dynamically, across fuzzy boundaries, with adjacent ecosystems.

- is a self-organizing dissipative system compromising between the Second Law of Thermodynamics and the biological imperative of survival and sustained identity.

This is an elegant expression of the concept of integrity but its relevance to the practical definition of sustainability and calculation of risk is not clear.

Resilience is a promising approach to objective integrity, the ability to recover from a specified stress. It is a context-dependent measurement; the meanings of recover and stress must be clear and testable in scientific experiments. Ecosystems are not a static integration of structure and functions. They continually evolve and change; this is the source of their resilience, which is the desired valuable behavior. Resilience may be gauged and interpreted by examining trends such as:

- the ecosystem has not changed, at a given stress level, from an original satisfactory natural condition.

- the ecosystem changes but returns to the original condition, even under continued stress (how much change? how long to return? is it stable on return?).

- the stress is reduced or removed and the ecosystem returns to the original condition (how much change? how long to return? is it stable on return?). and

- the ecosystem changes permanently (collapses? resumes original natural evolutionary pathway? Takes new but derivative evolutionary pathway? takes catastrophically different evolutionary pathway).

A site with high "biological integrity" is, supposedly, able to withstand natural or human disturbances.

The components of an Index of Biological Integrity are species abundance counts and ratios, water quality, habitat structure, flow regime, energy source, and biotic interactions. This is essentially a resilience measurement and, although valuable in EcoRA, does not relate directly to productivity or sustainability. Some quantitative indices purporting to measure integrity are solipsistic, self-referential, and constitute a pseudo-science exercise. Indicators of integrity that could be quantified and monitored include:

- general indicators like primary productivity, nutrient cycling, species diversity, population fluctuations, pest prevalence, spatial patchiness;
- threats, like increases in human population density, consumption rates of water, energy, renewable and non-renewable resources, wastes, infrastructure; and
- improvements in integrity, including increases in production, recycling, conservation and citizen involvement.

Sustainable development has multiple meanings with diverse roots in ecology (both "deep" and conventional), resources, carrying capacity, anti-technology, and ecodevelopment. Operational definitions and indicators of implementation achievement are required if sustainable development is to be anything more than an attractive, but empty, phrase. Most natural scientists who are managing ecosystems such as agriculture are sceptical about their capability to measure sustainability. Sustainability occurs where the productive potential of a managed ecosystem site will continue for a long time under a particular management practice. The utility, capacity, or potential of these natural systems for producing goods and services, is what is to be continued, and even enhanced.

Munn (in Regier, 1994 (14) offers a view of integrity more consistent with this definition of sustainability:

"An ecosystem with integrity should exhibit such properties as:

- strong, energetic processes, not severely constrained;
- self-organizing in an emerging, evolving way;
- self-defending against exotic organisms;
- reserve capabilities to survive and recover from occasional severe crises; attractiveness, at least to informed humans;
- productive of goods and services valued by humans."

9.9.3 Ecotoxicology

Where chemicals have been tested against animals or plants for exposure-response, a risk assessment procedure similar to that for human health is used. For example, the concentration of a toxin in water to a fish species that kills 50% of the population (LC50) is akin to a maximum daily intake (MDI) in that it may be adopted as an end-point to be avoided. The risk is then evaluated as acceptable when the "quotient" of actual measured concentration to LC50 is less than one. These single chemical-single species consequences seldom reflect the real world where several chemicals stress several species simultaneously. Fig. 9.16 shows how many different concentration levels of exposure (EC) can be examined. This is useful when the risk assessment outcome is not based on exceedence of a toxicity benchmark level.(*source:* U.S. Environmental Protection Agency, 1997) (13).

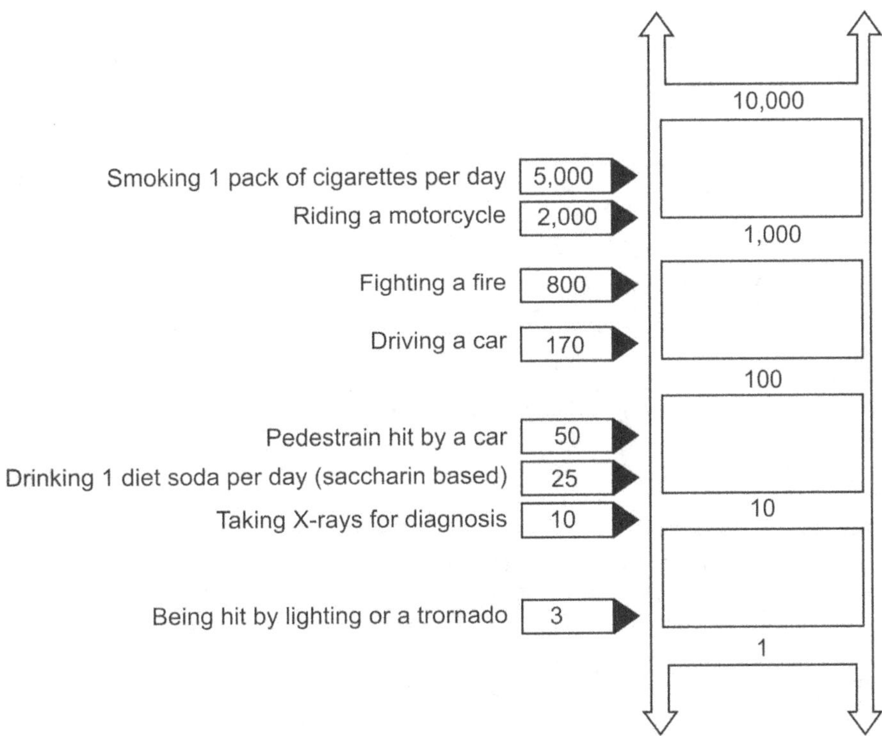

Fig. 9.16 Health risk ladder.

Models may make use of laboratory data to quantify biological and ecological processes and impacts, primarily at the species and community levels. This can be useful at site specific locations, but extrapolating the results to ecosystem and regional levels is more difficult, especially if two or more ecosystems and stressors are involved. A standard water column model comprising many biogeophysical parameters is used at Oak Ridge National Laboratory, "...to extrapolate the results of laboratory toxicity data into meaningful predictions of ecological effects in natural aquatic ecosystems." (Bartell et. al., 1992) (15).

Other methods evaluate structural and functional changes at the ecosystem and regional levels and are most easily applied where there is large-scale homogeneity in both the ecosystem and the stressor that affects it. Conversely, these methods break down when a region is a mosaic of many stressors and ecosystems. Normally there is a lack of sufficient data from a broad region to allow quantification. Fig. 9.17 (a) & (b) is the classical stress or response relationship, but even a qualitative estimate of the relationship is useful (for example, the dose at which about 50% of the organisms are killed).

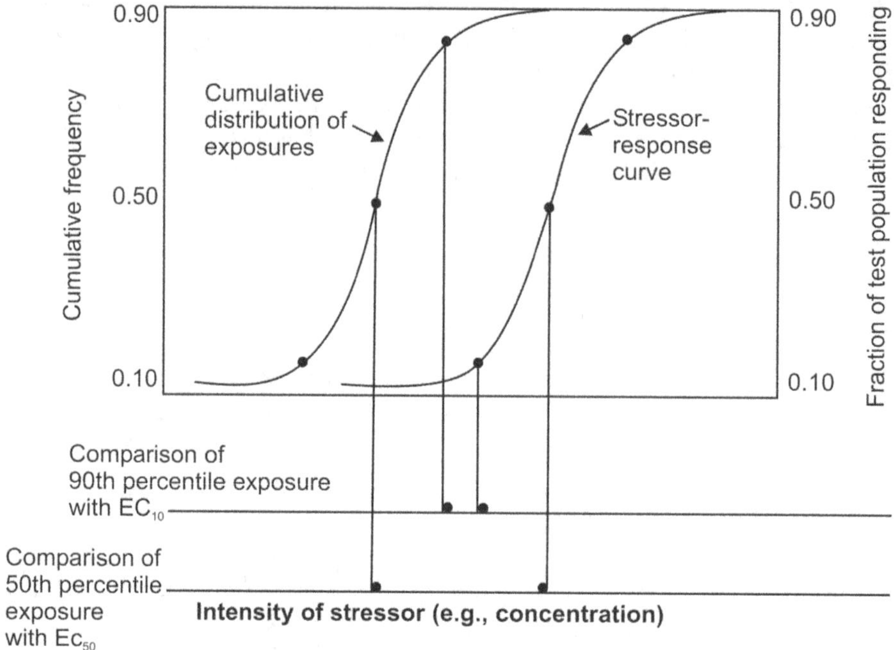

(a) Stressor-response curves
(e.g., dose–% mortality)

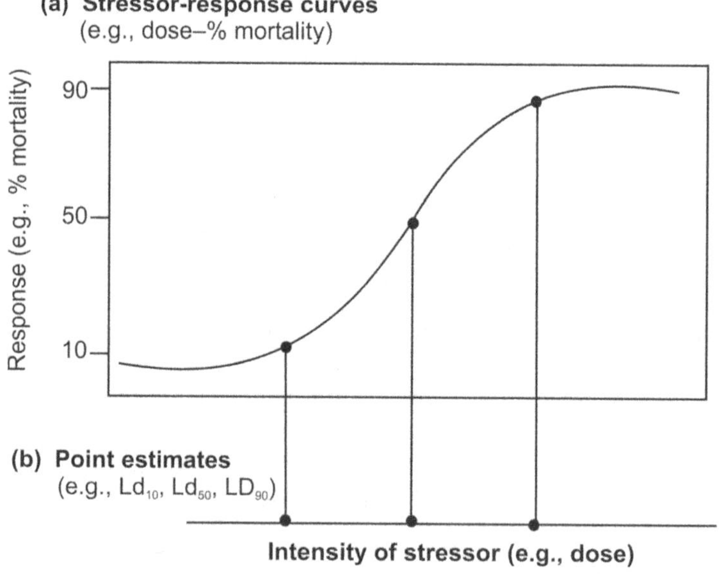

(b) Point estimates
(e.g., Ld₁₀, Ld₅₀, LD₉₀)

Fig. 9.17 (a) & (b)

Initial Screening Procedure-Steps include:

1. The contaminant data is sort by the medium (say groundwater, soil, surface water) for different types of contaminants

2. The mean and range of concentration values for all contaminants observed at the site are then tabulated

3. The reference concentration (for noncarcinogens) and slope factor (for carcinogens) for each potential exposure route, is then identified and

4. The toxicity score for each contaminant in each medium is then calculated as follows:

Formula for calculating toxicity scores for chemicals.

For non-carcinogens

$$TS = \frac{C_{max}}{Rfc} \qquad \qquad .....(9.2)$$

For carcinogens $TS = Cmax \times SF$ .....(9.3)

TS	Toxicity scores
C_{max}	Maximum concentration of a chemical
RfC	Reference concentration (chronic, i.e., for long-term exposure) (only for non-carcinogens) (source: www.epa.gov/iris/)
SF	Slope factor (or carcinogenic potency factor) (only for carcinogens) (source: www.epa.gov/iris/)

9.9.4 Practical Qualitative and Comparative Ecological Risk Assessment

Decision makers, politicians, and nonscientists have a need for practical comparative ecological risk assessments. Comparative EcoRA need not be quantitative; it may be preferable to keep it qualitative. A combination of best judgment of ecologists and professional land/water managers with on-site experience, and systematic evaluation of risks from available information is pursued. Effective communication to decision makers is accomplished through use of maps, simplified scoring systems, clearly defined evaluative criteria, and a manageable set of ecological stressors. Defining the specific problem areas and classifying the ecosystems of the study region are important early steps in this approach to comparative EcoRA.

Health risk assessments (with heavy emphasis on public health) differ from ecological risk assessments in several significant ways. For ecosystems, the ERA must consider effects beyond just individual organisms or a single species. No single set of ecological values and tolerances applies to all of the various types of ecosystems.

Stressors are not only chemicals or hazardous substances. They also include physical changes and biological perturbations. For public health purposes all humans are treated equally; with ecosystems, some sites and types are more valuable and vulnerable than others. Accommodating these factors complicates comparative ecological risk assessments and renders them more subjective.

9.9.5 Qualitative Methodology

Risks to ecosystems are based on the values (intrinsic and anthropocentric) of actual individual sites and the probability that stressors from human activities will significantly degrade these values in the near future.

Uncertainties about value, frequency of adverse impacts, and severity of response to stress are identified and evaluated as a part of the ERA. The ability of the ecosystem occurrence (site) to recover is also considered.

Just as the individual human being is the focus of health risk assessment, the individual ecosystem site is evaluated in ecological risk assessment. Ecosystems are bounded biotic communities in interaction with their physical surroundings of energy, air, water, minerals, and soil (and also other ecosystems). Usually, an ecological risk assessment treats only natural, more or less intact, ecosystems that are lightly managed or essentially undeveloped. Urban and agricultural areas that are substantially modified, intensively managed, and where economic value dominates all others are considered dedicated.

9.9.6 Risk Ranking Procedure

1. *Establish an Ecosystem Classification* - Define and select a manageable number of ecosystem types that are (a) identifiable (mappable) through currently available databases and reports, and (b) categorized by biophysical properties (climate, rainfall, topography and elevation, vegetation, geology and geomorphology, hydrology, soils, etc.). Examples of marine and terrestrial ecosystem types are: coral reef, freshwater stream, wetland, lowland dry scrub, monotone wet forest, and sub alpine dry grassland.

2. *Inventory and Map the Ecosystem Occurrences (Sites)* - Gather data about the location, extent and status of resources, degree of disturbance, and level of protection. Previous research and monitoring, and personal interviews may suffice for the inventory but new field studies are often necessary.

3. *Develop Criteria of Value for Each Ecosystem Type* - For each of the different ecosystem types, determine individual criteria for the components of value. Valued components include economic productivity, recreation, biodiversity, and cultural/aesthetic significance. Criteria include wetland classifications, the presence of endangered species, rarity, the ratio of native to alien species, and tourism visitor counts. Seek out previous valuation studies, measurable attributes, and changes to those attributes which degrade the resources.

4. *Estimate the Value of Each Ecosystem Occurrence* - Assign numerical scores to each value component at each site on the basis of a simple scalar using quantitative measurements of the criteria (where available) and professional judgments. The certainty of the score of each value component is recorded. Sum the component scores for an overall value score.

5. *Develop a List of Stressors* - Determine which consequences of, and perturbations from, human activities may plausibly cause unwanted, negative impacts on the natural ecosystems. Examples of stressors are alien species, toxic chemicals, excessive nutrients, erosion/sedimentation, water diversion, fire, and human crowding.

6. *Gather Data on Stressors and Estimate the Risk from Each* - Collect information on past, present and near-future human activities that affect the specific ecosystem sites chosen for the study.

   Environmental experts and site managers estimate the frequency (F) of occurrence and severity (S) of damage from stressors to each site with which they are familiar. The uncertainty of the estimates is also recorded. Scalars are used to roughly quantify these judgments and the product, $F \times S = R$, becomes a risk score for that stressor at that site.

7. *Map the Information* - Manually create map overlays or use computerized geographic information systems to display all data relevant to risk at each site. Data to be displayed may include location and site boundaries, values, stressors, risk scores, and geographic attributes such as present land use, native forest distribution, rare or endangered species habitat, historic/cultural sites, alien species distribution, public recreation areas, and concentrated fisheries.

8. *Rank Sites Comparatively According to Risk* - Compare a site's overall priority-for-attention score, which is the product of a site's value score and total risk score. Scores should differ by at least 20% of their absolute value to be regarded as different in priority.

9. *Rank Stressors and Ecosystem Types* - Compare the stressors as to importance in a region, and compare different ecosystem types as to degree of risk. Remedial actions can be guided by learning which stressors are widely felt and which ecosystem types are most susceptible to damage. These relative rankings can be used to inform public debate, set budget allocations, focus administrative attention, and establish site specific priorities for remediation, restoration, or protection. Since uncertainties are explicitly recognized and preserved in the assessment, areas where further research and monitoring would be worthwhile to decision makers are also illuminated. A multiple site comparative ERA can also reveal which stressors are the most common and damaging, which activities generate the most stress, and which types of ecosystems are most vulnerable.

9.10　Case Studies

9.10.1　Environmental Risk Assessment of Heavy Metal Contamination in the Industrial Area of Kattedan, India – A Case Study

Kattedanis an industrial area near Hyderabad with battery manufacturing, textile, electroplating and pharmaceutical industries. Twelve different locations when assessed in Kattedan, revealed high concentrations of heavy metals (Zn, Cu, Cr, Ni, Co, Pb, Hg, Cd, As) in soil, groundwater and surface water, and vegetation. A human exposure assessment conducted in the area revealed high amounts of Pb, Zn and Cr in the blood and urine samples of Kattedan residents (K. Chandra Sekhar et al., 2006). Concentration of heavy metals in blood and urine samples of residents shows a direct exposure pathway for the contaminants. Environmental hazard assessment was carried out by:

(i)　Sequential extraction of heavy metals from soil samples –

(ii)　Soil properties analyzed: - pH -Composition of organic matter in soil - Clay minerals - Redox potential -Fe, Mn, Al oxides and hydroxides

(iii) Speciation studies for contaminants present in higher concentrations - Groundwater and surface water collected from 12 locations and analyzed for heavy metals

(iv) Human exposure assessment was carried out by collecting urine samples and venous blood samples collected from about 100 residents across different age groups - Samples analyzed for heavy metal content using ICP-MS

The results of the case study can be condensed as below:

Table 9.3 showing Concentration of metals of ecotoxicological importance

Table 9.3 Concentration of metals of ecotoxicological importance.

Sample	As	Cd	Hg
Soil (μg/kg)	100–210	80–160	20–55
Ground Water (μg/L)	55–80	60–100	40–180
Surface Water (μg/L)	40–90	70–120	25–90
Forage Grass (μg/kg)	20–66	40–66	15–40
Blood (μg/L)	8–35	20–40	10–40
Urine (μg/L)	11–38	10–40	8–18
All the values are in ppb			

*The study concludes from the results obtained, that human health risks associated with exposure to heavy metal contamination is a pressing and relevant issue for the residents of Kattedan. Closure of the industries responsible for the pollution in the region needs to be initiated.

Ref:- Environmental Risk Assessment Studies of Heavy Metal Contamination in the Industrial Area of Kattedan, India—A Case Study, K Chandrasekhar, N Sridhara Chary, CT Kamala and Mariappanadar Vairaman, Y Anjaneyulu, V Balram, Jan Erik Sorlie **Human and Ecological Risk Assessment 12(2):408-422 · April 2006**

Summary

Both ERA and EIA are similar forms of impact assessment tools. While EIA is used as a tool for assessing the impacts of projects and proposals as per regulatory requirement, ERA can act as a supportive and complementary technique. Environmental Risk Assessment is developing rapidly and there is no one clearly superior approach to its performance. Adherence to probability theory is the one essential in adding this explicit presentation of uncertainty to the management information as EIA Health risks and ecosystem risks differ substantially in the endpoints chosen for risk characterization (the individual compared to the biological community), and in the uncertainties accompanying experimental data.

Human health risk assessment is far more advanced in methodology. However, both are dominated by concern with toxic chemicals. Both require close communication between environmental scientists and risk managers. A common but flexible framework for hazard identification, exposure pathway analysis, and hazard accounting is useful because the underlying treatment of uncertainty and the decision process is the same for both. ERA can help correct misperceptions of risk and avoid unnecessary public anxiety. Environmental risk assessment is maturing as a practical and valuable addition to the set of management and policy tools needed in a complex. There are a number of examples where ERA is used as a part of EIA and vice versa to provide comprehensive and holistic information to decision makers. As long as assumptions and limitations of ERA are made transparent it will be credible and very useful tool for the decision makers. Since EIA and ERA are developed in parallel and in isolation there is a great necessity and scope for cross-fertilisation of procedures and process between two methods.

References

1. Zhong ZL, Zeng GM, Yang CP. 1996. The development of environmental risk assessment research. *Advances in Environmental Science*; 4:17-21.

2. Lu YS. 1999 *Environmental assessment*. Shanghai: Tongji University Press, p. 531-58.

3. Royal Society 1992 Risk Analysis, perception and management. London: Royal Society.

4. Holling CS1978 Adaptive environmental assessment and management, Chichester: Willey.

5. DOE 1995 A Guide to Risk Assessment and Risk Management for Environmental Protection. London. HMSO.

6. DETR 2000 Guidelines for Environmental Risk Assessment and Management London. TSO.

7. Xie CQ. 1994 A brief introduction to environmental Risk Assessment. *Sichuan Environment*; 13:65-9

8. Steger-Hartmann T, Lange R, Schweinfurth H. 1999 Environmental Risk Assessment for the Widely Used Iodinated X-Ray Contrast Agent Iopromide (Ultravist) *Ecotoxicology and Environmental Safety*; 42:274-81.

9. Suter GW (2000) Generic Assessment Endpoints are needed for Ecological Risk Assessment. Risk Anal 20(2): 173–178.

10. Carpenter, R., C. Claudio, L. Habegger, and K. Smith. 1990. Environmental Risk Assessment: Dealing with Uncertainty in EIA. ADB Environmental Paper No. 7. Asian Development Bank, Manila.

11. Smith, Kirk R., Richard A. Carpenter, and Susanne Faulstich. 1988. Risk Assessment of Hazardous Chemical Systems in Developing Countries. Occasional Paper No. 5. East-West Environment and Policy Institute, Honolulu.

12. U.S. Environmental Protection Agency. 1997. Guidelines for Ecological Risk Assessment. Washington, DC.

13. Regier, H., J. Kay, and B. Bandurski. 1994. An Ecosystem in a State of Integrity. In Woodley, S., J. Kay, and G. Francis, eds. Ecological Integrity and the Management of Ecosystems, St. Lucie Press, Boca Raton.

14. Bartell, S., R. Gardner, and R. O'Neill. 1992. Ecological Risk Estimation. Lewis Publishers, Ann Arbor, MI.

Questions

1. What is meant by Environmental Risk Assessment? How it is different from EIA?

2. Explain the terms Hazard, Risk, Probability, and Risk Management.

3. EIA studies can identify linkages between injurious events and/or activities and their damaging consequences. Explain.

4. Compare the terms. Environmental Risk Assessment (ERA). Ecological Risk Assessment (EcoRA) and Human health risk assessment.

5. Explain the role of Fault Tree Analysis in Hazard Analysis.

6. Discuss the important aspects of Human Health Risk Assessment.

7. What are the main features of Ecological Risk Assessment? How it is different from Human Health Risk Assessment?

8. Risk assessment (RA) is the scientific method of confronting and expressing uncertainty in predicting the future. Explain.

9. Explain various key steps in performing an Environmental Risk Assessment.

10. Discuss five step sequence in performing ERA.

11. What are the major uncertainties in hazard analysis?

12. Write notes on i) Scoping choices and ii) Exposure Assessment in hazard analysis.

13. Discuss a. Risk characterization factors, b. Risk Management Strategies and Risk - Cost – Benefit analysis.

14. Explain Advantages and Limitations of Environmental Risk Assessment.

15. Discuss Human Health Risk Assessment Methods.

16. What are the objectives of any of ERA as applied to ecosystems?

17. Give details of Practical Qualitative and Comparative Ecological Risk Assessment approach.

18. Discuss the details of qualitative methodology of Risks to ecosystems.

19. Explain the procedure for Risk Ranking.

CHAPTER 10

Application of Remote Sensing (RS) and Geography Information Systems (GIS) for EIA

10.0 Introduction-Role of RS & GIS Technologies in EIA as a Decision making Tool

As a decision making tool, EIA is heavily influenced by the nature and structure of the local planning process and can be divided into 5 stages:

1. **Screening** - The step where the authorized body decides whether or not an EIA is needed. In most cases, the decision is based either on the lists in the EIA regulation, on the base of project type, or on the sensitivity of the project (1). Another widely used criteria state that EIA should be prepared for every project with "significant" impact on the environment. Of course, when using the last criteria, the main issue is the decision whether the proposed project will have a significant impact or not (2)

2. **Scoping** - Determining the scope of environmental issues to be scrutinized in the EIA, defining the scope for each issue.

3. **Impact Assessment** - Assessment of each topic selected in the scoping stage. This step takes up most of the EIA's time and resources. For each topic, the current status is delineated, and the predicted impacts are forecast by means of models. In cases where adverse impacts are identified, mitigation measures are proposed.

4. **EIS preparation** - At the end of the impact evaluation, a document (Environmental Impact Statement) is prepared. This document is passed on to the competent authority. Usually, after the authority makes its decision on the proposed project, the public can inspect the EIS and in many countries can resort to legal action. Hence, the EIS is a legal document.

5. **Post Project Analysis (PPA)** - In some countries, the environmental authorities continue to follow the proposed project in order to check that the project initiator is following the orders of the competent authority and also to improve the EIA system as an ongoing process.

Due to the diverse nature of the EIA process, ranging from EIA of small projects to large projects, Cumulative Effects Assessment (CEA), have to be carried out for a wide range of spatial and temporal variables for which Remote Sensing (RS) and Geography Information Systems (GIS) will be very effective, rapid, reliable and useful in all the five stages of EIA.

10.1   Remote Sensing Applications for Environmental Monitoring

"Remote sensing applications (RS) is the science of deriving information about the earth's land and water areas from images at a distance. It usually relies upon measurement of electromagnetic energy reflected or emitted from the features of interest." (Campbell, 1996)(3). Remote Sensing (RS) is one of the emerging technology for getting very useful information on Earth's land and Water Resources using images acquired from an overhead perspective using electromagnetic radiation in one or more regions of the electromagnetic spectrum reflected or emitted from the earth's surface

Using Remote sensing the earth's surface or the atmosphere can be observed with cameras or sensors installed in satellites as shown in Figure 10.1.

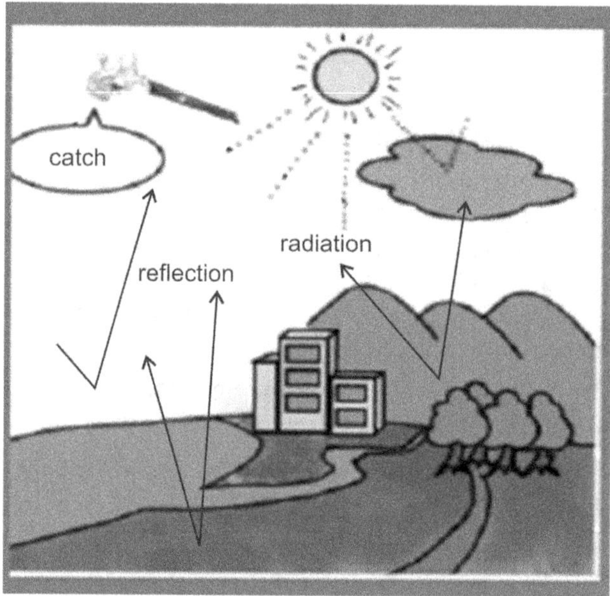

Source: Japan Aerospace Exploration Agency. 2003. Principles of Remote Sensing.
http://www.eore.jaxa.jp/en/hatoyama/experience/rm_kiso/whats_remosen_e-html

Fig. 10.1 Overview of Satellite Remote Sensing

Remote Sensing has the following advantages:

(i)   Very useful spatial data can obtained easily by remote sensing even for inaccessible areas.

(ii)   Specific, objective, wide range, frequent and periodic data can be at low cost per unit area and

(iii)   Historical spatial data can be retrieved

These advantages enable remote monitoring of project sites and cost-effective long-term periodic monitoring of the wide area necessary for development activities such as environmental monitoring and climate change impact assessment. Remote sensing is particularly effective in developing member countries (DMCs) that suffer from the lack of infrastructure for data acquisition.

10.2 Basic Principles of Remote Sensing

10.2.1 Electromagnetic Energy

Electromagnetic energy which is the source of remotely sensed information travels through space in waves which are characterized by wave length (distance between two crests of waves) and frequency (total number of equivalent crests that pass a reference point in a second).

As wave length decreases frequency increases The distribution of all radiation incident on Earth can be plotted using wave length or frequency in the electromagnetic spectrum (ES). The human eye can gather information from the visible portion of ES i.e., blue (0.4 - 0.5 um) green (0.5-0.6 um) and red (0.6-0.7 um) Fig. 10.2.

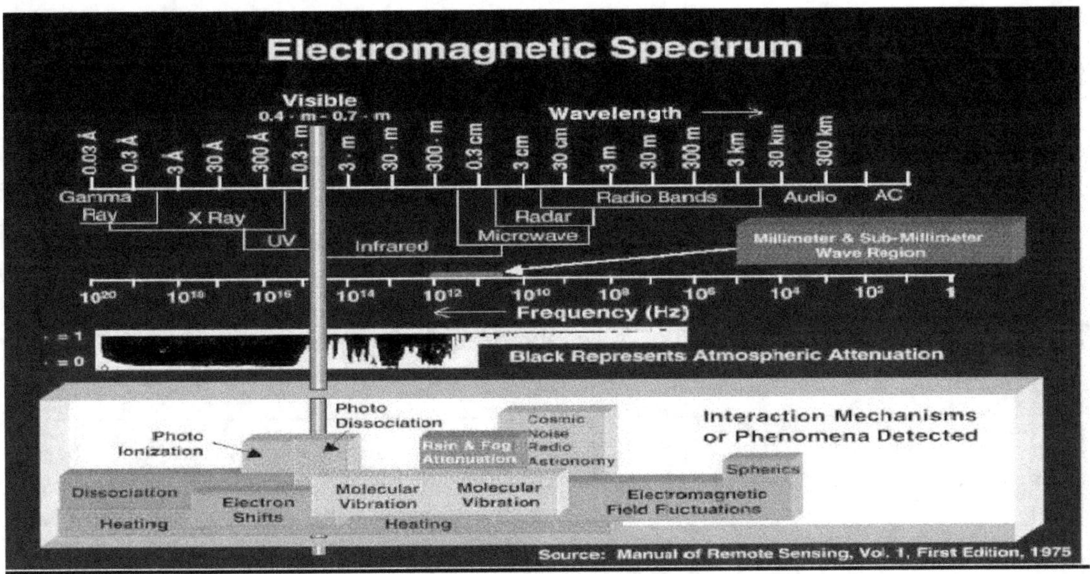

Fig. 10.2 Electro Magnetic Spectrum (Source: Manual of Remote Sensing Vol 1 First Edition)

Remote sensing sensors can extend human perception by collecting information below the blue portion of ES into ultraviolet (0.3-0.4 um) and smaller to X-rays. Likewise wave lengths larger than red light such as near infrared (0.7-3.0), known as NIR and far infrared (3–100 μm) known as FIR can be recorded and provide valuable information on vegetation patterns. Beyond FIR remote sensing sensors collect data in the micro wave portion of the ES. Micro waves can be divided into two different categories i.e. Passive and Active. Passive waves correspond to emissive radiation coming from Earth. Active Micro wave Remote Sensing is becoming an attractive source of data for scientists in many different disciplines as Microwaves can pass through clouds; valuable information will be coming from this RS for Tropical regions.

For monitoring primary impacts of any development project/program, Remote Sensing can be advantageously used to acquire spatial data over large area, and temporal frequency. By monitoring green vegetation and other spatial data RS can also be used to monitor atmospheric pollution, water pollution, contamination, reclamation of open cast mined areas and to assess impacts on environment (Latifovic et al, 2005) (4).

Table 10.1 presents the application of RS data for various types of EIA process.

Table 10.1 Application of Remote Sensing for EIA in different development Projects.(5)

Sector	Applications
Infrastructure	Satellite-based imagery, maps, land cover and/or land use maps in planning, monitoring, safeguard activities, etc.
Agriculture and food security	Satellite-based land cover and/or land use maps, crop yield estimation, cultivated area, vegetation index, evapotranspiration, soil moisture, precipitation, drought indices, etc.
Disaster risk management	Satellite-based rainfall data for flood forecasting, satellite-based drought indices for monitoring, Satellite-based disaster damage assessment (satellite imagery, inundation maps), disaster risk maps, etc.
Energy	Satellite-based solar irradiation maps, wind resource maps for solar and wind energy projects to identify suitable location, satellite right view map for understanding energy access.
Environment and climate change adaptation	Satellite-based imagery, impact area maps, land cover maps, forest and/or non-forest maps, water quality, river bank and coastal change maps, etc. for environmental and/or climate change monitoring projects or environmental and social safeguard activities.
Urban development	Satellite-based urban mapping of infrastructure and buildings, land cover and land use mapping, digital elevation models, land subsidence mapping for urban planning monitoring and assessment.
Water resources management and irrigation	Satellite-based land cover maps, water body maps, water quality information, temperature and water level, evapotranspiration to evaluate water productivity, soil erosion analysis, precipitation, soil moisture, drought indices etc.

10.2.2 Remote Sensing Elements and Present Technological Status

Remote sensing has acquired in the last two decades a special technological status for getting very useful environmental information in a systematic and reliable manner even from inaccessible areas.

Energy source, target (object) and sensor are main elements of a RS system.

The energy source is electromagnetic radiation provided by the sun or other sources. Further by using different regions of electromagnetic spectrum RS can help in getting information on various resources which go beyond human perception.

The area of interest on the terrain will be **target** while **sensor** is the device that receives the electromagnetic radiation which can be converted into signal and recorded either as digital data or image in RS systems.

Ground based Spectrometers, Aerial photography and Satellite Imagery are the three platforms through which Remote Sensing data can be obtained (Fig. 10.3).

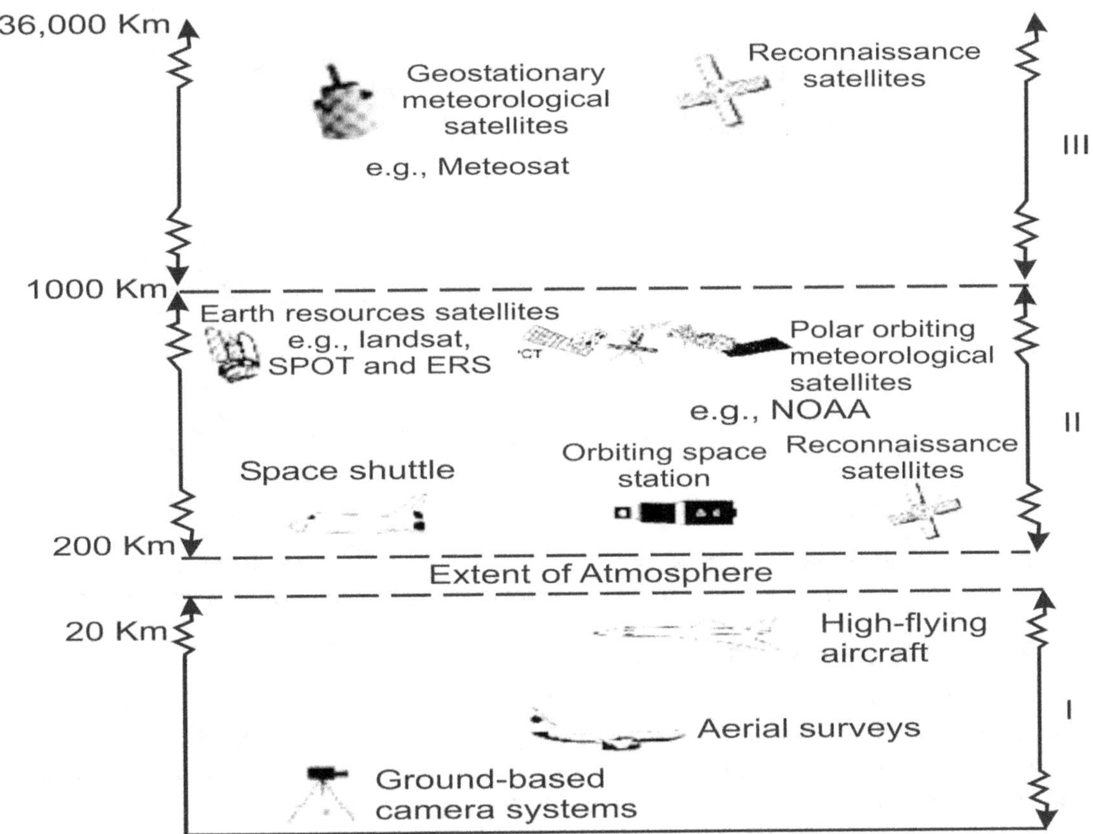

Fig. 10.3 Satellite, Aerial Photograph, Ground based Spectroscopy Flatforms for RS.

Ground based instruments such as Spectrometers will provide information on reflected radiation which will help scientists to create libraries for many types of land features, plant species, minerals, water resources etc., which can be used to assist classifying ortho imagery and satellite data. Aerial Photography has been under use for the last several decades for getting information on landscape and visual features and surveying data of various types of terrains which cannot be readily and speedily obtained from traditional surveys. Recently most modern Air Craft equipped with Sensors and high resolution cameras are under use for getting micro scale features.

The application of Satellite imagery which is of recent origin has revolutionized the Environmental Monitoring of Earth's Resources because of the ability of Satellites to scan large areas repeatedly. At present new satellites are sent into operation which have capacity of providing data that have spatial resolution under 1 meter extending the application of RS both for qualitative and quantitative studies (Figure 10.4). Air borne and satellite data are now finding extensive application to obtain information on geology, geomorphology of soils hydrology, vegetation, land cover land use studies and to produce Digital Terrain Models (DTM) for EIA studies.

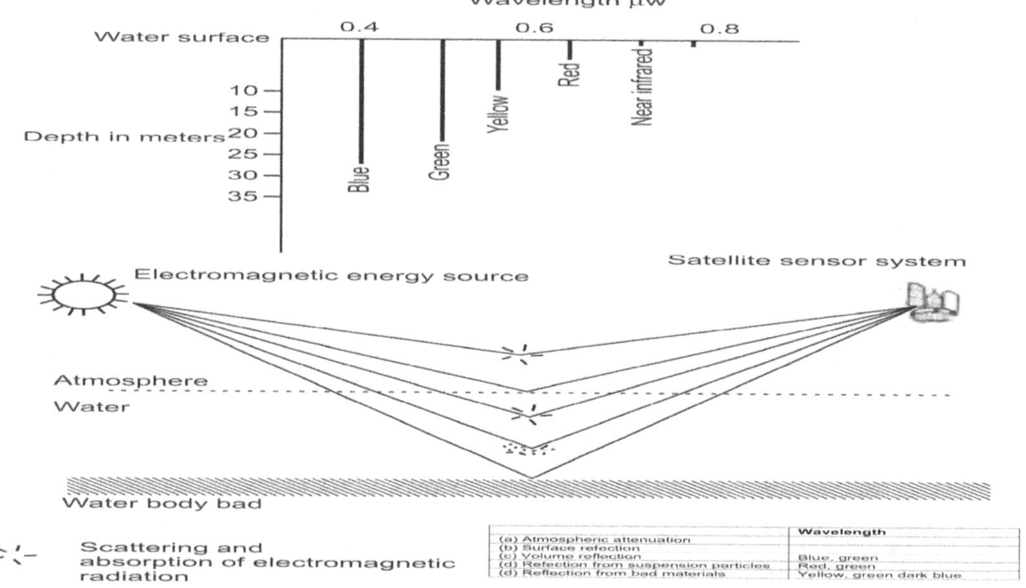

	Wavelength
(a) Atmospheric attenuation	
(b) Surface refection	
(c) Volume reflection	Blue, green
(d) Refection from suspension particles	Red, green
(d) Reflection from bad materials	Yellow, green dark blue

Fig. 10.4 Relative penetration of radiation of different wavelengths into a water body. The response of a water body for radiation and nature of changes a radiation undergoes

10.2.2.1 Photography and Optical Data

The images produced by photography and optical sensing instruments rely mainly on light reflected from the Earth's surface. However they may also utilize wavelengths beyond the visible band principally the near infrared band. This is partly because shorter wave lengths such as blue tend to be scattered by dust and water vapor in the atmosphere. Consequently images recorded with blue light energy can appear blurred or hazy and or not ideal for detailed resource observation. Green light is less affected by the atmosphere and provides useful images especially of vegetation. The same is true with red light and many studies use both green and red energy in combination with wave lengths beyond visible light (e.g., Infrared) to give a better overall understanding of vegetation trends.

Photography provides images using camera and photographic film which is now using optical sensors one can get the image in digitized mode. Again more useful option is infrared color photography which is less effected by atmospheric haze and can allow clearer discrimination of vegetation types and soil moisture variations (I of Figure 3).

Spatial data with ground resolution from 3m-30m with scales upto -5000 can be obtained using air- craft photography.

***Optical Sensors*:**

These are devices which can convert electromagnetic radiation into digital signal which can be recorded and displayed whenever required. Multi Spectral Scanner (MSS) is one of the versatile sensor with the ability to collect information from a larger portion of the Electro Magnetic Spectrum (ES) (0.3-14 um rather than the 0.3-0.9 um) available with photographic film.

In these optical sensors, an oscillating mirror will focus light energy through a filter. This filter will separate the light into different bands (blue, green, red etc.) and reaches the detectors and where it will be digitally recorded and processed by an image analyzer to produce an image. These optical sensors after being fixed to any space flatform can be used to collect and transmit digital data which can be converted to an image by computer assisted image processing system and this finds great application in earth resources analysis. By fixing to the satellites the sensors can collect land use, land cover data on any specific area over prolonged time which then depends on satellite orbital characters. The capacity of sensors fixed to any satellite to continuously collect spatial data and revisit an area can be used to monitor global hazards like forest fires, flood effected areas etc. American Land Sat and French Land Sat provide optical data which is very widely used in many countries. Land Sat earlier used to carry only MSS sensor which has limited band and low resolution (80 m). However present Land Sat-7 carries extended Thematic Mapper (TM) which has higher resolution 15m panchromatic band and seven spectral channels covering visible and NIR. Further it has also thermal sensor for one band in FIR region which can measure heat energy emitted from earth surface with a 60m spatial resolution at scales upto 1:50,000

The French SPOT (System Propatoire de l' Observation de la Terre) has two main advantages over LandSat1. It has better ground resolution of 10 m for one band in the visible (which can be used for scales up to 1:10,000) and of 20 m for two in the visible and one in the NIR 2). It can produce off centre images which allows the production of stereo images. However SPOT has reduced spectral coverage with four bands. It covers only part of visible spectrum and extends as far as the NIR.

The Indian IRS series have sensors similar to four of the Land Sat TM bands and improved ground resolution of up to 5 m. Some Russian satellites have similar ground resolution but more limited spectral resolution.

The Japanese JERS-1 sensor has good spectral resolution with additional bands in the IR.

Optical Sensors are also can be mounted on the aircraft. These include:

(a) DETR's airborne Thematic Mapper (ATM) which can achieve ground resolution.

(b) NASA's Airborne Visible Infrared Imaging Spectrometer. (AVIRIS) which has high resolution and a wide range of spectral bands.

(c)   Compact Airborne Spectrographic Imager (CASI) which also has a wide range of spectral range of 400-100 nm and can gather information at a spatial resolution of 40 cm depending on altitude.

CASI is found to be useful for the following applications:

(a)   Land cover and vegetation visualization and mapping.

(b)   Identification and tracking of dissimilar water bodies and mapping of mixing zones e.g. in estuaries.

(c)   Detection and monitoring of water pollution including suspended solids concentrations and eutrophication (Chlorophyll-a estimation).

(d)   Estimation of and changes in coastal morphology.

The success of CASI has prompted the development of CASI-2 which is an extremely compact and power efficient instrument. This allows the unit to collect visible and near infrared information from aircraft, land vehicles and terrestrial based flatforms. CASI and CASI-2 are being used by govt. and educational institutes, private service companies, international space agencies and the military.

10.2.2.2 Airborne Light Detection and Ranging System (LIDAR)

LIDAR uses a laser to measure the distance between the aircraft and the ground. LIDAR applies the same principles as RADAR remote sensing. The device transmits light to a ground based target. Radiation scattered by the target is collected by the instrument and processed to provide information about the target and or the path to the target.

There are three types of applications for LIDAR in remote sensing.

1.   LIDAR range finder methods can be used to calculate distances that can be converted to accurate Digital Terrain Models (DTM).

2.   Differential Absorption LIDAR (DIAL) uses two different laser wave lengths to record information about chemical concentrations.

3.   Doppler LIDAR can be used to determine the velocity of an object.

Though the versatile application of LIDAR is as a tool for creating extremely accurate Terrain Maps, e.g., for assessing flood risks it is now gaining importance in pollution monitoring in the atmosphere and specifically toxic emissions from large factory stacks.

10.2.2.3 Thermal Imagery

Thermal imagery uses measurements of temperature and heat and is commonly affiliated with infrared energy (IR). However this is only partially true because there is a difference between Near Infrared (NIR) energy and Far Infrared Energy (FIR). NIR like light energy that is initially part of incoming solar radiation and is reflected from Earths surface. For example inbound NIR energy is reflected by chlorophyll, making it an ideal portion of the spectrum to use for vegetation studies. In contrast FIR is the energy that has been absorbed by the earth or water body and reemitted as heat due to which FIR is often referred as thermal IR.

The difference between the two forms of IR energy marks a distinct change in the way sensors record information and thermal imagery which measures temperature strictly refers to FIR. Some satellites such as landsats carry thermal sensors but airborne sensors such as NASA's Thermal Infrared Multi Spectral Scanner (TIMS) can provide more information because it is nearer to the ground and is designed specially for recording thermal data in six separate channels.

During last decade thermal imagery is being extensively utilized for getting valuable information for environmental applications like monitoring open waters, surface waters, variations in ground water moisture, content, spring lines, leaks in pipelines or water borne pollution, thermal pollution in open waters from large factories and refineries based on the associated variations in temperature patterns.

10.2.2.4 Radar

Radar uses artificially created microwave energy in pulses at a predefined wave length which are reflected by objects on the ground and recorded as digital information by the sensor. Radar can provide images during day or night and or not effected by weather conditions as they can penetrate cloud covers. This is known as active Remote Sensing while when the emission of earths microwave energy is followed it is known as passive microwave Remote Sensing (Fig. 10.5).

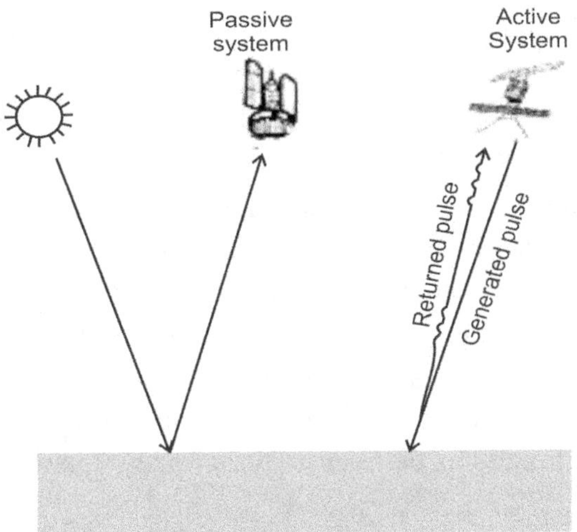

Fig. 10.5 Passive and Active Remote Sensing.

Surface roughness greatly effects the response of the radar signals and with increasing surface roughness the black scatter will increase resulting in the brighter surface of the image. The angle at which energy strikes the ground called incidence angle and as the incident angle increases the expected black scatter also increases. So radar can provide information on surface roughness and topography.

Four satellites are currently providing daily coverage of the earth's surface using Synthetic Aperture Radar (SAR) sensors. These are the European ERS-1 and ERS-2 the Canadian Radar and the Japanese JERS-1. Radar sat has most flexible angle which can change the incidence angle between 10-60 degrees. Shuttle Radar Topography Mission

(SRTM) uses two radar systems mounted on the space shuttle and to create a detailed DTM for entire globe using radar interferometer.

Airborne radar sensors are also used the most comprehensive coverage currently being provided by MS Intera. Airborne Radar can be broken down into two groups; Real Aperture Radar (RAR) and Synthetic Aperture Radar (SAR). Real aperture radar is often associated with side looking while Airborne Radar (SLAR) which is a sensor used for displaying back Scatter from surfacial objects. SLAR works by emitting and receiving microwave energy from high powered antenna mounted on the aircraft. As the antenna size is increased the spatial resolution is improved. To achieve greater spatial resolution without having antenna too large to mount on a aircraft, synthetic aperture radar systems have been designed. A small SAR antenna can achieve the same or greater spatial resolution as a SLAR with less power requirements. This makes SAR ideal for space borne platforms also.

10.2.2.5 Satellite Orbits

Satellites can be kept in motion in two types of orbits:

1. Equatorial which circle the earth near the plane of the equator and is also referred as geostationary as their orbital period is equal to rotating period of earth, and so the orbiting sensor appears to be motion less. However to achieve this satellite must maintain an altitude of 35000-36000 km which severely limit spatial resolution and so they are useful for meteorological and communication applications.

2. The second type of satellites are called polar satellites which circle the earth from pole to pole and hence at right angles to earths rotation and can be used for remote sensing applications of earths natural resources, due to following advantages

    (a) By offsetting the orbit slightly (oblique to the lines of longitude) the local sum time of each point along the orbital track will be the same. Most RS satellites including LandSat and SPOT are in this type of sun synchronous orbit and those with optical sensors collect information between 9.30 am and 11 am which is normally least cloud cover.

    (b) The earth's rotation at right angles to the orbit of the satellite allows complete coverage of earths surface during consecutive satellite orbits.

    (c) Polar orbiting satellites usually travel at relatively low altitudes (between 700 to 900 km) which allows better spatial resolution ranging from approximately 10 to 30 m orbit at altitudes as low as 30 km similar in space shuttle providing sub meter resolution. Unfortunately atmospheric drag becomes a serious problem at such low altitudes and the life span of low orbits will be limited by the need to maintain altitude and hence to carry fuel which will eventually burn out. New high resolution satellites are sure to make an impact on earth's observation of natural resources as projects like Quick bird and Orbview-3 are promising resolutions of less than 2m.

10.3 Remote Sensing as an Important Tool for Environmental Impact Assessment

Remote sensing for environmental impact assessment is an important tool for studying different aspects of ecosystems at local, regional and global scales. It provides an ability to observe and collect data for large areas relatively quickly, and is an important source of data for further analysis and inference.

World-wide, remote sensing is recognized as being a significant source of data for environmental analysis and assessment projects. Projects of this nature benefit from the unique advantages of remote sensing, many of which are fundamental to the successful outcome of environmental projects. Some of these advantages, pertinent to the Environmental Assessment are:

(i)   Regional coverage at adequate mapping resolutions. Landsat TM 180 x 180 km per scene at 30 m resolution and SPOT Panchromatic imagery at 10 m resolution 60×60 km per scene are recommended. These data sets provide the lowest cost per square km for base-mapping and for extraction of a multitude of derived data sets.

(ii) The multi-spectral nature of remote sensing data such as Landsat TM, Landsat MSS, or Spot XS (20 m resolution) permit image processing (IP) applications, allowing extraction of data from single data sets for multiple applications. Environmental applications that can benefit directly from or obtain evaluation data from these data sets are:

- Land use mapping and planning.
- Vegetation cover mapping and vegetation classification maps.
- Hydrological mapping and planning.
- Land degradation mapping.
- Infrastructure mapping, to update outdated infrastructural and topocadastral maps.
- Geological mapping and mineral potential identification.
- Agricultural production mapping, resource identification, and planning.
- Human settlement mapping and planning.

The extraction of applications data from remote sensing data for the above mentioned studies is automated via standard IP procedures in IP software packages.

(iii) Remote sensing data provides the most current information on an area available, as opposed to aerial and ground-based surveys or existing topocadastral data which is rapidly outdated. This function of remote sensing data has particular importance to the project in terms of gathering data about areas where access is denied, or difficult, due to political conflict or instability in certain regions covered by the project.

(iv) Continuous collection of remote-sensing data ensures that environmental planning is based on current data. Comparison of data obtained at different dates during the Environmental Assessment permits time domain analysis, enhancing the quality of:

- Vegetation studies.
- Land degradation monitoring projects.
- Infrastructural development assessments.
- Human resettlement/movement.

10.3.1 Summary of Sensors in Environmental Modeling

Table 10.2 presents an exhaustive overview of the various satellite sensors used in environmental applications. These sensors provide data in a wide range of scales (or pixel resolutions), radiometry, band numbers and band widths and also provide distinct advantage of consistency of data, synoptic coverage, global reach, cost per unit area, repeatability, precision, and accuracy. Long-time series of archives and pathfinder datasets (6, 7) that have global coverage are also included.

Several applications (8) in environmental monitoring require frequent coverage of the same area. This can be maximized by using data from multiple sensors (Table 10.2). However, since data from these sensors are acquired in multiple resolution (spatial, spectral, radiometric), multiple and width, and in varying conditions, they need to be harmonized and synthesized before being used (9, 10). This will help normalize for sensor characteristics such as pixel sizes, radiometry, spectral domain, and time of acquisitions, as well as for scales. Also, inter-sensor relationships (9) will help establish seamless monitoring of phenomenon across landscape.

Table 10.2 Satellite and Sensor Data Characteristics (Source:11)

Sensor	Spatial (meters)	Spectral (#)	Radiometric (bit)	Band range (µm)	Band widths (µm)		Irradiance (wm^{2}sr^3 µm^{-3})		Data Points (# per hectares)	Frequency of revisit (day)
A. Coarse Resolution Sensors										
1. AVHRR	1000	4	11	0.58-0.68	0.10		1390		0.01	Daily
				0.725-1.1		0.375		1410		
				3.55-3.03	0.65	0.38		1510		
				10.30-10.95	0.7		0			
				10.95-11.65			0			
2. MODIS	250, 500, 1000	36\7	12	0.62-0.67	0.05		1528.2		0.16.0.04, 0.01	Daily
				0.84-0.876	0.036		974.3		0.16, 0.04, 0.01	
				0.459-0.479	0.02		2053			
				0.545-0.565	0.02		1719.8			
				1.23-1.25		0.02		447.4		
				1.63-1.65		0.02		227.4		
				2.11-2.16		0.05		86.7		
B. Multi Spectral Sensors										
3. Landsat-1, 2, 3 MSS	56 × 79	4	6	0.5-0.6	0.1	1970		2.26	16	
				0.6-0.7	0.1	1843				
				0.7-0.8	0.1	1555				
				0.8-1.1	0.3	1047				
4. Landsat-4, 5 TM	30	7	8	0.45-0.52	0.07	1970			11.1	16
				0.52-0.60	0.80	1843				
				0.63-0.60	0.60	1555				
				0.76-0.00	0.14	1047				
				1.35-1.74	0.19	2271				
				10.4-12.5	2.10	0				
				2.08-2.35	0.25	80.53				

Table 10.2 contd…

Sensor	Spatial (meters)	Spectral (#)	Radiometric (bit)	Band range (µm)	Band widths (µm)		Irradiance (wm^{2}sr^3 µm^{-3})		Data Points (# per hectares)	Frequency of revisit (day)
5. Landsat-7 ETM	30	8	8	0.45-0.52	0.65	1970			44.4.11.1	16
					0.80	1943				
				0.52-0.60	0.60	1555				
				0.63-0.60	0.150	1047				
				0.76-0.00	0.200	2271				
				1.35-1.74	2.5	0				
				10.4-12.5	0.2	1368				
				2.08-2.35	0.38	1352.71				

Table 10.2 *contd…*

6. ASTER	15.30.00	15	8 12	0.52-0.63 0.63-0.60 0.76-0.85 0.76-0.86 1.60-1.70 2145-2.185 2.185-2.225 2.335-2.285	0.11 0.04 0.04 0.05 0.07 0.07 0.35 0.35 0.3.5 0.7 0.7 0.64 0.20 0.65 0.80 0.57 0.3.0 0.45 1.00 2.00	 0.06 0.1 0.1 0.1	1846.9 80.32 74.05 89.20 59.82 57.12 0 0 0 0 0	1546.0 1117.6 1117.6 282.5	44.4, 11.1, 1.23	16

				2.295-2, 365 2.360-2, 430 8.125-8-475 8.475-8.825 8925-9 275 10.25 10.05 10.05-11.65 048-0.69 (P) 0.435-0.453 0.450-0.515 0.425-0.505 0.633-0.690 0.775-0.590 0.845-0.890 1.200 1.200 1.550-1.750						
7. ALI	30	10	12	2.080-2.350	2.70		1.74/8600 1849.5 1985.0714 1.732.1765 1485.2308 1134.2857 948.36364 930 61905 233.30024 78.072727	11.1		10

Table 10.2 *contd…*

Sensor	Spatial (meters)	Spectral (#)	Radiometric (bit)	Band range (µm)	Band widths (µm)		Irradiance (wm²sr³ µm⁻³)		Data Points (# per hectares)	Frequency of revisit (day)
8. SPOT-1 -2 -3 -4	2.5-20	15	16	0.50-0.59 0.61-0.68 0.70-0.80 1.5-1.75 0.51-0.73 (p)	0.09 0.07 0.1 0.25 0.25 0.22	1858 1575 1047 234 1773			1000.25	3-5
9. IRSIC	23.5	15	8	0.52 0.50 0.62-0.68 0.77-0.86 1.55-1.70 0.5-0.75 (P)	0.07 0.06 0.09 0.15 0.25		1851.1 1582.8 1102.5 240.4 1627.8		12.1	16
10. DRS-1	23.5	15	8	0.52-0.59 0.62-0.68 0.77-0.80 1.35-1.70 0.5-0.75 (P)	0.07 0.06 0.09 0.15 0.25		1852.8 1577.38 1096.7 240.4 1603.9		18.1	16
11. IRS-P6-AWIS	56	4	10	0.62-0.68 0.77-0.86 1.55-1.70	0.12	1064.3		23.25 400.25		
12. CBERS-2 -3B -3	20 in pan 20 in MS 5 m pan 20m MS		11	0.31-0.73 0.45-0.32 0.52-0.59 0.63-0.69 0.77-0.89	0.22 0.07 0.07 0.06 0.12	1934.03 1787.80 1587.97 1069.21 1664.3		23.25 400.25		

Sensor	Spatial (meters)	Spectral (#)	Radiometric (bit)	Band range (μm)	Band widths (μm)		Irradiance ($wm^2 sr^3\ \mu m^{-3}$)		Data Points (# per hectares)	Frequency of revisit (day)
C. Hyper-Spectral Sensor										
1. Hyperton	30	196	19	190 effective calibrated bands VNDR (based 8 to 37)	10 mm wide (approx.) for all 190 bands	See data in Number and Labs (1984) Plot 4			11.1	1.0

10.4 Application of Geographic Information Systems (GIS) in EIA

Geospatial techniques such as geographic information systems (GIS) play a pivotal role in Environmental Impact Assessments (EIAs) and their various components such as ecological, social and hydrological impact assessment. The most common application of GIS is concerned with environmental issues. This prominent role has been attributed to the fact that GIS can answer questions that are central to the EIA process which according to ESRI (1995) (12) are

1. What is where?
2. What spatial patterns exist?
3. What has changed since?
4. What if?

The knowledge of "what is where" is pivotal to in conducting the screening, scoping and baselines studies while on the other hand data on the existing spatial patterns can help in developing an understanding of the baseline conditions as well as in impact prediction and mitigation. For instance the location of a wetland near to the proposed project site can identify impacts to do with effluent discharge while the impact of the effluent can be determined by the wetland's ability to self purify which depends on the soil type as well as flora and micro fauna.

An understanding of "what has changed" can then be relevant to impact prediction of changes in the absence of the project (i.e., business as usual alternative) and impact monitoring.

GIS assists in answering the "what if" question by assisting in the exploration of alternatives as well as in the formulation of mitigation measures. At the most basic level the GIS system can be used to produce map and visual aids that will be used by project managers, consultants and the EIA team. At a more sophisticated level, however, GIS functions such as overlays, spread and materials flow analysis can be used for the assessment of the significance of impacts (Figure 10.6).

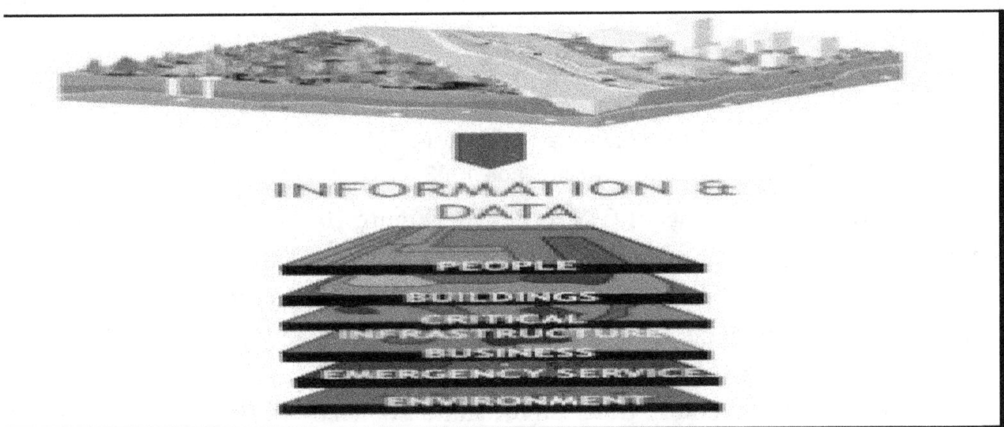

Fig. 10.6 Data and Information in set of layers.

10.4.1 Technical details of Geographic Information Systems (GIS) and its Application to EIA

10.4.1.1   Major Elements of GIS

GIS is a computer based information system which enables geographical data (13) both spatial and descriptive either simultaneously or separately for modeling, manipulation, retrieval, analysis and presentation in map form. Spatial data are two- or three dimensional coordinates of points (nodes), lines (arcs), or areas (polygons) representing one aspect of a geographic reality (coverage). Descriptive data, on the other hand, refers to the features or attributes of these points, lines, or areas (14).

Figure 10.7 depicts the main elements of GIS. Information which provides spatial relationships between similar geographical areas known as topological relationship like link connectivity, area contiguity (spatial features of the adjacent areas), can be obtained from GIS. Understanding topographical relationship helps for carrying functional analysis which facilitates faster data processing, area aggregation, route finding and overlaying geographic features (15,16).

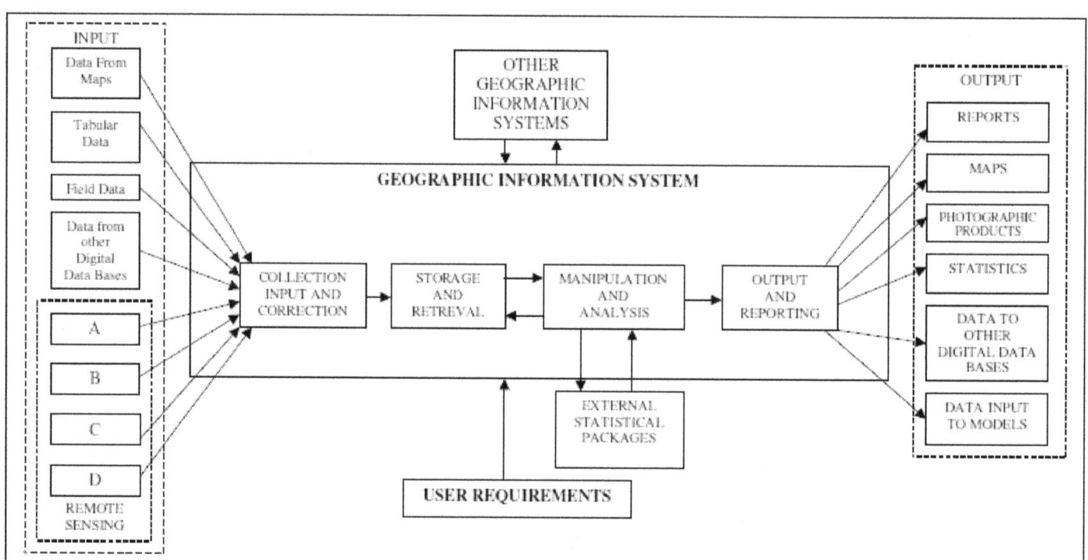

Fig. 10.7 Major Elements of GIS
(**Source** Marble and Arnundson 1988; Linden, 1990 16,17)

The following are major elements of GIS ((16, 17)) (Figure 10. 7) and their functions

1. Data input module: for collecting and/or processing spatial data derived from sources, such as existing maps, remotely sensed data and direct digital input.

2. Spatial data base module: location and shape data are stored and retrieved in the form of coverage (maps).

3. Attribute data base module: descriptive data associated with the spatial features are stored and retrieved.

4. Analysis module: in the form of a group of commands and functions which performs a number of tasks, such as changing form of the data through user defined aggregation rules or producing estimates of parameters for transfer to external analytical type models.

5. Output module: displays all the retrieved or selected portions of the spatial database in the term of standard reports or in a variety of cartographic formats and communicate with other systems.

User defined procedures/functions adopting special simulations and optimization models can also be incorporated into GIS either internally or externally to make it a very useful decision support system (18). For example Fig 10.8 shows the application of GIS integrated with external statistical package.

10.4.2 GIS Data Characteristics Spatial Data

Geographical features including their location and their spatial dimensions will be covered by spatial data which can be stored, either in vector or raster data models (Fig 10.8).

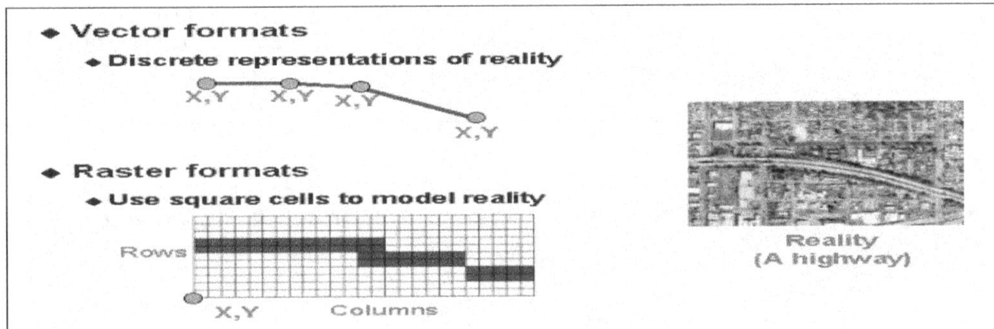

Fig. 10.8 Vector and raster data models for data storage in GIS

Geographic features are presented in vector data models as points, line or areas just as in maps. Locations which have no area like road intersections, telephone poles, buildings, mountain peaks etc. are represented by points while objects which have linear features length but no area like roads, streams, elevation contours are represented by lines. Areas such as states, counties, parcels, soil types, or land use zones are represented by closed figures. In GIS points are presented as x, ny coordinate system, while line as series of points with direction and area as a polygon with closed line.

In raster data model which is well suited for spatial analysis, geographical features instead of x, y coordinates are given some values to cells that cover their locations. This is more appropriate for the storage of data that is collected in grid format and the amount of detail you can show for a particular feature depends on the size of the cells in the grid. In-view of this Raster data will be in appropriate for applications where discrete boundaries must be known, such as parcel management.

The differences between Raster data and vector data are presented in Table 10.3.

Table 10.3 Differences between Raster and Vector Models

No.	Raster Data	Vector Data
1	Low spatial accuracy.	High Spatial accuracy.
2	Requires greater storage space on computer.	Requires less disc storage space.
3	Easy to be analyzed as it is easy for the computer to manipulate some complex number analysis.	Difficult to be managed as it is stored in a large list of coordinates.
4	Slow computation (analysis) and display	Quick computation (analysis) and display.
5	Not easy to be understood by the general public.	Easy to be understood by the general public.
6	Requires low technology and inexpensive systems.	Requires high technology and expensive systems.
7	Used in the applications which concern the continuous changes such as: environmental characteristics and change detection of the shore line.	Used in the applications which concern the stable conditions such as urban planning, site selection and crisis management.

10.4.2.1 Data Capture

The technology for GIS map data capture is quite varied and changing rapidly but the techniques can be divided into three categories that can be called primary, secondary and tertiary data capture.

Primary data capture techniques include:

(a) ground survey based on sampling, the traditional source of cartographic data.

(b) remote sensing based on classifying the pixels in the satellite infrared picture.

(c) Global Positioning System: a hand held system that can register positions. GPS is used today for all kinds of cartographic applications and is probably the most important advancement of recent times in the field of cartographic data input.

Secondary data capture techniques (from paper maps or aerial photographs) include:

(a) Digitizing (tracing) on a magnetic table the points on a map as well as the caricature of its lines (lines broken down into straight segments) - labor intensive and expensive.

(b) Scanning maps, using refined versions of the type of technology used in fax machines - cheaper but still prone to errors in the form of gaps in the scanned lines.

(c) So called overlay digitizing (or heads up screen digitizing) which combines the advantages of both: a map is scanned cheaply and then displayed on the screen and vector map features are derived from the image using a digitizer.

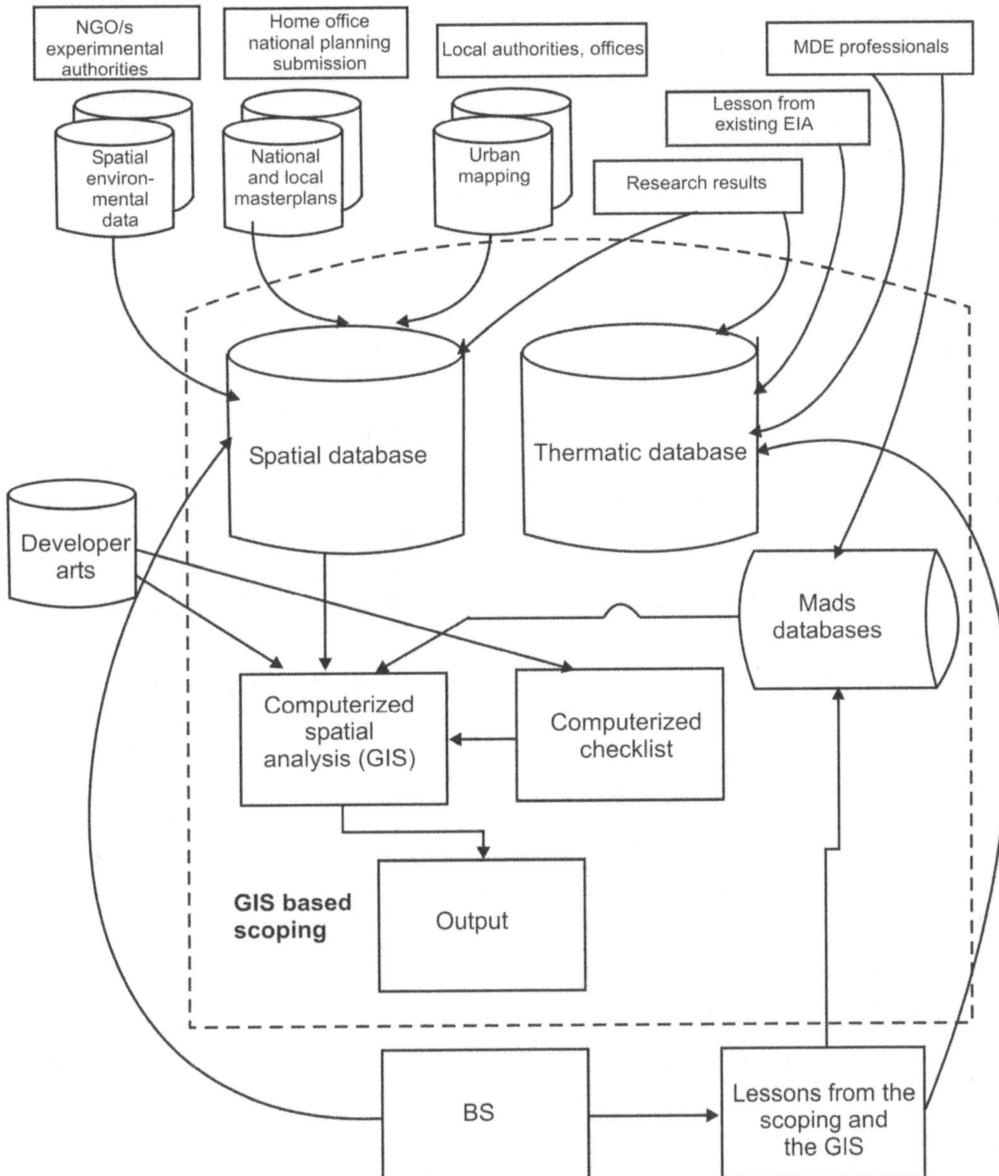

Fig. 10.9 Scheme for Database Collection and Storage System.

Tertiary data capture is based on importing data from existing sources already in digital form, provided by public or private organizations. This is currently an area of fast growth is not surprising, given the cost and difficulty of obtaining primary data and the labor intensive nature of secondary data capture. Digital data from airborne and satellite sensors is becoming increasingly available and many national cartographic and environmental agencies are now providing digital cartographic information which is GIS compatible, although this proliferation of data sources is increasingly raising the question of formation acceptability between them.

10.4.2.2   Data Sources

Using spatial data sets consisting of physical data (such as topographical data in the form of Digital Elevation Model - DEM), coverage data (buildings, infrastructures etc.), ecological data (sensitive species) and results of environmental studies (such as aquifer sensitivity) and relation ship between environmental issues and elements different thematic data bases can be prepared and stored. Figure 10.9 presents the general scheme for data collection and storage.

10.4.2.3 Data Storage

Raw map data become information when interpreted by conceptual data models, and the type of model used to store GIS maps is one of the clearest dividing lines between different types of systems.

Regular tessellation "raster models" store maps using more or less simplified dimensions of a matrix file, where the different square cells (rasters) are stored with their attributes. The advantage of a file of this kind is that it simultaneously defines the map and the values of particular attributes (one for each map) for every feature. First generation GIS belonged to this kind. They are easy to program and simple in terms of file structure, but are wasteful of repetitive information and their drawback is that their accuracy is ultimately determined by the size of the file they use.

Irregular tessellation "vector" models represent map features (point, lines) by the precise coordinates of their defining points and segment ends. This increases accuracy but has the problem of requiring two sets of files for each image one to store the position and shape of the map features, and another to store the attributes with those features. Vector data can be stored by layers (each containing one or several features), or by objects (the latest approach now being level) where the attention is on individual cartographic objects, their properties and their membership of different classes and sub classes with possibility of inheritance of properties between them. Well known vector based GIS software's include:-

ARC INFO -by ESRI http.www.esri.com

TIGRIS -by INTERGRAPH http www.intergraph.com

Some simple mapping programs with very limited GIS functional are now available for use with platforms such as windows or NT4.

Examples include:

- AutoCAD map 2000
- Geomedia and Geomedia professional
- Map maker pro
- Map sheets
- PAMAP

10.4.2.4 Data Manipulation and Analysis

Despite the cartographic sophistication of GIS, the tasks they can perform in terms of spatial analysis are quite limited and can be summarized as follows:

- map overlay superimposing maps to produce simple composite maps probably then single most frequent use of GIS functionality.
- clipping one map with the polygon of another to include (or exclude) parts of them, for instance to identify how much of the area of a proposed project overlaps with the sensitive area.
- producing partial maps containing only those features from another that satisfy certain criteria.
- combining several maps (weighted differently) into more sophisticated composite maps, using so called map algebra also referred to as cartographic modeling used for instance to do multicriteria evaluation of possible location for a particular activity, or calculating composite effect of a set of factors on an area.
- calculating the size (length, area).
- calculating descriptive statistics for the features of a map (frequency distributions, average size, maximum and minimum values etc).
- doing some multivariate analysis like standard correlations and regressions of the values of different attributes for different features of a map.
- calculating minimum distances between features (some systems measure straight line distances, others can also measure distances along network).
- using minimum distances between features on one map nearest to particular features on another map.
- using distances to construct buffer zones around features, which can then be used to clip other maps to include/exclude certain areas.

10.4.2.5   With a Third Dimension

- interpolating unknown attribute values for new points (a third dimension on a map) between the known values for existing points, using triangular irregular networks (TINs) to maximize the efficiency of interpolations.
- drawing contour lines using the interpolated values of an attribute.
- constructing digital terrain models (DTMs).
- calculating topographic characteristics of the terrain, like slope, and orientation aspect of different parts, their concavity and convexity etc.
- calculating volumes in a DTM e.g. to calculate water volumes in lakes or reservoirs.
- identifying areas of visibility of certain features of one maps from the features of another for instance to define the area from which the tallest building in a proposed project would be visible.
- Modeling, identifying physical geographic objects from maps like the existence of valleys, or streams forming a river basin, river networks etc.

10.4.2.6 Preparation of Results

The output of GIS is probably the best developed and most appealing aspect of 2D displays maps which are most common.

2.5D representations of Digital Terrain Models which uses third (z) dimension over an x-y map. Other maps can be superimposed on them so that they appear to be in 3D or the slopes and aspects of the different facets in these models can be used to calculate sunlight reflection and produce 'shaded' representations of the terrain.

3D models which are currently the object of considerable research, looking to the possibility of representing 3D objects as collections of sheets using the standard functions of GIS which are essentially two dimensional or may be incorporating into GIS some features of CAD or Virtual reality (VR).

A dominant current trend of GIS output when produced for the computer screen towards interactive multimedia output which combines maps, photographs, motion video images and even sound as part of the emerging approach of hypermedia in which the user can move between all those outputs by just zooming in and out between them.

10.5   Integration of Remote Sensing and GIS-New Software Modules

Though Remote sensing (RS) and GIS are developed independently. RS has the capacity to generate end products in the form of maps, tabular summaries, statistical graphs etc. (19) while GIS creates new base maps by digital map compilation in stereo-models and on building thematic databases by digitizing existing maps. Thus RS&GIS compliment each other. Air -borne/space borne sensors which produce data images will be the data sources for these techniques. GIS can extract vast information from RS images and process it by integrating with other information for building thematic data bases by integrated data analysis. Integrated GIS is finding very large applications particularly due to availability of submeter resolution satellite data and possible data management techniques like adaptation, modification, and extension (20) using advance computer technologies. However the new data produced require further processing to compare with standards (21).

10.5.1 Levels of Integration

Three levels of integration viz. i. separate but equal, ii. seamless integration, and iii. total integration. Proposed by Ehlers et al. (20) between RS & GIS for various earth resources evaluation applications.

Level I (separate but equal): Separate databases. Two software modules, GIS and image processing, are linked only by data exchange. The integration at this level should have the ability to move the results of low level image processing (e.g., thematic maps, extracted lines, and so on) to the GIS, and the results of GIS overlays and analysis to image processing software.

Level II (seamless integration): Two software modules with a common user interface and simultaneous display. Such a system will allow for a "tandem raster-vector processing" (20), such as incorporating vector data directly into image processing, entity-like control over remote sensing image components (e.g., themes). Moreover, the system will have abilities to accommodate hierarchical entities (e.g., "house" at one level, "block" at another, and "city" at another), and (spatially, radio-metrically, spectrally, and temporally) in heterogeneous data in a coherent manner.

Level III (total integration): A single software unit with combined processing. It is a long-term goal. In the fully integrated system, a single model will underlie all information in the GIS, which has the flexibility of handling both object- and field based space representations. Remote sensing will become an integral part of the functionality of the GIS

For Earth resources assessment and evaluation, integrated RS&GIS will be of great help as it will be able to process temporal and three-dimensional information. A hierarchical three-level integration system in which each successive level in the hierarchy deals with more detailed conceptual and operational factors and issues of remote sensing and GIS linkages was developed by Mesev (22).

In level I the issues of data unity, measurement conformity, potential integrity, statistical relationships, classification compatibility, and overall integration design will be addressed while these issues are more fully examined at Level II.

In level II more complex linkages among the various components is possible. Data unity, for example, is divided into factors such as information exchange, data availability, data accessibility, and data creation.

Factors are further refined to produce even more detailed levels at Level III. For example, data availability is subdivided into awareness, publicity, search, data type, age, quality, and access (Table 10.4).

While other schemes are proposed, Davies et al. (1991) (23), Mesev attempt to itemize the common linkages and reexamine the relationships, and thus provides a logical and structured framework for direct data coupling.

Table 10.4 Levels of Remote Sensing and GIS integration (Meesev 1997)(22)

Level II	Level III
Data unity (factors that bring RS and GIS data together)	
Information interchange	Definition of integration, type of information needed,
Data availability	Awareness, publicity, search, data type, age, quality, (access of create)
Data Accessibility	Cost, agreements, exchanges, sharing, proprietary, resistance, confidentiality, liability
Data creation	Digitizing, scanning, survey information encoding, sampling, data transformation. GPS
Measurement conformity (factors that link data beween RS and GIS)	
Data representation	Data structures (Vector, raster, quad-tree, etc). data type, level of measurement, field-based vs. object-based modeling, interpolation.
Database design	Type (relational, hybrid), schema, data dictionary, implementation (query, testing)
Data transfer	Formal, standards, precision, accuracy.
Positional integrity (factors that spatially co-ordinate data between RS and CIS)	
Generalization and Scale	Spatial resolution, scale, data reduction and aggregation, fractals
Geometric transformation	Rectification, registration, re-sampling, co-ordinate system, projection, error evaluation.

Table 10.4 *contd....*

Statistical relationships (factors that measure links between RS and GIS)	
Vertical	Boolean overlays, dasymmetric mapping, aerial interpolation, linear and non-linear equations.
Lateral	Time series, change detection spatial searches, proximity analysis, textural properties
Classification compatability (factor: that harmonize information between RS and GIS)	
Semantics	Classification schema, levels, descriptions, class merging, standardization
Classification	Stage (pre during post). Level (pixel, sub-pixel). Type (per-pixel, textural, contextual, neural nets, fuzzy sets). Change detection, accuracy assessment.
Integration design	
Objectives	Plan of integration, cost/benefit assessment.
Integration specification	Feasibility, alternatives to integration education system requirements (hardware, software computing efficiency)
Decision-making	Testing, Visualization, ability to replicate integration, decision-support, implementation or advocate alternatives, bi-directional updating and feedback into individual RS and GIS projects

10.6   Application of GIS as a Tool for Remotely Sensed Image Processing

GIS data can be used to enhance the functions of image processing at various stages: selection of area of interest for processing, preprocessing, and image classification. At the stage of geometric and radiometric correction, GIS data such as vector point, area data, and DEMs are increasingly used for image rectification (24). High-resolution topographic data play an important part in radar image interpretation (25). The impacts of varying topography on the radiometric characteristics of digital imagery can be corrected with the aid of DEMs (24). Perhaps, the most frequently used vector data sets are ground control points in image rectification, selected from an existing map with a defined coordinate system. Hinton (1996) (24) suggested that with the advance in pattern recognition and line following techniques, lines on images could be registered to roads, rivers, and railways in vector datasets for image registration. At the stage of image classification, the integration of GIS and remote sensing will facilitate the selection of training areas. Ehlers et al. (1991) (26) emphasized the need for a raster/vector intersection query in order to optimize the operation. This query is capable of providing image statistics within vector polygons without completing any data format conversion or raster masking. The emergence of such an intersection query, therefore, would allow for a better selection of training areas for image classification by examining whether the areas display the spectral response characteristics of a class. Likewise, it is possible to look at changes in image statistics within defined polygons directly without classifying the image and examine the raster result (Hinton, 1996) (24). Moreover, the integration between GIS and remote sensing will allow vector data to be rasterized as "image planes" and incorporated into traditional classification and segmentation procedures (Ehlers et al., 1989a) (20). A closer integration will make it possible to incorporate all GIS data layers in their native format into an image classification. Then, a two-way data flow between the raster image and the vector dataset comes true. This situation occurs only in "polygon classification" (Hinton, 1996) (24), in which image statistics are generated based on polygons, and then returned directly to the GIS database as attributes of the polygons. The full integration of raster and vector data processing could also be used to restrict the area of image to be processed. This indeed permits masking operations without raster masks, making image processing much more efficient.

10.6.1 Remotely Sensed Data as an Information Source to GIS

The automated extraction of cartographic information has been a major application of remote sensing imagery as data input to GIS. Lines and other geographic feature extraction have been achieved from satellite images by using pattern recognition, edge extraction, and segmentation algorithms (20, 24). A further development in the integration will produce smoother lines and boundaries, i.e., not stepped appearance, hiding their raster origin. Therefore, satellite images show a great potential in producing and revising base maps (Welch and Ehlers, 1988) (26). The production of base maps by means of remote sensing imagery will make it easier for tracking error propagation in GIS layers because of reliable map metadata. In addition, the extracted cartographic information can be used to improve image classification. Image segmentation polygons derived from optical imagery, for instance, could be very useful for stratifying radar data, which are traditionally difficult to digitally segment owing to their noise (24). Another area of the application of remotely sensed data as input to GIS is change detection and map updating. Ehlers et al. (1989b) (20) used SPOT data in a GIS environment for regional analysis and local planning at a scale of 1:24,000, and achieved an accuracy of 93 percent for growth detection. Brown and Fletcher (1994) (26) demonstrated that satellite images could be used to interactively update a land use database by comparing the database with the image statistics within areas defined in a vector database. In the future, a full integration between the two elements will allow for querying the raster pixels within vector areas (both vector topological query and raster/vector intersection querying), and performing analyses without format conversions and overlays. Image statistics within vector polygons are then used to examine change and update maps. Cartographic representation is the third area of application of remote sensing imagery as an input to GIS. Terrain visualization using satellite images in association with DEMs has been explored as a promising tool in environmental studies (28).

10.7 Application of Integrated Remote Sensing and GIS for EIA

10.7.1 Sources of Remote Sensing Applications

The following are internet sources where the information on application of **Remote Sensing** for Environmental monitoring can be found:

1. Indian Space Research Organization (ISRO) http/www.isro.org/sat.htm.irs.

2. National Remote Sensing Agency (NRSA), India, http/www.nrsa.org.in.

3. National Remote Sensing Center (NRSC) UK-http/www.nrsc.co.uk.

4. National Space Development Agency, Japan (NASDA) http/www.nasada.go.jp.

5. Netherlands Earth observation Net work-http/www.neronet.nl/.

6. Radarsat Canada http/wwww.rti.com.

7. USGS-EROS-http//edcwww.cr.usgs.gov/eros.home.html.

Satélites

IKONOS –space imaging USA, www.space imaging.com

 Landsat 7-NASA(USA) www.lansat.goc.nasa.gov

 Terra-NASA-USAwww.terra.nasa.gov

 Rdadarsat-Radarsat Internacional- Canadawww.rsi.ca

 Spot-Spotimage France, www.spotimage.fr

 Orbitvew-Orbimage (USA) www.orbimage.com

10.7.2 Integrated RS & GIS in Screening, Scoping and Baseline Studies

10.7.2.1 Screening & Scoping

Screening and Scoping is deciding whether a project requires EIA or not. This is usually based on:

(a) Characteristics of the project itself, e.g., the type of activity or construction it involves, the level of such activities, and whether they exceed certain threshold levels.

(b) The project's location and the sensitivity of this and the area nearby. Examples of how GIS can facilitate screening include:

- Certain types of projects e.g., industrial estates will require an impact assessment if they reach or exceed a certain area, and a GIS will be able to calculate this automatically from the map of the project.

- Often it has to be established if a project lies within an environmentally sensitive area in which case an EIA would be required. Although simple visual of a map will often suffice, using GIS to overlay the map of Industrial Project and the map of the relevant sensitive area will achieve the same result with increased accuracy, and with the additional advantage that the GIS may be programmed to do it automatically and report back.

- In some cases an EIA is required if the project is within a distance of the water stream.

10.7.2.2 Baseline Studies

For carrying out detailed analysis of base line environmental conditions for specific impact themes, in addition to initial scoping data, further specific environmental indicative factors data of the ecosystem will be required for improving the scoping in the EIA for which GIS will be useful for displaying and visualizing trends and patterns in spatial datasets.

Point type data which relate to a specific sampling location (e.g., a pollution monitoring station) can be displayed in the form of a proportional symbol map in where time series data are available, perhaps as a series of maps at various levels to reflect the dynamic nature of the environmental baseline.

Spatially continuous data (e.g., noise, rainfall, topography, groundwater and air pollution) can be used (given a sufficient spatial sample) to produce contour maps or in the case of topography, as a DTM to describe the baseline terrain.

Linear data describing features such as rivers or rocks can be represented using color coding or perhaps with variation of line width in proportion to the data values e.g., to illustrate traffic flow data along rocks.

Area data which relate to discrete spatial units (e.g., census data, designated and habitat patches) can be displayed as differently shaded maps, where the intention of shading is to reflect the data values.

For detailed environmental analysis in any EIA process, GIS is specifically suitable for organizing and storing multi disciplinary monitoring datasets into a framework which can be analyzed, queried and displayed interactively. For example the spatial query capabilities intrinsic to GIS can be used to highlight potential hotspots (e.g., locations with pollution levels of prediction and assessment of significance where comprehensive spatial data sets are available.

GIS are also ideal for determining the extent to which hotspots and sensitive locations are spatially concentrated across a variety of different environmental parameters.

10.7.2.3 RS & GIS in Impact Prediction Reflections

Introduction impact prediction lies at the core of EIA and is intended to identify the magnitude and other dimensions of likely changes to the environment which can be attributed is a development proposal. GIS are obviously most suited to dealing with the spatial dimension of impacts, and at the simplest level of analysis that can be used to make quantitative estimates of aspects such as:

- the land loss caused by development (e.g., the total area of agricultural land, grass land or wetland habitat which may be lost)
- the length of road or pipeline which passes through a designated landscape area such as an area of outstanding natural Beauty
- the number/importance of features, such as archaeological or ancient monuments lost to the development.

More sophisticated predictions will require some form of modeling to represent or simulate the behavior of the environment, and two broad ways in which GIS may be used for modeling in impact prediction can be identified.

1. The entire process of developing and implementing a model takes place within the GIS software, i.e., GIS is used for data input and preparation, modeling, and finally for the display and spatial analysis of model output.

2. While GIS may be used in data preparation, the actual modeling is undertaken outside the GIS software using an independent computer model, the output from which is imported back into the GIS for purposes of display and further spatial analysis.

10.7.2.4 RS & GIS in Mitigation

Out of the most effective uses of GIS technology in terms of mitigation in the broadest sense relates to the identification and evaluation of alternative locations for a development project. Given a comprehensive spatial database and a series of clearly defined constraints or preferences, GIS overlay analysis can be used to good effect to identify and compare potential sites (or route alignments for linear development).

The results of comparing the issues identified through the GIS based scoping effort with those identified in the comprehensive EIS are summarized in Table 10.5. These results show that the GIS based scoping study identified several issues not addressed in the EIS, and provided clearer and more explicit guidelines for several issues (in particular for the visual effects) than were set before the consultants who prepared the EIS. While the scoping study did not identify some impacts (such as noise) as accurately as the comprehensive EIS, it did not overlook any of the issues identified in the EIS.

Table 10.5 Comparison of GIS and EIS

	GIS Based Scoping Outcome	EIS Results	Notes
Effect on Land Use Plans	The proposed plan split the municipal area of two rural settlements.	The EIS states that a proper passageway will be defined in due course.	
	For the part of the plan where detailed mapping exists, all buildings that should be removed were identified	The buildings that should be removed due to the proposed plan were identified using an aerial photo.	
Effect on Open Land	A landscape reserve was identified through a national level master plan	The landscape reserve identified through a regional master plan	
	A nature reserve was identified. Furthermore, uncertainty about its precise border was discovered.	The nature reserve is not mentioned in the current EIS.	The EIS editor claimed that it should be dealt within another EIS for a different part of the road.
	A man-made plantation site was identified.	In the same area, a small grove of eucalyptus was identified in the flora research.	
Land and Soil	Soil types were identified based on	Soil types were identified during a local	Though the types match, there is a shift

Table 10.5 *Contd..*

	existing survey material.	survey.	in borders.
	Cut and fill sites were identified in a simplified model.	Cut and fill sites were planned in the engineering part of the EIS.	All but one site were matched.
Hydrology	Possible contamination of a local reservoir.	The EIS ignores this subject.	The EIS editor stated that the road is not a contaminating body.
	The local watersheds were identified using a Digital Terrain Model (DTM)	The local watersheds were identified in a local survey	The DTM forecast was inaccurate, due to the poor quality of the DTM data source.
Noise	Public buildings (school, hospitals etc.) in the vicinity of the proposed road were identified.	The guidelines order probe of each site. By mistake, the guidelines did not include one settlement.	
	Noise levels were identified using the IUCZ model (Canter, 1996)	Noise levels were predicted with a special purpose software.	A deviation of 5 to 10 db(A) between the two model, as result of simplification of the scoping model.
Visible Value	Several sites were suggested. The sites are characterised by being sensitive to micro alignment changes in the road scheme.	Several arbitrary sites were chosen to depict the view of the road.	Current guidelines do not force the EIS to give the view of the road from a specific site.
Flora	Several protected species were identified on the base of ecological data.	Several protected species were identified in a local survey.	Survey results match the GIS data.

10.7.2.5 Application of RS and GIS for Linear Projects

Roads affect habitats at a range of spatial scales and usually impinge upon a number of habitats across a region as well as fragmenting individual habitat patches and act as barriers to species movement. The impact of roads is thus difficult to assess using conventional techniques while RS and GIS can be of great help

10.7.2.6 Coastal Zone Studies

The coastal zone has a similar linearity to that of roads and pipelines and poses the associated problems with measurement and monitoring. Coastal processes in large spatial scales and over both short and long term temporal scales, and consequently monitoring techniques are difficult to apply.

Remote sensing can monitor both short and long term changes at the coast. For example, historical aerial photographs can be used to measure changes at coastal geomorphology resulting from variation in the long term sediment balance; aerial satellite imagery can be used in conjunction with current nautical charts to detect gross sediment transport. More dynamic water based processes can be analyzed by studying the evolution of temperature patterns.

10.7.2.7 Estuaries

Estuaries can be monitored for changes in their sedimentation regions. Both the gross sediment load and more subtle indices such as the distribution of suspended solids, turbidity, temperature, salinity and amount of chlorophyll and phosphorus can be measured to check the behavior of an estuary. Some coastal developments have a direct effect on smaller sections of the coast line but Remote Sensing is an invaluable data gathering tool due to the extensive nature of the coastal processes.

Remote sensing is particularly useful in Fresh water studies for measuring the extent of coastal zone waters because of their inaccessibility to measurement. Fresh water bodies such as lakes, reservoirs, and wastewater stores are similarly difficult to measure other than by time consuming and costly direct sampling.   Remote sensing can be used to.

(a)   Measure wastewater bodies, which can be classified by combining remote sensing and laboratory analysis

(b)   Estimate the volumes of wastewater held in reservoirs

(c)   Monitor reservoirs for effluent quality by studying water transparency and color due to the organic matter, suspended solids and chlorophyll

(d)   Assess micro pore distributions in reservoirs

(e)   Target the optimum location for mitigation measures such as establishing buffer zones of vegetation to control runoff

The estimation and monitoring of runoff can be useful index of land use change. Runoff as a function of soil moisture can be estimated using remote sensing using SAR imagery. Similar techniques can be used for watercourses whereby field sampling is used in conjunction with remotely sensed land use data and climatic, hydro-geological and geomorphologic data to monitor their chemical composition.

10.7.2.8 Land Use and Land Cover Studies

Landfills have been extensively classified and their status assessed using current and historical aerial photographs, visible – FIR imagery and multi-spectral images. The classification allows for the prediction of impacts from new or expanding sites, but more specific scoping is possible when the geology of surroundings for water bearing fractures are analyzed.

The scoping and monitoring of wastelands is important and they can be identified with thermal RS in having relatively high soil moisture content and associated low temperature. Remote sensing will be ideally suited for mapping and classification of wasteland sites at the regional scale. In the rural –urban fringe, remotely sensed mapping is especially useful as information becomes rapidly outdated. This allows for scoping over large areas in zone where there is conflict between the demands of urban or industrial development, and there is agricultural or nature conservation. But for this it is essential that the remote sensing data are at a scale both temporally frequency of repeat observations and spatially (both extent and resolution) appropriate to the subject under investigation.

10.8   Key problems or Disadvantages of using RS and GIS for EIA

There are a number of issues/concerns/problems of use of RS & GIS technologies for wide range applications and hinder exploit their full potential.

(i)   Time and cost are some important constraints in the application of RS&GIS in Environmental Impact Assessment. For installing the GIS server/clusters, to build data base and to process the data the key constraints are time and costs which are limiting the widespread use of GIS in EIA. For example for many projects EIA preparations will have strict time schedules which may not match with the time required for GIS data generation. Further these sophisticated data bases involve huge costs and are for one time use only which will not be worthwhile.

(ii)   The initial start-up costs for hardware and software acquisition, and the costs incurred for data collection and conversion to digital form are another important drawback in the application of GIS. Further the cost incurred for data collection and conversion to digital form is very high and the cost of converting data to digital form exceeds by far the cost of any other component of the system (between 50 and 80% of the total cost of the project) (29).

(iii)   New staff with expertise in GIS operating skills need to be appointed additionally or existing staff need to be trained in the GIS skills to acquire data in digital format (29).

(iv)   The two problematic concerns in the use of GIS for environmental modeling are the accuracy of the input data into the GIS system and tracking down the error propagation due to the GIS manipulations.

10.9   Conclusions and Recommendations

Advantages in using RS and GIS for EIA lie in their capacity in improving environmental assessment effectiveness: better analysis, efficient storage and access to spatial digital data, and good visual display capabilities.

Summary

For rapid and efficient Environmental Impact Assessment (EIA) Process RS &GIS are used as effective tools. The basic principles of Remote Sensing and its capabilities in monitoring environmental resources and its potential applications in EIA with examples are discussed. Remote Sensing data from Satellites, aerial photographs can be used for monitoring environmental /resources data over very large areas. Air borne satellite data at a 1m spatial resolution which are now available from new satellites are of great help for qualitative and quantitative studies, geology, geomorphology of soils, hydrology, vegetation, land use land cover changes etc. Further they are also useful to produce Digital Terrain Model (DTM) for detailed EIA studies on Coastal zones, Estuaries, linear projects etc.

GIS is a computer based information system which enables geographical data (13) both spatial and descriptive either simultaneously or separately for modeling, manipulation, retrieval, analysis and presentation in map form. These capabilities of GIS particularly in manipulating with spatial properties of any area of study will be of great help for analyzing environmental issues which are spatial in nature.

Integrated RS&GIS is finding very large applications particularly due to availability of submeter resolution satellite data and possible data management techniques like adaptation, modification, and extension (20) using advance computer technologies. New software modules for integrating RS and GIS at different levels are emerging and their application for different process of EIA are discussed.

A thematic database, which stores the links between environmental issues and elements and potential impact of the proposed project and a spatial database, which contains the spatial data sets can be established. The salient features of various GIS softwares and their manipulation capabilities are discussed with examples.

References

1. Brachya V., and Marinov U. 1995. Environmental Impact Statements in Israel and Other Countries: A Comparative Analysis. *Horizons in Geography*, 42-43, 71-78 (in Hebrew).

2. Gilpin, A. 1995. Environmental Impact Assessment (EIA): Cutting Edge for the Twenty-first Century, Cambridge: Cambridge University Press.

3. Campbell, J. B., (1996); "Introduction to Remote Sensing", 2nd Edition, the Guilford Press,

4. Latifovic R., Kostas Fytas, Jing Chen, Jacek Paraszczak, 2005, Assessing Land Cover Change Resulting from Large Surface Mining Development, *International Journal of Applied Earth Observation and Geoinformation*, 7, 29-48

5. Asian Development Bank. 2014 Space technology and Geographic Information Systems Applications in Asian Development Bank projects. Mandaluyong City, Philippines: Asian Development Bank.

6. Tucker, C.J.; Grant, D.M.; Dykstra, 2005 J.D. NASA's Global Ortho rectified Landsat Dataset. *Photogrammetric Engineering & Remote Sensing*, 70(3), 313.

7. Agbu, P.A.; James, M.E. *NOAA/NASA Pathfinder AVHRR Land Data Set User's Manual;*

8. Greenbelt: 1994 Goddard Distributed Active Archive Center, NASA Goddard Space Flight Center.

9. Thenkabail, P.S.; Enclona, E.A.; Ashton, M.S.; Legg, C.; Jean De Dieu, 2004. M. Hyperion, IKONOS,ALI, and ETM+ sensors in the study of African rainforests. *Remote Sensing of Environment*, 90, 23-43.

10. Thenkabail, P.S. 2004 Inter-sensor relationships between IKONOS and Landsat-7 ETM+ NDVI data in three ecoregions of Africa. *Int. J. Remote Sensing*, 25(2), 389-408.

11. Thenkabail, P.S.; Biradar, C.M.; Turral, H.; Noojipady, P. (1999); Li, Y.J.; Vithanage, J.; Dheeravath, V.; Velpuri, M.; Schull M.; Cai, X.L.; Dutta, R. *An Irrigated Area Map of the World derived from Remote Sensing. Research Report # 105; International Water*

Management Institute, 2006; pp.74. Also, see under documents in: http://www.iwmigiam.org.

12. Webb, T. 1995. New Methodology for the Identification and Assessment of Environmental Impacts: An Application that Integrates ArcView 2 with Rule-based Decision Support System. In: *Proceedings of the Fifteenth Annual ESRI User Conference,* Redlands: ESRI

13. Remote Sensing Sensors and Applications in Environmental Resources Mapping and Modelling Assefa M. Melesse Qihao Weng, Prasad S. Thenkabail and Gabriel B. Senay; *Sensors* 2007, *7*, 3209-3241.

14. Worboys, M. F., (1997), "GIS: A Computing Perspective"; Taylor & Francis Inc., Bristol, PA, 376 pp.

15. Huxhold, W.E., (1991), "An Introduction to Urban Geographic Information Systems", Oxford: University press.

16. Marble, D. F., and Arnundson, S. E., (1988); "Micro-computer Based Geographical Information Systems and Their Role in Urban and Regional Planning: Environmental and Planning", B. 15, p. 205-324.

17. Linden, G., (1990), "Education in Geographic Information Systems", Dordrecht, Kluwer Academic Publishers.

18. Clarke, M., (1990), "Geographical Information Systems and Model Based Analysis: Towards Effective Decision Support Systems", Dordrecht, Kluwer Academic Publishers

19. Derenyi, E.E. (1991). "Design and Development of Heterogeneous GIS". CISM Journal of ACSGS, Vol. 45, No. 4, pp. 561-567.

20. Ehlers, M., Edwards, G. and Y. Bedard. 1989a. Integration of remote sensing with geographic information systems: a necessary evolution. Photogrammetric Engineering and Remote Sensing 55(11): 16 19-1 627

21. Poulter, M. (1995). "Integrating remote sensing and GIS: designs for an operational future". In Proceedings of a Seminar on Integrated GIS and High Resolution Satellite Data, eds, Palmer, M. Farnborough: Defense Research Agency (Ref No: DRA/CIS/ (CSC2)/5/26/8/1/PRO/ 1).

22. Mesev, V., (1997). "Remote Sensing of Urban Systems: Hierarchical Integration with GIS, Computer, Environment, and Urban Systems" Vol. 2 (3-4): 175-187.

23. Davies, F. W., Quaitrochi, D. A., Ridd, M. K., Lam, N., Walsh, S. J. Michaelsen, J. C., Franklin, J., Stow, D. A., Johannsen, C. J. and C. A. Johnston. (1991); Environmental Analysis using Integrated GIS and Remotely Sensed Data: Some Research Needs and Priorities". Photogrammetric Engineering and Remote Sensing 57(6): 689-697.

24. Hinton, J. C., (1996); "GIS and Remote Sensing Integration for Environmental Applications"; International Journal of Geographic Information Systems 10(7): 877-890.

25. Kwok, R., Curlander, C. and S. S. Pang, (1987); "Rectification of terrain induced distortions in radar imagery"; Photogranmetric Engineering and Remote Sensing 530: 507-513.

26. Ehlers, M., Greenlee, D., Terrence, S. and S. Star. (1991). Integration of remote sensing and GIS: Data and Data Access. Photogrammetric Engineering and Remote Sensing 57(6): 669-675.

27. Brown, R. and P. Fletcher, (1994); "Satellite images and GIS: making it work". Mapping Awareness 8: 20-22.

28. Gugan, D. J., (1988), "Satellite Imagery as an Integrated GIS component"; Proceedings of GIS/LIS '88, San Antonio, Texas, pp. 741-750.

29. Huxhold, W.E., (1991), "An Introduction to Urban Geographic Information Systems", Oxford: University press.

Some useful References on the application of RS & GIS for EIA

1. Johnston, C. A., Detenbeck, N. E., Bonde, J. P., Niemi, G. J. 1988. Geographic Information Systems for Cumulative Impact Assessment, *Photogrammetric Engineering and Remote Sensing*, 54(11): 1609-1615.

2. Schaller, J., 1990, Geographical Information System Applications in Environmental Impact Assessment. In: *Geographical Information Systems for Urban and Regional Planning*, Edited by Scholten, H. J., and Stillwell, J. C. H. Dordrecht: Kluwer Academic Publishers:107 - 117.

3. Scott, D. R., and Saulnier, T., 1993, ARC/Info for Large and Small EIS Applications: Interstate 93, NH; Garcia River, CA; Sakhelin Island, Russian Federation. In: *Proceedings of the Thirteenth Annual ESRI User Conference* Redlands: ESRI inc.: 119 - 124.

4. Shopley, J., Sowman, M., and Fuggle, R. 1990. Extending the Capability of the Component Interaction Matrix as a Technique for Addressing Secondary Impacts in Environmental Assessment. *Journal of Environmental Management*, **31**(3):197-213.

5. Webb, T. 1995. New Methodology for the Identification and Assessment of Environmental Impacts: An Application that Integrates ArcView 2 with Rule-based Decision Support System. In: *Proceedings of the Fifteenth Annual ESRI User Conference,* Redlands: ESRI. *Geographical Information Systems* 8: 1–12.

6. Chapal S E, Edwards D 1994 Techniques for visualisation and quality control. In Michener W K, Brunt J W, Stafford S G (eds) *Environmental Information Management and Analysis: Ecosystem to Global Scales*. London, Taylor and Francis: 141–59.

7. Federa K, Kubat M 1992 Hybrid Geographical Information Systems. *EARSel Advances in Remote Sensing* 1: 89–100.

8. Goodchild M F, Haining R, Wise S 1992 Integrating GIS and spatial analysis: problems and possibilities. *International Journal of Geographical Information Systems* 6: 407-23.

9. Ji W, Mitchell L C 1995 Analytical model-based decision support GIS for wetland Resource Management. In Lyon J G, McCarthy J (eds) *Wetland and environmental applications of GIS*. New York, Lewis Publishers: 31-45.

10. Maguire D J 1995 Implementing Spatial Analysis and GIS Applications for Business and Service Planning. In Longley P A, Clarke G (eds) *GIS for Business and Service Planning.*

11. Cambridge (UK), Geo Information International: 171–91.

Schroeter H, Olsen L 1996 The evolution of a space/earth information system. In Lautenschlager M, Reinke M *Climate and Environmental Database Systems.* Boston (USA), Kluwer: 25-35.

12. Steyaert L T, Goodchild M F 1994 Integrating geographic information systems and environmental simulation models. In Michener W K, Brunt J W, Stafford S G (eds) *Environmental Information Management and Analysis: Ecosystem to Global Scales.* London, Taylor and Francis: 333-57.

13. Thorkilsen M, Dietrich J, Larsen L C, Nielsen A 1997 Environmental and Geographical Information System for Dredging Control. *Proceedings CEDA dredging days 1997.* Central Dredging Association: 109-17.

14. Warren I R, Bach H K 1992 MIKE 21: A Modelling System for Estuaries, Coastal Waters and Seas. *Environmental Software* 7: 229-40 GIS in Environmental Monitoring and Assessment 1007.

15. Antunes, P., R. Santos, L. JordBo, P. GonGalves, and N. Videira. 1996. "A GIS based decision support system for environmental impact assessment." Proceedings of the 1996 international IAIA conference. Improving Environmental Assessment Effectiveness: Research Practice and Training, 17-23 June 1996, Estoril, Portugal, pp. 451-456.

16. Carver, S., M. Blake, and I. Turton. 1996. "Open Spatial Decision Making? Evaluating the potential of the Internet." Proceedings of GIS Research UK (GISRUK) conference, 10-12 April 1996, University of Kent, Canterbury.

17. Eedy, W. 1995. "The use of GIS in environmental assessment." **Impact Assessment** Epstein, E. 1991. "Legal aspects of GIS." In **Geographical Information Systems: Principles and Applications,** vol. 1, edited by D. Maguire, M. Goodchild, and D. Rhind, 489-502.

18. Harlow: Longman. Faber, B.G., W. Wallace, and R. Miller. 1996. "Collaborative modeling for environmental decision making." Proceedings of the 10th Annual Symposium on Geographic Information Systems (GIS'96), February 1994, Vancouver, BC, March 1996. GIS World Inc. (published on CD).

19. Fonseca, A., C. Gouveia, A. C. L. mara, and J. Silva. 1995. "Environmental Impact Assessment with Multimedia Spatial Information Systems." **Environment and Planning B: Planning and Design** 22: 637-648.

20. Goodchild, M., L. Steyaert, and B. Parks. 1996. **GIs and Environmental Modeling: Progress and Research Issues.** Cambridge: Geo Information International.

21. Agrawal M.L. and Dikshit A.K.; (2002); "Significance of Spatial Data and GIS for Environmental Impact Assessment of Highway Projects; Indian Cartographer, Convergence of Imagery, Information & maps, Volume 22.

22. Ahern, F. J., R J. Brown, J. Cihlar, R Gauthier, J. Murphy, R A. Neville, and P. M. Teillet, (1987). Radiometric Correction of Visible and Infrared Remote Sensing Data at

the Canada Centre for Remote Sensing. International Journal of Remote Sensing. 8: 1349- 1376.

23. Anderson, J.R., Hardy, E.E., Roach, J.T., and R.E. Witmer (1976)."A Land Use and Land Cover Classification System for Use with Remote Sensor Data." Washington, DC: U.S. Government Printing Office.

24. Armstrong J. M. and Khan A. M.; 2004 "Modelling Urban Transportation Emissions: Role of GIS", Computers, Environment and Urban Systems, Volume 28, Issue 4, July, Pages 421-433.

25. Ban, Y., and P. J. Howarth, (1996). "Integration of ERS-1 SAR and Landsat TM Data for Agricultural Crop Classification."Proceedings for the 26th International Symposium on Remote Sensing of Environment, the 18th Symposium of the Canadian Remote Sensing Society: Information Tools for Sustainable Development, Vancouver, B. C., Canada, 25-29.

26. Booz, Allen, and Hamilton (2003); "Environmental Information Management and Decision Support System: Implementation Handbook"; National Academy Press Washington, D.C.

27. Rashmin, G.: Use of GIS for Environmental Impact Assessment: An Interdisciplinary Approach Interdisciplinary Science Reviews, 2004 (March) 29:1, 11-1.

28. Bajracharya SR, Maharjan SB, Shrestha F (2014) The Status and Decadal change of Glaciers in Bhutan from the 1980s to 2010 based on satellite data. *Ann Glaciol* 55: 159-166.

29. Alan Gilpin (1995). Environmental Impact Assessment. Cutting Edge for the Twenty first century. Cambridge University Press. London.

30. Birhanu. M. (2005). Remote Sensing and GIS techniques in land use/land cover mapping, change detection and evaluation of environmental degradation in the main Ethiopian rift (MER) between Koka and Ziway.

31. Campo, A. G. D. (2012). GIS in Environmental Assessment: A Review of Current Issues and Future Needs, Journal of Environmental Assessment Policy and Management Vol. 14, No. 1, 1250007 (23 pages).

Case Studies

Application of Remote Sensing (RS) and Geographic Information Systems (GIS) to Environmental Impact Assessment (EIA) for Sustainable Development.

Idowu Innocent Abbas and J.A. Ukoje Department of Geography, Ahmadu Bello University, Zaria, Nigeria.: Research Journal of Environmental and Earth Sciences 1(1): 11-15, 2009

Case Study: The proposed dredging of Lower Niger River (Warri to Bifurcation at Onyia segment) using Remote Sensing and GIS techniques.

The proposed dredging of lower Niger River from Warri (Delta State) through the Forcados /Warri River to Baro (Niger State). The 592 km have been divided into five segments:

• Warri to Bifurcation at Onyia 155 km

- Bifurcation at Onyia to Onitsha 120 km

- Onitsha to Idah 115 km

- Idah to Kotonkarfe 105 km

- Kotonkarfe to Baro 097 km

The EIA of the Forcados segment dredging on the area was conducted using Landsat TM (1994), topographic maps, vegetation and landuse/landcover maps of the area. The image was georeferenced, processed and resampled for visualization in a GIS environment. The tonal values recorded in the image with the features on the ground were validated by ground truthing. Arch/info 3.5.1 software was used for the analysis

After the analysis, the following result was obtained: From Table 10.6, A total of 17 settlements of about 6.03048 km^2 (25%) of the total built up area will be impacted; 7 major roads covering 36.078 km^2 (67%) out of 54.3616km^2 will be impacted; about 634.60708 km^2 of various vegetation types will be impacted in the area. The analysis also showed that 7 fishing camps will also be lost, thus an important means of livelihood will be destroyed.

Table 10.6 GIS Analysis of lower Niger River

Feature Class	Impact (km^2)
Built up Areas	6.03048
Cultivation	215.13124
Different Types of Vegetation	634.60708
Roads	36.078

Source: GIS analysis

Rapid environmental impact assessment using remote sensing and geographic information systems: A case study of river Ib Barrage, Odisha

Debiprasad Taudia# and Sudha Goel*#School of Water Resources and *Civil Engineering Department, IIT Kharagpur, Kharagpur - 721302, West Bengal, India Email: sudhagoel@civil.iitkgp.ernet.in

Journal of Geomatics -Vol 7 No.1 April 2013

A barrage project has been proposed by Odisha State government across river Ib, downstream of the confluence of rivers Basundhara and Ib, near village Deogaon of Jharsuguda district, Odisha, India. The major objectives of the proposed project are to provide water for irrigation for approximately 800 ha, drinking water to three urban bodies, i.e., Jharsuguda, Brajrajnagar and Belpahar, and water for various industries in Jharsuguda and its periphery. Like all water resources projects, this project also has wide-ranging impacts on the environment. The government-sponsored Environmental Impact Assessment (EIA) based on conventional ground-based methods is still in progress since 2005. The objective of this study was to conduct a rapid EIA using remote sensing and geographic information system (GIS) to demonstrate the utility of these tools in providing high-quality survey results which is not possible with conventional ground based methods.

An EIA requires comprehensive baseline data for the area that is likely to be impacted. All data regarding the study area were obtained from Ib Investigation Division, Sundargarh,

Department of Water Resources, Govt. of Odisha, unless stated otherwise. Remote Sensing data were collected from various sources on the Internet, and Census of India data from the Registrar General of India.

Population data used in this study are based on Census of India 2001 (Registrar General of India, 2001).

Geomorphological data: Elevation data were required for watershed delineation, marking river cross–sections, submerged areas, etc. These data can be freely downloaded from the Internet in the form of Digital Elevation Model (DEM) from http://srtm.csi.cgiar.org/. The downloaded DEM file of the study area was processed in Arc-GIS and Soil and Water Assessment Tool (SWAT) was used to delineate the basin boundary, sub-basins or watersheds and drainage networks.

To extract geomorphologic input parameters of the basin, SRTM (Shuttle Radar Topography Mission) data of 90 m × 90 m resolution was used. SRTM data sets provide terrain information which can be used for handling regional environmental problems, and can be applied in river network and terrain analyses. Land-use/land-cover data for the study area were available in the form of Landsat Enhanced Thematic Mapper (ETM) images which can be freely downloaded from Global Land Cover Facility (GLCF: http://www.glcf.umd.edu/index.shtml). The land use/land cover map of the entire study area and catchment area were derived from Landsat ETM images using ERDAS IMAGINE software.

Hydrological data: Rainfall and temperature data were collected from India Meterological Department (IMD) website for Jharsuguda. Evaporation data for the region are available in CWC reports (2006).

Yield data: CWC has collected flow data at Sundargarh (north of Jharsuguda on the River Ib) for the period 1987-88 to 2007-08 (CWC, 2009). Department of Water Resources (DOWR), Government of Odisha is also gauging the flow in river Ib near village Deogaon, Jharsuguda and data have been collected for the period 1985 to 2010 (Ib Investigation Division, 2007).

Calculation of submerged area and storage volume: Digital Elevation Model of the study area was used in Micro DEM and ArcGIS software for different heights of the barrage. Model outputs allow calculation of storage volume and submerged area and corresponding changes in land use/land cover.

Table 10.7 Distribution of submerged areas by land use for different barrage Heights

Elevation (m)	198		199		200		201	
Agriculture	311.13	36.66	528.53	46.91	786.83	54.12	956.25	52.44
Vegetation	29.7	3.50	63.09	5.60	85.33	5.87	148.68	8.15
Settlement	18.9	2.23	36.63	3.25	46.26	3.18	69.57	3.82
Total useful land	359.73	42.38	628.25	55.76	918.42	63.17	1174.50	64.41
Water bodies	115.83	13.65	120.06	10.66	130.68	8.99	168.75	9.25
Barren land	373.23	43.97	378.43	33.59	404.82	27.84	480.15	26.33
Grand Total	848.79		1126.74		1453.92		1823.40	

Satellite data have been used as input parameters for demarcation of water spread and computation of barrage storage capacity, and land-use/land-cover assessment for various heights of the barrage. The total annual water demand estimated for fulfilling the project objectives is 94.8 million m^3/y, while water availability in the river Ib catchment area is estimated to be 50 million m^3/y. Three design alternatives were evaluated and included barrage heights of 4, 5 and 6 m. The storage capacity of the barrage with a height of 6 m is close to 14 million m^3 and can satisfy only a fraction of the total annual water demand, i.e., none of the alternatives evaluated is capable of satisfying the total annual water demand of the area. A barrage height of 4.5 m was chosen after comparing the impacts for each alternative. The results of this study highlight the accuracy and effectiveness with which RS and GIS can be used to conduct a rapid EIA. Further, it is clear that alternatives to the proposed project need to be examined in greater detail with similar tools.

Questions

1. Explain how EIA can be useful as a decision making tool for Development Projects Discuss the advantages of using remote sensing data in environmental impact assessment (EIA). What is GIS? Discuss its special advantages in EIA.

2. Explain how Airborne Light Detection and Ranging System (LIDAR) works.

3. What kind of data can be obtained thermal imageries? Explain the various technical aspects in analyzing thermal imageries for land use land cover studies.

4. Discuss the state-of-art of satellite sensors widely used in environmental monitoring.

5. What are the Major elements of GIS? Explain their functions.

6. Discuss GIS Data characteristics of Spatial Data.

7. Explain the differences between Raster and Vector Models.

8. Discus the scheme for Database Collection and Storage System in GIS.

9. How is Data Manipulation and Analysis carried out?

10. Explain why Integration of Remote Sensing and GIS is necessary for Ecological EIA studies. Discuss salient features of different Levels of Integration.

11. Discuss various applications of GIS as a Tool for Remotely Sensed Image Processing.

12. Explain how Remotely Sensed Data is used as an Information Source to GIS.

13. Discuss the application of Integrated RS and GIS in Screening, Scoping and Baseline Studies in EIA.

14. Explain how integrated RS and GIS are useful in Impact Prediction Reflections and in Mitigation.

15. Discuss the advantages of integrated RS & GIS in Coastal Zone Studies and Land Use and Land Cover Studies

16. What are the Key problems or disadvantages of using RS and GIS for EIA?

17. Discuss the basic principles of Remote Sensing. How it is useful for Environmental monitoring?

18. What are the sources for RS data? What is meant by passive and active Remote Sensing? and LIDAR.

19. Discuss how Remote Sensing can be advantageously used for EIA on Coastal Zones, Estuaries, Transportation projects, Land use and land cover changes.

20. What is GIS? Discuss the basic concepts and techniques used in GIS.

21. Discuss the external and internal approaches in the application of GIS for EIA.

22. Discuss use of GIS in screening, scoping and baseline studies in EIA.

23. Discuss the advantages of GIS in impact prediction and development of mitigation strategies in EIA.

24. What are major application areas for GIS in EIA? Compare the relative merits and demerits of GIS and general EIS methods in EIA.

Chapter **11**

EIA Case Studies

11.0 Environmental Impact Assessment of Industrial Projects

11.01 Introduction

To take a decision for starting a major industry or a cluster of industries in any particular area it is necessary to carryout EIA through which the environmental, social and economic impacts likely to occur if the projects are started can be predicted/assessed. Thus EIA can be used as a Decision Support Tool for selecting Industrial sites, for implementing proper design, planning, mitigation measures to reduce the intensity of adverse impacts and to suit to local environment which will greatly help in achieving both environmental and economic benefits. Critical examination of environmental impacts and mitigation measures from the early stages of project planning cycle greatly helps in proper utilization of resources, saving time & costs and overall protection of environment. Thus EIA if conducted professionally helps decision makers for selecting proper sites for starting industries and also by promoting public participation lessens conflicts.

Industrial promotion is a major priority for the governments of many developing countries. Industrial development can make significant beneficial contributions to a country's overall economic development by providing jobs, promoting socio-economic infrastructure and so on. However, by its very nature, industrial development can also have profound impact on the environment (Fig. 11.1). All industries require use of natural resources, many of which are limited, such as water, and so can directly affect local ecosystems. The conversion of natural resources to finished or semi-finished products results in residues that are often discharged as wastes. These wastes in solid, liquid and gaseous forms, can be detrimental to the quality of life by adversely affecting land, water and air resources.

The type, size, and location of industrial development projects will determine the extent of potential environmental impact. Generally, industrial development can be divided into two kinds of activities: (a) establishment of individual industries in a certain locality and (b) establishment of industrial estates containing industries with differing functions. Industries can also be classified according to the type of operation, extraction, mineral processing and manufacturing; manufacturing can be subdivided up into light, medium (sometimes) and heavy industry. The impact of selected industries on major environmental parameters are given in Table 11.1. Table 11.2 presents various environmental parameters to be analyzed for evaluation of the impact of various industries.

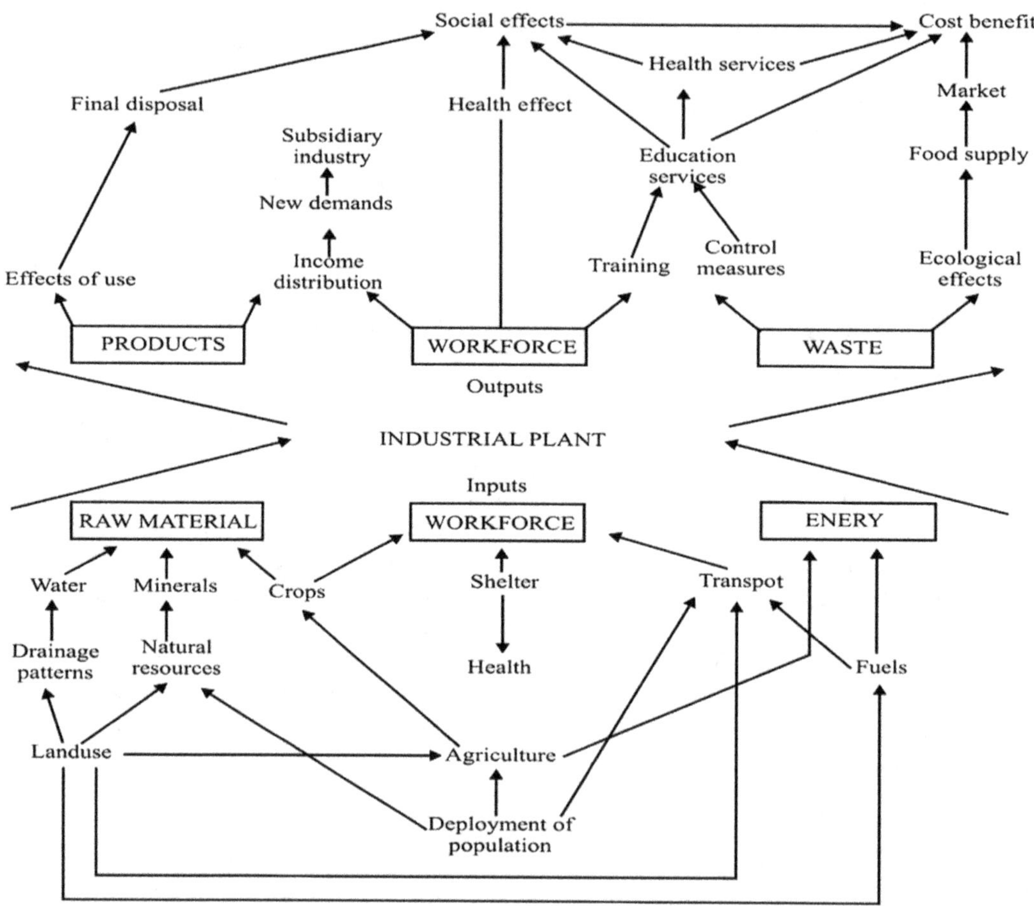

Fig. 11.1 Input-Output Model Showing Relationships between and Industrial Plant and its Socio-ecological Environment.

Source: Ecological and Social Evaluation of Industrial Development, Environment Conservation Vol. 3: 4, Marstand, P.K., (1).

Table 11.1 Major Impacts of Industrial Sector Development on Environment

Development projects \ Environmental parameters	Surface water hydrology	Surface water quality	Ground water hydrology	Air quality	Land quality (pollution)	Fisheries	Vegetation	Forest (resources depletion)	Mineral resources	Aesthetic	Socio-economic	Public health
Food Processing	O	⊕	⊕	•	O	•	•	•		•	•	O
Sugar Refining	O	⊕	O	•	O	•		⊕		•	•	O
Pulp and paper	O	⊕	•	•	O	⊕		⊕		•	•	O
Fertilizer	•	⊕	•	⊕	O		⊕		•	•	•	•
Cement	O	•	•	•	⊕	•			⊕	•	•	•
Tannery	O	⊕	O	⊕	•	⊕				⊕	•	O
Pharmaceutical	O	⊕	O	•	•	⊕				O	•	⊕
Steel and Iron Manufacture	•	⊕	•	⊕		•			⊕	•	•	•
Electroplating	•	⊕	O	O	O	•	⊕			O	•	O

⊕ Significant impact

• Moderate to significant impact

O Negligible impact

Source : Technology for Development Environment Aspects,

Source : Technology for Development Environment Aspects,

Table 11.2 Environmental Parameters for Analysis of Industrial Projects.

Environmental Resources		Physical Resources									Ecological Resources				Human Use Values											Quality-of-Life values			
	Type of Projects	Surface Water	Surface Water Quality	Ground Water	Ground Water Quality	Climate	Air Quality	Land Quality; Pollution	Mineral Resources	Geology Resources	Fisheries	Aquatic Biology	Forests/Vegetation	Terrestrial wildlife	Industrial wildlife	Highways / Railways	Navigation	Water supply	Power	Agriculture	Housing	Recreation	Flood Control	Sewage disposal	Slid Waste Disposal	Aesthetic	Public Health	Socio - EconomicHealth	Industrial Safety &
Impact on Environment	Light industry	1	–	1	–	–	1	1	1	–	–	–	–	–	3	1	1	1	1	1	1	1	–	1	1	1	1	1	2
	Heavy Industry	3	3	2	2	–	3	3	3	–	2	2	2	1	3	2	2	2	2	1	3	1	–	2	2	3	3	3	3
	Mineral processing	3	3	2	2	–	3	3	3	–	2	2	2	1	3	2	2	2	2	1	3	1	–	2	2	3	3	3	3
Impacts of Environment on project	Light Industry	1	1	1	1	1	–	–	–	1	–	–	–	–	1	3	3	1	2	1	1	1	2	3	1	1	–	1	1
	Heavy Industry	3	3	3	3	2	–	–	–	2	–	–	–	–	1	3	3	3	3	1	1	1	3	3	3	1	–	1	1
	Mineral Processing	3	3	3	3	2	–	–	–	2	–	–	–	–	1	3	3	3	3	1	1	1	3	3	3	1	–	1	1

Notes : Numbers indicate usual magnitudes of significant impacts : (3) = major, (2) = intermediate, (1) = significant.

Source : Manual of NEB : Guidelines for preparation of Environmental Impact Evaluations, National Environment Board of Thailand,

United Nations Economic and Social Commission for Asia and the Pacific (ESCAP) 2,3,4

The primary function of an EIA in industrial sector development is to identify the constraints which arise from impact on the environment and to do so early enough for the constraints to be incorporated into the industry's planning, including site selection for new industries, and the decision-making process.

It is important that the report brings out the following to help decision makers:

- Predict environmental impact of projects.
- Suggest means to reduce adverse impacts.
- Identify and align projects to suit local environment. and
- Present the predictions and options to the decision-makers to facilitate better environmental protection and achieve sustainability in the region.

11.1 Key factors for Environmental Impact Assessment of Industrial Development Projects

The key functions which should be considered in evaluating the impact of industrial development projects are discussed below.

11.1.1 Physical Resources

(a) ***Resources depletion***: The raw material requirements may cause depletion of the natural resources, for example, bamboo used in paper production. Existing conditions of the natural resources that will be used as raw materials for industrial production and the projected impact of use should be evaluated to determine if and how the resources can be used on a sustained basis.

(b) ***Surface water hydrology***: Many industries rely on large volume of water for industrial processes or for cooling. This could significantly affect surface water hydrology, particularly if many industries are to be located in the same vicinity, such as the industrial estates. It is equally important to consider secondary developments related to industrial sitting, especially development of new towns or major population increases which could affect the water supply and demand and consequently surface water hydrology to a more significant extent than actual industrial operations. The type and extent of the existing and projected water demand and use by the industry, municipality, agriculture and hydropower should be described in relation to the water sources, whether surface or ground water. This requires description of water flow rates, volumes, seasonal variations, normal, flood and drought year flows, and related ecology.

On occasions the water supply is not reliable enough to ensure uninterrupted industrial operations. Consequently, a reservoir may be built and this will itself pose potential environmental problems. This must be described and analyzed as part of the industrial development project assessment.

(c) *Surface water quality*: Water quality and related beneficial uses are often most affected by industrial development. The impact of industrial construction and operation on water quality will depend on the volume of flow and concentrations of Bio-chemical Oxygen Demand (BOD), Total Organic Carbon (TOC), Suspended Solids (SS), Total Dissolved Solids (TS), heavy metals and other toxics, turbidity, acidity and pollutants to waterways. This requires preparation of mass balances for each toxic or hazardous material entering or produced in the plant and for all waste emissions.

(d) Groundwater: The effects of industrial development on groundwater quality and hydrology can be especially pronounced in regions where groundwater is the major source of water supply and few or no alternatives exist. Both highly concentrated waste water and solid wastes, if improperly disposed off, can percolate or produce leachates that could contaminate ground water. The EIA should describe the extent of major existing groundwater development and on developed aquifers and their rates of recharge and/or depletion.

(e) *Soils*: Soil types should be evaluated accounting to various parameters like sustainability to erosion settlement, earthworks, beating capacity, soil structure and slope. The assessment should consider possible effects from the use of sand and gravel for construction and soil erosion from uncovered cut/fill areas. Special soils analyses should be undertaken to determine suitability for land disposal of wastes and, in addition to geologic analysis, to determine suitability for filling.

(f) *Geology*: An understanding of local geology is also important to describe the loss of unique features, potential disasters from geologic activity, loss of mineral development, and potential subsidence.

(i) *Air quality*: Emission of gaseous pollutants can have adverse health or productivity effects on man, animals and vegetation, and can degrade man-made structures. An example of the last is deterioration of structures owing to acid rain. Significant air pollution can also result from the increases in automobile traffic and dust and fumes that accompany major industrial development. The assessment requires an adequate understanding of wind direction and speed, precipitation, temperature, humidity and existing ambient air quality. All significant air pollutants and their sources should be described including raw materials used and their types and the amounts used during both the construction and the operation phases. This information is used to prepare the "projected emission inventory" (i.e., emissions from combustion, manufacturing, storage and transport) and the "emission inventory of the surroundings" (i.e., emission from all sources).

11.1.2 Ecological Resources

(a) *Fisheries and aquaticecology*: The most significant impacts of Industrial Development Projects on local fisheries are caused by the discharge of effluents. Impacts can be of three major types, (i) immediate fish mortality owing to toxic substances, (ii) reduced productivity caused by degradation of water quality(either physical or chemical), and (iii) degradation or alteration of the aquatic habitat. The EIA should describe the aquatic habitat and productivity, and project future conditions as related to the description of the future water hydrology and quality.

(b) *Wildlife*: The EIA should include a description of the significant wildlife resources in the area, indicating if the rate of endangered species will be affected by the Industrial Development Project. This includes description of their behavioral and ranging patterns and habitat requirements in order to indicate the projects expected effects on the species and measures for offsetting any detrimental effects.

(c) *Forests*: The forest resources existing in the area which may be exploited or degraded as a result of the Industrial Development Project should be described. Impacts on forests from Industrial Development Projects will mainly take the form of cutting for use as construction/raw material and clearing forests prior to construction. Some industries require forest products as raw material, such as, bamboo for pulp of paper industry, and for these types of industries a plan for sustained exploitation should be prepared.

(d) *Environmentally sensitive areas*: The existing conditions and the nature of potentially significant impacts from the Industrial Development on environmentally sensitive areas should be described if not mentioned elsewhere. Examples include: (i) prime agricultural land (ii) wetlands/coastal zones/shorelines and (iii) tourism resources.

11.1.3 Human use Values

(a) *Water supply*: The supply of drinking water down stream and ground water resources likely to be effected by industrial water usage and generation of effluents which need to be evaluated.

(b) *Land use*: With introduction of industry, land- use patterns in the surrounding area often experience substantial changes. This may be particularly pronounced in relation to changes in socio-economic infrastructure, such as, urbanization, which accompanies siting of industrial estates. Existing and projected land-use for "with" and "without" development scenarios should be described.

11.1.4 Quality-of-life Values

(a) *Socio-economics*: Beneficial effects of the industrial development projects in the local vicinity can be substantial, particularly in the establishment of large industries and industrial estates. An obvious example is job creation. However, it is important to describe carefully and analyse the present and projected socioeconomic effects of the industrial development projects, particularly for major industries, such as, fertilizer and petro-chemicals. Such an evaluation can help to identity potential social conflicts and suggest measures for mitigation or resolution. For example, there may be a shortage of skilled labour and the need to bring them from elsewhere. Local residents may resent this loss of potential income and cultural conflicts could arise between the locals and the outsiders. The problem might be resolved by offering training to local candidates. Another example is a reduction in land values adjacent to the industrial plant.

The EIA should include assessment of the socio-economic conditions for before/after and with/without scenarios. Some parameters to be assessed include population structure, population dynamics, land use/settlement patterns, labour and employment structure, economic production and distribution, income distribution, social organization, cultural characteristics, and social institutions.

(b) ***Occupational and public health***: Industrial development projects can produce wastes that are deleterious to humans within the project vertically and beyond, particularly in relation to air and water quality. Hazards to employees and nearby residents can also originate from fires, hazardous wastes, and spills from petrochemical industry. The potential for accidents or spills should be measured using historical records and the judgement of engineers. Risk analysis is often required including the identification of dust and fumes which can cause serious health problems to employees. Increased automobile traffic can affect the health through exhaust fumes and accidents. Therefore, a critical component of any industrial development project EIA is to describe in detail the potential health hazards and their impacts, and plans for preventing, and if necessary, reacting to hazardous incidents.

Noise pollution is another major factor for industrial development project impacts on public health. It occurs in three forms: (i) general audible noise, (ii) special noises such as infra-sound, ultra-sound and high energy impulse noise, and (iii) noise induced vibrations. Vibration transmissions may be structural or generated by airborne noise. Again, it is important to identify clearly the sources and the receptors during both construction and operation.

(c) ***Aesthetics***: Industrial development projects can have high visual impact, particularly when the industrial plant is sited in areas of scenic beauty. The effects could extend to recreational use in some areas. The project area's visual aesthetic values and recreational use and/or potential should be described along with plans to minimize such impacts.

(d) ***Archaeology and historical resources***: Any archaeological or historical/cultural resources which may be affected by the industrial development project should be identified, and if necessary, plans to be developed to protect or relocate valuable artifacts, buildings, and so on.

11.1.5 Project Siting

Of particular importance when assessing the environmental management needs of a proposed industrial development project is the site of development. Site selection by the project developers is generally based on the availability and/or costs of land, raw materials, transport and utilities infrastructure, labour and markets. Developers will often assume that the site location plays a minimal role in the impact of waste disposal on environmental resources. In actuality, the capacity to assimilate the wastes can vary widely, and so the project site can play a major role in determining the costs and methods of pollution control. An example of this is the prevailing winds that blow gaseous emissions towards communities rather than away from them. In the above case the industrial plant may be required to add more features to its emission control procedures in order to obtain a level of environmental protection equal to that of prevailing winds flowing away from communities. The EIA for new industrial projects should therefore determine whether the proposed site is an environmentally good choice and if not, whether the proposed waste system can be expected to function adequately. Table.11.3 lists important site selection factors for industrial development.

Table 11.3 Environmental Site Selection Factors

A. Water Supply

	1.	**Water needs** : process -- cooling -- potabale -- fire protection.
	2.	**Water availability** : public water supply -- private water supply -- ground water -- surface water.
	3.	**Ground water :** geological potential (onsite)
	4.	**Water characteristics :** chemical -- bacteriological -- corrosiveness
	5.	**Water distribution :** amount available -- pressure -- variations -- proximity of site -- size of lines.
	6.	**Cost of water supply :** extension of existing service -- development of new supply -- cost per 1,000 gallons
	7.	**Water treatment requirements :** process -- cooling -- boiler feed water -- potable -- others.
	8.	**Special considerations :** restriction on use -- future supplies -- compatibility for use in process.
	9.	Applicable government regulations.

B.   Ecological Considerations

	1.	**Discharges :** gaseous -- liquid -- solid wastes. What are the ecological considerations?
	2.	Existing area ecological relationships (use available background data and augment as necessary)
	3.	Control measures to minimize ecological effects
	4.	Wildlife propagation areas
	5.	Terrestrial and aquatic areas
	6.	**Physical tolerance levels :** -- ambient air quality standards -- glare and/or lighting standards
	7.	Nutrients
	8.	Detrimental and beneficial development
	9.	Buffer zones and green belts
	10.	Applicable government regulations

C.   Air Pollution Control

	1.	Air pollution enforcement regulations and ordinances
	2.	Meteorological conditions : -- wind direction and velocity variability inversion frequency intensity and height, and other microclimatology factors.
	3.	Proximity to population/employment centers
	4.	Local topography
	5.	Effect other area industrial emissions may have on the quality of new plan environment of allowable emission rates.

D. Wastewater Disposal

	1.	Sewerage system -- storm water -- cooling water -- process wastewater
	2.	Anticipated mode of occurrence, flow and characteristics of plant wastewater discharges
	3.	Proposed pollution loadings
	4.	Toxic materials present
	5.	Variations in flow and strength
	6.	Variations of wastewater treatability
	7.	Implant control measures

Table 11.3 *Contd...*

8.	Onsite wastewater treatment and disposal possibilities
9.	Nearby watercourses which may be considered for wastewater disposal
10.	Existing stream quality
11.	Water uses to be protected
12.	Stream quality standards
13.	Wastewater effluent standards
14.	Government regulatory agencies concerned : -- permit requirements
15.	Stream flow characteristics : -- design critical flow
16.	Development of treatment design parameters
17.	Availability of a public sewerage system
18.	Pretreatment requirements if discharged to public sewers
19.	Sewer service charges and surcharges for industrial wastewaters
20.	Onsite underground disposal system : -- percolation rates
21.	Scavenger hauling of liquid wastes
22.	Emergency operation : -- electrical power dependability
23.	Performance reliability requirements
24.	Applicable government regulations
E. Solid Waste Disposal	
1.	Applicable government regulations
2.	Disposal facilities available : -- incineration -- sanitary landfill -- other
3.	Local contract pickup and disposal : -- municipal control -- competition between haulers
4.	Costs of solid waste disposal
5.	Dependability : -- lifetime of disposal facility -- probability of flooding
6.	Onsite disposal : -- incineration -- landfill
7.	Responsibility : -- if public collector -- if private collector -- if other disposal
8.	Special handling and disposal practices required for industrial wastes

11.2   Factors to be considered in making Assessment Decisions

There are six factors that should be taken into account when assessing the significance of an environmental impact arising from a project activity. The factors are interrelated and should not be considered in isolation. For a particular impact some factors may carry more weight than the others but it is the combination of all the factors that determines the significance.

1. *Magnitude*: This is defined as the probable severity of each potential impact. The impact can be reversible or irreversible.

2. *Prevalence*: It is the likely eventual extent of the impact as for example the cumulative effect of a number of actions. Each one taken separately might represent a localized impact of a small importance and magnitude but a number of such activities could result in a widespread effect.

3. ***Duration and frequency***: The significance of duration of frequency is reflected in the following questions. Will the activity be long-term or short-term? If the activity is intermittent, will it allow for recovery during inactive periods?

4. ***Risk***: Risk is the probability of serious environmental effects. To accurately assess the risk, both the project activity and the area of the environment impacted should be known and understood.

5. ***Importance***: This is defined as the value that is attached to an environmental component in its present state. The impacted component may be of regional, provincial or even national importance.

6. ***Mitigations***: Are solutions to problems available in the existing technology provide a solution say to a silting problem expected during construction of an access road or to a bank erosion?

The possible decisions, using the above criteria are:

1. No impact

2. Unknown and potential or adverse impact

3. Significant impact

Checklists of project activities for industries

1. ***Investigation change***: Access roads and tracks, site surveying, burning, engineering investigation, raw materials survey, abandonment.

2. ***Site preparation and construction***: Access roads and tracks, site clearing, burning, stripping, earth work contouring erosion control, drilling and blastings, demolition, belching relocation, drainage alteration, reclamation, installations, equipment, utilities, services, labour force, transportation, suspension of works, landscaping, revegetation, waste disposal.

3. ***Operation and maintenance***: Raw material handling, and transportation, storage of raw materials, mechanical processing, chemical processing, operation failure, process water, cooling water, energy requirement, atmospheric heat discharge, gas/vapor emissions, waste disposal and recovery (Liquids), waste disposal and recovery, (solids) product handling and storage, accidents, labour force utilities, services, amenities, dust control, abandonment.

4. ***Future and related activities***: Ancillary pipelines and transmission lines, industrial development, urbanization.

Saving time and money: A feature of environment management often overlooked by industrial project developers is that a well- prepared EIA incorporated into planning and design can save the developer and regulatory agency valuable time and expense. If the IEE/EIA is performed early enough to be considered during the decision-making phase, delays in regulatory procedures can be minimized. Improper planning or design that will lead to unacceptable levels of environmental deterioration may require costly rectification or replacement.

Operation and maintenance: A final major consideration when dealing with environmental management of ID projects is the operation and maintenance (O & M) of equipment. Problems with this can be broadly categorized as (a) paucity or absence of monitoring system, (b) low salary for O & M personnel, which will negate pollution abatement problems.

11.3 Guidelines for Preparations of TORS for Life of Industrial Development Projects for Initial Environmental Examination

1. The initial environmental examination, as indicated in the checklist if it shows that the project involves potentially serious adverse environmental impacts, which must be given careful attention in the planning, design, construction, operation and monitoring of the project in order to minimize and offset the adverse effects, and therefore a follow-up EIA is required.

2. The feasibility study for the project, to be done by the Project Consultant, should include an EIA (ES/EIA). The ES/EIA should include, inter-alia, study of each of the environmental effects found by the IEE to be significant. For each of the following items, the consultant will conduct a study, as part of the overall EIA,

    (a) to make an assessment, which delineates the significant environmental effect of the project.

    (b) to describe and quantify the effects.

    (c) to describe feasible mitigation measures for minimizing, eliminating, or offsetting unavoidable adverse effects and

    (d) to recommend the most appropriate mitigation and/or enhancement measures.

3. The selected significant environmental impacts (SEIs) to be studied as part of the overall EIA, are the following.

    (a) Environmental problems caused by project location

    (i)   --------------------

    (ii) --------------------- etc. as appropriate

    (b) Environmental problems related to a system design (including assumptions on O &M)

    (i)   --------------------

    (ii) --------------------- etc. as appropriate

    (c) Environmental problems during construction phase

    (i)   --------------------

    (ii) --------------------- etc. as appropriate

    (d) Environmental problems resulting from operations

    (i)   --------------------

    (ii) --------------------- etc. as appropriate

4. The estimated cost of the overall EIA is approximately-------man-months of professional input. Of which -------per cent should be allocated for use of expartiate EIA expertise for guiding and supervising the EIA and for transferring technology to the local staff. This estimate assumes that the EIA will be done as part of the overall project feasibility study.

5.   The estimated optimal time required for the EIA is ----------------

6.   The total estimated cost of the recommended EIA is approximately __________ including approximately ------- per cent for foreign exchange.

11.4 Management Requirements for the Preparation of EIA for Industrial Projects and Timely Preparation of an EIA

Generally, the industrial development projects proceed in four stages: (a) Strategic planning, (b) Feasibility studies, (c) Design and construction, and (d) Project operation. The environmental management problems very often occur at the planning and feasibility stages when most of the information relevant to decision- making is gathered and analyzed. These problems can be traced to a lack or paucity of consideration given to the effects that the proposed project will have on the environmental resources, particularly resources in which the project developer has little or no interest. In addition, in areas where the pre-project trends have already been towards environmental degradation, for example, deforestation by villagers, project developers may justify potential adverse environmental effects as simply accelerating an ongoing process. However, the major issue from the environmentalist's perspective should be how the industrial development project can help reverse or minimize these deleterious trends.

Effective environmental management of the industrial project must therefore rely first and foremost on identifying the major environmental impacts of plant operations and incorporating constraints into the initial stages of the proposed development, i.e., during the planning and feasibility stages. This will allow environmental concerns to become part of the early planning and decision making. The EIA's purpose is to provide for appropriate planning and design so that project impacts on any sensitive environmental resources or values will be minimized or mitigated.

11.5 Environmental Clearance (EC) for Industrial Projects - Indian Context

11.5.1 Environment Clearance Process for Industrial Projects – Ministry of Environment and Forests Govt of India Guidelines

In India the Ministry of Environment & Forests, Govt of India made Environmental Clearance (EC) for certain development projects mandatory through its notification of 27/01/1994 under the Environment Protection Act, 1986. Keeping in-view the experience gained in environment clearance process over a period of one decade, the MoEF came out with Environment Impact Notification, So 1533(E), and dt.14/09/2006. It has been made mandatory to obtain environmental clearance for 32 different kinds of developmental projects (Schedule-1 of notification).

The notification has classified projects under two categories - A and B. Category-A Projects (including expansion and modernization of existing projects) require clearance from Central Government (Ministry of Environment and Forest, Govt. of India) while category-B projects should be considered by State Level Environmental Impact. Assessment Authority (SEIAA), constituted with the approval of MoEF.

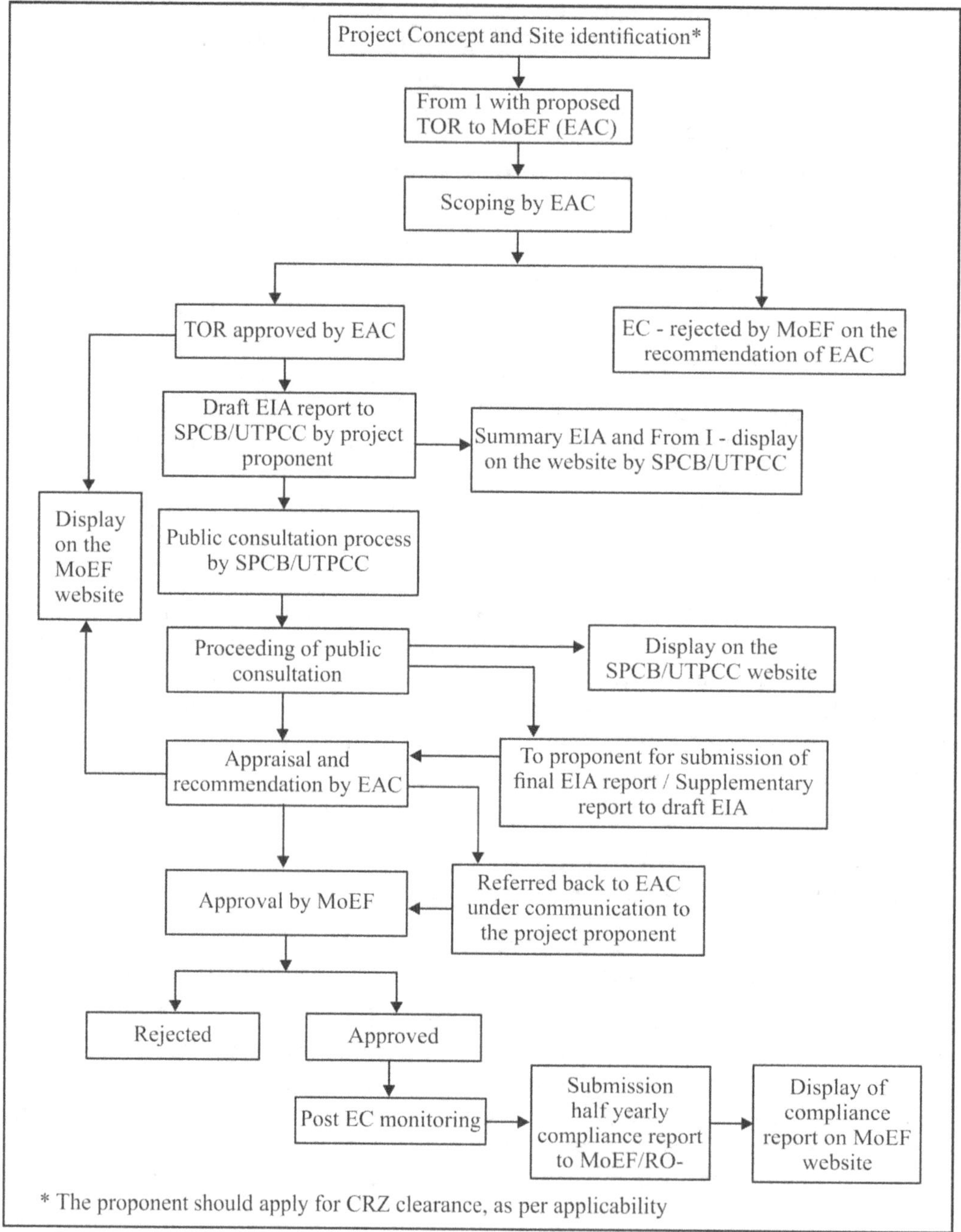

Fig. 11.2 Prior Environmental clearance process for category 'A' projects

Source: MoEF Environment Impact Notification, So 1533(E)

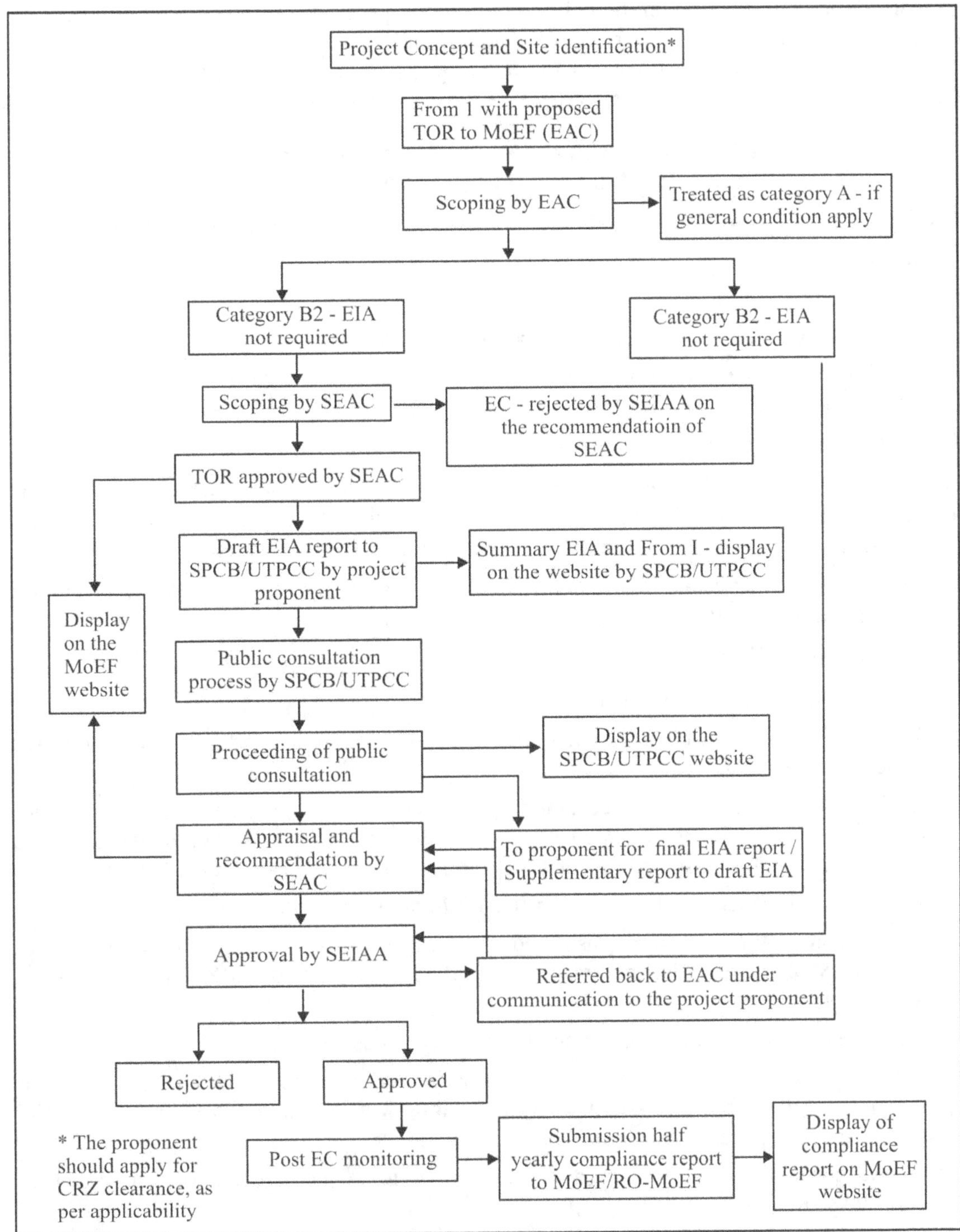

Fig. 11.3 Prior Environmental clearance process for category 'B' projects

Source: MoEF Environment Impact Notification, So 1533(E)

Considering EIA report and the public opinion, the expert appraisal committee, approves or rejects the projects. All the steps of environmental clearance have been time bound. Validity of environment clearance of mining projects is for a maximum of 30 years. The environmental clearance procedure is presented in Fig. 11.2 and Fig. 11.3.

11.5.2 EIA for Starting Industrial Estates- Guidelines for Essential Points to be covered in preparing EIA

The following are the essential points need to be covered in an EIA for Industrial Estates:

- Generic points
- General
- Locations specific concerns and corresponding components
- Any other component not covered but felt appropriate by the expert appraisal Committee of the Central Govt./ State level

11.5.2.1 Generic Points

For any common facilities such as CETPs, municipal solid waste management, common incinerators, TSDFs, coming-up as a part of the industrial estates, then respective developmental activity-specific generic guidance points may be considered.

Besides, generic coverage of points for EIA studies for Industrial Estates may include but not limited to the following:

11.5.2.2 General Features

- Executive summary of the project.
- Details of the industries, for which the estate is being planned and their proposed capacities of installation.
- Land requirement for the project including 50 meters of the peripheral green belt inside the boundary, and immediate 200 meters to 500 meters based on composition of air pollution potential industries, outside the boundary of estate for restricted land use.
- Justification for selecting the proposed size of the industrial estates.
- Comparison of alternate sites considered and the reasons for selecting the proposed site. Conformity of the site with the prescribed guidelines in terms of CRZ, rivers, highways, railways etc.
- Strategy being followed for development of industrial estate.
- Layout map of estate indicating processing zones, admin area, roads, plots, green belt, common utilities area etc., shall be shown along with contour map. Landscape plan including open spaces may be described.
- All the coordinates of the Industrial Estate site to be demarcated on the toposheet (1:10000 scale).
- The project study area for EIA studies include 10 km radius from the boundary of the proposed industrial estate.

- Land use of study area should include data about the residential/institutional/nearest village/township/locality/housing society, etc., based on the satellite imagery.

- Topography of the area should be given clearly indicating whether the site requires any filling. If so, details of filling, quantity of fill material required, its source, transportation etc., should be given.

11.5.2.2.1 Project-Related

- Each industrial activity shall be defined with respect to its manufacturing process, product, material balance, waste generation, treatment and its disposal. (Classify the industries to the extent possible with the A1 to A4 categories and/or W1 to W4 categories – Please refer Guidance Manual for classification).

- The mass balance for each type of industry giving material in and out etc., shall be taken into account. Planning of industries with respect to flow of goods and services in sequential order and workout production figures with respect to utilization of automatic and labour intensive technology.

- Backward and forward linkages of the industrial estates (availability of input resources and markets for the products/by-products and anticipated benefits for the regional development).

- Details of Infrastructure Development within the Industrial Estate and in the region.

- Examine in detail the proposed site with reference to possible impact of infrastructure covering water supply, pipelines, roads, storm water drainage, sewerage, power, temporary waste storage facilities, treated wastewater disposal (land/sewer/surface water bodies), common facilities etc.

- Each known industrial activity-specific proposed activities, resource consumption and rejects assessment.

- Anticipated pollution loads from each of the known composition of industrial units. Cumulative wastewater quantity and pollution load, point source–specific details for air pollutants and their loads, total solid/hazardous waste generation etc.

- Quantity of fuel required, its type and nature, source and transportation.

- Details of rainwater harvesting and how it will be used in the Industrial Estate and outfall.

- Details of water balance taking into account conservation measures, reuse and re-cycling of treated effluents.

- Individual and/or common facilities for waste collection, treatment, recycling and disposal (all effluent, emission and refuse including MSW, and hazardous wastes)

- Evaluate alternative disposal modes of effluent and solid wastes, from the point of view of disposal points and associated impacts.

- Commitment from the concerned authorities regarding availability of power, water and sewerage network.

- Details of Solid Waste management. Arrangements for hazardous waste management including e-waste may be described.

- Provisions made for safety in storage of materials, products and wastes may be described.

- Describe the application of industrial ecology concept for planning of estates. Explore possibility of utilizing waste of one unit as raw material for the other units.

11.5.2.2.2 Baseline Studies

- Identification of existing potential sources of pollution like industries in the study area.

- Present and projected population, present and proposed land use, planned development activities, issues relating to squatting and relocation, community structure, employment, distribution of income, goods and services, recreation, public health and safety, cultural peculiarities, aspirations and attitudes shall be explored in study

- The historical importance of the area shall also be examined in the study. While this analysis is being conducted, it is expected that an assessment of public perception of the proposed development be conducted.

- Details regarding availability of social infrastructure and future projections, details of facilities such as sanitation, fuel, restroom etc., to be provided to the labour force during construction as well as to the casual workers including truck drivers during operation phase.

- Detailed study of the hydrological and geo-hydrological conditions of the project area including a contour plan indicating slopes and showing drainage pattern and outfall.

- Information regarding surface hydrology and water regime and impact of the same, if any due to the project.

- One season site-specific meteorological data shall be provided.

- Examine soil characteristics, topography, rainfall pattern and soil erosion.

- Water quality of nearby river, if any, source of water supply and nearby water ponds shall be analyzed.

- Climatic conditions of the study area shall be monitored for hourly wind speed, wind direction, relative humidity, ambient dry and wet bulb temperatures and precipitation.

- One complete season AAQ data (except monsoon) to be given along with the dates of monitoring. The parameters to be covered shall include SPM, RSPM, SO_2, NOx and Ozone (ground level). The location of the monitoring stations should be so decided so as to take into consideration the pre-dominant downwind direction, population zone and sensitive receptors including reserved forests. There should be at least one monitoring station in the upwind direction and one in down wind direction where maximum GLC falls.

- Fuel analysis to be provided (sulphur, ash content and mercury). Details of auxiliary fuel, if any including its quantity, quality, storage etc, should also be given.

- Noise levels monitoring at about 15 locations within the study area.

- Details of Groundwater Quality around the Industrial Estate.

- Examine entry/exit of the project including the crossings from the highway and provision of service roads on the basis of traffic density studies and analysis

- Examine water quality with reference to Persistent Organic Pollutants, if relevant.

The guidelines for Base line Environment Monitoring are presented in table 11.4

Table 11.4 Guideline for Sampling, Frequency & Method for Baseline Environment Monitoring

Attributes	Sampling		Measurement Method	Remarks
A. Air Environment	**Network**	**Frequency**		
Meteorological • Wind speed • Wind direction • Maximum temperature • Minimum temperature • Relative humidity • Rainfall • Solar radiation • Cloud cover • Environmental Lapse Rate	1 site in the project area	1hourly continuous	Mechanical/automatic weather station Max/Min Thermometer Hygrometer Rain gauge As per IMD specifications As per IMD specifications Mini Sonde/SODAR	IS 5182 Part 1-20 Site specific primary data is essential Secondary data from IMD CPCB guidelines
Pollutants • SPM • RPM • SO$_2$ • NO$_x$ • Lead in SPM	Nos. of sampling location to be decided	24 hourly twice a week @4 hourly. Twice a week, One non monsoon season 8 hourly, twice a week 24 hourly, twice a week	As per CPCB guidelines	**Monitoring Network** • Minimum one location in upwind side, two sites in downwind side / impact zone • All the sensitive receptors need to be covered for core zone and buffer zone

11.5.2.2.3 Impacts

- Environmental condition scenarios shall be developed based on industrial activities and pollution potentials.

- Details of traffic density vis-à-vis impact on the ambient air.

- Impact of the developmental activity on drainage of the area and the surroundings.

- Impact of the project on the AAQ of the area. Details of the model used and the input data used for modeling should also be provided. The air quality contours may be plotted on a location map showing the location of project site, habitation nearby, sensitive receptors, if any. The wind roses should also be shown on this map.

- Mathematical modeling for calculating the dispersion of air pollutants and ground level concentration along with emissions from boilers.

- Cumulative impact on regional supportive capacity shall be studied in terms of population density, water supply, sewerage, storm water drainage, power supply, educational facilities, medical facilities, public transport, traffic, housing for EWS, and communities facilities etc.

- All kind of resources, both renewable and non-renewable shall be taken into account.

- The impacts shall be distinguished between significant positive and negative impacts, direct and indirect impacts.

- Project activities and impacts shall be represented in matrix form with separate matrices for pre and post mitigation scenarios.

- Public hearing should be conducted as per the prescribed procedure.

11.5.2.2.4 Management Plans

- Use of local building materials should be described. The provisions of fly ash notification should be kept in view.

- Details of greenbelt - details of species, width of plantation, planning schedule etc., within the boundary of industrial estate shall be given.

- Detailed plan of treated water disposal, reuse and utilization/management.

- Detailed R&R plan/compensation package for the project affected people shall be prepared in accordance with the latest national resettlement and rehabilitation policy, taking into account the socio economic status of the area, homestead oustees, land oustees, landless labourers, fishermen, reduction in fishing yields, etc. As far as possible, quantitative dimension to be given.

- Risk assessment and corresponding on-site and off-site emergency management plans.

- Specific chemical emergency response and proposed rescue system shall be established.

- Traffic management plan including parking and loading/unloading areas may be described. Traffic survey should be carried out on week days and weekends and also analyse the anticipated traffic increase. Back to back parking shall not be permitted.

- Odour mitigation plan may be described. Also make provision of green belt as a measure for mitigation of dust and noise and buffer between habitation and industry.

- EMP to mitigate the adverse impacts due to the project along with item-wise cost of its implementation.

- Elaborate corporate social responsibility proposals.

- Rain water harvesting proposals should be made with due safeguards for ground water quality. Maximize recycling of water and utilization of rain water.

- Temporary plans for the housing of construction labour within the site with all necessary infrastructure and facilities such as fuel for cooking, mobile toilets, mobile STP, safe drinking water, medical health care, crèche etc.

- Compliance management plan, if diesel power generating sets are proposed, as source of backup power for elevators and common area illumination during construction/operation phase.

- Identification of the industries, which should be avoided in the industrial estate, in view of the anticipated impacts (w1 to w4 and/or A1 to A4 as given in the Manual ref. CPCB).

- An outline-monitoring programme for construction and operation stage shall also be developed. The monitoring programme shall include the parameters to be monitored with frequency, locations and reporting.

- In case of any scheduled fauna, conservation plan should be provided.

- Points raised in the public hearing and commitment of the project proponent on the same, including necessary allocation of funds for the same.

- Impact of the project on local infrastructure of the area such as road network and whether any additional infrastructure would need to be constructed and the agency responsible for the same with time frame.

- Any litigation pending against the project and/or any direction/order passed by any Court of Law against the project, if so, details thereof.

Additional Conditions, if the components of Industrial estate are connected to CRZ area/marine ecology:

- Identification of CRZ area. A CRZ map duly authenticated by one of the authorized agencies demarcating LTL, HTL, CRZ area, location of the project and associate facilities w.r.t. CRZ, coastal features such as mangroves, if any. The route of the pipeline, conveyor system etc., passing through CRZ, if any, should also be demarcated. The recommendations of the State Coastal Management Authority for the activities to be taken up in the CRZ should also be provided.

- Provide the CRZ map in 1:10000 scale in general cases and in 1:5000 scale for specific observations.

- Impact of the activities to be taken up in the CRZ area including jetty and desalination plant etc., should be integrated into the EIA report. However, action should be taken to obtain separate clearance from the competent authority as may be applicable to such activities.

- Make provision for guard pond of adequate capacity before conveying effluent to marine outfall and similar provisions for safety against failure in the operation of wastewater treatment facilities. Identify acceptable outfall for treated effluent.

- Explore the possibility of expansion of Port facilities instead of Copy of the MOU for Port facilities shall be provided.
- Capital quantity of dredging material, disposal and its impact on aquatic life.
- Common marine outfall design, features, impacts on marine biology etc.
- Fisheries study should be done with respect to Benthos and Marine organic material and coastal fisheries, bathymetric studies.

11.5.2.3 Location-Specific Coverage of Points for EIA Studies

Based on environmental sensitivity, location-specific concerns for EIA study may broadly include the following points given in table 11.5.

Table 11.5 Location-Specific Coverage of Points for EIA Studies Based on Environmental Sensitivity

S. No.	Type of concern	Attribute	Whether any attribute is falling within 10 km radius from the plant boundary? (Yes/No)	If yes, proponent shall describe the sensitivity (distance, area and magnitude) and propose the additional points/ conditions based on magnitude for review and acceptance by the EAC/SEAC
1	Incompatible land use areas	Public water supply areas from rivers/surface water bodies, from ground water		
2	Incompatible land use areas	Ground water recharge areas		
3	Incompatible land use areas	Scenic areas/tourism areas/hill resorts		
4	Incompatible land use areas	Religious places, pilgrim centers that attract over 10 lakhs pilgrims a year		
5	Incompatible land use areas	Protected tribal settlements (notified tribal areas where industrial activity is not permitted)		
6	Incompatible land use areas	Coastal Regulatory Zone (CRZ)		

Table 11.5 contd...

S. No.	Type of concern	Attribute	Whether any attribute is falling within 10 km radius from the plant boundary? (Yes/No)	If yes, proponent shall describe the sensitivity (distance, area and magnitude) and propose the additional points/ conditions based on magnitude for review and acceptance by the EAC/SEAC
7	Incompatible land use areas	Monuments of national significance, World Heritage Sites		
8	Incompatible land use areas	Flood prone areas (based on flood in 1 in 25 years)		
9	Incompatible land use areas	Agricultural research stations		
10	Incompatible land use areas	Air port areas		
11	Incompatible land use areas	Any other feature as specified by the State or local government and other features as locally applicable (including prime agricultural lands, pastures, migratory corridors etc.)		
12	Biodiversity concern	National parks		
13	Biodiversity concern	Wild life sanctuaries		
14	Biodiversity concern	Game reserve		
15	Biodiversity concern	Tiger reserve/elephant reserve/turtle nesting ground		
16	Biodiversity concern	Breeding grounds concern		
17	Biodiversity concern	Core zone of biosphere reserve		

Table 11.5 Contd...

S. No.	Type of concern	Attribute	Whether any attribute is falling within 10 km radius from the plant boundary? (Yes/No)	If yes, proponent shall describe the sensitivity (distance, area and magnitude) and propose the additional points/ conditions based on magnitude for review and acceptance by the EAC/SEAC
18	Biodiversity concern	Habitat for migratory birds		
19	Biodiversity concern	Mangrove area		
20	Biodiversity concern	Areas with threatened (rare, vulnerable, endangered) flora/fauna		
21	Biodiversity concern	Protected corals		
22	Biodiversity concern	Wetlands		
23	Biodiversity concern	Botanical gardens		
24	Biodiversity concern	Zoological gardens		
25	Biodiversity concern	Gene Banks		
26	Biodiversity concern	Reserved forests		
27	Biodiversity concern	Protected forests		
28	Biodiversity concern	Any other closed/protected area under the Wild Life (Protection) Act, 1972, any other area locally applicable		

Table 11.5 *Contd...*

Ministry of Environment and Forests Govt of India issuedguidelines for preparing TORs for various types industries which are presented in Appendix I

11.6 Case Study

11.6.1 Bulk Drug Industry - Environmental Impact Assessment

Introduction

A bulk drug manufacturing industry has to be set up in Mukthayala village, Jaggiapeta Mandal, Krishna district of Andhra Pradesh.

The details of the proposed products to be manufactured along with the production capacities are shown in Table 11.6.

Table 11.6 Proposed products and their capacities

S.No	Products proposed	Production kg/day	Production kg/batch
1.	Trichloro acetone (20%) TCA	4325	8650
2.	Folic acid (FA)	255	85
3.	Butyl Cyano Acetate (BCA)	421	632
4.	Tri amino pyrimidine sulphate (TAPS) wet	800	400
5.	Para amino BenzoylGlutamicAcid (PABGA) wet	800	800

The production will be in phases. The first two products TCA and FA will be manufactured during phase – 1. TAPS and PABCA will be taken up after stabilization of FA production during phase 1. In early stages along with FA, TCA is proposed to be manufactured where as other products will be produced only after stabilization of FA. Till then BCA, TAPS, and PABGA will be procured from the local market.

Folic acid is one of the constituents of B complex and is very much essential for arsenic and iron deficiency cases. Folic acid is one of the most important hemotopoetic agents necessary for proper regeneration of the blood forming elements and their functioning. It acts as coenzyme in intermediary metabolic anemia, gloss it is, diarrhea, weight loss and macrocytic anemia in elderly persons.

Need for EIA Studies

In all chemical, pharmaceutical, drug, dye manufacturing industries, the plant activities must co-exist satisfactorily with its surrounding environment so as to reduce the environmental impact of these activities. This requires sound and safe environmental management plan to be implemented by the proponents, which makes environmental protection an essential requirement along with production and profits.

Justification of the Project

During the last two decades, the Indian Pharma industry has reached a level of maturity, whereby the country has become self reliant in terms of production and distribution of over 98% of the country's demand for bulk drugs and formulations. The growth of vitamin industry including Folic acid is in pace with other bulk drug industries.

Uses of Folic Acid

Addition of folic acid to women's diet during child bearing years can reduce brain and spinal chord defects in new born. Intake of folic acid along with Vitamin B_{12}, and B_6 could help the risk of cardiovascular disease including stroke. Folate deficiency due to inadequate intake, absorption or utilization or increased excretion or need as in sprue, nutrition algaloblastic anemia, and megaloblastic anemia of infancy and pregnancy or associated with anticonvalescent therapy. It is also used in animal nutrition as a feed supplement where a dietary deficiency is encountered.

Market in India

Government of India has set up a steering committee to give among things the demand estimates for bulk drugs in order to formulate a drug policy in 1984. The demand estimates for the periods 1990 to 2000 has varied from 18.7 MT to 60 MT. The figures include the demand in the veterinary usage and food industry.

EIA Methodology

The EIA study encompasses 10 km radius area with the proposed plant as its centre.

Scope of EIA

The scope of the study includes a detailed study of the environment in an area of 10 kms radius with the proposed plant as its centre for environmental components, air, noise, water, land, and socio-economic environment.

To assess the present status of air, noise, water, land, eco-system and socio-economic components of environment.

To identify and quantify significant impacts of processing operations on environmental components.

To evaluate proposed pollution control facilities.

To delineate future environmental quality monitoring program to be pursued by the proponents.

Methodology for Environmental Impact Assessment:

It was observed from past experiences that the impact on the environment is felt mainly up to a distance of 10 km from the plants site. Therefore under the scope of EIA studies, an area of 10 km radial zone around the plant is studied for detailed characterization of various environmental component viz. air, noise, water, land, and socio-economic.

(a) Study Period

> *Micrometeorology*: The micro climatic parameters were recorded using manual weather station for the study period. Wind speed, wind direction and relative humidity

was recorded on hourly basis. Minimum and maximum temperatures were also recorded.

(b) Air Environment

The baseline status of the existing ambient air quality within the study region has been assessed by establishing a monitoring network of appropriate ambient air quality monitoring stations based on the available climatological norms of predominant wind directions and wind speeds of the study region.

The baseline status of air environment were monitored for Suspended Particulate Matter (SPM) and gaseous pollutants like SO_2, NO_x. All the pollutants were measured on 24 hours average. Pre-calibrated High Volume Samplers (HVS) has been used for monitoring all the air pollutants.

(c) Noise Environment

High dBA noise levels causes adverse effect on human beings and associated environment, including land, structures, domestic animals and natural ecological systems. A detailed survey on noise environment in and around the project site was carried by measuring spot noise levels at residential areas, schools, hospitals, bus stands and commercial centers etc., using a precision sound level meter.

(d) Water Environment

Information on water resources was collected during the study period. The parameters of prime importance were selected under physical, inorganic, organic and heavy metals groups. Water samples from ground water and surface water sources were collected within 10 km radius around the existing site. Due care was taken during sampling, and transportation of these samples.

(e) Land Environment

Soil samples were collected within 10 km, radius in order to assess the cropping patterns field infiltration rates and limitations of soil for growth of appropriate plant species around the site. Plant species for development of green belt were identified taking attenuation factors into consideration of air pollutants.

(f) Eco-system

Information on eco-system within 10 km radius was collected from the state agricultural and forest departments. The important flora species native to the area was enumerated. A test check survey was also under taken to judge the correctness of the data collected.

(g) Socio-economic environment

As any development activity will bring about changes in socio-economic pattern. Data on demographic pattern, population characteristics, employment, income, health status, land use pattern, transport and recreation facilities were collected from mandal offices and national informatics centre at Hyderabad. A cross-check survey was also conducted in some of the villages.

Site Location and Site Description

The proposed project is located adjacent to Jaggaihpeta – Huzurnagar road. The site is located at a distance of 8 km from Jaggaihpeta. The site falls under Jaggaihpeta mandal of Krishna district. It is located on the border areas of Krishna and Nalgonda District. The site is located between the geographical coordinates 16" 49' latitude and 80" 02 E longitude. The altitude ranges between 70 to 71 MSL (Table 11.7).

The nearest human settlement is Mukthyala which is about 2.0 km radially, however, a small hamlet called Venkteshnagar is present around 1.0 km distance of the plant site in west direction. Vedadri is an important pilgrim centre in the study area. There are cement industries located in the study area.

River Krishna is the major important surface water body flowing at a distance of 3 km in the south east direction with respect to the plant site. It is also the intake source for the proposed industry. Paleru river is another surface water body flowing in the NE direction at distance of 8 km and joins river Krishna at Ravirala.

The nearest railway station is Jaggiahpeta on the Motammari – Jahhiahpeta line to cater to the requirements of the cement industries. It is not open for passenger traffic. National Highway No.9 is located at distance of 9.5 km from the site. The area around the proposed plant site mostly represents rural nature with a few important industries. The study area falls under Krishna district including the project site. The area on the other side of the River Krishna comes under Guntur District and rest of the area falls under Nalgonda district.

Table 11.7 Site Selection

Selection Criteria	Details
Latitude and Longitude	$16^0 49'$ north and 80^0 02' 30" East
Climate conditions	Mean Annual Max Temp. 34^0C Mean Annual Min. Temp. 23^0C Average Annual Rainfall 795 mm
Land acquired for the plant	10 Acres
Land use and major crops	Dry land with scattered shrubs
Predominant wind direction	SE followed by E
Nearest Town	Jaggiahpeta around 8 to 9 km from plant site
Major urban settlement	Vijayawada city
Water bodies	River Krishna is flowing around 2 km south east, and river Palleru in the east side of the project area
Selection Criteria	Details
Hills and mountains	None
Ecological sensitive zones	No wild life sanctuaries No migration route auifavana
Reserved forests	Baluspada, Budavada in north, Jaggiahapeta Extension in East, Kuntimaddi, Venkatayapalem in South.east and Chintapalem is south side of the plant.

Table 11.7 *Contd...*

Selection Criteria	Details
Historical places	No major historical places, Vedadri pilgrim centre around 9 km east of the plant site.
Socio-economics	No rehabilitation required
Locational advantage	Availability of basic industrial, social infrastructure necessary for the bulk drug and intermediate manufacturing industry.

11.6.2 Process Description for the Present and Proposed Products

The drugs that are proposed for production are taken up after considerable evaluation of various factors like domestic market and availability of the drugs in India, export potential etc. (Table 11.8 (a) to (e)).

Table 11.8 (a) Material Balance for TCA

Input	Weight (kg)	Output	Weight (kg)
Water	200	TCA + Isomers	1870
Acetone	1000	HCl (100%)	630
Chlorine	2000	Aqueous layer for recycling	600
		Cl_2 (trapped in)	100
Extraction stage			
TCA + Isomers (org. layer)	1870	TCA – 20%	8650
Water	6920	Resin	140
Total	8790	Total	8790

11.6.2.1 Folic Acid

Folic acid is produced by condensation of para amino benzoyl glutamic acid, triamino pyrimidine sulphate and trichloro acetone in the presence of sodium bisulphate. This is transformed to Pharma grade by purification with HCl. Sodium hydroxide, zinc chloride and calcium hydroxide are later on precipitated with acetic acid.

Table 11.8 (b) Material balance for folic acid

Condensation			
Input	Weight (kg)	Output	Weight (kg)
Trichloroacetone (20%)	1138	Reaction mass	9246
Water	7500		
Sodium bisulphite	93		
Sodium bicarbonate	236		
Condensation			
Input	Weight (kg)	Output	Weight (kg)
p-amino benzoylglutamicacid	114		

Table 11.8 (b)*contd…*

Tri amino pyrimidine sulphate	165		
Total	9246	Total	9246
Filtration			
Reaction mass from condensation	9246	Product with 50% moisture	2000
Total	9246	Mother liquor	7246
		Total	9246
(a) Mother liquor is separated and subjected to pretreatment followed by evaporation			
(b) The cake is washed with 6000 kg of water and the wastewater so obtained is sent to ETP			
Purification			
Cake from stage 2	2000	Reaction mass	5317
Hydrochloric acid (30%)	530		
Water	2787		
Total	5317	Total	5317
a. Mother liquor is separated and subjected to pretreatment followed by evaporation			
b. The cake is washed with 5000 kg of water and the wastewater so obtained is sent to ETP			
2nd Purification			
Cake from previous Stage	1000	Reaction mass	2313
Water	1278		
Hydrogen peroxide	1		
Calcium hydroxide	16		
Sodium hydroxide	16		
Zinc Chloride	2		
Total	2313	Total	2313
Filtration			
Reaction mass from previous stage	2313	Wet solid (waste)	50
Total	2313	Filtrate containing product	2263
		Total	2313
Precipitation / centrifugation			
Filtrate from previous stage	2263	Product	100
DM water	909	Filtrate	3262
Acetic Acid	190		
Total	3362	Total	3362
(a) Mother liquor is separated and subjected to pretreatment followed by evaporation			
(b) The cake is washed with 4000 kg of DM water which will find its way as wastewater to ETP			
Drying			
Product wet	100	Final product dry	42.5
Total	100	Moisture	57.5
		Total	100

11.6.2.2 Butyl Cyano Acetate (BCA)

Sodium cyanide reacts with sodium monochloro acetate to give sodium cyano acetate in the form of reaction mixture. This on treatment with butanol and sulphuric acid gives butyl cyano acetate.

Table 11.8 (c) Material balance of BCA

Stage 1			
Input	Weight (kg)	Output	Weight (kg)
Sodium cyanide (32.3%)	885	Reaction mass	2591
Sodium monochloro acetate solution (41.38%)	1706		
Total	2591	Total	2591
Stage II Distillation			
Reaction mass	2591	Reaction mass	1491
Total	2591	Aqueous distillate for recycle	1100
		Total	2591
Stage III Esterification			
Reaction mixture	1491	Reaction mixture	3791
Butanol	1500	Total	3791
Sulphuric acid	800		
Total	3791		
Stage 1			
Input	Weight (kg)	Output	Weight (kg)
Filtration			
Reaction mass	3791	Cake (sodium sulphate)	1500
Total	3791	Filtrate	2291
		Total	3791
Cake (sodium sulphate) for disposal by sale			
Stage V			
Filtrate	2291	Aqueous layer	1000
Soda ash (5%)	600	Organic layer	1891
Total	2891	Total	2891
Aqueous layer for pretreatment followed by evaporation			
Stage VI			
Organic layer	1891	Butanol recovery	1251
Carbon	6	Reaction mass	646
Total	1897	Total	1897
Stage VII: Filtration			
Reaction mass	646	Butyl cyano acetate	632
		Spent activated carbon as solid waste	14
Total	646	Total	646

11.6.2.3 Tri Amino Pyrimidine Sulphate (TAPS)

Guanidine nitrate is condensed with sodium methoxide. Butyl cyano acetate to give an intermediate, which on treating with sodium nitrate in acidic medium gives a nitroso compound. This on reduction with hydrazine hydrate followed by acidification gives tri amino pyrimidine sulphate.

Table 11.8 (d) TriAmino pyrimidine sulfate (TAPS)

Input	Weight (kg)	Output	Weight (kg)
		Stage I	
Sodium methoxide	945	Reaction mass	2769
Water	2000	Methanol (recovered)	850
Guanidine nitrate	320	Total	3619
Butyl cyano acetate	354		
Total	3619		
		Stage II	
Reaction mass	2769	Reaction mass	3435
Sodium nitrite	216		
Hydrochloric acid (30%)	450		
Total	3435	Total	3435
Stage III: Filtration			
Reaction mass	3435	Wet cake	600
		Filtrate	2835
Total	3435	Total	3435
Input	Weight (kg)	Output	Weight (kg)
Stage I			
(a) Mother liquor is subjected to pretreatment, followed by evaporation			
(b) The Cake washed twice with 200 kg of chilled water, which will find a way as wastewater to ETP			
Stage IV			
Wet cake	300	Reaction mass	1491.5
Water	1000	Total	1491.5
Caustic lye	109		
Hydrazine hydrate	80		
Nickel catalyst	2.5		
Total	1491.5		
		Stage V Filtration	
Reaction mass	1491.5	Solid waste containing nickel catalyst with impurities for sale	20.5
Total	1491.5	Filtrate containing product	1471

Table 11.8 (d)*contd…*

Input	Weight (kg)	Output	Weight(kg)
		Total	1491.5
Input	Weight (kg)	Output	Weight(kg)
Stage VI			
Filtrate from stage V	1471	Reaction mass	1796
Dilute sulphuric acid (40%)	325		
Total	1796	Total	1796
Stage VII Centrifugation			
Reaction mass	1796	Wet cake (product)	400
Total	1796	Filtrate	1369
		Total	1796

a. The wet cake is washed with 100 kg of water, which will find a way along with filtrate for pre-treatment followed by evaporation.

11.6.2.4 Para Amino Benzoyl GlutamicAcid (PABGA)

Para nitro benzoic acid reacts with thionyl chloride to give an acid chloride, which on treatment with mono sodium glutamate gives a nitro-derivative. This on hydrogenation with hydrazine hydrate gives PABGA.

Table 11.8 (e) Material balance for PABGA

Stage 1			
Input	Weight (kg)	Output	Weight (kg)
Toluene	600	Hydrogen chloride	80
Thionyl chloride	270	Sulphur dioxide	141
p-nitrobenzoic acid	368	Reaction mass	1017
Total	1238	Total	1238
In the 1st stage water will be in circulation and scrubbant contains dilute HCl and traces of SO_2			
In the 2nd stage alkali solution will be in circulation and the scrubbant contains sodium bisulfate with traces of sodium chloride			
Stage II			
Reaction mass	1017	Reaction mass	1767
Caustic Lye	400	Recovered toluene	590
Mono sodium glutamate	350	Loss	2
Water	600	Pot. Residue (solid waste)	8
Total	2367	Total	2367

Table 11.8 (e) *Contd...*

Stage III			
Input	Weight (kg)	Output	Weight (kg)
Reaction mass	1767	Reaction mass	1917
HCl	150		
Total	1917	Total	1917
Stage IV Centrifugation			
Reaction mass	1917	PNBGA	520
		Mother liquor	1397
Total	1917	Total	1917

a. Mother liquor is subjected to pretreatment, followed by evaporation

b.   PNBGA cake is washed with 150 kg of chilled water, which will find a way as wastewater to ETP

c. In the 2nd stage alkali solution will be in circulation and the scrubbant contains sodium bisulfate with traces of sodium chloride

Stage II			
Reaction mass	1017	Reaction mass	1767
Caustic Lye	400	Recovered toluene	590
Mono sodium glutamate	350	Loss	2
Water	600	Pot. Residue (solid waste)	8
Total	2367	Total	2367
Stage III			
Reaction mass	1017	Reaction mass	1917
HCl	150		
Total	1917	Total	1917
Stage IV Centrifugation			
Reaction mass	1917	PNBGA	520
Total	1917	Mother liquor	1397
		Total	1917

(a) Mother liquor is subjected to pretreatment followed by evaporation

(b) PNBGA cake is washed with 150kg of chilled water which will find its way into ETP

Stage V			
Input	Weight (kg)	Output	Weight (kg)
Water	600	Reaction mass	1270
PNBGA	520	Total	1270
Alkali	150		
Total	1270		
Stage VI			
Reaction mass	1270	Reaction mass	1800
Iron	150	Total	1800

Input	Weight (kg)	Output	Weight (kg)
Water	250		
Sodium chloride	30		
Total	1800		
		Stage VII	
Reaction mass	1800	Iron sludge	200
Washing	100	Filtrate	1700
Total	1900	Total	1900
		Stage VIII	
Filtrate	1700	Reaction mass	1950
Hydrochloric acid	250	Total	1950
Total	1950		
		Stage IX centrifugation	
Reaction mass	1950	p-amino benzyolglutamicacid	800
Water for washing	700	Mother liquor	1150
Total	2650	Washing	700
		Total	2650
a. Mother liquor along with water used for washing is subjected to pretreatment			
b. The wastewater from the cake washing will be sent to ETP			

Water Source

The total water required for the plant (Table 11.9) during phase I and II will be 171 m^3/day. For production of TCA and folic acid in phase I water required is around 163 m^3/day, during phase II the consumption will be increased by 8 m^3/day. The entire water quantity required would be drawn from River Krishna, which is 2 km from the plant. The water required for greenbelt development in the plant site would be taken along with fresh water from River Krishna.

Table 11.9 Water Requirement

S.No.	Process / Use	Present m^3/day	Future m^3/day	Total m^3/day
1.	Process water	110	7	117
2.	Floor washings	5	1	6
3.	Chilling plant	3	0	3
4.	DM Plant regeneration	10	0	10
5.	Boiler makeup	30	0	30
6.	Service and potable water	5	0	5
	Total	163	8	171

11.6.3 Baseline Status of the Study Area

Introduction

The prime objective of baseline environmental study is to delineate the prevailing conditions inand around the proposed project site, relating to important environmental components viz. air, noise, water, land, and socio-economics. Depending upon the size and nature of the industry a suitable area is designed with the proposed project as the nodal centre. The area is designed keeping in tune with the guidelines formulated by the regulatory authorities. In the present project keeping in view the size of the industry an area of 10 kms with the project site as the centre is taken for the baseline environmental study.

Land Features

Topography

The topography of the study area is undulatory in general with a mild slope. The site exhibits rugged terrain with wild shrubs and bushes spread all over.

Climatology

The climate in the study area is generally hot. Higher temperature may be due to the local factors, as the area is situated in a Limestone belt. Like other places in the state the study area has three seasons viz. summer, winter and monsoon. The area in general experiences rainfall from the Southwest monsoon. The annual mean minimum temperature, maximum temperature and rainfall based on the average of 5 years data are:

Mean annual minimum temperature	22^0C
Mean annual maximum temperature	34^0C
Average annual rainfall	795 mm

Geology and Hydrogeology

The area is rocky in general, comprises of proterozoicBalnadu Basin with mild slopes. The ground water table varies from 15 to 20 m in the study area, however, it has been observed that, at few places the formation of perched aquifers exist which has limited yield of ground water explorations. The depth of the ground water table normally abides by the ground level of the region. Drainage pattern is observed trending west to east.

Meteorological scenario exerts a critical influence on air quality as the pollution arises from the confluence of atmospheric contaminants, adverse meteorological conditions and certain topographical conditions.

A manual weather station was fixed near Mukthyala and important micrometeorological parameters like wind speed, direction, temperature and relative humidity were recorded on hourly basis for twenty four hours for four weeks. The data was recorded during the months of December 2000 to February 2001. The meteorological station was installed at a height of 8 m from the ground level. Wind speeds, direction and relative humidity were collected on hourly basis and wind rose pattern.

Air Environment

The ambient air quality was monitored at all the eight locations for four weeks on 24 hourly basis for SPM, SO_2, and NO_X. The monitoring locations including plant site, Multhyala, and Bugga Madhavaram reflect the regional background (upwind) because of their relative locations with respect to the proposed project site.

The limits as per CPCB standards for this pollutants in industrial/mixed areas are SPM 500 $\mu g/m^3$, SO_2 and NO_x are 120 $\mu g/m^3$, whereas in residential/rural areas SPM 200 $\mu g/m^3$, SO_2 and NO_x are 80 $\mu g/m^3$.

The measurements recorded in Ramapuram and Budawada villages located in the directions of NW and N winds respectively are SPM 96.6 $\mu g/m^3$ and 79.6 $\mu g/m^3$, SO_2 values are 7.8 $\mu g/m^3$, and 6.6 $\mu g/m^3$ respectively, for NO_x values are 9.0 $\mu g/m^3$ and 7.0 $\mu g/m^3$ respectively.

At plant site SPM, SO_x, NO_x minimum and maximum values are 59 $\mu g/m^3$, 86$\mu g/m^3$, 5.8 $\mu g/m^3$, and 7.75 $\mu g/m^3$ respectively. The AAQ levels observed at all sampling locations were within limits specified by CPCB for industrial/mixed use and also industrial/mixed.

Noise Environment

The environmental impact of noise can have several effects varying from Noise Induced Hearing Loss (NIHL) to annoyance depending upon decibels of noise levels. Taking into consideration various factors can carry out the EIA of noise from the proposed industry. The assessment of noise pollution on neighborhood environment due to the proposed activity was carried out keeping in view all the considerations mentioned (Table 11.10).

Table 11.10 Equivalent day–night noise levels–study area (10 km)

S.No.	Location	Equivalent day – night levels dB(A)		
		Day "Ld"	Night "Ln"	Equivalent "Ldn"
1.	Plant site	46	41	49
2.	Budhavada	48	41	49
3.	Ramapuram	55	50	58
4.	Dondapadu	57	52	59
5.	Vajinapalli	46	41	49
6.	Bugga Madhavaram	53	42	53
7.	Madipadu	57	52	59
8.	Mukthyala	61	46	60
9.	Koutavari	44	39	47
10.	Ravirala	58	44	57

Water Environment

Samples were collected to evaluate the surface and subsurface water quality in and around the study area. Necessary caution was exercised during sample collection. The sampling locations were selected as per Global Environmental Monitoring norms.

In the surface water sources, pH is ranging from 7.84 to 7.96. Total dissolved solids are ranging from 405 to 491 mg/*l*. The minimum TDS is observed in River Krishna near

Vedadri. The fluoride content is ranging from 0.81 to 1.18 mg/l. The sulphate content of River Krishna is slightly higher 78.1 mg/*l* when compared to the other two sources.

The dissolved oxygen in the collected surface water samples are in the range of 6.0 to 6.4 mg/*l*, whereas BOD is ranging from 2.3 to 2.7 mg/*l*.

In Groundwater sources pH is ranging from 6.86 to 7.71 whereas hardness is ranging from 193 to 1100 mg/l. A minimum value is observed in Jayanyhipuram water source whereas maximum values are observed in Budawada source. Remaining samples are in the range of around 250 mg/l, except sample collected from Kautavari Agraharam i.e., 550 mg/l. Total dissolved solids are ranging from 506 to 3931 mg/l. Fluoride content in the samples collected from the study area are ranging from 0.90 to 1.37 mg/l, whereas nitrate content is ranging from 7 to 113 mg/l. The highest value was observed in the sample collected from the village Budhawada whereas minimum values are observed in the sample collected from Ravirala source. Sulfate content in the samples collected in the study area are ranging from 36 to 783 mg/l. In the overall observation of the analyzed ground water samples, Budhawada source may be contaminated because of high electrical conductivity, high TDS, chlorides and nitrates.

Land Environment

In EIA studies, the land and biological aspects of ecosystem are important for identifying sensitive issues and taking appropriate action by maintaining ecological homeostasis in the early stages of development of the project.

Soil Quality

The soil in the study area is slightly reddish to brown and black with loamy texture, because presence of this soil water quickly percolates into the ground without causing any marshy conditions. The pH values of the soil is an important property, plants cannot grow in low and high pH value soils. Most of the essential nutrients like N, P, K, Ca, Mg are available for plant at the neutral pH except for Fe, Mn, and Al which are available at low pH value. The pH values in the study area are varying from 6.01 to 7.63 showing neutral to slightly acidic pH in the entire area.

The other important macro nutrients for characterization of soil for irrigation are N, P and K. The nitrogen value is varying from 0.10 to 0.13% and phosphorus is varying from 0.54 to 0.059% whereas potassium in ranging from 0.013 to 0.055%. The organic matter is varying from 0.04% to 0.93%.

The other macro and micro nutrients like Ca, Mg, Fe, Mn, Zn etc., are adequate and all sources of soil samples are suitable for irrigation.

Flora and Fauna

There is no animal kingdom in the study area at present but it is learnt that the area was rich in wild life about eight decades earlier. The main fauna attraction of that time was tigers, leopard, cheetah, bears, hyenas and spotted dears. It is said that the fauna has migrated due to construction activities.

Socio Economic

The population density of the study area is 245 persons per sq. km. The percentage of literacy is moderately low about 30.7%.45.5% of the population in the study area come under the category of main workers, which include cultivators, agriculture laborers, those engaged in household activities, construction, forestry, and allied activities. Most of the houses in the villages are electrified while some villages have benefited by the subsidized electric supply scheme for agriculture by the state government. The main sources of water available are river, canals, tube wells, hand pumps, and taps.

People in general aquifers to be healthy, about fifty percent of the villages don't have the basic medical facilities and villagers have to travel 1 to 10 km for proper medical aid. Communication facilities in the villages are quite good with all villages having post offices, Grameena bank and cooperative societies. Bus services are available for approaching the villages. There is no place of tourist attraction in the study area.

Agriculture

The cultivators yield two or three crops in a year with paddy as a major cereal crop. Paddy is grown in Kharif (June – September), Rabi (October – March). Due to adequate irrigation facilities, the productivity of land is fairly good.

11.6.4 Identification and Prediction of Environment Impacts

Identification of Impacts

Based on the baseline environmental data at the proposed project site, the environmental factors may be affected (impacts) are identified. Both positive (beneficial) and negative (adverse) impacts are considered.

11.6.4.1 Prediction of Impacts on Air Environment

Construction Related Impacts

In early stages of the new projects, some construction activities will take place, like earth moving, laying roads, construction of sheds, etc. All these activities and increased traffic at proposed access road will increase the concentration of particulate matter in the air, however, this being an small scale industry not much change will be observed. However, to control the particulate matter, regular water sprinkling will be carried out. The site being nearer to existing industries the study area has black topped roads in some places and kutcha roads in some parts of area so the particulate matter generation therefore will be negligible due to this industry. Ambient air levels of SO_2 and NO_x are also not going to get affected much because there is no operation of construction related equipment such as generators, bulldozers, pay loaders, trucks etc.

Operations Impact

Manufacturing process of the proposed drug industry is based on indigenous technology using raw materials available locally. All production blocks proposed to be provided with very efficient ventilation leading to several air changes per day. During chlorination of acetone in aqueous medium HCl gas is liberated, which is a byproduct. This is scrubbed in a two stage scrubber to get 30% HCl, which is used for captive consumption. Sometimes during chlorination there are changes for unreacted chlorine getting out along with HCl. This change is remote as the acetone is taken in excess than the requirement. However, an acetone trap is provided to trap any unreacted chlorine.

Boiler Emissions

The plant is proposed to have one boiler of one ton capacity for generation of steam. Coal is used as fuel, at peak load the coal consumption is around 3.5 MTPD. The main pollutants of significance that are released on burning of coal are particulates, sulphur dioxide and to some extent nitrogen dioxide. Particulate matter emissions are calculated based on the assumption that the ash content in coal as 40% and fly ash content as 80% of the total ash. The SO_2 emissions have been calculated assuming sulphur to be 0.5% of the total coal. The sulphur dioxide emissions from the boiler stack were calculated as 0.40 g/sec. The NO_x emissions have been calculated in accordance with the emission factor 8 kg of NO_x emission for every one ton of coal burnt. Based on this factor, the NO_x emission rate from the boiler stack was found to be 0.33 g/sec.

DG Set Emissions

The plant has proposed to keep one DG set of 225 kVA as standby to use in case of power failure. The fuel (diesel) consumption for the DG set will be in the range of 200 litres/day. The main pollutants of significance that are released on burning of fuel are sulphur dioxide and to some extent nitrogen oxide.

Since DG will be used only during power break down which will be around 2 to 3 hours, the sulfur dioxide and nitrogen dioxide emissions will be negligible as the fuel consumption is minimum.

The All Terrain Dispersion Model (ATDM) is a hybrid Gaussian Dispersion model that calculates concentrations from point, area, and volume source emissions in simple, intermediate and complex terrain.

ATDM is used to calculate the plume rise and plume centerline elevation for a given source to determine whether a receptor is located in which terrain with respect to that source. Depending on the terrain regime of the receptor, the model then uses one of the approaches.

Predictions have been carried out for these emissions considering the following points:

Predicted Results

- Predictions are carried out for 100% load where the maximum emissions are emitted and which would be the worst case scenario.
- Predictions have been carried out for sulphur dioxide only.
- No half-life time of the pollutant is considered.

The maximum predicted results for SPM, SO_2 and NO_x emissions for proposed plant are given in Table 11.11.

Table 11.11 Predicted results for SPM, SO$_2$ and Nox.

Pollutants	Emissions rate gm/sec	Predicted max. value µg/m^3
SPM	2.6	16.8
SO$_2$	0.4	2.20
NO$_x$	0.3	1.94

The maximum concentration of 16.80 µg/m^3 of SPM is found in W direction of the plant site at a distance of about 1.0 km the isopleths of SPM are shown in Fig. The maximum predicted concentration when superimposed over the baseline value (109 µg/m^3) results in an ultimate ambient SPM levels of about 125.8 µg/m^3, which is well within the limits of CPCB.

Similarly the maximum concentration of 2.20 µg/m^3 of sulphur dioxide is found in W direction of the plant site at a distance of about 1.0 km. The maximum predicted concentration when superimposed over the baseline value (11µg/m^3) results in an ultimate ambient sulphur dioxide level of about 13.2 µg/m^3 which is well within the limits of CPCB.

11.6.4.2 Predictions of Impacts of Water Environment

Water Sources and Requirement

The estimated water requirement for the proposed industry is 171m^3/day, of which 117 m^3/day is used for process. The remaining 64 m^3/day amount is for other utilities like floor washings, chilling plant blow down, DM plant regeneration, etc.

The water required for the plant is met from the River Krishna flowing around 2 km east of the project site.

Wastewater Generation

The various sources of wastewater in the plant are process, floor washings cooling blow down and domestic wastewater. The quantities expected from various processes are in Table 11.12.

Table 11.12 Quantities expected from various process

S.No.	Process	Phase – 1 m^3/day	Phase – II m^3/day	Total m^3/day
1.	Process water	112	11	123
2.	Floor washing	5	1	6
3.	Chilling plant	1	0	1
4.	DM plant regeneration	2	0	2
5.	Boiler blow down	6	0	6
6.	Service water and potable water	4	0	4
	Total	130	12	142

Effluent characteristics of all process being sent to Effluent Treatment Plant (Table 11.13).

Table 11.13 Effluent characteristics of all process

Stage no.	Water kg	In. org. kg	Org. kg	Wastewater Kg	Fixed solids mg/l	Volatile solids mg/l	TCOD Mg/l	BOD mg/l	'pH
FAII B	17860	60	80	18000	3330	4440	6670	1990	6.0
FAIVB	14837	70	93	15000	4670	6200	9300	2780	6.2
FAVIIIB	23946	54		24000	2250				6.5
Phase– 1				57000	3227	3033	4553	1360	
TAPSIIIB	190	1	9	200	5000	45000	67500	20250	6.5
PABGAVIB	146	1	3	150	6670	20000	28000	8400	7.2
PABGAIXB	695	5		700	7142				7.1
Phase- II				1050	6666	11428	16857	5057	
Phase 1&II				58050	3289	3184	4775	1426	

Treatment of Wastewater Streams Coming from Other Utilities

The effluent coming from auxiliary units like Chilled water blow down, floor washing, boiler blow down and DM plant regeneration 2nd and 3rd wash are sent to collection pit and later used for green belt development within the plant premises. The domestic waste is sent to septic tank followed by soak pit. The DM plant regeneration 1st wash wastewater will be sent to multiple effect evaporator for further treatment.

Prediction of Water Impacts

Hydrogeology

No effect on the level of the water table is envisaged as water required is only 171 m^3/day, and most part of this is acquired from River Krishna. Part of the treated effluent is used within the plant premises for greenbelt development after meeting the onland disposal standards. However, for green belt development the water pumped from the borewell along with treated wastewater is used. The effect on the ground water due to percolated water from green belt development will be insignificant as it contains only nutrients and fertilizers used for green belt development.

EMP during Construction Phase

In the present project though the potential for environmental pollution during construction phase is meager but control of pollution is of considerable importance, the following factors require control during construction phase.

Sanitation

The construction site should be provided since it is a repetition with sufficient toilet facilities for workers and the waste should be sent to a septic tank to ensure minimum environmental impact.

Air Environment

The engine exhausts from construction traffic, dust and other sources of emission may affect air quality during construction phase, but they will be kept to minimum levels. Both gasoline and diesel driven construction vehicles should be properly maintained to minimize exhaust emissions.

Construction Equipment

Proper care should be taken in the installation of equipment for heavy foundation jobs and movement of heavy construction machines.

Noise Environment

The noise effect on the nearby inhabitants due to the construction activity will be negotiable. However, it is advisable to use noise protection equipment like ear muffs etc., by workers when operating such equipment.

Socio-economic

Employing the local labour from adjoining villages during construction phase will have significant beneficial impact.

EMP during post construction phase

The major pollutants during the operation of the plant will be in the form of particulates, liquid effluents, noise and solid wastes.

Air Environment

The source of air pollution from the plant is due to the emissions SO_x and SPM from the coal fired boilers. Based on the mathematical modeling a stack height of 30 m has been proposed. The results of modeling predict that for a stack height of 21 m the ground level concentrations of SO_2 are well within the limits.

In order to control suspended particulate matter (SPM) the management has proposed to install a multi cyclone system (Table 11.14). An amount of Rs. 2.5 lakhs has been allocated for the cyclone and ID fan system.

Table 11.14 Control of Suspended Particulated Matter (SPM)

Pollutant	Base line environment $\mu g/nm^3$	Increase due to proposed activity $\mu g/nm^3$	Resultant environment $\mu g/nm^3$
Suspended particulate matter (SPM)	130		
Without pollution control equipment		78.3	208.3(500)
With pollution control equipment		17.5	147.5 (500)
Sulphur dioxide	20	4.28	24.28 (120)

All the above values represent the 8 hours average concentrations while the value in parentheses is that of limits prescribed by CPCB for industrial area (24 hrs. avg. values).

11.7 Preparation of EIA of Land Clearing Projects

11.7.1 Introduction

Land clearing (LC) projects by their very nature have profound environmental effects, normally covering extensive areas. The key to proper control of the adverse environmental effects of land clearing is appropriate land use planning seen as a multidisciplinary activity. It is therefore essential that whatever plans are proposed for land usage, they must take into account the actual physical character of the land itself. Plans and decisions must be based on factual data concerning all landforms present in the proposed project area and the environs as each relates to the feasibility and efficiency of the project operations; "armchair" planning without sufficient attention to site conditions is likely to result in project failure.

An essential element in land use planning is site selection. It is important that site selection should include a narrowing-down process that involves use of increasingly detailed data/criteria based on land capability so that effective use of the selected site is optimized. Closely related to this is the need for a detailed benefit/cost analysis that clearly shows whether realistic projections of "after project" benefits outweigh the expected project costs and the "before project" beneficial uses of the area to be cleared.

Furthermore, in many cases there is a preconceived notion on the part of project proponents that any piece of land set aside for land clearing for agriculture must be usable for food crop production. In such cases it is often not realized that the less productive a land area is, the more technical, managerial and financial inputs will be required. This can lead to unexpected and sometimes unrealistically high burdens on the government.

In the following discussion of the environmental impacts of land clearing for agriculture, emphasis is given to two broad land classifications: upland forests and swamp land. However, the general impacts will be similar for all forest sites.

The advent of LC projects, of course, open the doors for a multitude of agricultural activities like irrigation and fishery projects.

11.7.2 Environmental Impacts of Land Clearing in Upland Forests

Discussed below are some of the typical effects of land clearing on environmental parameters in upland forest conditions.

Physical Resources

(a) *Adequacy of data*: A check should be made on the adequacy of data concerning soil conditions, climatic variables, terrain, socio-economic conditions and so on to provide the basis for proper site selection and for long-term project operations particularly as concerns the ability to sustain agricultural activity. If critical data is absent it may be necessary to delay the project implementation until such data is in hand.

Special attention should be paid to the impacts of the previous operations owing to proper or improper planning, design and operations, and modifications made to the project based on the information. Provision should be made to store all information in a data base of indicators that the government and others can use for planning and decision-making for this and future projects.

(b) *Water hydrology*: A drastic change in the ecosystem from forest to agriculture will cause significant changes in the hydrology of surface and groundwater.Extensive road construction for new communities and for access to the new agricultural land can significantly affect both the surface and groundwater hydrology, as can water consumption for domestic use and irrigation. The "before" and "after" water flow rates, volumes, seasonal variations and normal flood and drought year flows should be described.

(c) ***Water quality***: Adverse impacts on water quality may result from logging operations, including road construction, clearing of ground vegetation, disposal of human and domestic wastes in the new communities, and application of fertilizers and pesticides during agricultural operations. Special attention should be given to riparian zones with provision of buffer areas along the zones where disturbance of existing vegetation is minimized or prohibited. The existing and expected water quality in waters to be affected by the project should be described.

(d) ***Soil fertility, erosion and sedimentation***: This parameter is most affected by land clearing and subsequent agricultural operations. Soil classification, erosion, stability, texture, bulk density, waterholding capacity, porosity, soil chemistry and fertility should be described. Soil management/conservation measures that will be taken during land clearing operations and when agricultural activities are operational, should also be described.

Soil suitability assessment should clearly state what land clearing methods are assumed: mechanical semi-mechanical or manual clearing. For example, upland rain forest sites that are initially assessed as marginally suitable for sustained agriculture may be wholly unsuitable if the project planning and design call for mechanical clearing.

Full mechanical clearing of fragile upland sites is not recommended because severe erosion, compaction and often removal of the humus-rich topsoil can severely affect inherent soil fertility and thus crop yields. When heavy equipment is used, work should normally be restricted to the dry season. Appropriate drainage is another main factor to be considered in soil fertility. Construction of minor diversion ditches to carry run-off to natural drainage channels and construction of contour ridges are often necessary to minimize soil erosion. Provision should be made for planting and adequate maintenance of a leguminous crop cover immediately following the land clearing to prevent the soil erosion and weed development.

Ecological Resources

(a) ***Aquatic biology and fisheries***: Fisheries specifically, and water ecology generally, can be affected by (i) erosion/siltation both during the land clearing and during agricultural operation, (ii) introduction of pesticides and fertilizers to the waterway from crop land run-off, and (iii) heating of streams where adjacent vegetation has been removed. The existing fisheries, aquatic ecology and conditions of other aquatic fauna/flora as well as the anticipated effects of the project on their values should be described. The impacts of large and clearing projects in upland areas may reach estuarine systems for downstream, and so the effects on these systems and estuarine fisheries will also need to be described.

(b) ***Wildlife***: The land clearing for agriculture must consider the wildlife parameter in two ways (i) the project's impact on existing wildlife population, and (ii) wildlife's potential impact on agricultural activity. In the former case, wildlife species likely to be affected by the land clearing should be listed and those species that are of regional/national/international significance should be identified. For significant species, habitat requirements, their behavioral characteristics, and the effects the project will have on these parameters should be described. Potential mitigation measures like declaring surrounding forests as wildlife reserves should also be described. In the other case, wildlife species such as rats, wild pigs and birds that could become agricultural pests should be identified and measures be taken to decrease the potential for crop depredation by creating buffer zones between new crop land and remaining forests.

(c) *Forests*: The forest resources existing in the area along with their regional and national importance should be described. It should be determined whether the project siting has taken into account regional and/or national plans for forest conservation and utilization, and whether the project will contravene plans for minimum forest area/types that should be maintained for long-term regional welfare. A description should be made of how timber is to be harvested and whether the proposed timber processing will make maximum economic use of this resource. Careful attention should be given to how second holdings, i.e., those forests not initially cleared by the concessionaire, will be handled; if the new land holders clear these areas themselves, valuable timber may be lost to burning. The land clearing for agriculture may open access to nearby forests not targeted for clearing, and so provisions for protection of these forests from encroachment should be described. The provision of appropriate training for farmers in efficient soil conservation and cultivation techniques should also be described.

Human Use Values

(a) *Water supply*: The loss of forest cover may adversely affect water supply to the new settlers as well as to established downstream users. The impacts may include: deterioration of water quality owning to erosion, addition of pesticides and other chemicals and human/domestic wastes from the new communities, and disruption in the periodicity of water flow. Conditions before and after the project should be described.

(b) *Navigation*: The log transport or improper felling operations can block traditional navigation routes.

(c) *Flooding*: The loss of forest cover can result in hazards from increased flood peaks and overland flows.

(d) *Land uses*: Particular attention should be paid to whether the project will infringe on other existing or planned dedicated land- uses such as, mining, recreation areas, parks and wildlife preservation zones. The kinds and levels of traditional forest uses by members of the existing nearby communities should be described.

Quality-of-life Values

(a) *Socio-economics*: For the land clearing projects that will be sited near the existing communities, the socio-economic conditions around the project site should be fully described and analysed based on before/after and with/without project scenarios. Some parameters to be assessed for both the existing and the new communities include population structure, population dynamics, land-use/settlement patterns, labour and employment structure, economic production, income distribution and social organization, cultural characteristics; and social institutions. Such an evaluation can help to identify the potential social conflicts and suggest measures for mitigation or resolution. If the indigenous forest dwellers are to be relocated in view of the project development, a description will be needed to show how this will be done and what rehabilitation measures have been provided. A description will also be required of expected infrastructural problems owing to the remoteness of the project site, high transportation costs for marketing of products, and processing of farm products, so they can be stored for long periods.

Major socio-economic impacts can be expected to include one or more of the following (i) social conflicts between the existing communities and the immigrants for the new agricultural land, (ii) infrastructural development for new communities, (iii) losses of traditionally-utilized forest products, such as, firewood, medicinal/food plants, and food from wildlife hunting, and (iv) beneficial impacts on income production from employment during land clearing and from agricultural production.

(b) *Human health* Threats from insect vectors of human diseases are usually minimal when jungle is completely cleared of vegetation. However, if jungle areas remain along the perimeter of the project area, new communities could be faced with serious threats of diseases like malaria. Plans for control of vector-borne diseases should be described along with plans for provision of adequate community sanitation facilities.

(c) *Recreation*: The impacts on existing recreational use of the project area should also be identified.

11.7.3 Environmental Impacts of Land Clearing in Swap Lands

The above discussion concerning the land clearing impacts on upland sites will generally apply to swamp lands as well. Presented below additional considerations of particular concern when reclaiming the swamp lands for agriculture.

Physical Resources

(a) *Water hydrology*: The effects of the swamp land clearing on water hydrology can play a significant role in the success of subsequent agricultural activity. Lowering of the water table because of land reclamation and drainage can cause oxidation of the potential acid sulphate soil often found in swamp lands, thus lowering the pH value, and may also deprive crops of water requirements fluctuations. The water table may be very severe causing floods during the rainy season and low water table levels during the dry season.

(b) *Navigation*: Generally, land clearing for agricultural development of swamp lands will increase navigational opportunities for local communities. However, remote farms may be cut off from main transit routes and therefore consideration should be given to construction of small transit harbour facilities along the rivers/canals and development of arterial waterways to connect these with remote farms.

(c) *Flood control*: In reclaimed swamplands, areas can be flooded by stagnant rain water, river floods and high tides. Land level lowering owing to peat soil subsidence could impede drainage of low-lying back swamps. Proper dredging and planning of drainage canals and reclaiming the area around canals will reduce the possibility of flooding.

(d) *Aquaculture*: The clearing of mangrove forest can affect breeding stocks on which aquaculture projects rely.

Quality-of-life Values

(a) *Socio-economics*: Constraints to crop management (and also income to be derived from the project) in tidal swamp land result from problems of soil acidity, soil salinity, low diversification of vegetation, lack of fresh water, drainage and so on. The development pattern should consider (i) The diversification of crop plans to accelerate the reaching of equilibrium of components in the agricultural system, (ii) The choosing of crops in accordance with their suitability to chemical, physical and environmental characteristics of the land, and (iii) avoiding the land difficult to cultivate. Consideration should be given to introducing agroforestry for diversification.

(b) *Public health*: Swamp insects are vectors of human diseases particularly mosquitoes which transmit malaria. Well water during the dry season can become acidic or contaminated by bacteria from human wastes. The expected impact of the project in altering hazards of water-related diseases, such as, dysentery, diarrhoea and skin diseases should be described, as there should be plans for provision of adequate community sanitation facilities and plans for control of vector-borne diseases.

11.7.4 Planning and Management Requirements for EIA in Land Clearing Projects

11.7.4.1 Review of Overall Project Size and Purposes

(a) *Optimum area extent*: Feasible alternative area extents/sites should be described and the process of analysis leading to area extent, including a summary of environmental impacts associated with each alternative site (see item (c) below), should be reviewed.

(b) *Alternative uses of storage*: Feasible alternative uses of the agricultural/irrigation waters to be made available by the project should be described and the rationale for the final selection should be reviewed.

(c) *Alternative project locations*: A description of feasible alternative locations for the LC project, and the reasons for selection of the recommended site, including considerations of environmental effects, should be made.

11.7.4.2 Management for Achieving Comprehensive Multiple Use

One of the most difficult problems with LC projects, from the environmental point of view, is the difficulty in achieving the project planning design and management, which will achieve the optimal balance of beneficial uses, irrigation water supply, aquaculture, watershed management/reforestation/regreening/wildlife protection, use of the project for stemming out migration and so on. The primary reason for this is that the implementation of most of the LC projects are assigned to a single implementing agency (such as that for agriculture) which simply is not equipped to handle all the other issues. This means a new approach is needed to planning, designing and implementing of the LC projects in the developing countries. The new approach is expected to establish co-ordinating mechanisms to make the projects multipurpose with optimal benefits to the people. The EIA can serve a very valuable purpose by delineating the needs for co-ordination, so that the decision-makers can grasp the importance of these mechanics and realize their aims.

The importance of achieving understanding by decision-makers of the optimal multipurpose potential can scarcely be over-emphasized. It is precisely when a major investment project is being formulated that attention must be paid to funding for all "secondary" but essential economic-cum-environmental parameters, in addition to the primary objectives which stimulated initial interest in the project. This is the "golden opportunity" from the environmentalist's point of view. Trying to get attention for funding after the fact is almost always very difficult and perhaps hopeless.

11.7.4.3 Environmental Management Measures and Monitoring

The major objective and benefit of utilizing an EIA in project planning is to prevent avoidable losses of environmental resources and values as a result of environmental management. Environmental management includes protection/mitigation/enhancement measures as well as monitoring. Environmental management may require revision of the project site or operation to avoid adverse impacts. More often environmental management requires additional project operations sometimes not incorporated in the conventional operations. An example is incorporating watershed management as part of a land cleaning and dam/reservoir project for agricultural development. Monitoring is required to evaluate the success or failure (and consequent benefit or losses) of environmental management measures and subsequently to re-orient the management plan. Regardless of the quality of an EIA and consequent environmental management measures, they are of limited value unless implemented. Even with the authority to delay a project until approval of an EIA is obtained, there is frequently no assurance that the environmental management measures prescribed will be implemented. It is essential that detailed monitoring programmes be designed for appropriate projects (this design should be prepared as a part of the EIA study and should be presented as a major component of the report including the detailed monitoring work plan, reporting procedure and manpower and cost budgets) and that regular monitoring reports be submitted to environmental agencies. In many cases the environmental agency will have to rely on its own monitoring, to "monitor the monitoring" as well as to monitor implementation of management measures. Monitoring procedures are well developed for most environmental resources and if carried out and analysed correctly, the results will allow the determination of the level of compliance with environmental management requirements and, among other important benefits, will allow evaluation of the cost effectiveness of the requirements.

When these procedural needs are fulfilled the EIA planning tool is put to use in a much more effective manner, and a benefit analysis will be possible, which will determine how successful the EIA process is in preventing or minimizing environmental degradation. Even then, the measurement will be in terms of environmental values, most of which can be quantified in monetary values. However, the evaluation of costs and benefits of environmental values is important for decision-making, especially in regard to the decision for the requirement of environmental management measures based on whether or not their costs justify their benefits. The EIA report should provide, along with well-defined management measures, work-plans and budget requirements, a concise summary of evaluation which displays to the decision-maker the savings in environmental values that will be gained by expenditures on environmental management.

Due to increased experience in using the EIA process for environmental planning in many developing countries, the need and justification for continuous monitoring for establishing meaningful databases have become very obvious. The importance of performing adequate baseline surveys during the EIA studies, and project post-construction monitoring to check actual impacts on environment, implementation and effectiveness of management measures, is gradually gaining recognition. It is also increasingly recognized that the effective way to get funds for the management measures including monitoring, is to include these funds as an integral part of the project budget. This should be done at the time when the project is being approved and funded because the funds are seen as a small part of the overall project budget, but important for maximizing benefits from the project. Trying to get funding after budget allocations are made, is usually very difficult.

11.7.4.4 Mitigation Measures

The use of mitigation measures to offset unavoidable impacts of the LC projects is a practice which is frequently applied in the industrialized countries but of which there is little experience in the developing countries. It is important to understand that planning for mitigation measures should not be an activity independent of the EIA process but rather that mitigation measures should be planned within the context of the overall environmental management plan. Therefore, when one discusses the environmental management planning for a development project, mitigation measures, as well as enhancement measures and monitoring, are automatically components of the planning activity. Once appropriate mitigation measures have been identified, the cost of implementation must be incorporated in the project cost/benefit analysis (along with the environmental benefits).

Mitigation measures vary widely for the LC projects depending on the types of impacts and value of affected resources. Some fairly typical mitigation measures applicable to the LC projects (based on measures for water resources development) are described below:

(a) If the storage reservoir for agricultural purposes, as well as cleared area, results in loss of valuable downstream fisheries and this loss will not be offset by the future reservoir fisheries, then a downstream fisheries development scheme including provision of technology, infrastructure and marketing may be required to mitigate the loss.

(b) If clearing the area involves destruction of valuable wildlife habitat, it may be necessary to designate an area away from the project as a wildlife sanctuary. The mitigation measures would not only include the designation of the new sanctuary but also provide for technology and budgets required to manage the area so that equal wildlife benefits can be achieved.

(c) If the LC results in degradation of water quality (owing to pesticides run-off), or reduced flows of traditional downstream water supplies, it may be necessary to mitigate the loss by providing groundwater development or alternative water supplies.

11.7.4.5 Guidelines for Evaluating Typical Impacts of LC Projects in Forest Areas

Physical Resources

Hazard of soil erosion loss without proper refacing, resulting in impairment of downstream water use values as noted below:

(i) Hazard of soil fertility loss from physical stresses in clearing and leveling.

(ii) Loss of rainwater infiltration which normally occurs under forest conditions.

(iii) Micro-effects on increasing temperature (importance for resort areas).

Ecological Resources

(i) Loss of forest resources associated with the wildlife habitat.

(ii) Encroachment hazards for nearby forests stemming from agricultural development.

(iii) Hazards from pesticides and other agricultural toxics of forest ecosystems in vicinity.

Human Use Values

Impairment of downstream water quality and of beneficial water uses from lilt runoff, including community water supply, fisheries, etc., besides sedimentation and flooding hazards.

Quality-of-Life Values

(i) Loss of forest tourism/aesthetic values.
(ii) Hazards of impairment of downstream water quality/aesthetic values.
(iii) Disruption of local forest is hazardous to population and socio-economics.
(iv) Insect vector disease hazard to farmer population.
(v) Increased disease hazards due to increased population densities.

11.8 Assessment of Impacts of Traffic and Transportation

Introduction

The evaluation of traffic and transportation impacts is closely interrelated to the assessment of land use, social, economic, air quality and noise effects.

Road development can have wide-ranging environmental impacts compared to many other development projects. This is because roads extend over long distances and, by promoting rapid communication, they can catalyze dramatic changes in land- use patterns not only in the immediate vicinity but also in adjacent hinterlands.

Projects or Actions required to be studied for Impacts on Transportation

The following are examples of types of proposed projects or actions where traffic impacts may become a key issue in the environmental impact assessment:

- Land-use or comprehensive plans.
- Proposed highway or transit improvements.
- Projects that attract large volumes of traffic such as shopping centers, amusement. parks, schools, convention centers, parking structures, or municipal buildings.
- Major event venues or employment centers.
- Housing developments.
- Changes in bus or parking rates in major urban areas.
- Individual projects that may block or render unsafe pedestrian and bicycle travel or access for the handicapped.

Projects with geographically extensive or long-duration construction periods may also create adverse traffic impacts. In cases where an impact is expected, a maintenance of traffic plan should be prepared by the project designers. The maintenance of traffic plan will describe staged construction activities, detour routes, signing, and other measures to lessen the impact on traffic during construction.

For projects requiring construction-related or permanent truck or heavy equipment access, haul routes should be identified in advance. Mitigation measures, such as, dust control, air quality control, or excise control through limiting hours of operation should be identified and assessed for successful mitigation of impacts to acceptable levels.

Development projects or other activities frequently have impacts on local and regional traffic patterns and transportation systems. The conceptual approach depicted in Figure.8.A.1 can be applied.

Step 1 Identification of Potential Traffic and Transportation System Impacts

The first step is to determine the potential impacts of the proposed project on local traffic and/or the transportation system in the ROI. Examples of the key transportation impacts which might occur, include 1. increases or decreases in local-area or regional traffic situations, 2. temporal changes in local-area or regional traffic situations (daily, weekly, monthly, and/or seasonally), 3. construction phase disruptions of existing local-area or regional traffic patterns, and 4. increases or decreases in commuting times and congestion in the local area and/or region.

Quantitative information should be aggregated on expected local and regional traffic change (increases and decreases) which might occur as a result of the construction and/or operation of the proposed project. Particular attention should be given to the timing (daily, weekly, monthly, and/or seasonally) of the expected changes. It is anticipated that the project proponent (or contracted proponent) would have such information, or, if no such data exists, this information could be developed during discussions with the project proponent.

Step 2 Documentation of Baseline Traffic Information

Certain basic information on the traffic and the transportation system in the vicinity of a proposed project or activity is necessary for describing the affected environment or baseline conditions. Key information includes the following: (1) the type of transportation network and its use, (2) the type and purpose of traffic using the network, and (3) the character of traffic flow for example, periods of maximum and minimum use. This information can be assembled for the majority of cases by:

1. Procuring from the appropriate governmental engineering staff the necessary maps showing the locations of all paved and unpaved roads in the study area. In addition, traffic count information, if available, should be procured from the appropriate governmental engineering staff and local, regional, or national transportation agencies.

2. Making site visits to the study area and collecting ad hoc data on traffic counts for pertinent roads, streets, and highways, such counts should be focused on the peak and minimum periods of usage of the network.

The information necessary to accomplish step 2 is assumed to be readily available or easily obtainable.

Step 3 Procurement of Pertinent Standards or Criteria

Traffic Analysis

Vehicular traffic on streets and highways can be assessed by using various standard traffic analysis procedures. Most projects expected to produce traffic impacts will, at the least, require a description of existing, and perhaps historic, traffic volumes. Flow characteristics will change in the future both with and without implementation of the proposed project or action. In other words the analysis should discuss in equal measures of detail the future traffic characteristics of all proposed alternatives including the no-built alternative, which will be used as the baseline for comparison with the proposed build alternatives.

Volumes and Levels of Service

Traffic volume data is normally available from the state department of transportation, local planning agencies, or regional metropolitan planning organizations. Raw data is gathered by actually counting traffic volumes throughout the hours of a day at a particular point of a street or highway. Traffic volumes are normally reported as average daily traffic (ADT) and morning and evening peak-hour traffic.

Certain physical characteristics of streets and highways dictate a calculated traffic capacity for that particular facility. Examples of the features entering into the capacity analysis include the lane width, number of lanes, shoulder width, grade (slopes of hills), radii of curves, type of access permitted, and distance between ramps of traffic signals. The type of access permitted can be divided into several categories. There are three basic types of access:

1. *Free access*: at-grade intersections, adjacent joining driveways, and left or right turns possible.

2. *Controlled access*: streets or highways with medians that only permit crossing at designated places, and only right turns on to or off the street permitted except in designated areas.

3. *Limited access*: the freeway, expressway, or turnpike facility where crossings of other highways and streets are grade-separated and access is limited to free–flow interchanges.

An at-grade intersection is the typical stop sign or signal light at ground level. A grade-separated crossing occurs where the crossing street is carried over or under the expressway via a structure (bridge). A free-flow interchange provides ramps onto and off a freeway or expressway without requiring the vehicle to stop at the freeway or expressway. Common interchange designs include the diamond and cloverleaf, but there are many varieties, each with specific advantages and disadvantages retarding traffic flow and capacity.

Capacity can be determined for the mainline of the expressway or street and for intersections and interchanges. Capacity analyses for intersections and interchanges are more complicated because characteristics, such as, the number of left-turn lanes, timing of signal red and green cycles, and timing of nearby signals must be factored into the analysis.

Level of service (LOS) is a qualitative measure to describe the flow or operational characteristics of traffic, as perceived by the level of congestion or delay experienced by the motorist. The level of service is a result of the transportation facility's capacity, or ability to accommodate the volume of traffic on the facility. Also, the LOS can be affected by the characteristics of the traffic itself, such as, the percentage of trucks in the total traffic. The levels of service are as follows:

LOS A represents a free flow of traffic. Individual users are unaffected by others and have the freedom to select desired speeds and to maneuver within the traffic stream.

LOS B is in the range of stable flow, but the presence of other users begins to be noticeable. Freedom of speed is unaffected, but there is a slight decline in freedom to maneuver.

LOS C is in the range of stable flow, but marks the beginning of the range where individual users become significantly affected by interactions with others, hindering selection of speed and maneuverability. The general level of comfort and convenience declines noticeably at this level.

LOS D represents high-density but stable flow. Speed and freedom to maneuver are severely restricted, and the driver experiences a generally poor level of comfort and convenience. Small increases in traffic volumes will generally cause operational problems at this level.

LOS E represents operating conditions at or near capacity. Speeds are reduced to a low but relatively uniform level, and maneuvering is extremely difficult. Comfort and convenience levels are extremely poor, and driver frustration is generally high. Operations at this level are usually unstable because small volume increases or minor fluctuations will cause breakdowns.

LOS F represents forced or breakdown flows, and these flows exist when the amount of traffic approaching a point exceeds the amount which can traverse the point. Queues form behind such locations, and the extremely unstable operations are characterized by stop-and-go waves.

Level of service is governed by traffic density, measured in passenger cars per mile, per lane. The density is converted to passenger cars per hour, per lane using the average speed of the traffic stream. Generally, levels of service A, B, and C are considered good operating conditions with only minor delays. LOS D represents fair to below-average operating conditions, but is sometimes acceptable in urban areas. Levels of service E and F represent extremely congested conditions.

Level of service for traffic analysis of signalized intersections is defined in terms of delay. LOS criteria are stated in terms of the average stopped delay per vehicle for a 15 minute analysis period.

Table11.15 summarizes information on a six category "Level of Service" (LOS) delineation used by the U.S. Transportation Research Board. The LOS for a highway, for example, is a qualitative measure of the effect of a number of factors, including speed and travel time, traffic interruptions, freedom to maneuver, safety, driving comfort and convenience and operating costs. If impacts on local or regional highways are anticipated, it would be appropriate to determine the LOS classifications for the highways in the study area. In addition to the LOS system, local roads and streets in the study area may have been classified by local or regional traffic or transportation authorities, or even by the engineering section of a military installation. The delineation of these classifications would also be appropriate in step 3.

Table 11.15 Levels of Service

The level of service concept

The definition of "level of service" is "a qualitative measure of the effect of a number of factors, which include speed and travel time, traffic interruptions, freedom to maneuver, safety, driving comfort and convenience, and operating costs." It goes on to indicate that "in practice selected specific levels are defined in terms of particular limiting values for certain of these factors." Service level A through F represent the best through the worst operating conditions.

Level of service A represents virtually free-flow conditions, in which the speed of individual vehicles is controlled only by the driver's desire and by prevailing conditions, not by the presence of interference of other vehicles. Ability to maneuver within the traffic stream is unrestricted.

Level of service B, C and D represent increasing levels of flow rate with correspondingly more interference from other vehicles in the traffic stream. Average running speed of the stream remains relatively constant through a portion of this range, but the ability of individual drivers to freely select their speed becomes increasingly restricted as the level of service worsens (goes from B to C to D).

Level of service E is representative of operation at or near capacity conditions. Few gaps in traffic are available. The ability to maneuver within the traffic stream is severely limited, and speeds are low (in the range of 30 millions). Operations at this level are unstable, and a minor disruption may cause rapid deterioration of flow to the level of service F.

Level of service F represents forced or breakdown flow. At this level, stop-and-go patterns and waves have already been set up in the traffic stream, and operations at a given point may vary widely from minute to minute, as also operations in short, adjacent highway segments, as congestion waves propagate through the traffic stream. Operations at this level are highly unstable and unpredictable.

Steps 4 and 5: Prediction of Traffic and Transportation – System Impacts and Assessment of Impact Significance

Step 4 requires the consideration of the changes in terms of increase or decrease of timing in the baseline traffic conditions in the ROI as a result of the construction and operational phases of the proposed project. The basic mathematical relationship for this step is as follows:

Percentage change in baseline conditions =

Percentage changes can be calculated for each pertinent local or regional road or highway and for each project or activity phase. For example, assume a local road has a baseline average daily traffic (ADT) of 1,000 vehicles, with the peak-hour traffic being 250 vehicles. Further assume that the project-construction phase of 6 mo will add 200 (vehicles) to the ADT, with 150 being associated with the peak hour. The project operational phase will add 75 to the ADT, and none of these vehicles will be associated with the peak hour. The percentage changes are calculated below.

Construction Phase

Percent change in ADT =

$$= 20\%$$

Percent change in peak hour =

$$= 60\%$$

Operational Phase

Percent change in ADT =

$$(100) = 7.5\%$$

Percent change in peak hour =

$$= 0\%$$

Forecasts

Future traffic volumes can be predicted in a variety of ways. Regardless of the method used, it is extremely important to the environmental analyst that the exact assumptions and methodologies be documented. There should be a review and an agreement by local and state traffic specialists that the applied approach is acceptable and reliable. Results of traffic studies are subsequently used in the analysis of land-use, neighborhood, air quality, and noise impacts. If there is a justified question concerning the development of forecast future traffic volumes, conclusions in these other areas of impact analysis also are in doubt. Methods for forecasting future traffic volumes have become significant issues of controversy in a good many projects.

A relatively simple method sometimes used to predict traffic volumes involves reviewing historic data on traffic growth rates for a particular transportation facility or area and then predicting the future growth rates. The predicted future growth rates may depend on predicted employment or population growth contained within a regional or local comprehensive plan. The future growth rates are then applied to existing traffic volumes to arrive at future volumes for the target year of analysis.

More detailed methods can include the systematic division of an area into sectors or zones and then application of population and employment growth predictions to each sector. Travel origin and destination studies can be done for existing traffic through questionnaires and surveys, and future conditions can be estimated. Predictions are then made to the number of trips originating in a particular zone and traveling to another zone. Computer models are used for the complicated data input required.

For individual site analysis, estimates of potential generated traffic can be made by gathering information on the number of future employees, expected number of shoppers at retail centers, number of potential attendees at major sporting events and so on so that total traffic volumes in future years may be predicted.

For the evaluation of comprehensive and land-use plans, the permitted density and dispersion of various land- uses will determine the generated traffic and should yield estimates of rates of growth.

Since the basic output for step 4 is the percentage change in information in relation to baseline traffic conditions, the next step is focused on how to interpret this percentage change in information (step 5). No transportation criteria or standards provide a delineation of appropriate interpretation method; however, the absolute changes and the LOS should be given consideration.

Burchell has described a five- component traffic-impact-analysis methodology for development projects (5-7). The components are (a) introduction, (b) analysis of existing conditions, (c) traffic characteristics of the development site, (d) future demands on the transportation network, and (e) impact analysis and mitigation recommendations.

The introduction of the traffic-impact analysis should contain a complete project description, including the proposed land use (or uses), the extent of the development, proposed-site-access points, and phasing plans (9).

The next component, the analysis of existing conditions, should address the current volume of traffic using the roadways and the LOS provided to current traffic. Consideration should be given to specific analysis periods, such as, existing peak-traffic times and the times of peak traffic generated by the development project. Recent traffic counts should be procured from agencies covered. New traffic counts should be taken to fill in those critical locations without acceptable historical counts. New roadway counts are taken with an "Automatic Traffic Recorder" (ATR) for a one-week period (including the weekend). The data is tabulated on hourly bases (by direction of travel) with a 24 hour volume, or "Average Daily Traffic" (ADT) shown for each roadway in the study area (5). In addition to ADT, turning-movement counts at key intersections may need to be taken and monitored.

Results

Predicted future traffic volumes are applied to the future transportation network, which may or may not be changed over existing street and highway characteristics in view of the type of project being assessed. Normally future traffic volumes are given in average daily traffic (ADT) and AM and PM peak hourly volumes for comparison with existing volumes. For each proposed alternative, including the no-build alternative, future volumes should be presented in either tabular or graphical form. Often a line drawing of the street network is used to represent traffic data with volumes shown at particular locations.

The level of service can be calculated by applying future volumes to the transportation network. Traffic impacts of each proposed build alternative can be compared to those of the no-build alternative. The difference will be the impact of the individual project or action.

For example, a particular arterial street operates at LOS C. Future traffic volumes are expected to grow, and with the future year (say, 10 years from the present) no-build alternative, the street is projected to operate at LOS D. The proposed project being assessed is a major employer, with thousands of new jobs located on this particular stretch of street. When the projected employee traffic is added to the total forecast peak-hour traffic volumes, the street will operate at LOS E, indicating a severe traffic congestion impact caused by the proposed project.

Caution is required in interpreting the above example results as generated traffic. The actual impact is generated traffic on that particular piece of street, leading to increased congestion during peak hours. It could be described as a change in traffic patterns. Or perhaps a particular project will "generate" traditional traffic in a particular town. At the regional level, however, there is debate often over whether a major employer, or a roadway improvement, or other types of traffic impact projects actually generate a net increase in traffic, or just redistribute existing traffic into different patterns throughout the region.

The next aspect of the analysis of existing conditions is to determine the capacities and LOS within the study area. "Capacity" is defined as the maximum number of vehicles that can be expected to travel over a given section of roadway, or a specific lane, during a given time-period under the prevailing roadway and traffic conditions. Details on information can then be integrated: for example, Table 11.16 summarizes average capacity of a two lane road expressed as maximum ADT volumes for three levels of service.

Table 11.15 Capacity of two-line road in relation to level of service level of service.

Table 11.16 Capacity of two-line row

Terrain	C	D	E
Level	7,900	13,500	22,900
Rolling	5,200	8,00	14,800
Mountains	2,400	3,700	8,100

Assumes: Peak hour traffic = 10%; 60:40 split; 14% trucks; 4% recreational vehicles; 25 percent no passing (level terrain); 40% no passing (rolling terrain); 60% no passing (mountainous terrain).

Addressing the traffic characteristics of the development site, the third component in the methodology involves developing answers for two questions (9): (1) How much traffic will the proposed site produce? (i.e., what is the trip distributions?) and (2) which roadways will use site-generated traffic? (i.e., what is the trip distribution?)

There are three approaches for aggregating the trip generation information namely, (1) the use of local rates (2) the use of estimates based on the type and characteristics of the project or activity and (3) the use of national rates.

After the site-generated (project-or activity-induced) traffic is estimated, the next activity is to determine the directional distribution of the traffic. For small sites, it is reasonable to assume that the traffic will arrive and depart in a manner similar to the existing travel patterns. Calculations for large sites often require the formation of a detailed distribution model combining elements of population, employment, travel times, highway network characteristics, and competing uses (9).

The three most typically used methods for estimating trip distribution are based on the use of (1) the existing data, (2) the origin – destination data and (3) a trip – distribution model. The first two methods are self-explanatory. A "trip distribution model" (referred to as a "gravity model") assumes that the number of trips between two zones is proportional to the size of the zones and inversely proportional to the square of the distance between the two zones.

Determination of future demands on the transportation network is the fourth component in the traffic-impact-analysis methodology. A "horizon year" must be determined for each phase of proposed development, as well as the subsequent completion, or "buildout" year. The determination of future volumes without the site development (or project or activity) is calculated through the use of (1) growth rates (or trends), (2) the buildup method and (3) the area-transportation plan. The growth-rate method is the simplest to use, and so is most often utilized for relatively small developments or for developments with a buildout for no more than five years into the future. Growth rates (or trends) are determined from historical traffic counts maintained by the appropriate traffic or transportation agencies. In the absence of specific historical traffic counts, growth rates are often indexed to area population growth. For each phase of the development (including final buildout), the existing base volumes are factored upwards by the appropriate growth rate to determine future without-site traffic columns.

The "buildup" method is most appropriately used in an area experiencing moderate to rapid growth. The buildup method combines elements of the growth-rate method with a detailed analysis of approved and anticipated developments within the study area. For each horizon year, the existing volumes are increased by the applicable growth rate. Furthermore, the trip-generation and distribution characteristics of approved and anticipated development, estimated and added to the area transportation plan unusually project traffic volumes on major streets 20 years into the future (this is analogous to the without-project condition). If the proposed development is on one of these roadways, future volumes may be interpolated to the horizon years.

The next activity involves assigning the site- generated traffic to the study-area roadway and intersections, that is, with-project conditions. Finally, for each analysis period being studied, totals for future nonsite and site-related traffic volumes are calculated for the study area. Separate graphics and tabulations of the various components of total future traffic are useful in illustrating site-related changes.

The final component in the methodology constitutes the actual impact analysis and the development of appropriate mitigation recommendations. This component should focus on the LOS with and without the site development. The first activity involves a calculation of the future without project LOS for the analysis periods and horizon years described earlier. After this calculation, a comparison is made of the results with the "acceptable standard" of the community. For those developments not expected to meet the extant community standard, a determination of recommended improvements necessary to achieve the desired LOS should be developed.

The second activity involves the calculation of the future LOS with the development-site traffic. The results should once again be compared with the community standard and with the results of the future-project analysis to identify changes in the LOS caused by the development and additional improvements that may be required. As an alternative to additional capacity improvements, demand-reduction strategies (mitigation measures) may need to be seriously considered. Examples of these strategies include utilization or development of public transportation, car pools, and van pools, implementation of modified work schedules (flextime or staggered working hours), and parking limitations.

Secondary and Cumulative Impacts

As with many other types of impacts, traffic congestion impacts can cause secondary effects, such as the following: Increased noise and air pollution -Adverse visual effects - Delays in emergency vehicle services 1. Loss of patronage to restaurants or retail establishments due to inconvenience of access 2. Increase in motorist accidents and decrease in pedestrian safety 3. Reduction of ability of an area to keep major businesses or to attract new business 4. Changes in travel patterns of through traffic into the neighborhood as motorists attempt to avoid congested portions of major streets. 5. Inconsistency with goals and objectives of local land-use or comprehensive plans. It is important to carefully review the results of traffic studies so as to rectify the possible secondary impacts.

Step 6: Identification and Incorporation of Traffic and Transportation -System Impact Mitigation Measures

Numerous mitigation measures can be employed to offset traffic and transportation impacts. The range of possible mitigation techniques can extend from a regional level to very specific design characteristics of a particular proposed project or action.

"Mitigation measures" in this context are steps that can be taken to minimize the magnitude of the increases in traffic in the RIO. The key approach is either to reduce the traffic or to change the timing of the traffic anticipated to be emitted from the project (or activity). Mitigation measures (1) the use of car or van pooling or buses from residential areas for travel to and from military installations, (2) scheduling construction-equipment movement during nonpeak periods in the local area and (3) scheduling troop movements related to training exercises during nonpeak traffic.

Mass Transportation Systems

Proposed new mass transit systems or bus service, or changes in existing systems, can have effects on vehicular traffic patterns on roadways. Traffic impacts may occur at major transit stations or at park-and-ride lots. If trips are transferred from individual vehicles on the roadway network to trains or buses, the system will work more efficiently and effectively. Proposed projects should be evaluated for design features to encourage mass transit use, such as, location near transit stations or special incentives for employees who carpool.

Changes in parking policies or rates, bus and train schedules or fares also can cause transportation-related impacts. Particularly in urban areas, a portion of the population will be transit-dependent. Raising fares can sometimes unfairly impact special social population groups, such as the elderly, the low-income and the handicapped. Changes in schedules and fares could cause adverse accessibility effects for transit-dependent employees in addition to affecting medical facilities, shopping and visiting trips.

Pedestrian and Bicycle Travel

All projects should be assessed for possible adverse impacts on pedestrian bicycle access and safety. Local and regional land-use and comprehensive plans should include information on existing designated pedestrian trails or walkways and on bicycle routes. A review of the potential area of impact of the proposed project or action can be conducted to identify major routes for nonmotorized traffic.

Traffic Congestion

A direct means to reduce traffic congestion is to increase capacity on the highways or streets operating at poor levels of service, as adding lanes to the roadway. Ramp meters can be installed to control the timing of the flow on to limited-access highways. Synchronization of signals can improve flow on arterial streets, and intersection operation can be improved with specific traffic control measures, such as left turn lanes or signal cycle timing.

Traffic congestion can also be reduced by changing the characteristics of the traffic. They includes staggered work hours at major employers, incentives for carpools or use of mass transit, or provision of high-occupancy vehicle (HOV) lanes on ramps and lanes of major expressways (for use of only cars with two or three passengers and/or buses). Message systems that warn motorists of congested areas to avoid and a program for rapid emergency service for broken-down vehicles also assist in mitigating congestion on major routes. Transportation management techniques can be applied at a local level or within a large region.

11.9 Environmental Impacts of Highway/Road Development Projects

Directions

(a) Physical resources

(i) *Water hydrology*: Highway/Road (H/R) projects that cross waterways can have significant impacts on both the surface water and groundwater hydrology. For example, without the provision of adequate drainage a road can act as a dike separating waters in a stream or swamp, and can possibly lead to increased flood water levels. A change in water hydrology may affect surface water quality as well. Sediment transport, water quantity including alternations in the water, water logging of wells, change in infiltration rates and present stream hydrographies should be described.

(ii) *Surface water quality*: Water quality can be affected during construction and operation of the road. Examples of the former are pollution from runoff and sanitary wastes from construction. Pollution can occur during H/R operation through accidents or spills of transported materials. The effects of pollution on the water's beneficial uses such as community, industrial and agricultural, should be described and evaluated.

(iii)*Air quality*: There are two main sources of emissions during construction namely, mobile sources and fixed sources. Mobile sources are vehicles involved in construction activities. Fixed-source emissions include non-mobile construction equipment like compressors, and demolition/excavation/grading activities which produce dust. During the operation phase, air quality is affected primarily by vehicular exhaust; those pollutants of primary concern include suspended particulate matters No_x, CO, hydrocarbons and lead. Expressways can significantly alter air pollution distribution patterns, resulting in polluted "air tunnels". The existing and expected air pollution patterns along the H/R route should be described according to daily average and maximum conditions. Special attention should be given to "sensitive" areas such as hospitals and adjacent residences.

(iv) *Soils*: Soils are mainly affected through cut-and-fill operations and soil erosion. Inadequate protection of cut and fill areas (for example, with vegetation), inadequate culvert capacity for streams and poor drainage from the road can result in serious erosion problems. This in turn can damage the road, lead to flooding problems and degrade water resources. The type and origin of soil materials to be used in cut-and-fill operations should be described and the amount of soil involved estimated. The amount of erosion expected, its impacts on resource values and erosion control methods during and after construction should be discussed considered in detail.

(v) *Roadways in mountainous terrain*: The construction of roadways in mountainous terrain presents special problems concerning the physical environment. Highly unstable geological conditions are worsened by roadway construction on steep slopes. Landslides can destroy sections of newly-completed roadways. Conversely, roadways require reshaping of land both up and down-slope; consequently erosion potential is increased. Both surface-water hydrology and geotechnical factors should be addressed as critical issues governing the construction of roadways in mountainous regions.

(b) *Ecological resources*

(i) *Fisheries*: Fisheries specifically, and water ecology generally, can be affected by: (a) erosion during both construction and operation of H/R projects, (b) runoff from highways containing petroleum drippage and spilled materials, (c) spills of toxic and hazardous material, and (d) alterations of water hydrology. Increased accessibility may lead to depletion of listing fisheries. Aquatic ecology and conditions of other aquatic fauna/flora, and the anticipated effects of H/R construction and operation on their items should be described.

(ii) *Forestry*: The effects of H/R projects on forestry are primarily caused by (a) site clearance for the road-bed and right-of-way; and (b) improved accessibility leading to encroachment by people. Encroachment may involve villagers searching for farmland or firewood, businessmen in fields such as logging and mining, and illegal operators (especially loggers), and so on. The forest composition, the types and number of trees to be cut down during construction, the estimated loss of forest productivity and the estimated impacts of this loss on sub-national and national levels, should be described.

(iii) *Wildlife*: Wildlife will be affected in a manner similar to forestry, that is, through habitat loss and encroachment (mainly hunting). The wildlife species likely to be affected by the project should be listed, and those species that are of sub-national/national/international significance should be identified. For significant species, habitat requirements and their behavioral characteristics should be described besides showing the effects of the H/R project on these parameters. If possible, there should be an assessment of the intrinsic value of the wildlife resources in the overall national resource context to determine whether alternative routing can be given to preserving wildlife travel routes, especially for such susceptible species as arboreal animals and deep-forest birds.

(c)   Human use values

(i) *Navigation*: H/R projects have beneficial effects on navigation by allowing access to navigable waters. Adverse effects could include blocking traditional navigation routes, that might occur when a road bisects a large swamp. The effects of the project on inland/marine navigation and any compensatory measures should be described.

(ii) *Flood control:* Road development can adversely affect flood control existing flood patterns, and flood control systems. The project's effects on these parameters and possible measures to mitigate adverse effects, should all be described.

(iii) *Land-use*: Land-use patterns can be vertically altered by the effects of H/R projects mainly owing to improved accessibility. For example, forest details may be converted for agricultural use and agricultural areas utilized for industry. The following are the project's effects on the existing land transport patterns:

- Existing agricultural conditions including types and amounts of crops, irrigation practices and marketing practices;

- The types, production capacities, raw materials and markets of industries to be affected by the project, as well as the potentials of new types of industries to be attracted by the H/R projects;

- The types and locations of existing mineral development operations and the project's expected effects, including potentials for new operations.

For all land-use types, the H/R project's effects on environmental and socioeconomic conditions and methods for offsetting any adverse effects should be detailed.

Landscape is a subjective concept that cannot be precisely quantified. It includes a large number of parameters. A study of the relief, vegetation, buildings, hydrograph (water courses), and land division system makes it possible to identify several different landscape units on the site. Each unit is defined as a part of the territory with its own special characteristics (relief, forms of land use, vegetation, buildings, color, etc.) which can be perceived by the eye and enjoyed by the senses. Land-scape units are homogeneous parts of the land-scape which can be defined by such criteria as coherence, readability, hierarchy, harmony, and stability.

Coherence: A landscape is coherent if its various components (e.g., relief, vegetation, buildings) harmonize – if they are aesthetically in keeping with one another. This is a strong feature of truly vernacular land-scapes. Contemporary structures, on the other hand, rarely attempt to relate to their natural setting.

Readability: A landscape is readable if it is easy for the observer to comprehend.

Hierarchy: A landscape with hierarchy is one with a predominant feature.

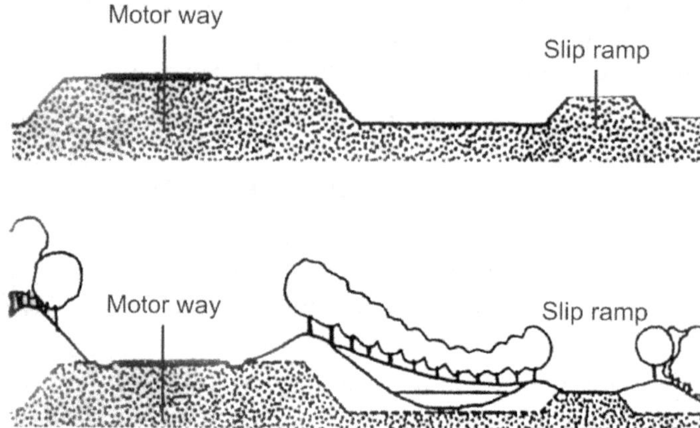

Fig.11.5 Using vegetation to improve harmony between a road and terrain.

Source: Handbook on Impacts of Transportation and Highways. Published by World Bank (1997)

Harmony: A landscape exhibits harmony if there is a relationship in terms of mass and scale between the various components making up the landscape. It aims for maximum overall coherence compatible with the widest possible diversity (Fig. 11.5).

Stability: A stable landscape is one which, although dynamic, retains the same characteristics and qualities through time and space.

Landscape analysis must consider the overall route and integrate sections which have been studied separately, in order to avoid creating a project which appears splintered and lacking in cohesion.

Remedial Measures

Prevention

It is not possible to prevent the presence of a road from affecting the surrounding landscape. Even maintenance and rehabilitation works can change the appearance of a road, for example through the use of vegetation and shaping of the roadside.

Mitigation

The regional landscape design principles should provide guidance in resolving major issues relating to alignment, landscaping maintenance, and the provision of user services.

Alignment

Vertical and horizontal alignment should follow the natural relief as closely as possible within technical constraints such as slopes and radius of curvature.

Curves can accentuates views, while ensuring adequate safety for passing, coming into close proximity slopes on either side of the road can be varied to match the site's natural topography.

Bridges, viaducts, and tunnels can be used across steep terrain rather than high cuts and embankments.

To preserve the landscape's visual and physical continuity. computer landscape illustration may help the road agency to visualize the completed road project within the landscape.

Views from the road can be revealed, composed, or reinforced by road layout and design, but should also take road speed into account

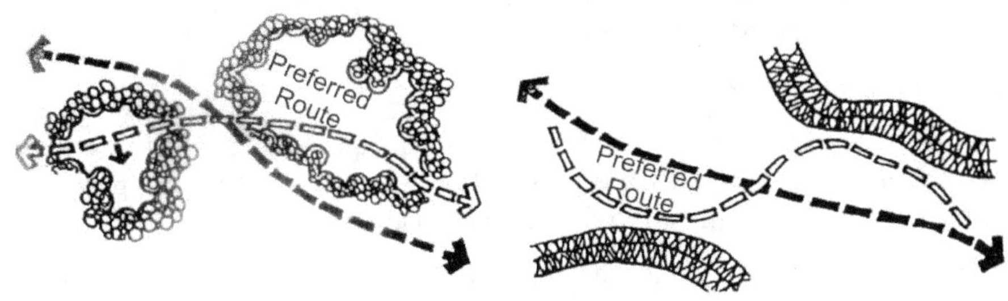

Fig. 11.6 Making the most of landscape features.
Source: Handbook on Impacts of Transportation and Highways. Published by World Bank(1997)

Landscaping proposed for the route should:

Fit in with local vegetation (trees, shrubs, avenue trees, hedges).

Make use of vegetation to harmonize with or improve the existing landscape.

Be representative of the road's category and function.

Take advantage of natural openings in the existing vegetation.

Frame and underscore the various landscape units crossed.

Suit and underscore the various engineering structures.

Ensure user safety by using the landscape to signal changes in the route, for example, by decreasing the space between avenue trees before entering a curve or village; and

Maintenance

Maintenance of roadside vegetation, slopes, and structures can greatly affect visual appearance and can be enhanced by involving maintenance workers in the planning and management of the roadside environment.

Plant indigenous wildflowers and grasses for a low maintenance ("no mow") roadside.

Avoiding the use of too many different types of noise barriers.

Establishing regulations or fines for littering. and

Regulating billboard and storefront advertising along roads, especially at the entrance to cities or towns, to prevent unsightly proliferation and protect road user safety.

User Services

User services made available to motorists along the roadway can help to avoid concerns such as littering or vehicles making indiscriminate stops along the roadway. They also contribute to road safety by allowing drivers to rest or check vehicles and loads during a trip.

The guidelines and essential points to be covered in preparing TORs for EIA of Highway projects issued by Ministry of Environment Govt of India are presented in Appendix II.

11.10 General Guidelines for Preparation of TORs for EIA of Transportation Projects

Some Management Considerations

Transportation projects can have very disruptive effects on forests and wildlife as described in the earlier sections of this book. Careful attention must be paid to site planning so that these disruptive effects are minimized to the maximum extent possible.

Consideration should be given to the maximum use of enhancement and protection measures funded by the project to offset unavoidable degradation. Enhancement and projection measures may include (a) establishing forest reserves to minimize the effects of encroachment; (b) fencing off and/or policing roads and (c) promoting new rural occupations so that villagers will have economic incentives to protect the forest. An example is a rural development project in Thailand that increases village income and enhances forest protection by promoting nature tours. A similar step can be taken with swamp lands, a further alternative is to use engineering techniques to create a new swamp to replace the old. Whenever special enhancement or protective measures are to be recommended for funding by the project, they should be clearly justified in terms of economics and resource conservation, including projections of the forest/wildlife/swamp status with and without the recommended measures.

1. The feasibility study for the project, to be done by the Project Consultant engaged by the Government, should include an EIA (FS/EIA).

2. The FS/EIA should include, inter alia, study of each of the environment effect found by the Banks IEE to be of significant importance.

3. For each of these items, the consultant will conduct a study, as a part of the overall EIA, sufficient:

   (a) To make an assessment, which delineates the significant environmental effects of the project.
   (b) To describe and quantify the effects.
   (c) To describe feasible mitigation measures for minimizing, eliminating, or offsetting adverse effects and
   (d) To recommend the most appropriate mitigation and/or enhancement measures.

4. The selected Significant Environmental Impacts (SEIs) to be studied, as a part of the overall EIA, are the following:

   (a) Environmental problems for the major H & R rehabilitation projects.

      (i) Does review of experience with existing project indicate any significant environmental projection problems? If so, list and grade them.

    (ii) Construction stage (new project):

        1. Hazards of silt run-off during construction

        2. Hazards of continuing silt run-off from areas not properly resurfaced

        3. Other construction hazards (Annex III/I), and

        4. Provision of appropriate construction monitoring.

    (iii) Post- construction operations monitoring:

(b)   Environmental problems for major new highway projects

    (i)   Encroachment on precious ecology

    (ii)   Encroachment on historical/cultural/monument/areas

    (iii)   Impairment of fisheries/aquatic ecology and of other beneficial uses

    (iv)   Erosion and siltation

    (v)   Environmental aesthetics

    (vi)   Noise and vibrations

    (vii)   Air pollution hazards

   (viii)   Highway run-off pollution

    (ix)   Highway spills of hazardous materials

    (x)   Construction stage problems

1. Erosion and silt run-off

2. Other construction hazards

3. Monitoring

    (i)   Post-construction monitoring

    (ii)   Environmental problems for rural roads

    (iii)   Encroachment into precious ecology

    (iv)   Encroachment into historical/cultural values

    (v)   Impairment of fisheries on other beneficial water uses

    (vi)   Erosion and silt runoff

    (vii)   Dust nuisances

   (viii)   Construction stage problem

    (ix)   Post – construction monitoring

5. The above analysis shows that this project has many significant impacts, hence a full-scale EIA is needed.

6. The estimated cost of the overall EIA is approximately man-months of professional input, or which percent should be allocated for use of expatriate EIA expatriate for guiding and supervising the EIA and for transferring technology to the local staff. This

estimate assumes that the EIA will be done as a part of the overall project feasibility study.

7. The estimated time required for the EIA is to months. (From 2 to 12 months, depending on the size and complexity of the study).

8. The total estimated cost of the recommended EIA is or, which about percent is foreign exchange.

11.11 Case Study

11.11.1   Impact Assessment on Soil Erosion due to Highway Construction

Source: Mohan Lal Agrawal, Indian Institute of Technology, Kharagpur, Anil Kumar Dikshit, Center for Environmental Science and Engineering, Indian Institute of Technology, Bombay, Mrinal Kanti Ghose, Regional Remote Sensing Service Center, Kharagpur - journal.

Introduction

The EIA carried out on National Highway (NH-60) from Balasore to Kharagpur, between latitude of 210 22'24''N to 22028'24''N and longitude of 86047'30''E to 87030'04''E, with an approximate length of 109 km. To generate soil erodibility map, roughly 10 km on either side of the highway was taken into consideration.

The methodology used for soil erosion assessment for the study area is based on the USLE (Wiscmeir and Smith, 1978). The USLE is defined as follows.

$$E = R\ K\ LS\ C\ P\ \text{(tons/ha/year)} \qquad \qquad \text{.....(11.1)}$$

where

E = soil loss

R = rainfall erodibility factor

K = soil erodibility factor

LS = slope-length factor

C = cropping management factor

P = conservation practice factor

This model was developed for soil erosion prediction based on empirical research and statistical analysis of field experiments. In this study each factor has been considered as a thematic layer in the GIS. The procedure for determination of different factors considered for generating thematic layers are as follows:

Land-use/Land-Cover Map

The satellite images (LISS III) of the study area were taken and rectified with respect to topographic sheets. The rectified images were mosaiced to extract the study area and classified in different categories of land-use/land-cover, using Maximum Likelihood supervised Classification techniques (MLC). From the classified image, land-use/land-cover map was composed using ERDAS IMAGINE 8.3. and is shown in Fig. 11.7.

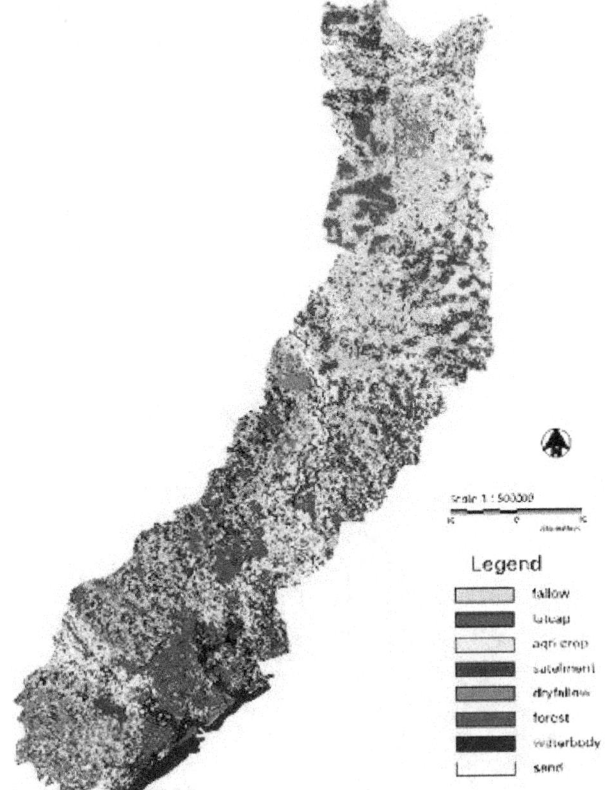

Fig. 11.7 ERDAS imagine 8.3

Digital Elevation Model

Contour lines of different elevations and some spot heights available from topographic sheets of the area were digitized to make vector layers of line and point features respectively and were used as input to ERDAS IMAGINE 8.3 to prepare digital elevation model (DEM) of the study area.

R Factor Layer

Rainfall factor (R) is the factor that depends upon the energy of rain drops and the intensity of rainfall. R is computed as:

$$R = \sum E/30 \qquad\qquad .....(11.2)$$

where

E = Total kinetic energy of rain,

I_{30} = peak 30 min intensity.

The energy table given by Wischmeier and Smith (1958) (13) was used to calculate R. A vector layer of R value was prepared and it was converted to raster map as shown in Fig. 11.8.

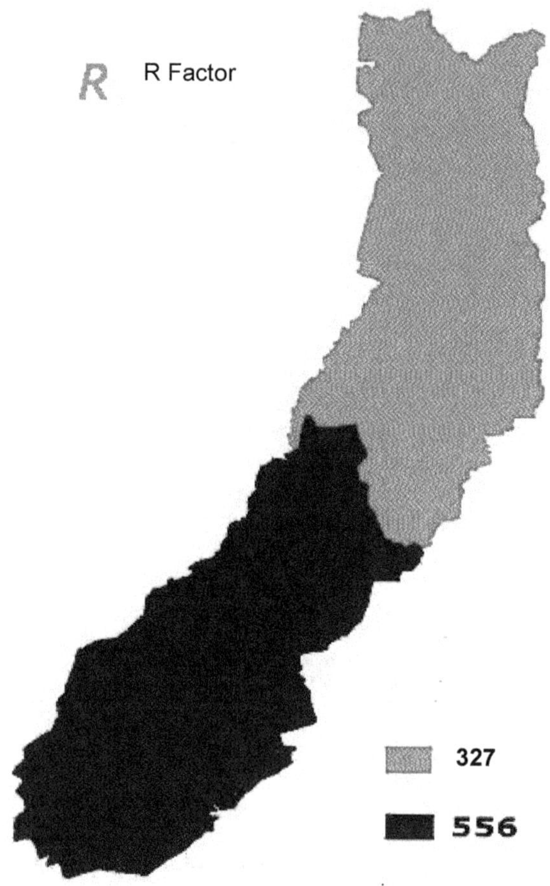

Fig. 11.8 R factor map of the study area

K Factor Layer

Soil erodibility factor (K) gives an idea about the resistance of the soil to detachment and transport caused by rainwater. The soil map from NBSS&LUP (India) of the study area at scale 1:500,000 was referred. A vector coverage having polygon features of the soil classes was digitized. Detailed classification of soil were noted as per soil texture, structure, permeability given in the map. K values (as given by USEPA, Agricultural Research Service, U.S. Department of Agriculture, 1975) were assigned to different classes of soil and a layer was generated. This layer was then rasterised and converted into a K value map shown in Fig. 11.9 The organic matter content of the soil in general is taken between 0.5% to 2.0%.The values of K for different soil types taken for this study are shown in Table 11.17a.

Table 11.17(a) Soil erodibility factor (K).

Soil Type	K factor
Loamy fine sand	0.20
Very fine sand	0.36
Loamy very fine sand	0.38
Silty loam	0.42
Sandy clay loam	0.25
Clay loam	0.25
Silty clay	0.23

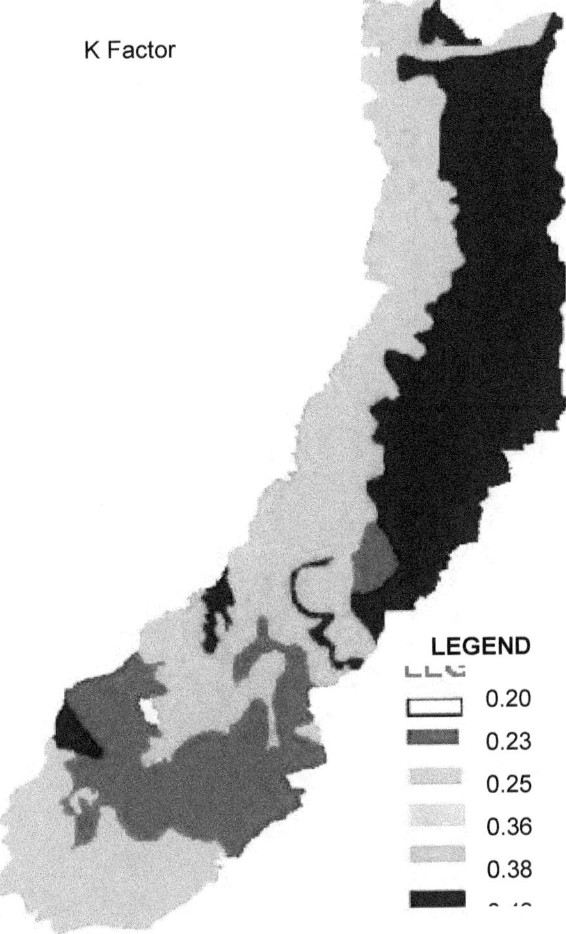

Fig. 11.9 *K* factor map of the study area

LS Factor Layer

Slope-length factor (*LS*), depends on percentage slope and length of the slope. Percentage slope layer was derived from digital elevation model (DEM) of the study area having values;

for each pixel and slope length assumed to be fixed as 23 m for each pixel. *LS* factor was calculated using Eq.

$$LS = (L/22.1)^{0.5} (0.065 + 0.045\ S + 0.0065\ S^2)$$

where

L = slope length in m

S = slope gradient in percent

A raster map having *LS* factor values is shown in Fig. 11.10.

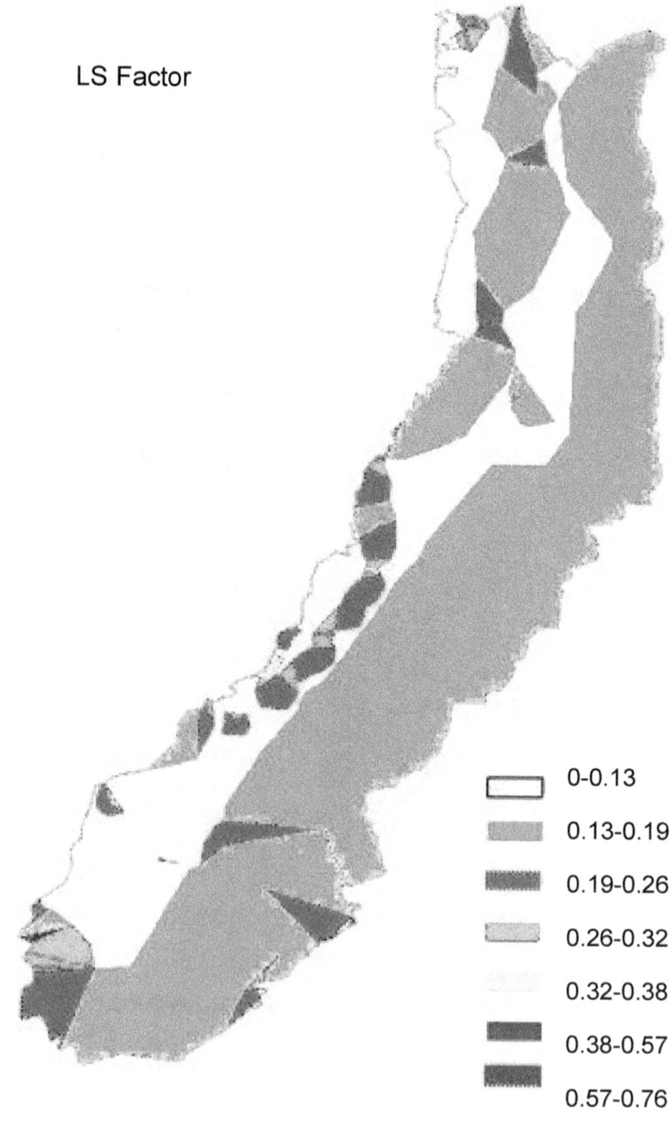

Fig. 11.10 *LS* factor map of the study area

C **Factor Layer**

Cropping management factor (*C*) depends on vegetation cover. Vegetation cover dissipates the kinetic energy of the rain drops before reaching to ground surface. *C* values were decided according to the type of land cover, using land-use/land-cover map and Table 11.17b. This layer was converted to *C* layer through reclassification of each cover type into its corresponding *C* values and is shown in Fig. 11.11.

Fig. 11.11 *C* factor map of the study area

Table 11.17(b) *C* and *P* factor values

Land-use/land-cover type	C factor	P factor
Fallow	1.0	1.0
Laterite cap	1.0	1.0
Agricultural crop (rice)	0.1	0.03
Settlement	0.1	1.0
Dry fallow	1.0	1.0
Open forest	0.8	0.8
Water bodies	0.1	1.0

P Factor Layer

Conservation practice factor (*P*) is to incorporate the erosion control management practices. In the study area, no erosion control practice is specifically adopted (Table 11.14b). So, only inherent control due to growing of rice crop and other land cover is taken into account (Fig.11.12).

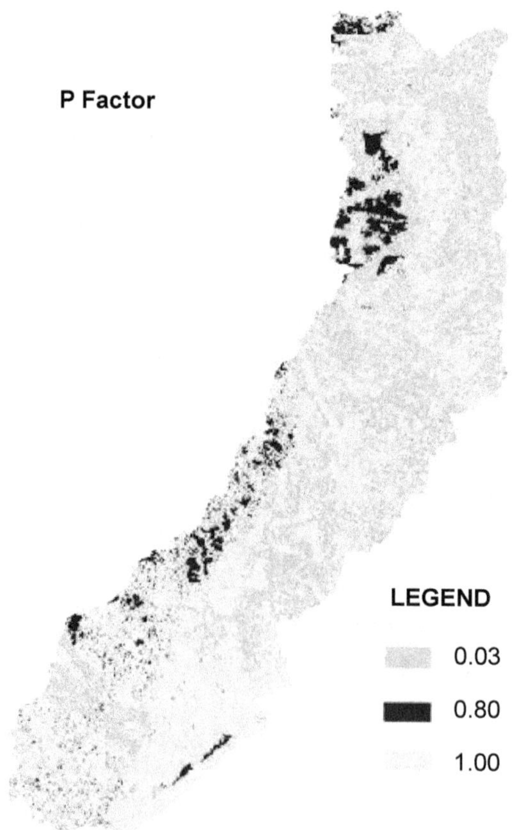

Fig. 11.12 *P* factor map of the study area

Assessment of Impact on Soil Erosion Due to Highway Project

Using USLE model different raster layers mentioned earlier (R, K, LS, C, P) were integrated using model maker option of ERDAS IMAGINE 8.3 software and the results were stored in an output file. This output image has the erosion loss value for each pixel. Soil erodibility map was composed with 7 classes of erosion loss as shown in Fig.11.13.

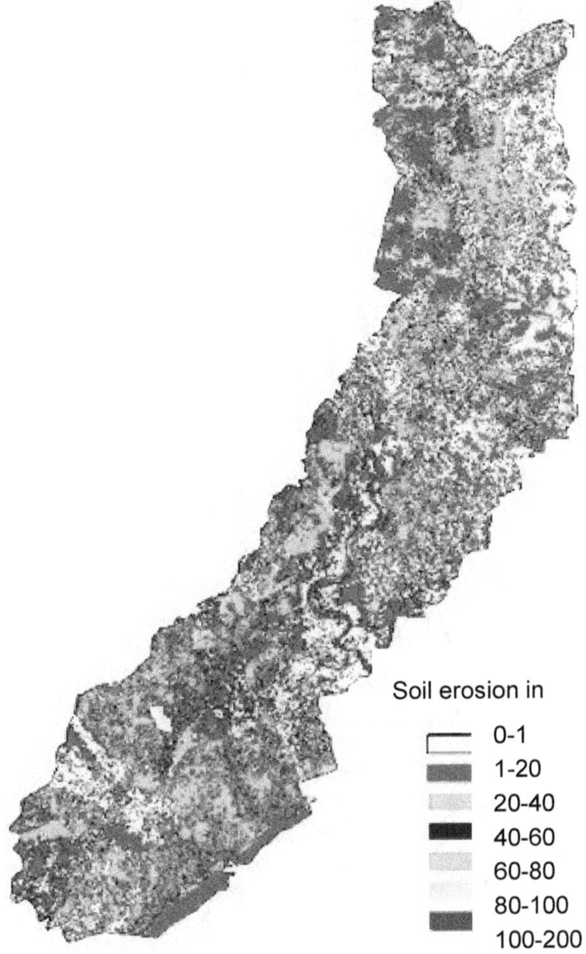

Fig. 11.13 Soil erodibility map of the study area

Since highway is a linear feature having 109 km length and a buffer of 100 meter on either side of road is considered as directly affected zone by the project, C factor will change to 1.0 for the affected zone. For assessing the impact of highway project on soil erosion an image of erodibility was prepared by taking the new values of C factor (1.0), and a new map was composed taking same 7 classes of erosion.

Result and Discussion

Soil erodibility map of the study area is shown in Fig. 11.13. It is evident from the soil erodibility map that most of the area has an erosion potential of less than 60 tons/ha/yr. It may be due to the slope of terrain which is varying only in the range of 0 to 5%.

To get the visualization of impact due to highway construction on the soil erosion, a map only with the area which is likely to be affected directly as mentioned above (a buffer of 100 meter on either side of the road center line) before and after the construction of highway are shown in Fig. 11.14.

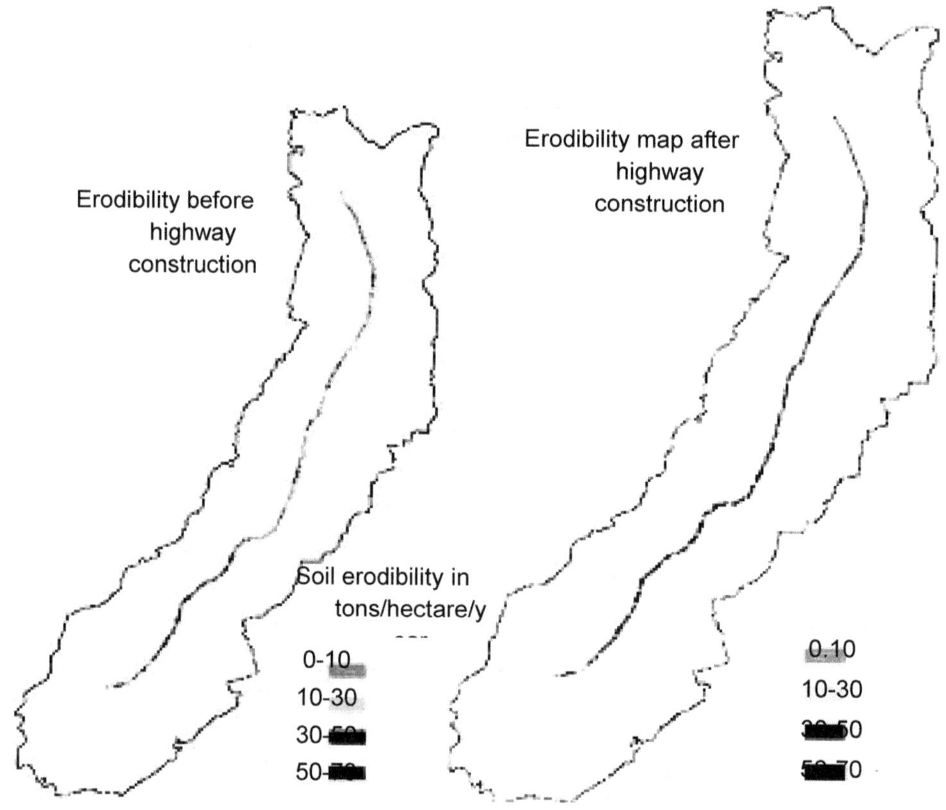

Fig. 11.14 Soil erodibility map along the road before and after highway construction.

Erosion potential before and after construction of highway are also plotted along the length of highway in segments of 30 km each and are shown in Fig.11.15(a) to (d). It is evident from these plots that there is substantial impact on most of the portions of road. At many portions, it is changing abruptly from very low values to high values such as: 4-10 km, 34-35 km, 39-44 km, 47-52 km, 55-60 km, 63-68 km, 75-76 km, 82-87 km and 104-109 km. It is mainly due to reason that these portions were having the value of $C = 0.1$ (Agricultural crop) before construction and it changed to 1.0 after construction. Some portions like: 68-75 km, 78-81 km, 88-105 km are unaltered implying that the impact is almost nil. This was may be attributed to the fact that C value was already maximum before construction.

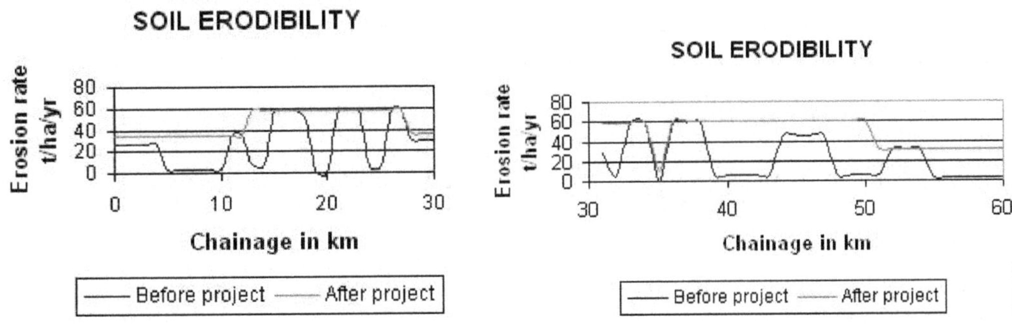

Fig. 11.15(a) Change in soil erodibility between 0-30 km (b) Change in soil erodibility between 30-60 km.

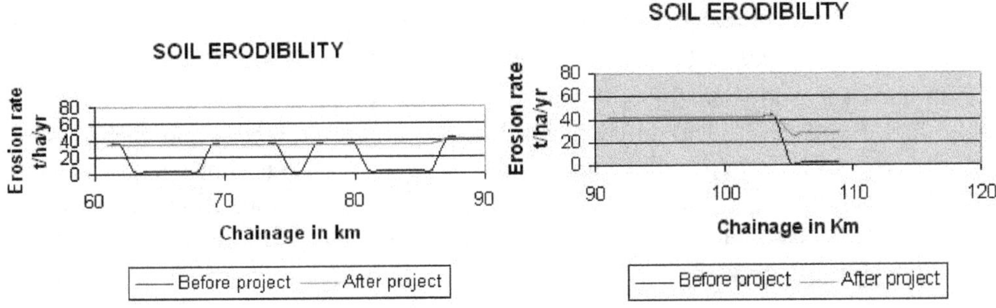

Fig. 11.15(c) Change in soil erodibility between 60-90 km (d) Change in soil erodibility between 90-110 km.

11.12   EIA for Irrigation Projects

Development of irrigation schemes is explicit objective of the national policy. These schemes comprise of two phases. In the first phase the water is either stored, diverted or lifted by construction of a dam/reservoir, diversion work or wells. The second phase consists of conveying water to fields through a canal network. Completion of irrigation schemes require heavy expenditure, time and workmanship. Hence it is essential to study the impact of these structures on the environment in planning stage itself, so that by proper management, its adverse effect on the environment can be reduced and an overall reduction in the cost of development can be achieved without creating a large ecological imbalance.

The implementation of large scale water resources management projects, especially dams, has been a matter of wide dispute between environmentalists, sociologists, developers, and governmental agencies. This debate is expected to continue regardless of where, or when the intended project is going to be carried out. Usually the situation evolves a confrontation between two points of view. The first includes calls for preserving the natural ecological balance of the untapped water resources and the region of influence of the project, protecting the legitimate rights of local residents, safeguarding against deterioration of water quality and threats to biodiversity, changes in soil fertility etc., whereas the second opinion, on the other hand, diverts the attention to the benefits of flood control, drought management, clean energy generation, expansion in agriculture sector, flow regulation, increment in GDP. etc. Highly diversified concerns encompass the situation, not to mention the political will, which usually over-rules in many instances.

Conflicts and disputes related to environmental issues may arise over strategies for resource management, expected environmental impacts of the new project or development plan, operation of existing projects (such as dams), and environmental restoration efforts for degraded resources.

Although agriculture is usually associated with its positive impacts on human life, irrigation practices may be associated with adverse impacts on environmental conditions, which may eventually curtail the sustainability of irrigation projects. For this reason, Environmental Impact Assessment (EIA) has been recognized as an integral part of the early planning studies of irrigation projects in-order to identify any expected negative impacts and suggest the necessary actions to curb these impacts. In the process, EIA can demonstrate the positive and negative impacts of different design alternatives for the project as an essential step for better decision making in implementing irrigation projects.

EIA studies evolved primarily from cost-benefit analysis due to the growing necessity of including social and ecological impacts while planning for development projects. In this context, EIA is meant to predict the future changes in environmental quality and to evaluate these changes. Ultimately, this aims at protecting the environment including the human welfare and health from any foreseen negative impacts as well as to elaborate on the positive impacts of the project.

Available EIA techniques for irrigation projects range from simple checklists to sophisticated simulation models. In fact, the accuracy and scope of each technique varies according to the intended purpose as well as the targeted user. A fundamental objective of any method remains; however, to determine all potential positive and negative impacts in a manner that facilitates the comparison between available project alternatives. In addition, it should identify those impacts that need an evaluation by a specialist. Advanced techniques provide wider scope of objectives that may support the decision making by weighing environmental effects on common basis with economic costs and benefits. Accordingly, targeted users for the different techniques vary to include specialists and non-specialists planners, designers and decision-makers.

11.12.1 Implication of Irrigation Schemes

The implication of irrigation schemes on the environment are presented from the view point of physico-chemical, ecological and socio-economic aspects. These effects may be divided as long term or short term, beneficial or non-beneficial and reversible or non-reversible so as to aid in evaluating the degree of impact.

1. ***Physico-chemical aspects***: This aspect can be studied with respect to the impact of irrigation scheme on atmosphere, water and land ecosystems.

2. ***Atmosphere***: The construction and consequent operation of dams and canal networks influences the values of meteorological phenomena, in relation to the original state. Irrigation structures convert bare soil land into large flooded lands. These water bodies

act as a cooler part of the environment during rapid increase in air temperature. Similarly they function as a warmer part of the environment during rapid fall in ambient temperature. This influences the thermiczonation of the air on the reservoir shore. The air flow depends on the changes of temperature. The low roughness of the reservoir surface promotes horizontal air movement. The influence of reservoir surface therefore changes the shape of the relevant wind rose increasing the occurrence of strong winds from the reservoir to the shore. The change of climate depends on frequency of the wind occurrence, its intensity and change of the solar radiation input derived from the change of the area surface.

The increase in the extent of free water surface in the catchment owing to construction of reservoir often results in reduction in precipitation. This decrease in precipitation is a consequence of low temperature above the surface in the summer season in comparison with original temperature above non-wetted surface. An inversion often occurs above an open water surface causing vertical air motion decreasing the ratio of saturation and hence probability of rainfall occurrence.

Flooded land due to impoundment, increase evaporation rate from water surface when compared to the original bare soil. The increase in evaporation causes a rise in air humidity and also raises the concentration of suspended and dissolved matters. This fact is extremely important in arid and semi-arid areas where evaporation is very high.

The extent of the measurable effect on climatological factors depends on local conditions. It is a compensating effect, occurring more noticeably in areas with a rough climate than in areas with a mild climate. The distance of the climatological influences of irrigation structures and its network depends on its size. It does not exceed 100 m in small reservoirs whereas in big structures, to around 1-3 km, form their shore.

1. **Water**: Water is stored in reservoir so as to convey it to the field mainly for agricultural purpose through canal network. Water required for agricultural purpose should have standard quality or else its use would have an effect on vegetational growth.

Due to impounding, the self purification process in reservoir with the exception of sedimentation, is negatively influenced resulting in low oxygen and low temperature in comparison with original conditions of river bed. An increase in detention time leads to increase in bioactivity resulting in the mineralization of organic matter and production of new organic matter. The decay of this organic matter can cause pollution, decrease in oxygen content and also disrupts biological balance. The change of water quality also occurs due to creation of a comparatively heavier/cooler layer at reservoir bottom due to thermal stratification by external forces. Human activities also contribute to the occurrence of zone with oxygen deficit especially, pits/dike at the bottom of reservoir.

Dams and weirs inhibit and disrupt bed load, suspended load and wash load transport. They also change the course of erosion process at upstream and downstream. The sedimentation occurs due to decrease of velocity and kinetic capacity. The abrasion and land sliding of shores is a process of their destruction caused by the effect of water, wind and by fluctuation of ground water table.

2. *Land*: The extent and intensity of changes in the landscape caused by the irrigation schemes depends not only on the topography, character and accessibility of the area and the density of population and communication lines but also on the type and size of irrigation structures.

The most adverse change that occurs is flooding which may be a cultivated land, urbanized land, forest, communication lines, buildings, engineering works, landmarks, historical places, mines, mineral deposits and national parks and sanctuaries. The inhabitants of submerged area may be people, flora, fauna and rare species etc. These may lead to depletion of some resources which may be rare and of high value.

Second great impact of water flooding or submergence is rise in ground water table of the surrounding area which may result in water logging and salinity, which is a serious defect from agricultural point of view. Similarly, fluctuations in water table result in acidity or alkalinity of soils, increase or decrease in salination rate and lastly escape of nutriments of plants especially nitrogen.

Irrigation projects required heavy foundation deep into the earth crust. This may effect the land stability and increase stress and strain within. Such regions have high probability of earth quake and land slide occurrences.

3. *Ecological aspect*: As soon as the filling of a reservoir begins, the original eco system is gradually being replaced by a new eco system of stagnant water. Trees, shrubs, vegetation, natural resources, wild life, forest ecosystem completely disappears with the new eco system. Some of this loss may be of medical and economical value.

The aquatic life of manmade structure is greatly influenced by frequently draw down and rise in water table. This results in the poorer heterogeneity of ecosystems in upper littoral zones in comparison to natural ones. A number of current species are disappearing as a result of the action of fluctuating water level. The occurrence of aquatic species depends upon availability of nutrients and the extent of the tolerance of the relevant organism to occurrence of unwanted components of living environment. The existence of relevant fish species depends on water quality, temperature, oxygen, water depth, rate of flow, morphology material at the bottom and the bank and occurrence of flora. The construction of reservoir changes all this condition. A dam or weir forms an obstacle for the draw of migratory fishes. The water level fluctuations reduces surface production and destroys zooplankton basic to fish food, resulting in depletion of fishes.

Socio-economic aspects: Irrigation schemes provide water for agriculture, opportunity of huge employment and other subsidiary economic activity like pisciculture, tourism etc. The economic impact of the dam construction and reservoir operation causes a constructional boom as the equipment used on the building site offer possibility for further utilization and the construction of new communication links and increase the accessibility for recreation purpose which lead to modernization of the life style of the population. The social group and personal interest of the population are affected by the creation of new shore. The increase in dwelling values results in increased density of habitation. The change in the dwelling value has an important impact on the life style supporting its recreational aspects. The reservoir creates or supports favourable condition for fishing, camping hunting and other type of weekend and vocational recreation.

The increased density of habitation has an effect of breakdown of natural vegetation and increase in erosion rate, a rise in the transport density, an increase in noise density, a gradual pollution of environment with the need for water supply and waste handling.

Depending on water quality and prevailing sanitary conditions, the reservoir operation can create or strengthen the condition for the dissemination of germs or their bearers specially in tropical and sub tropical areas. Negative circumstances may result in economic, cultural and social losses either permanently or temporarily especially in the period during and shortly after the construction.

Some of the expected effects may not be achieved due to various planning, financial, political, organizational obstacles or due to unexpected reaction of population. The main objection of habitants may be regarding rehabilitation, employment, loss of flora and fauna, loss of cultivable or forest land and wild life. The check list of probable impacts of irrigation scheme have been developed.

There is no absolute or unique judgment that could apply for all water resources management projects, but the judgement is rather case dependent and site specific. The decision of whether the project is on the whole beneficial or detrimental should be based on a process that identifies all positive and negative impacts, estimates their significance, and considers mitigation measures. A technique that prioritizes the relative importance (weights) of the different factors, should then be followed, in-order to reach a rational and objective decision. The criteria that has to be considered while judging a Dam project are given below:

11.12.2   Typical Impacts of Dams and Reservoirs

1. Change in quality of impounded water (seasonal).
2. Water loss due to evaporation (seasonal).
3. Downstream effects: decreased (and more uniform) flow into estuaries, thus causing changes in saltwater intrusion patterns and changes in estuarine fisheries.
4. Increased cultivated lands and agricultural produce.
5. Hydropower generation, cleaner energy production.
6. Changes in local groundwater levels and quality.
7. In-reservoir landslides, increased regional seismic activity due to water pressure.
8. Changes in microclimate of area - more wind, humidity, and/or precipitation.
9. Inundation of mineral resources.
10. Flood control.
11. Drought management.
12. Changes in number and types of fish-from cold water to warm water fishery.
13. Preclusion of movement of migratory fish.
14. Fish destruction in turbines and pumps (use protective screens).
15. Possible creation of "new reservoir fishery" as positive impact.

16. Increase areas or breeding of mosquitoes and related insects – and their public health implications (e.g., malaria and schistosomiasis).

17. Promote growth of aquatic weeds such as water hyacinths.

18. Changes to habitat in inundated area and wildlife associated with habitat.

19. Changes to waterfowl habitat from shallow, flowing habitat to deeper lakes; possible impact on migratory birds.

20. Impacts on rare, threatened, endangered, unique flora and fauna.

21. Decrease in waste assimilative capacity of river segment.

22. Inundation of historical, cultural, archaeological, or religious resources.

23. People relocation-resettlement (and possible change in style of life).

24. Influx of construction workers, associated social, infrastructure, health impacts.

25. Increased tourism around reservoir.

26. Downstream effects on traditional floodplain cultivation; reduced flood delivery of nutrients to downstream fields.

27. Developments in catchment area resulting from roads and from other associated increases in sediment and nutrients into reservoir.

28. Grazing land capacity and livestock breeding.

29. Reduction of energy generated from fossil fuel.

30. Transboundary impacts.

31. Direct and indirect effects on GDP.

11.12.3   Environmental Impact Assessment Techniques

Checklists are known to be the simplest method for the evaluation of any project impacts on the different components of the environment. With respect to irrigation projects, the two most widely acclaimed checklists are those prepared by the World Bank (1991) (13) and the International Commission for Irrigation and Drainage (ICID) (Mock and Bolton, 1993) (14).

The first of the two lists is a comprehensive inventory of potential negative environmental impacts associated with irrigation projects and the corresponding mitigation measures. However, the second list is a descriptive checklist that includes more details about the environmental effects of irrigation projects and identifies the necessary data for an assessment study. As a matter of fact, both checklists do not provide an absolute measure for environmental risk or relative magnitude for rating different project alternatives.

A developed form of simple lists is the scaling checklist since it allows the relative rating of each impact and guides the evaluation of the different criteria. The Battelle Environment Evaluation System (ESS) is a typical example of scaled checklists for water resources projects. It consists of 78 parameters describing the different environmental components together with their corresponding fixed importance weights. The user has to assign a value between 0 and 1 that reflects the environmental quality of each item. The given scores are eventually transformed into one environmental impact unit for each specific alternative. This makes the ESS a useful tool for comparison between the project and no project options and meanwhile detecting the serious adverse impacts.

Another form of scaling checklists is the multi-attribute utility functions checklist, developed for measuring the relative environmental quality of the parameters in the checklists. This involved fixing a scaling value for each parameter to reflect its importance, and by combining the utility functions a total utility is generated for various project alternatives which can form basis for comparison. However, this method was criticized for its complexity and involvement of certain subjectivity.

Interaction matrices, unlike checklists, adopt an objective procedure for environmental impact assessment that enables rating, weighing, and consequently can assist in decision making. Leopold matrix is the most well known example of an interaction matrix, which consists of 8800 cells. The matrix is a more complex and time-consuming technique than a checklist. Meanwhile, it does not include any description of measurement strategies or guide for assignment of the impact scores (Hyman and Stiffel, 1988) (15). As a result, matrices provide no guarantee that the applied weights and ranks represent the actual impacts and are not biased values.

Networks and system diagrams techniques are the approaches that link the main impacts to subsequent indirect impacts. They have the advantage of incorporating several alternatives into their formats and also including mitigation measures in the planning stages of the project. Similar to matrices, networks are very time-consuming processes and in addition expensive to use. Moreover, networks inability to include socio-economic impacts is considered as a major shortcoming that limits their usage.

The overlay mapping method is simply the superimposition of a set of transparent maps representing each environmental attribute on a base map of the area. The boundaries and the density of the color indicate the extent and value of impact. This spatial distribution of the project effects provides a powerful visual representation of the potential impacts that can be easily displayed and comprehended by audience. However, this presentation always lacks any probability or duration considerations and ignores the impacts arising from interaction between two or more factors. In fact, these problems can be tackled by Geographic Information Systems (GIS) technology when used for environmental impact assessment. GIS provides a tool for collecting, storing, retrieving, transforming and displaying data in a manner that allows the data to be modeled and analyzed (Burrough, 1986) (16).

Simulation modeling is a powerful interactive tool that can demonstrate and predict the response of the environmental components to the project activities. They are normally based on mathematical formulations that can be quickly processed using computer capabilities. The simulation results can guide impact assessment studies including environmental, social and economic aspects even in cases of few available data or considerable uncertainty of the dynamic relationships. Moreover, they can swiftly generate several project scenarios for comparison with various initial assumptions and help greatly in conducting sensitivity analysis. Nevertheless, a team of professional specialists indifferent fields is always required to develop and verify simulation models. Also models has to be used cautiously within context otherwise it may produce misleading results.

Thus the first step in any EIA for irrigation projects is to identify the environmental criteria that are susceptible to change due to irrigation projects construction or operation, and then categorize these criteria under main headings. The physical factors that are known to be responsible for changing the criteria conditions have to be then identified. All possible conditions of these factors have to be determined and compiled in a multiple-choice questionnaire. The expected positive and negative impacts to be relatively associated with the different factors conditions have to be integrated in the EIA to represent the initial assessment. The conditions of each factor in the initial assessment should indicate its individual positive or negative impact on the criteria. The cumulative effect of the factors impact determine the impact on each criterion. The impact on each main environmental category will be then calculated assuming equal weights for all criteria impact. However, the overall impact for the project alternatives have to be calculated according to the given importance weights for each of the five main categories.

In this context, the five main environmental categories to be considered are, natural resources, biological life, socio-economic, political and economic impacts Table (11.18 a&b). Environmental criteria relevant to irrigation projects can be categorized carefully under the appropriate main headings in order to avoid replication or double counting. The criteria and the physical factors influencing them have to be compiled from the available literature. They are believed to be covering all possible criteria of impacts of irrigation projects and all factors in irrigation projects that may be affecting these criteria.

Table 11.18 (a) Impacts criteria of irrigation projects

1.Natural Resources Impacts	2. Biological Life Impacts	3. Socio-Economic Impacts
Soil	2.1 Wildlife	3.1 Public Health
1.1 Soil Erosion	2.2 Fish	3.2 Land Use
1.2 Soil Fertility	2.3 Aquatic Life	3.3 Tourism & Recreation
1.3 Soil Salinity	2.4 Eutrophication	3.4 Resettlement
1.4 Soil Pollution	2.5 Pests and Rodents	3.5 New Communities
Water		3.6 Sites of Special Importance
1.5 Surface Water Quantity		3.7 Job Opportunities
1.6 Surface Water Quality		
1.7 Groundwater Level		
1.8 Groundwater Quantity		
1.9 Groundwater Quality		
Air		
1.10 Gas Emission		
1.11 Dust Pollution		
1.12 Local Climate		

Table 11.18 (b) Impacts criteria of irrigation projects

(Les critères de l'impact de projets d'irrigation)

4.Political Impacts	5.Economic Impacts
4.1 National Security	Feasibility Indicator
4.2 Foreign Affairs	5.1 Benefit/Cost Ratio
4.3 Public Opinion	Productivity Ratios
	5.2 Net Production/Irrigated Area
	5.3 Net Production/Water Volume
	5.4 Net Production/No. Full Time Workers
	5.5 Return/Investment (Foreign Currency)
	5.6 Net Production/Invested Capital
	Investment Ratios
	5.7 Jobs Created/Capital Invested
	5.8 Irrigated Area/Capital Invested
	5.9 Water Volume/Capital Invested

As an example, Table 11.19 shows the factors affecting the four soil impacts criteria in the project and neighboring area. It may be noticed that one factor, such as the drainage system, can be affecting more than one criterion.

Table 11.19 Factors that control the soil impacts criteria

Soil Erosion	Soil Fertility	Soil Salinity	Soil Pollution
Irrigation Method	Applied Fertilizers	Existing Soil Salinity	Agro-Chemicals Application
Irrigation Water Quality	Irrigation Method	Irrigation Advisory Services	Irrigation Water Quality
Irrigated Land Slope	Land Use	Drainage System	Type of Agro-Chemicals
Irrigation Advisory Services	Irrigation Advisory Services	Soil Texture	Drainage Water Disposal
Drainage Water Disposal	Irrigation Water Quality	Irrigation Water Salinity Downstream	Irrigation Water Quality Downstream
	Drainage Water Disposal	Irrigation Water Salinity	
		Salt Leaching	

Sufficient information to describe the baseline conditions and the general design for several project alternatives. In this sense, the selected answers by the user reflect the impact of the physical factors on the different criteria in the project area as well as the neighboring area. Consequently, a positive score is assigned for each factor indicating a positive impact and conversely a negative score is given for every anticipated negative impact. The scores have to be given to reflect the expected impact of each factor relative to its all possible impacts. The scores of the impacts on the project area and neighboring area are then summoned for each criterion separately. Depending on the sign of the total impact score of each criterion, it will be divided by the maximum possible positive or negative score of that criterion to give a normalized +/- impact on a scale from "-100%" to "+100%". The normalized impacts for all criteria have to be then averaged over the criteria for each environmental impact category to obtain the impacts on the main environmental categories. The pre-defined importance weights for each environmental category can be used to calculate the overall environmental impact for each project alternative. For detailed method of computation for economic impacts, one has to insert estimated values for some of the economic parameters required to calculate the indicators listed in Table 11.20 The indicators included in the evaluation have to be based on the prime economic objectives of the project and the scarcest resources in the economy as identified by the user (Bergmann and Boussard, 1976) (17). The economic indicators enter into the final evaluation as a ratio relative to the highest values among the project alternatives. The economic impacts category, similar to the other environmental categories, have to be weighed according to their importance by the decision maker and added to the overall impact (Table 11.20).

Table 11.20 Overall impacts *(Ensemble d'impacts)*.

	Importance Weights (set by the user)	Alt. 1	Alt. 2	Alt. 2
Natural Resources Impacts	I1	N1	N2	N3
Biological Life Impacts	I2	B1	B2	B3
Socio-Economic Impacts	I3	S1	S2	S3
Political Impacts	I4	P1	P2	P3
Economic Impacts	I5	E1	E2	E3
Overall Impact	100	N1*I1+B1*I2+...	N2*I1+B2*I2+...	

Case Study

11.13 Environmental Impact Assessment for Dam Construction Using GIS/Remote Sensing (by Dr. Shibani Maitra) – Journal- GIS & Remote Sensing (2002)

11.13.1 Study Area

This is a EIA study on a dam proposed to be located on the Man river in the Sidhumber village in Gujarat, India. This dam shall impound gross storage of 174,430 cubic meters of water and is expected to irrigate 15170 hectares of land. This region of Gujarat, India is backward and underdeveloped. The irrigation potential is heavily underutilized, with only one Daman Ganga reservoir project. This study shall prepare a detailed environmental impact analysis of the proposed dam and environmental management plan in the catchment and the command areas of the Man River for clearance by the Ministry of Environment.

The watershed of the Man River has a semi-arid climate. The scarcity of water is acute. In spite of having a rich black cotton soil, the area is backward because of dependency on rain fed agriculture. The increased irrigation potential due to the proposed dam shall have a great impact on the socio-economic upliftment of the region.

The construction of the dam is aimed at improving the land productivity in the command area. Proper utilization of the irrigation water from the dam is suggested to prevent increase in the land salinity in the command area. The study shall assess the status of erosion and land degradation in the catchment area to prevent siltation and to suggest the future plan of treatment.

The Man River originates in the Nasik district of Maharashtra, India and flows through Valsad district of Gujarat, India (Fig. 11.16) and discharges directly into the Arabian Sea. The area is located between 73° 15' - 73°36' E longitude and 20° 25' - 20° 40' N latitude, and covers parts of the Survey of India topographic sheet numbers 73H/2,3,6,10. The catchment area is demarcated into sub-watersheds and consists of 55 villages covering an area of 261.0sq.km. The command area covers an area of 192.6 sq.km. consisting of 37 villages.

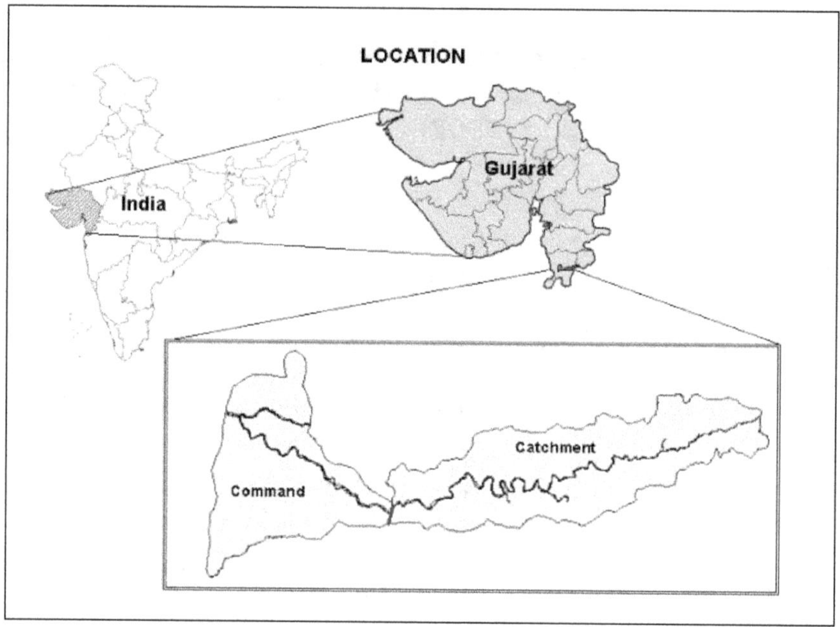

Fig. 11.16 Location of study area

Methodology

The following tasks have been undertaken:

- Generate thematic maps of various natural resources
- Integrate the thematic maps
- Define the plan of implementation

  The input data from all of the above diverse sources are translated into the thematic maps by the methods of:

- interpretation,
- classification,
- manipulation,
- integration,
- editing and
- analysis.

The data translation into thematic maps employed the GIS software Arc/Info and Arc View and the remote sensing software ERDAS. The multi layer thematic maps generated necessary information to provide detail insight to suggest the necessary improvement of the command and catchment areas.

11.13.2 Physiography and Drainage

The catchment area of the proposed dam consists of *flat topped, highly dissected plateaus* and dyke ridges running in a west to east direction (Fig. 11.17). The Man River in the catchment area flows in incised meanders forming steep 'V' shaped valleys with steep sides. The entire catchment has a rectangular and trellis pattern of drainage depicting the influence of the structures on the drainage development. The *piedmonts* at the base of the steep plateaus and dyke ridges are covered with thin soils, which supports agriculture in very few areas. The *river valley*, wherever flat, has good quality soil and is mostly cultivated based on the availability of water. The *valley fills* with thick alluvium provides the only area for cultivation.

The command area of the proposed dam is surrounded by *moderately dissected plateaus* and *piedmont* slopes. The slope of the land along the piedmont and the nature of flow of the streams (e.g., in the upper part of the command area near the village Panikhadak) provide ample proof of scarp retreat. The rivers in the piedmont slope area show parallel pattern, which are partly controlled by the lineaments. Along the river valley the *flood plain* consists of good quality soil, suitable for cultivation.

The largest portion of the command area is the *alluvial plain*, which has been formed by the river Man. The alluvial plain is studded with number of *residual hills* with degraded forests. The river Man in the command area also flows in straight channels, which shows that the river is structurally controlled.

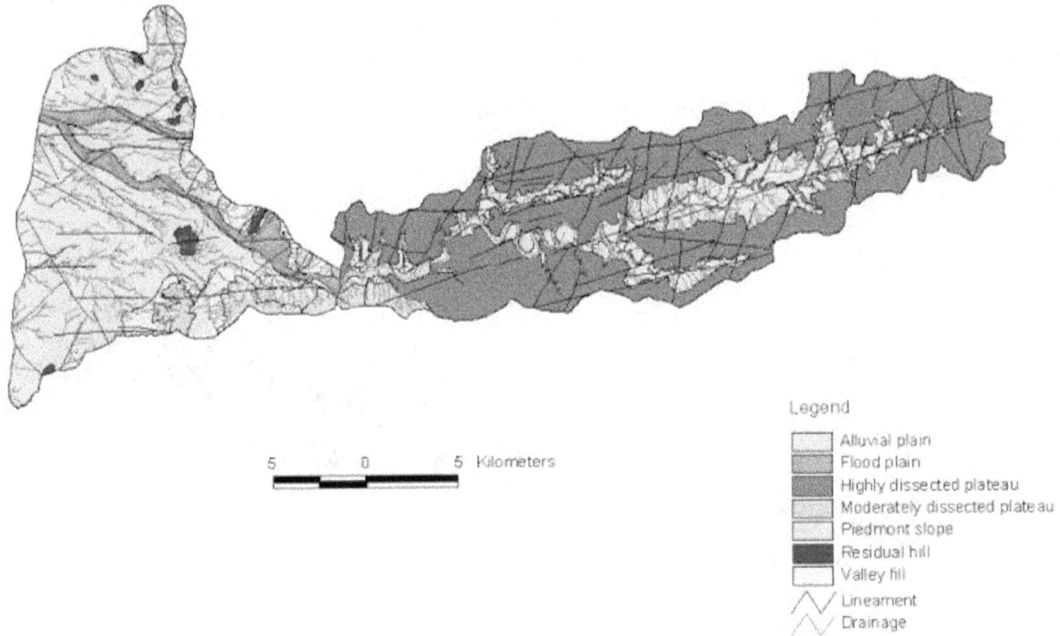

Fig. 11.17 Physiography, Drainage and lineaments.

11.13.3   Geology and Structure

The catchment and command areas are occupied by various Deccan trap lava flows viz. Amygdaloidal basalt and Porphyritic basalt and agglomerate of the Cretaceous – Eocene age. These are traversed by dolerite dykes. Except in the river bed the rock is highly weathered and exhibits spheroidal weathering.

Big cavities filled with secondary quartz are also seen. The agglomerates separate the lava flows and occur in the form of lenses of thickness varying from 1-2 m.

Top of the plateaus consists of porphyritic amygdaloidal basalt, which is highly fractured and jointed. The joint patterns as recorded in the rocks show a W-SW – E-SE and W-NW – E-NE direction (Figure-11.19).

11.13.4   Climate

The area has a semi-arid type of climate. It records a maximum temperature of 42°c with a mean annual temperature of 27°c. April and Mayare the hottest months. In winter the temperature drops to 7°C. January is the coldest month.

About 95% of the rainfall comes from the South-West monsoon. The average annual rainfall is 2465 mm is concentrated in a few months and the remainder of the year is dry.

11.13.5 Soil Series

The area, being of basaltic formation, falls under the broad soil group of red loams and black clayey soils. The transmission of water through similar parent material seems to have influenced the development of different physiographic characteristics of the soils in the area.

Below is given a brief description of the physiographic characteristics of the soils of the area (Fig. 11.18).

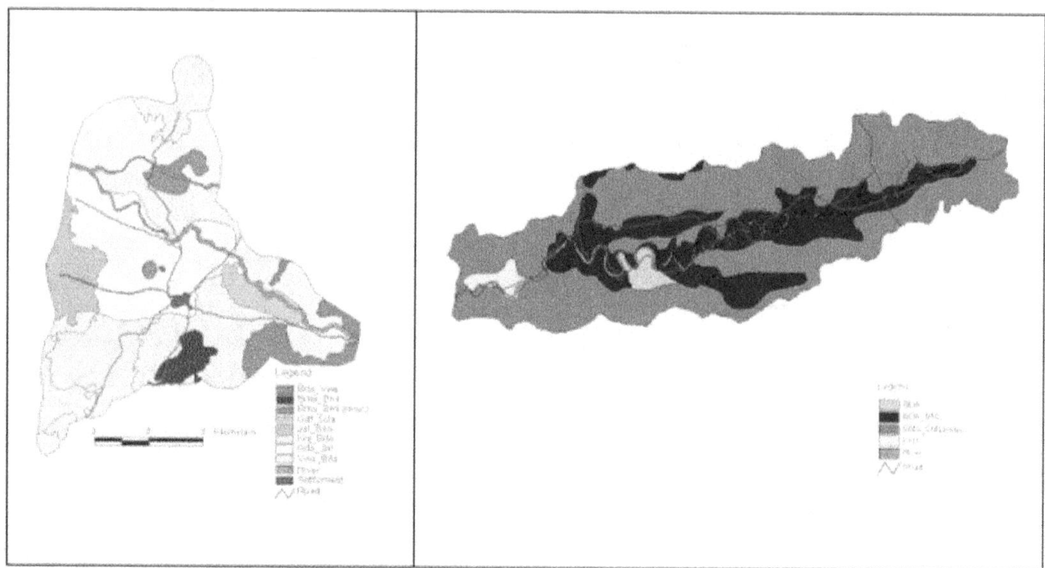

Fig. 11.18 Soil series.

Soils of the high lands: The characteristics of the dissected plateau with a slope above 8°
have gravelly, shallow to moderate thickness, loamy, non-calcareous soil, which has eroded
and derived from intra-trappean basalt rock in situ condition.

The area in between the hills with sloping lands, contains dark yellowish brown to very
dark grayish brown gravelly clay loam to clayey soils of shallow to moderate thickness.

The dissected hill and steep slopes suffer from severe erosion hazards. The steep hill
slopes are almost devoid of soil. These are marked as miscellaneous land type with
Billimoda-Bedmal soil association, having very poor and sparse forest growth.

Soils of piedmont slopes and river valleys: The relief decreases towards the west and the
materials washed out from the upper reaches are deposited in the downstream areas. The
area is usually having 3-5% slope gradients. The soils in these area are of shallow to
moderate thickness, dark reddish brown to very dark grayish brown, clay loam to clayey,
non-calcareous of the Kanjod-Baldha and Vadwania-Baldha soil series associations.

Soils of the alluvial plain and flood plain: The piedmont slope gradually merges into the
alluvial plains with gently sloping land. The soil becomes deep to very deep, clayey
calcareous and non-calcareous, dark grayish brown, very dark grayish brown to dark
yellowish. The soil cracks vertically during dry season due to montmorillnitic type of clay.
Jalalpur and Sisodra soil series occur on this physiographic unit.

The flood plain consists of moderately deep to very deep alluvial soils. The soil series
identified are Gadat-Sisodra. They are deep soil formed by river weathering of basaltic
parent material. Gadat is reddish brown to dark reddish brown, silty clay loam to clay loam,
non-calcareous soil. Sisodra is moderately drained, dark brown to brown soils, formed by
alluvium.

11.13.6   Land use/Land cover

Digital interpretation of IRS LISS-III FCC on 1:50,000 scale for two season dates is done in
ERDAS for identification of different land use land cover classes based on the image
characteristics (Fig. 11.19). The multidate imagery are interpreted for the details of the crop
land in the two harvest seasons known as the Kharif and Rabi seasons. Based on ground
truth verification, the boundaries are finalized which synchronizes well with the
physiography, slope and soil of the area. Twelve land use/land cover classes have been
identified in the command area and eight classes have been identified in the catchment area.

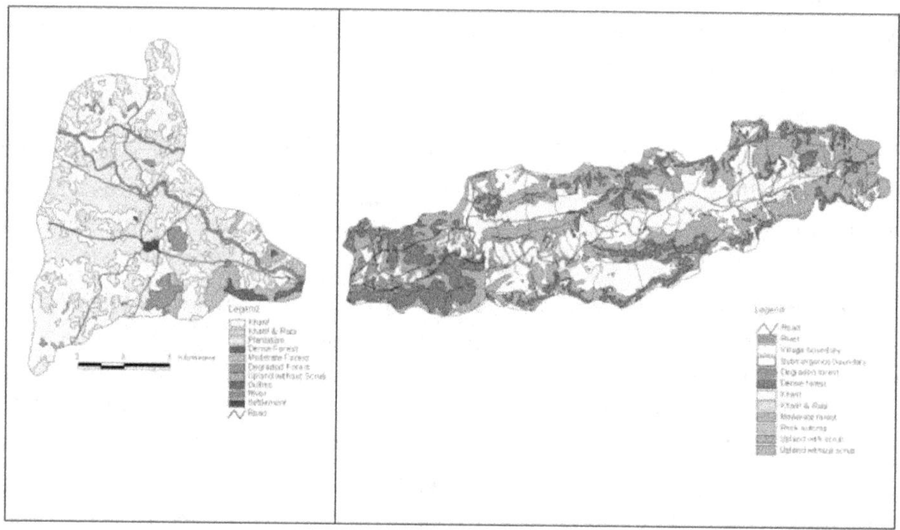

Fig. 11.19 Landuse

As a result of the proposed dam, 5 sq. km of forest area and 7 sq. km. of kharif land are going into submergence. Remainder of the total submergence area of 17.6 sq. km. is mainly wasteland.

11.13.7 Preparation of the Secondary Overlays

The secondary layers are derived from the above data sets in Arc/Info by the various overlay functions. Polygons below a threshold limit eliminated to generate the final layers based on which the decisions can be made.

11.13.7.1 Slope

The slope map is derived by using the GRID and TIN features of Arc/Info. The input data are the contours from the Survey of India topographic sheets. After converting from raster to vector layer it is processed for generating the secondary overlays.

In the catchment area, the plateau tops have slopes of 0-3% and the steep hillsides are above 8%. The piedmonts comprise of slopes mainly between 5 – 8%. In the river bed and the valley fills, the slope remains below 5%.

The command area, comprising mainly of the alluvial plain and the flood plain, has a slope ranging between 0-3%. In the surrounding dissected plateau and piedmont, the slope varies between 3 – 8%.

11.13.8 Generation of Final Overlays for Decision Making

11.13.8.1 Hydrogeomorphology and Groundwater

The hydrogeomorphological map is prepared by overlaying geomorphology, lithostratigraphy, structure and land use. The hydrogeomorphologic conditions for each landform type are identified based on the above layers. Groundwater prospects are assigned to each unit (Fig 11.20). A total of five classes of groundwater prospect areas have been identified in the catchment area. The groundwater status in most of the area varies between poor to moderate, especially in the dissected plateaus and dyke ridges. Good groundwater prospect exists in few places in the flood plain near the Sidhumber reservoir area, along the lineaments and in the weathered zones.

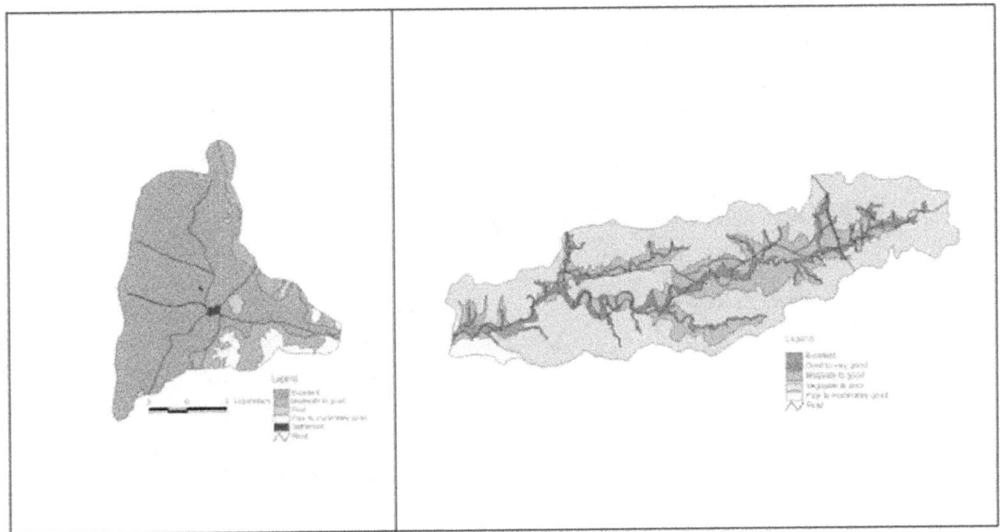

Fig. 11.20 Groundwater prospect.

11.13.8.2   Land-Irrigability

In the command area, based on the texture, structure, permeability, of the soil, soil-irrigability classes are assigned and each type of soil irrigability class is given a unique code (Fig. 11.21). This soil-irrigability layer is unionized with the slope layer to derive the land-irrigability classes. Based on the percent slope and soil irrigability classes, four land-irrigability classes have been identified.

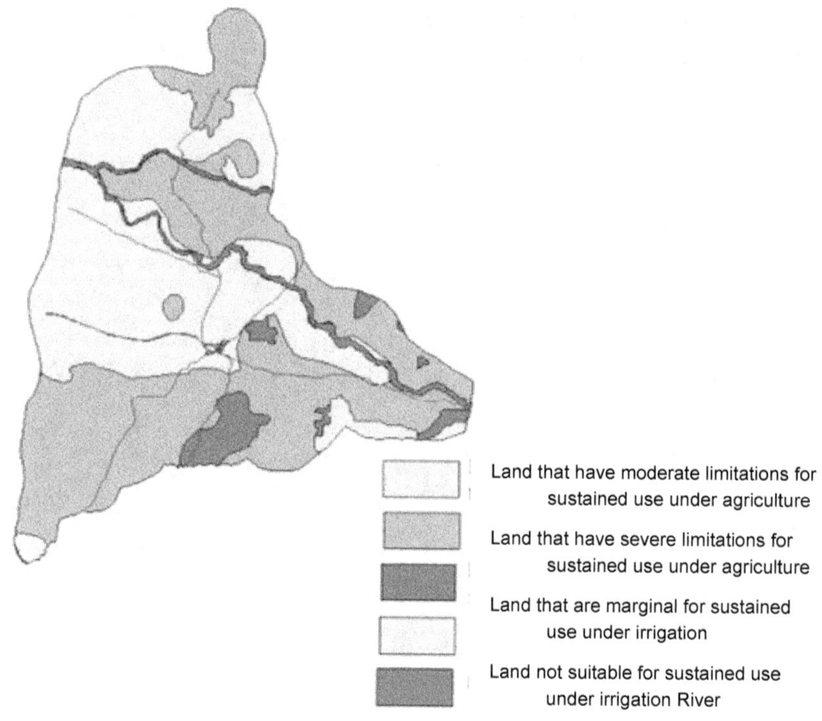

Fig. 11.21 Land irrigability

11.13.8.3   Composite Erosion Intensity Units/Composite Land Development Sites

Overlaying of the slope, soil and land use, in Arc/Info for both the catchment and command areas, has generated the Composite Erosion Intensity Units (CEIU)/Composite Land Development Units (CLDU) respectively. Total 47 unique CEIU and 42 unique CLDU have been generated for the catchment and the command areas respectively.

The area of each unique CEIU is estimated for each sub-watershed. These CEIU/CLDU have been used for decision making.

11.13.8.4   Land Capability

Overlaying the slope, soil, land use and environmental factors of each CEIU/CLDU, land capability classes are generated. Each land capability class is identified by a unique characteristic, having similar hazards of the soil to various factors, which causes soil damage, decreases soil fertility, and its potential for agriculture (Fig. 11.22).

In the command area the land-capability has been assigned for the development of the area. In the catchment area land capability has been assigned for future treatment.

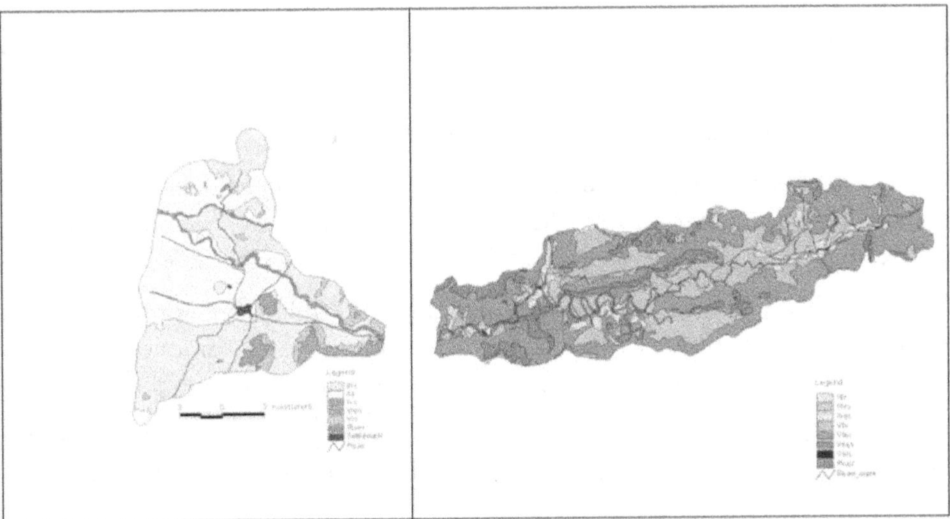

Fig. 11.22 Land capability

11.13.9 Sediment Yield Index

Based on the characteristics of the 47 units of CEIU, each CEIU are assigned weightage and delivery ratio. The Sediment Yield Index is assigned to each CEIU. The sub-watersheds have been identified based on the area eroding more in time and space and have been prioritized based on the Sediment Yield for future treatment (Fig. 11.23).

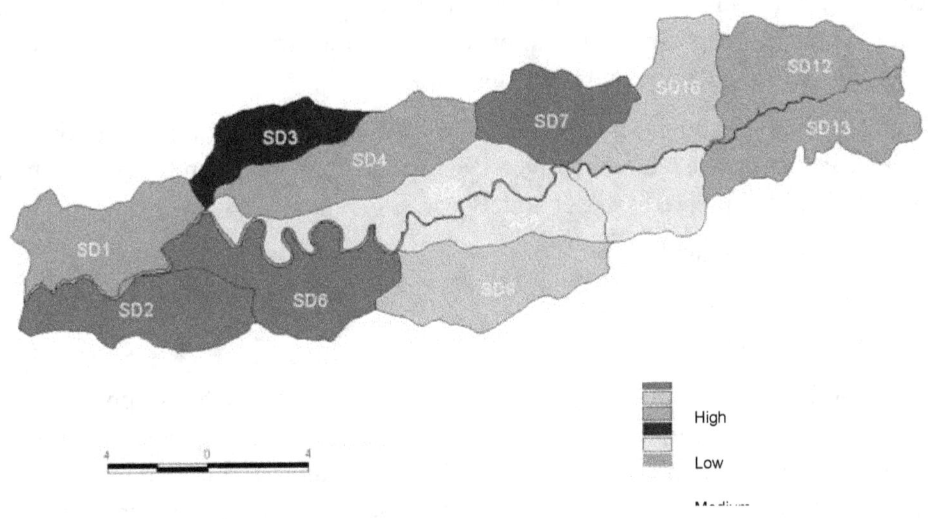

Fig. 11.23 Priority sub-watershed.

11.13.10 Treatment Plans

The treatment plans have been customized based on the composite final overlays for the Sidhumber Dam on the Man River.

11.13.10.1 Development Plan for the Command Area

Based on the land irrigability of the command area, it has been found that around176.8 sq.km. of the total command area of 192.6 sq. km or 91.8% is expected to produce better crop by irrigation water from the dam, provided proper drainage facilities are present.

The land capability also shows that approximately170 sq.km. area of the command area has II and III land capability, which indicates that the area can produce better crops by improving the quality of soil with appropriate treatment of nutrients and fertilizer.

Based on the soil, land use, slope, land irrigability etc., of the command area, suggestions have been made to change the land use pattern and undertake reforestation in some of the areas.

Treatment plan for the Catchment Area

The longevity of the proposed dam can be extended by checking soil erosion in the catchment area. After a detail assessment of the above overlays, it is recommended to treat each sub-watershed from different aspects.

Based on the Sediment Yield Index, the sub-watersheds were prioritized for treatment.

The land capability of the catchment area shows that 16.4 sq.km. of the total catchment area of 261.0 sq. km. can be treated with soil and water conservation methods to increase the moisture retention capacity, thus improving the fertility of the soil. Development measures for the command area have been suggested.

The drainage line treatment is also necessary in the catchment area to prevent soil erosion. The catchment area having diversified slope, soil and land use, the treatment suggestions carried out in three parts separately:

- For forest and watershed
- For agricultural land
- Drainage line treatment

The major drainage lines are assigned stream order and divided into upper, middle and lower reaches based on the slope of the terrain. The length of each drainage line in each type of land use is calculated. Different treatments based on the length of the drainage lines in the upper middle and lower reaches, in each type of land use, different soil conservation measures have been recommended.

Case Study

11.14 Environmental and Ecological Aspects of Polavaram Project

General : Source

11.14.1   Study Area for Polavaram Dam

Present Environmental and Ecological Status of the Project Area

The Andhra Pradesh State Govt. has planned Polavaram project as a multipurpose project to provide irrigation benefits to the up land areas, water supply to the industries in Visakhapatnam city including the steel plant, generation of hydro power, development of navigation and recreational facilities. The Polavaram Vijayawada link canal as conceived by NWDA will make use of Polavaram dam as its headwork. The length of the link canal is about 174 km. The Polavaram dam is located about 42 km upstream of Godavari barrage at Dowlaiswaram. The power house with an installed capacity of 720 MW is also envisaged on the left flank of the Polavaram dam.

The various features of the project area are:

(a)   No major industries or thermal power house is located in the project area.

(b)   The area likely to be inundated by the Polavaram project includes outcrops of Barakar rocks at Vinjaram and north of Tummalakunta, but their resource potential of workable coal will be estimated at DPR stage. The only other mineral occurrences known in the inundated areas are of graphite at Kavilkunta and Bollapalle.

(c)   At present the ground water is being use for agriculture purposes. The ground water quality in the command area is generally within the permissible limits for irrigation. The depth of water table varies from near surface to 25 m below ground level.

(d)   There is no record of fish sanctuary in and around Polavaram reservoir area.

(e)   The forest area coming under submergence are generally dense with jungles, open scrubs of thorny bushes and other trees. The important timber species are teak, eppa, and bijasal or yegi. Common fuel species are tella tumma, maredu, udaga and Korier. Commercially important species found in the forest are teak, neem, kalan, sandalwood, sailaichar, cattle grass, gum etc. There is also good bamboo growth, which are being exploited by paper mills.

(f)   The downstream area of the project has a long history of floods. To protect the area from floods, flood banks have been formed considering the maximum flood level of year 1986.

11.14.2 Favourable Aspects of the Project

Irrigation: The Polavaram-Vijayawada link will irrigate an area of 5.82 lakh ha in A.P. and provide 2265 Mm^3 of water to Krishna Delta. Besides this, the project will provide 1236 Mcum of water for stabilization of existing command area under Krishna Delta.

Power generation: A power house on the left flank of Polavaram dam is proposed with an installed capacity of 720 MW. 14.3.3.

Pisciculture: The reservoir can be utilized for development of fisheries. However, pre and post impoundment surveys have to be undertaken to work out the steps needed for development of fishery in the reservoir. Fish ladders will be provided to allow movement of important migratory fish population.

The Pisciculture development proposed to be created in the project area will also provide additional work to the local fishermen and revenue to the Government.

Water Supply: The project will also provide about 162 Mm^3 of water for meeting the domestic and industrial water requirement of areas.

Industrial development: Coming up of such a multipurpose project in the area may encourage setting up of some medium and small scale industrial units in and around the project area and will be helpful for the overall development of the area economically.

11.14.3 Need for Impact Assessments on Environmental and Ecological Aspects of the Project Area

The major environmental and ecological aspects of a inter basin water transfer project mainly pertain to the areas of the reservoir site, downstream river course below the dam, link canal enroute and command area of the project. Different types of environmental and ecological impacts may be observed in the areas due to the coming up of the project. It is, therefore, necessary to anticipate the possible adverse impacts along with the positive aspects from the relevant areas of the project. This will help to incorporate adequate control measures on the adverse effects from the project planning phase to various other stages of developments, such as implementation and management to accrue optimum benefits from the project. Relevant aspects on environment and ecology of Polavaram-Vijayawada link canal project and the possible impacts along with mitigative measures thereon are discussed.

Reservoir Site

The site of the dam on Godavari river near Polavaram village in West Godavari district of Andhra Pradesh has been proposed by the State of Andhra Pradesh on topographical, geological and economic considerations giving due relevance to submergence and rehabilitation aspects. The total area of submergence at FRL 45.72 m is 63691 ha, out of which 3705 ha is forest, 30650 ha are culturable land, 12688 ha are unculturable land and 16648 ha area is under river bed. 250 numbers of villages with about 16207 families having a population of about 1,44,812 are coming under submergence. Available information on population and properties affected are given in Tables 11.21 and 11.22.

Table 11.21 State-wise number of villages and families coming under submergence*.

Sl.No.	State	Number of Villages	Number of families	Population (1991 census)
1.	Andhra Pradesh	233	15235	135449
2	Madhya Pradesh	10	680	6620
3	Orissa	7	292	2743
	Total	250	16207	144812

Table 11.22 Properties affected due to submergence*

Sl. No.	State	Permanent houses	Semi permanent houses	Kutha houses
1.	Andhra Pradesh	1350	2300	18800
2.	Madhya Pradesh	42	-	300
3.	Orissa	13	-	290
	Total	1405	2300	19390

Environmental Impact

The National Council of Applied Economic Research (NCAER), New Delhi was entrusted with the studies of socio-economic and environmental implications of 6 inter-basin water transfer proposals of NWDA, and the present link is one among them. The present section on the Environmental Impact of the link project is mainly based on the conclusions drawn in their report.

Surface Water Regime

The link canal is designed as a contour canal and will interfere with natural surface drainage of the area. As such adequate cross-drainage works are provided in the project.

Impact on Groundwater

Provision of canal irrigation in the proposed command area causes additional recharge to the groundwater. As a result the groundwater levels will rise gradually year after year. Part of this augmented groundwater reserves find its way into the stream. To avoid likely rise in water table with consequent harm to crop pattern, the drainage system will have to dispose of the surplus recharge along with surface drainage.

Natural Resources

The area likely to be inundated by the Polavaram project includes outcrops of Barakar rocks at Vinjaram and north of Tummalakunta, but their resource potential of workable coal is not known. The only other mineral occurrences known in the inundated areas are of graphite at Kavilkunta, Velagapalle and Bollapalle. However, these are reported to be minor and not of economic importance. The main purpose of the Polavaram project is to provide water for irrigation to the ayacut upstream of the Godavari barrage, to supply drinking water to the Visakhapatnam steel plant and also to provide water to the chronic drought prone Cheepurupalle tract in which the manganese belt is situated. The mineral resources likely to be lost or their use precluded as a result of inundation will be very negligible. Even for these mineral commodities, plentiful and alternate sources are available in the vicinity of appropriate manufacturing centers.

Public Health Aspects

The formation and use of the water body is not likely to result in introduction or enhancement of water borne diseases provided no heavy industrialisation around the area takes place.

Aquatic Weeds

The chances of impounded reservoir leading to noxious aquatic weeds and intermittent host are remote in the given circumstances. The nature of existing aquatic weeds in submergence area and their impact on fisheries development due to the formation of the reservoir are studied in detail.

Impact on Seismicity

The Godavari river flows along a faulted graben, with the highest recorded earthquake in the region having occurred in 1968 near Bhadrachalam. The Director (Seismology), India Meteorological Department, has opined that since the height of dam is less than 100 m, seismological observations are not necessary. However, keeping in view the past history of earth tremors in the region, proposals are under consideration in consultation with the India Meteorological Department to monitor the pre and post project seismic activity. The dam site falls in Zone – III as per the map of India showing the various seismic zones (IS code: 1893 – 1975 "IS criteria for earth quake resistant design of structures").

The horizontal and vertical inertia co-efficients worked out and adopted for various components are as given in Table 11.23.

Table 11.23 Seismic co-efficients adopted for various components

Component	Horizontal	Vertical
Spillway	0.12	0.06
Earth dam	0.08	0.04
NOF dam	0.12	0.16

Sedimentation

The Government of Andhra Pradesh has anticipated a sedimentation rate of 1.25 Ac. ft/sq.mile/year (0.0595 ha.m/km^2/year) for the free catchment against the IS code recommended rates of 0.048 to 0.096 ha.m/km^2/year. However, the actual rate of sedimentation, based on observed data of 22 years between 1969-70 and 1992-93 at Polavaram G&D site, maintained by CWC, is 0.0299 ha.m/km^2/year.

Frequency of Cyclones

Number of occurrences of severe cyclone storms that affected the area during past seventy years are given in Table 11.24.

Table 11.24 Number of occurrences of cyclonic storms.

Month	Near Vishakhapatnam	Near kakinada
May	4	1
Jun	1	5
Jul	0	2
Sep	2	3
Oct	13	11
Nov	8	2
Dec	0	1

Source : Polavaram project report (Vol-I), July 1982.

Archaeological Centers

No archaeological researches were attempted in the past on account of the presence of impregnable forest with wild animals, though it is well recognised and acknowledged that the two banks of the river Godavari and adjoining areas have been treasure houses of undiscovered cultural, archaeological and historical sites. Detailed survey of the adjoining areas covered has however been proposed by the state Archaeological Department. Before the reservoir is formed the detailed survey is proposed to be completed and any archaeological finds will be retrieved to safer places.

11.14.4 Adverse Impact of the Project

Though the implementation of any irrigation project helps in upliftment of general prosperity in the region, there are bound to be some adverse effects, which should be mitigated through suitable remedial measures. Some of the adverse impacts could be listed as below:

(i) Resentment of the displaced people in the project area as well as submergence area, since most of the benefits of the project are for the people living in the command area: A proper R&R package for resettlement of Project Affected Peoples (PAPs) in the vicinity and similar climate with better civic amenities will be evolved. The PAPs will be resettled before the commencement of the project work.

(ii) Submergence of forest area may have environmental and ecological impact: Proper Environmental Management Plan (EMP) will be evolved to reduce the impact on the environment due to the project. Also, to minimise the loss of forest additional afforestation programme will be taken up. Necessary provision has been made in the estimate for compensatory afforestation.

(iii) Waterlogging and salinity due to increased irrigation in the command area: Proper drainage systems will be provided in the command area.

Labour Requirement

As per the norms of Central Water Commission, the employment generation per crore of rupees of the cost of the project is 155 persons in case of a major project. This 155 comprise 10 engineers, 11 other technical, 12 skilled, 93 unskilled and 29 clerical personnel. The expenditure towards the manpower in case of a major irrigation project would be 23%, of the total expenditure of the project. This expenditure on manpower includes expenditure on pay and allowances, bonus, social security, office expenses and traveling expenses. Total estimated cost of Polavaram-Vijayawada link project is Rs.148391 lakh (1994- 95 price level). The manpower required for the construction of the project will be 230020 persons considering the cost of the project as Rs.1484 crore. This manpower of 230020 comprise 14840 engineers, 16324 other technical, 17808 skilled, 138012 unskilled and 43036 clerical personnel. The expenditure for the engagement of manpower will be Rs.34130 lakh.

Socio-Cultural Aspects

Population Density

The catchment area of the Godavari basin at the Polavaram dam site is 306643 km^2. The unintercepted catchment area below the existing Sriramsagar dam up to Polavaram dam site is 215249 km^2. The population density in the catchment area of Polavaram dam site is 166 persons per km^2, in the submergence area 227 persons per Km^2 and in the command area of the whole project is 497 persons per km^2.

11.14.5 Rehabilitation and Resettlement of Project Affected People

The proposed Polavaram-Vijayawada link project involves the creation of storage reservoir at Polavaram, which submerges large areas including forests, cultivated land, villages etc. The prospects of submergence leading to loss of homes and means of sustenance will have a traumatic effect on the affected population. A proper and timely step for resettlement and rehabilitation of these persons is essential to minimise their suffering. However, the problems relating to resettlement and rehabilitation (R & R) are quite complex. It is essential that the contents of R & R package should be very attractive. However, an efficient institutional arrangement for implementing the entire programme of R & R effectively is equally important. A humane approach during implementation is required, as it is a very sensitive issue. Active co-operation of the affected persons will be beneficial for successful implementation of the project. The attractive R & R packages along with effective implementation in reasonable time schedule has become an essential input for construction of major projects.

The primary objective of a good rehabilitation and resettlement strategy should be to reinforce the traditional ethos and aspiration of displaced people to develop a society living in perfect harmony with nature. Besides, the main thrust of the rehabilitation strategy also should aim at providing fair and equitable treatment of the persons displaced from their homes, professions, farms etc., due to construction of a project. This may require a detailed analysis of the cost involved in providing houses, land and civic amenities to the displaced people.

Housing

The total 16207 families are likely to be affected due to creation of Polavaram reservoir. These families would need to be resettled in different villages in the nearby areas. To expect a displaced person to embark upon the task of constructing a house by him is perhaps expecting too much from him. Therefore, a modestly constructed house needs to be allotted to each of the affected family that would facilitate their prime need.

Land

There are considerable variations in the norms prescribed by different States and agencies in respect of land compensation to be provided to the affected persons. In some cases, the norms differ from project to project within the same State. The policy of providing land for land is commendable. However, complications may arise when the choice of land is also given to the affected families. To avoid dispute and problems, the selection of suitable agricultural land in the command area and its division into required sizes and its distribution by draw of lot with the control of a High Level Committee comprising senior officers of concerned departments should be performed. In the case of Polavaram-Vijayawada link project, 30650 ha of culturable area is coming under the submergence of the proposed reservoir at Polavaram. Therefore, at least an equivalent area of land has to be acquired, suitably in the command area of the project for encouraging to carry out the normal agricultural activities by the affected families.

Basic Amenities

Facilities for health, education, water supply, market, sanitary, communication, community park, panchayat ghar etc., are to be provided to make the life in resettlements more adaptive and comfortable.

Tourism

The area can be developed as a tourist resort after the formation of reservoir. The Papikondalu Gorge about 5 km upstream of dam site in particular is a scenic spot and can be developed as a tourist resort.

11.15 Case Study: EIA-Carrying out Exploration, Development and Production of Coal Bed Methane at Sohagpur (N) Block [SP(N)-CBM-2005/III] at Madhya Pradesh

11.15.1 Introduction

The Consortium of Geopetrol International Inc., Reliance Natural Resources Ltd. (RNRL), Reliance Energy Ltd. (REL) have been awarded the Sohagpur (N) Block [SP(N)-CBM-2005/III] at Madhya Pradesh for Production Sharing Contract (PSC) for carrying out exploration, development and production of Coal Bed Methane. Geopetrol is the operator of the block.

11.15.2 Objective and Scope of REIA Study

The objective of EIA study is to meet the regulatory environmental clearance criteria as well as to ascertain a sustainable development through the assessment of likely impacts due to project related activities on the surrounding environment. The study envisages toassess likely negative impacts and alleviation of these negative impacts, to such extent so as to avoid any harm/permanent changes in the naturally existing environment.

The scope of the REIA study includes detailed characterization of the existing status of the land, water, air, biological and socio-economic environment within the block, identification of the potential environmental impacts of the project, and formulation of an effective Environmental Management Plan (EMP) to prevent, control and mitigate the adverse environmental impacts, and ensuring the environmental compliance. Apart from suggesting mitigation measures to the negative impacts, the report reserves implementation of various positive and enhancement measures as a part of project benefit program to people of the nearby areas.

11.15.3 Legal Framework

As per the notification S.O. 1533, dated 14th September 2006, the Ministry of Environment and Forests (MoEF), Govt. of India, has made an amendment in the EIA notification and directed that all projects "Offshore and onshore oil and gas exploration, development &production" will require "prior environmental clearance from the central government in the Ministry of Environment and Forest (MoEF) on the recommendations of and Expert Appraisal Committee (EAC) to be constituted by the Central Government for the purpose of this notification".

Apart from Environmental Clearance, Consent to Establish and Consent to Operate from the State Pollution Control Board and other statutory permissions from concerned state/local authorities are required before commencement of exploration activities in the said area.

The standards applicable to oil and gas exploration projects are given in Table 11.25.

Table 11.25 Standards Applicable to Oil and Gas exploration

Issues	Applicable Legislation
Hazardous Substances & Wastes	1) The Environment (Protection) Act, 1986, and Rules framed there under a) Hazardous Wastes (Management and Handling) Rules, 1989 and amendment Rules 2000 and 2003; b) Guidelines for disposal of solid wastes by Oil Drilling and Gas Extraction industry as notified, vide notification dated GSR 176 (E) April 1996; c) Manufacture, Storage and Import of Hazardous Chemicals 1989 and amendment Rules 2000
2) The Public Liability Insurance Act, 1991 and Rules 1991	
3) Central Motor Vehicles Act, 1988 and Rules, 1989	
4) The Petroleum Act, 1934	
Water	5) The Water (Prevention and Control of Pollution) Act, 1974, amended in 1988
6) The Environment Protection Act, 1986 - Standards for liquid discharge by Oil Drilling and Gas Extraction industry as notified vide notification dated GSR 176 (E) April 1996	
Air	7) The Air (Prevention and Control of Pollution) Act, 1981, amended in 1987
8) The Environment Protection Act, 1986 – Guidelines for discharge for gaseous emissions by Oil Drilling and Gas Extraction industry as notified vide notification dated GSR 176 (E) April 1996	
9) The Environment (Protection) Second Amendment Rules, 2002 – Emission Standards for New Generator Sets	
10) The Factories Act, 1948, amended in 1987	
11) The Motor Vehicles Act, 1938, amended in 1988 and Rules, 1989	
Noise	12) The Environment (Protection) Second Amendment Rules, 2002 (Noise Limits for New Generator Sets)
13) The Noise (Regulation & Control) Rules, 2000	
Safety and Protection against Pollution of Environment	14) Oil Mines Regulations, 1984

11.15.4   Brief Project Description

During Phase I of exploration, Geopetrol plans to drill eight information core holes and Two production test wells in Block [SP(N)-CBM-2005/III], to determine the presence of coal bed methane within the geological formation starting at a depth of about 800 m to 1000 m below ground level. The exploration project envisages selection of drilling locations within the allotted project block using a truck mounted diesel land rig equipped with top drive system. Drilling process includes drilling fluid or "mud" is pumped through the drill strings down to the drilling bit and returns between the drill pipe –casing annulus up to surface back into the circulation system after separation of drill cuttings/solids through solids control equipment. Only water will be used as mud system in core hole. KCl will be added in core of caving shale section. Air + Water or air + foam + water will be used in production well drilling and will be environmental friendly. If sufficient quantity of methane is found during the exploration, then the site will be selected for operation purpose (i.e. coal bed methane production).   Other secondary activities associated to the drilling operation includes construction of roads, store houses, site office building, other utility structures, site construction, cementing program, well evaluation (well logging and well testing), rig demobilization, restoration and rehabilitation of the site.

The salient features of the project have been summarized in Table11.26.

Table 11.26 Salient Features of the Project

Number of Wells	Eight information core hole and fifteen production well
Block Area	609 sq km
Depth of each well	Ranges from 800 m - 1000 m below ground level
Total Estimated Drilling Period for each well	15 days
Total Estimated Testing Period for each well	2 days
Type of hydrocarbon expected	Coal Bed Methane
Proposed Drilling Fluid for each well	Water in core hole; air + water/foam in production well
Anticipated Volume of Cuttings for each well	2 to 4 cum in core holes 20 to 25 cum in production test wells

11.15.5   Brief Baseline Environmental Status

The study area of EIA study for this project encompasses all areas falling within the project site. The existing/baseline environmental set-up of the study area, has been studied as described in subsequent sections.

Topography, Climate and Meteorology

Topography: The block shows gently undulating topography with few prominent hillocks in northern part. Elevation usually varying between 420 to 500 m above MSL. The maximum elevation is 760 m near the block boundary in the northern part of the block.

Climate & Meteorology: The area experiences a typical tropical climate. The summer months (April to June) are very hot with temperature shooting up to 40 °C. Monsoon usually sets in June and continues up to September. The average rainfall around this area is about 110 cm. Winter starts from November and continues up to January. In these months the temperature lies between 4-10 °C.

Air Quality & Noise

Air Quality: Air quality is monitored at nine different locations within the study area. Analysis were conducted for Suspended Particulate Matter (SPM), Respirable Particulate Matter (RPM), Sulphur Dioxide (SO_2), Oxides of Nitrogen (NO_X) and Hydrocarbon (HC). The 24-hourly average SPM level at all monitoring stations varies 61 $\mu g/m^3$ (at AQ5, Malya and AQ 7 Dhaurai) and 125 $\mu g/m^3$ (at AQ4, Chuhiri and AQ6, Gohparu). RPM values at all nine locations are found within the prescribed limits. The analytic result of SO_2 reveals that the concentrations of SO_2 are below stipulated limit at all nine locations. Maximum NO_X levels from all the nine locations were found to be 28 $\mu g/m^3$ at AQ3, Khodri which is within permissible limits.

Noise: Noise intensity at nine locations within the study area has been collected. Out of these nine locations, five locations are under residential areas and four locations are under commercial area. Maximum noise level observed in the region is 60.2 dB (A).

Soil

Soil quality is analyzed from ten different locations within the study area. The data obtained from analysis of the obtained soil samples reveals that:

- Soil samples are sandy loam to loamy and grey to black in colour.
- pH of the soils varies from 7.41 to 8.99.
- Electrical Conductivity was found to be low, ranging from 0.103-1.214 millimhos/sec.
- Other elements in the soil samples are found within satisfactory level .

Land Use

A broad study of the land use in the study area indicates that most of the study area is unutilized (open space) and agricultural field. All the core holes locations are out side of forest land and surrounded by agricultural fields. Large portion of natural land is forest cover with dense forest area and sparse vegetation cover. Forest areas are mostly predominant in the northern and western part of the block. The eastern part of the block is mainly covered under cultivation along with scattered village pockets. Land use pattern of the study area is given in Table 11.27.

Table 11.27 Land Use Pattern

S.No.	Land Use	% of Study area
1.	Dense Forest	16
2.	Open mixed forest	26
3.	Sparse Vegetation	17
4.	Agricultural land	19

Table 11.27 *contd...*

S.No.	Land Use	% of Study area
5.	Settlement	16
6.	Open land	4
7.	Water Bodies	2
Total		100

Flora and Fauna

The exploration site is covered by forest in its southern, northern and western part of the block. Patches of forest also occurs at the central part of the block. About 42 percent of the block is covered under forest area and accounts to 253.76 sq km. Sparse forest cover is mostly associated with the thick vegetation cover within the block and forms boundary of the dense forest cover and constitutes 17 percent (105.29 sq km) of the entire block. Agricultural land, which forms 19 percent of the block area (113.87 sq km), is predominant in the peripheral portion of the block.

The forests in the study area are under severe threat as it is being gradually destroyed over the years. This has mainly been due to increasing human activities. The common domestic animals observed in the study area are *Bos indicus* (Cow), *Babalus babalis* (Buffalo), *Capra domesticus* (Goat), *Felis domesticus* (Cat), *Canis familiaris* (Dog), *Sus domesticus* (Pig) and *Equus cabalus* (Horse). Wild animals in the region are less, however, wild animals found in the region include Bear, Hyena, Monkey, Leopard, Cheetal, Sambhar, Chinkara, Elephant, Fox and Jackal. No rare or endangered animal species are found in the forest under this block.

The avifauna of the study area include Pigeons (black and blue), Parrots, Peacock, Neelkanth, Purki, Crow, Titahri, Woodpeckers, Eagles, Vultures and Owls. No endangered or rare species of birds have been found in the study area.

No Ecologically Sensitive Areas such as National Parks, Sanctuaries or Biosphere Reserves exist within the block. No well defined corridor of wildlife movement or routes of wildlife migration has been reported by the Forest Department of concerned State Governments around the project site.

Socio-economic Profile

The total population of Shahdol district and Sohagpur tahsil according to the 2001 census is 908148 and 469242 respectively. The decadal growth rate of population in Shahdol district and Sohagpur tahsil are 17.61% and 5.22% respectively. The sex ratio in Sohagpur tahsil is 900 females per 1000 males. Population density of the district is 264 persons/Sq km.

11.15.6   Proposed Environmental Mitigation Measures and Required Actions

The environmental management/mitigation measures and net final environmental impacts on various environmental parameters during construction and operation phases have been given in Table 11.28.

Table 11.28 Proposed Mitigation Measures and Required Actions

Hazard & Effect(s)	Proposed Mitigation	Required Actions
Land Acquisition In Case of govt land necessary permissions will be obtained from concerned departments.	• Ensure that all necessary protocols are followed and legal requirements implemented. • Ensure that appropriate legal requirements have been met with regard to land occupancy, land ownership or usage rights, notice and compensation, etc. • Establish and clearly document all land agreements with owners, users and state authorities (Forest Department) and mark out site boundaries. • Acquiring necessary approvals from State Pollution Control Board in a timely manner.	• Geopetrol to initiate interaction with the concerned officials in the Forest department/land revenue department to identify necessary permits and approval mechanism. • Apply for approval for Land acquisition with proper maps and prescribed fees with due involvement of the Forest department/revenue departments at the field level. • Preliminary site survey to be carried out by Geopetrol's civil works consultants to mark the road & site requirement on ground and the land owner and revenue department to be approached with exact location maps in hand for Land acquisition. • Geopetrol's Drilling & Permit team to meet the local Pollution Control authorities to apprise them of the plan and to identify and apply for necessary permissions prior to drilling of well.
Soil Erosion	• Minimize the extent of site clearance area, by choosing best layout with respect to existing topography.	• Detailed contour maps of the site to be prepared with big trees marked on it to work out the best layout to minimize cut & fill & avoid cutting of trees.

Table 11.28 contd...

Hazard & Effect(s) Proposed Mitigation Required Actions

• Minimize removal of trees at site • Collect topsoil removed during road development/construction, site preparation, etc., and stockpile the same at edge of site to be used to the extent possible for site restoration later.	• To see that arrangement is in place for collection. • Plan to minimize tree cutting prior to site construction and ensure implementation on ground during site construction phase.

Site Protection Effluent pits at site may pose a threat to local people and animals.	• Mark out road & site boundaries. • All bulldozer operators and other manual laborers involved in road and site preparation will be trained to strictly confine to their works within the defined site boundaries. • Pits for containment of cuttings and liquid effluents will be dug at site after fencing is in place.	• To ensure that clear boundary marks are in place. • To ensure that integrity of boundary markers is maintained by the workforce at all times. • Ensure that fencing of site is in place prior to cutting of pits at site. • To appoint guards to be stationed at site during construction phase.
Waste and Effluent Management Poor planning and execution might pose a threat to environment	• Geopetrol to identify different type of waste anticipatedduring operations, work out estimated quantities, and lay down procedures for collection, handling, treatment and disposal of each type of waste. • Waste Management Plan to be implemented during operations.	• Finalizing Waste Management Plan (draft plan given in EMP report). • Waste management plan to be implemented during drilling and be made available for inspection at site to all regulatory bodies.
Contamination of rain/storm water runoff with rig wash water & waste mud	• Detailed drainage design will be developed as a part of the site design. It will be ensured that mud and associated drainage system is isolated from the rain/storm water drainage system. • Pits must have adequate capacity to prevent flooding during high rains (maintain free board) and should be fully bunded.	• Geopetrol to work with Civil works consultants /contractors to develop detailed drainage system addressing concerns outlined here. • Geopetrol to work out required pit volumes based on maximum case scenario including rainwater.

Table 11.28 contd…

Hazard & Effect(s) Proposed Mitigation Required Actions

Wastewater & cuttings may contain trace amounts of drill fluid and residual chemicals. CBM Produced water	• All wastewater, which will be generated from washings & spent mud will be contained in HDPE lined (1 mm thick) pits. The wastewater will be treated if necessary to achieve SPCB compliance for discharged into a nearby nullah/stream. • Proper estimation of quantity and characterization of its quality is necessary to device a disposal plan.	• Site design will include adequately sized pits to contain wastewater & also treated water prior to discharge. • Geopetrol dispose of the produced water either by making evaporation pond or if water is of desirable quality can be disposed in nearest stream so that it can be used for irrigation or pisciculture.
Fuels, Lubricants and Chemicals Management pose threat of major, moderate & minor spills	• Keeping all fuels, lubricants and chemicals in well-designed storage facility with regular inventory checking. • Used and unused chemicals will be stored in a lined & bunded area. • Executing delivery of fuel to drilling site under strict supervision and carrying out refueling operations in an area with impervious flooring and surface drainage with oil interceptor. • Use of suitable delivery trucks.	• Checklist of all drums and containers located within footprint of the storage area • Live risk assessment trainings and awareness rising among all workers associated with mock exercises. • The lined & bunded area for the diesel tank will have extra space to contain used and unused lubricants in drums. • Keeping an inventory of all fueling and refueling operations. • Check all delivery trucks for suitability & ensure that they meet safety requirements

Table 11.28 contd…

| Contamination by way of oil/lubricant spills and leaching. | • Impervious liners in place for fuel, lubricants storage area. Fuel/lubricant containment & generator area to have drains with oil entrapment provision.

• Effective bunds capable of containing 110% of the volume of the largest container within and enclosing all potentially contaminating materials. To be used for fuel/lubricants storage area. | • Impervious liners to be installed in the fuel & lubricant storage area. Fuel/lubricant storage area & generator area to have drains with oil entrapment mechanism.

• Site design to incorporate bund requirement for the fuel/lubricant storage area.
• Ensure separate runoff routes during site design for rainy season. |

Hazard & Effect(s) Proposed Mitigation Required Actions

• Non-contaminated and potentially contaminated runoff will be kept separate. Non-contaminated runoff will be routed to off-site area. Potentially contaminated runoff will be treated. • Oil drip pans will be used wherever there is significant potential for leakage. • All spills/leaks to be contained reported and cleaned up immediately.	• Drip pans will be used. • Soil contaminated will be scraped and sent to for proper disposal. • Oil spills will be contained and controlled using shovels, sands and native soil. These equipment and materials will be made available at camp sites and well site during the operation. The contaminated soil will be excavated and stored in a bunded area lined with an impermeable base.	
Noise and Vibration	• Checklist of all machineries with record of date of procurement, installation and age. • Regular maintenance of all equipments. • Implement good working practices to minimize noise. • Wearing of ear protector when appropriate.	• Inventory of all machineries to be prepared and submitted to Geopetrol for review. • Maintenance Log Book to be prepared and submitted to Geopetrol for review. • No machinery running when not required. • Geopetrol to distribute noise protection equipment and ensure utilization by the work force. • Geopetrol to ensure that all drilling locations will be at least one km away from the

Table 11.28 contd...

		human habitations to avoid disturbance due to noise.
Air Emissions	• Operate all equipment within specified design parameters. • Store all dry, dusty material (chemicals, etc.) in sealed containers. • Minimize dust generated from truck movement	• Ensure proper Equipment maintenance. • Ensure absence of stockpiles or open containers of dusty materials.
Solid Wastes Wastes will include organic wastes, scrap metal, waste oil & surplus chemicals, sacks, broken wooden pallets, medical	• Litter and debris not to be discarded at site and to be segregated at a segregation pit on the well site •Non-toxic biodegradable waste to be buried during operations and at decommissioning, ensuring that local water resources are not contaminated in any way. • Bulk supply of materials to be preferred for minimization of packaging wastes. Unused materials to be returned to supplier.	• Pre operation inspections to ensure waste disposal facilities are in place. • A special segregation pit to have waste types segregated into separate compartment/drums at the well site. • All biodegradable waste at the drilling site is to be collected and disposed off into two small humus pits (each of 2m x 2m x 1.5 m) within the drilling site area away from common use by rig crewmembers. The humus pits are to be covered with soil on daily basis to avoid any odor nuisance due to putrification and check any contact with the flies or insects.

Hazard & Effect(s) Proposed Mitigation Required Actions

wastes etc.	• Material such as scrap metal, waste oil & surplus chemicals will be disposed of in a controlled manner through authorized waste contractors.	• Geopetrol to include in tender requirements wherever possible. • Geopetrol to arrange for proper disposal and waste recycling contractors.

Table 11.28 *contd...*

Non-routine events and accidental releases. (Well kicks, blow out)	• Emergency Response Plan (ERP), Well Control Plan & keep it updated. • Maintain state of readiness for quick response including plan awareness, training and regular exercises • Proper drilling program design to ensure selection of properly rated BOP equipment. • Ensure that the Geopetrol's supervision team & Rig contractor's relevant operating personnel are trained to handle well, control situations and hold relevant well control training certificates. • Ensure advanced detection system is in place and BOP equipment is well maintained.	• Geopetrol to monitor strict compliance with the provisions of ERP & Well Control Plan. • Records of interaction between the management and the work force. Records of training and drills. • Ensure all available offset data is examined for proper design parameters. • The CBM being low pressure drilling operation up to a maximum depth of 1000 m. The CBM is adsorbed in coal beds and required to be pumped out along with water. Hence well kick and well blow out scenarios are not expected. • Well monitoring equipment to detect influx from reservoir. Pressure detection service provided by Mud-logging contractor. Blowout preventers tested on installation and routinely. • While at the drilling location, any spill will be reported promptly.

Table 11.28 contd...

Socio-Economic Impacts	• Ensure no water (surface or ground) contamination occurs from drilling operations.	• Implement waste management plan and undertake water quality monitoring before, during and after the operations.
	• Dust emissions on access road to be minimized.	
	• All manual labor and other jobs for which local skills are available are recruited from local people.	• Regular monitoring of the access road and deployment of water tankers to minimize dust.
	• Undertake social welfare projects for the local communities through well thought out CSR strategy and data provided in the EIA	• Geopetrol to keep a record of all jobs provided to locals reporting on each job profile.
		• Develop a CSR strategy for the area and implement one social welfare project during each drilling well.

11.15.7 Environmental Management Plan

Geopetrol will have an environmental advisor who will train the site staffs, drilling contractors and other staffs in environmental management. Proper training will be provided at all staff levels from the management to supervisory levels to skilled and non skilled workers. Scope of training will cover the requirements of EIA and EMP, with special emphasis on sensitizing the project staff to environmental, social, ethnic, and tribal context of the area.

The Environmental Advisor will conduct on-job live risk assessment trainings to the Geopetrol staff (including HSE coordinator & Company man) and the contractor staff to better appreciate environmental risks and their mitigation measures.

Apart from training, suitable measures will also be taken for handling hazardous wastes, segregation of various types of solid wastes, availability of onsite first aid facilities and most important, restoration and rehabilitation of the abandoned area.

After well testing and evaluation, a decision on whether to abandon or develop the well will be taken. If no indications of a commercial quantity of methane are encountered either before or after testing, the well will be declared dry, accordingly plugged and abandoned, and the site restored in line with local regulations and good industry practice. As a minimum, the following steps will be undertaken to restore and rehabilitate the area:

- All concrete structures will be broken up, and the debris disposed off as per the regulatory requirements.

- All other waste products, solid and liquid, will be disposed of in accordance with the requirements of the EIA and will be treated to render them harmless.

- All fencing and access gates will be removed.

- All pits whose contents would show regulatory compliance for on-site disposal, at the time of site closure, will be backfilled and closed out as per the legal requirements.

- That portion of the access track likely to be of no use for other exploratory wells in the reserved forest will be restored by removing cross drainage structures.

- Waste products, solid and liquid, will be disposed of in accordance with the waste management plan.

11.15.8   Risk Assessment and Disaster Management

Geopetrol is committed to maintain high standards for health and safety at all times. However, on rare occasions, an unplanned event can have the potential to jeopardize the safety of the crew and cause environmental damage. Potential non-routine events that may occur whilst drilling the exploratory wells includes Well Kicks, Oil Spills, Well Testing Fallout and Well Fire.

Specific procedures and training will be carried out to ensure that the correct action would be taken on the rig in the event of a kick occurring. The operating personnel will be trained for such an eventuality and the key responsible people will be required to hold relevant well control certifications. The rig will be equipped with Blow out Preventers of suitable ratings while drilling different sections to ensure safety of equipment and personnel in case of a blow out.

Geopetrol has well outlined emergency response and contingency plan to:

- Obtain an early warning of emergency conditions so as to prevent a negative impact on personnel, the environment, and assets.

- Safeguard personnel to prevent injuries or loss of life by either protecting personnel from the hazard and/or evacuating them from the facilities.

- Minimize the impact of such an event on the environment and the facilities by mitigating the potential for escalation and, where possible, containing the release.

Different staffs will be given different scale of responsibilities to take care during the emergency situation. All care is being taken to avoid any kind of hazard to human health and property, environmental components and local plants and animal species.

11.16   Case Study: EIA of Vindhyachal Super Thermal Power Project-STAGE-IV (2 × 500 MW) NTPC Executive Summary of EIA (By Envirotech East Pvt. Limited Kolkata-June, 2008) Executive Summary

11.16.1 Project Description

Vindhyachal Super Therrnal Power Prolect (VSTPP), owned and operated by NTPC Limited, is located on the North West bank of Rihand Reservoir (Gobind Ballabh Pant sagar) at Vindhyanagar in District Sidhi of Madhya Pradesh

Stage-I (6×210 MW), Stage-II (2×500 MW) and Stage-III (2×500 MW) are already under commercial operation totalling to the present capacity of 3260 MW. lt is now proposed to augment the capacity of the station by addition of two units of 500 MW capacity under Stage-IV, leading to an ultimate capacity of 4260 MW.

Site and the Surroundings

Vindhyachal Super Thermal Power Project (VSTPP) is geographically located at 24°6 N latitude and 82°40' E longitude. lt is located on the north-western bank of Govind Ballabh Pant Sagar (GBPS) at Vindhyanagar of Sidhi district in the state of Madhya Pradesh. Vindhyachal site is adjacent to Singrauli STPP of NTPC Renukoot town is about 50 km away from the project site. Varanasi, the nearest major town, is about 220 km by road.

The site is confined within the boundaries of Singrauli STPP discharge channel towards south, Ballia nalla towards east, VSTPP township towards west and power corridor of SSTPP and Jayant mine township towards north. The proposed plant site is a piece of vacant land located within the existing boundary of the VSTPP.

Land Requirement

No additional land is proposed to be acquired for Main Plant and Township for Stage-IV. However, approximately 500 acres is proposed to be acquired for ash disposal. A R & R plan for the Project Affected Person (PAP) will be prepared in consultation with State Govt. and PAP as per policy.

Fuel Requirement

The total coal requirement would be approximately 6.132 MTPA at 90% PLF. The source of coal is Northern Coalfields Ltd. (NCL).

Water Requirement

Make-up water requirement of the expansion project is estimated as 35 Cusecs, which will be met from Mp's share of water in Rihand basin. The make up water requirement will be met from surplus water available within existing commitment.

11.16.2 Description of the Environment

In order to identify the environmental inpacts due to the construction and operation of Stage-IV of the project and associated facilities, an Environmental Impact Assessment (EIA) study has been undertaken.

Baseline environmental status has been established for various environmental attributes within a study area of 10 km radius with the plant site as centre. The major environmental disciplines covered in the EIA report includes ambient air quality, water quality, noise, soil, ecology (terrestrial and aquatic), land use, geology, hydrology and demographic and socio-economic conditions, The EIA report consists of field data generated for a period of three months frorn (January 2008-March,2008). Environrnental Management Plan (EMP) including pollution mitigation measures, afforestation plan and environmental monitoring plan during the construction and operation of the project, occupational health and safety and disaster management plan have also been incorporated in the EIA report.

Meteorology

The rnonthly temperatures recorded on-site varied between 10.1 to 35.4°C. The maximum rainfall was recorded in the month of January, which was 44.01 mm. The total rainfall received during the three months study period was found to be 49.58 mm.

The maxirnum value of the mean Relative Humidity at 8:30 hours was observed to be 82.2%, which was recorded during the month of January, 2008. The corresponding minimum value at 17:30 hours was observed to be 26.1%, which was recorded during the month of March, 2008.

Ambient Air Quality

The ambient air quality has been monitored at six locations in the study area for three months. The overall mean for the 6 stations for SPM, RPM, SO_2 and NOx are 179.3 $\mu g/m^3$ 62.9 $\mu g/m^3$, 25.0 $\mu g/m^3$ and 34.7 $\mu g/m^3$, respectively.

Water Quality

To assess the physical and chernical properties of water in the region, water samples from Rihand Reservoir, Surya Nala and ground water were collected on rnonthly basis for 8 (Eight) locations. No significant impact on water quality of Rihand Reservoir due to VSTPP discharge was perceived. The results indicate that ground water in general meets the drinking water quality standards.

Soil

Soil samples from 10 (Ten) locations within 10 km radius were collected and analyzed for physical and chemical constituents. The soils were observed to be slightly alkaline at most of the locations with the pH ranging between 6.6-7.6 The nitrogen, potassium and phosphorous levels were found in the range of 6-30 mg/kg, 52-75 mg/kg and 8-45 mg/kg, respectively. The soils in the area are generally good for cultivation and agricultural use.

Noise

Ambient noise levels were measured at 10 (Ten) locations in the study area, once during March, 2008. The day and night Equivalent Noise Levels varied between 52.4 to 66.2 dB(A) and 41.5 to59.2 dB(A), respectively. lt was observed that the day time as well as night time arnbient noise levels in the industrial as well as at the commercial areas were mostly within the prescribed standards.

Biological Environment

The study on Biological Environment was conducted in March, 2008. The vegetation in the study area and around Rihand reservoir is mainly deciduous and degraded forests, The forests are characterised by the presence of Sal (*Shorea robusta*). The aquatic birds are found abundant in the Rihand reservoir. At present, it appears that these birds are not much affected by the various discharges into the lake. No National parks and Wild life Sanctuary exist within the 10 km radius of study area.

Social Environment

The study area includes 52 villages and 4 urban areas. The urban areas are Morwa, Vindhya Nagar, Amlori colliery and Vaidhan.

These towns are included in Singrauli tehsil of Siddhi district. The total population of study area is 117739 (as per 2001 Census).

Infrastructure facilities including road, communication, education, health facilities etc. are available in the study area. The region is well linked by paved roads and rail transport.

The baseline status of major environmental disciplines and the findings are presented in Table 11.29.

Table 11.29 Baseline environmental status of the study area

Discipline	Main Parameter	Average Level observed during the study	Standard (24 hourly)
Ambient Air Quality (Periodic Average)	SPM	179 pq/ms	500 pg/rnr
	RPM	63 rrq/m3	120 rrq/rn^3
	SO_2	25 pLg/rn3	120 ;rgirn^3
	NOX	35 uq/rn3	120 pg /mt
(a) Surface Water	As specified in 1S:2296 and IS:10500 (1991)	Values of different parameters within specified limits.	-
(b) Ground Water	As specified in IS:10500 (1991)	Values of different parameters within specified limits.	

Noise Levels	L_{eq} for IndustrialArea	52 4-65.7 dB(A)	75 dB(A) for daytime.
	L_{eq} forCommercial Area	55-66.2 dB(A)	65. dB(A) for daytime.
	L_{eq} for ResidentialArea	56 3-64.6 dB(A)	55 dB(A) for day time

11.16.3   Anticipated Environmental Impact and Mitigation

Measures

Additionat power generation of 2×500 MW would help to reduce the demand and supply gap. lmproved power availability would encourage industrial and overall development.

This project would also provide the necessary stimulus for better infrastructure facilities in this region by its own presence as well as expected growth. Infrastruclure facilities like telecommunication, water supply, sanitation, health care, education etc. are expected to improve further.

Land Use

No additional land is proposed to be acquired for Main Plant and Township for Stage-lV. However, an additional land of about 500 acres is proposed to be acquired for disposal of ash. Most of the proposed land is private non-irrigated land. So, it will not have any significant impact on the land use pattern of the area. In addition, there will be some ternporary changes in land use pattern due to construction activities. However, this shall be temporary and restricted to construction site only.

Water Use and Hydrology

The additional water requirement of the project will be met from the cooling water discharge channel of Singrauli STPP. Therefore, no impact on the hydrological scenario of the natural water bodies in the study area is expected.

11.16.4 Demography and Socio Economics

The non-workers in the study area constitute about 57.39% of the total population, which indicates the availability of sizeable manpower required for the construction activity. Migration of workforce to project site and increase in floating population may create additional strain on civic amenitres like road, transport, communication, drinking water, sanitation and other facilities to meet the work force requirement. However, such impacts will be ternporary and restricted to the period of construction only. Operation phase of power plant will require work force of non-technical and technical persons. Migration of persons with better education and professional experience will result in rninor increase of population and literacy in the area.

Cornmissioning of power ptant will result in considerable growth of service sector and will also generate new industrial and business opportunities in the area. As the power plant and its ancillary facilities would act as an active nucleus for new industries and business activities, shift of population towards this centre and peripheral area is likely to increase.

Soils

The impacts on soil during construction phase shall be mainly due lo loss of topsoit in the construction areas and contamination of the soils of surrounding area due to construction materials such as cement, sand, oils, etc. However, it shall be temporary and shall be confined to the areas of construction only. Appropriate soil conservation measures associated with improved construction techniques would minimize such local impacts,

The impact on the soil during operation of the project could result due to deposition of residual particulate matter and gaseous emissions on the soil. However, these irnpacts would be negligible.

Surface Water Quality

Effluents from the construction area rnainly contain suspended solids while the sanitary waste from the labour colonies contains suspended as well as organic matter. Routing of effluents from construction area through sedimentation basins and provision of proper sanitary facilities with treatment will eliminate these problems of water pollution. Moreover, these irnpacts will be temporary in nature.

The impacts of a coal based thermal power project during operation phase could result from several activities such as discharge of hot cooling water, discharge of main plant effluents and sanitary effluents and discharge of ash pond overflow, However, in case of Vindhyachal STPS, Stage - lV, all these impacts are envisaged to be insignificant, due to the adoptation of the following mitigative measures:

- The project will have a closed cycle cooling system with cooling towers, hence there will be no thermal pollution of the receiving water body.

- Ash handling system will also have a re-circulating system and there will be no discharge of overflow from ash disposal area. Only treated blow down, conforming to standards shall be discharged.

- Each of the effluent streams emanating from the project shall be individually treated and then routed through a Central Monitoring Basin (CMB), which will also act as an equalisation chamber. Only treated eflluents, conforming to regulatory standards shall be discharged.

- The sanitary effluents from the main plant and township areas shall be treated in the existing Sewage Treatment Plant (STP) and the treated effluent from STP conforming to regulatory standards shall be discharged.

All the effluents shall be treated to conform to regulatory standard and recycled/reused to the maximum possible extent and only a rninimum quantity of effluents shall be discharged which is negligible as compared to the volume of water in Rihand Reservoir.

Ground Water Quality

A detailed geo-hydrological study of the impacts of VSTPP ash pond area was conducted by Department of Earth Science, University of Roorkee in 1993-95. The study included detailed geological, geophysical and geohydrological investigation, soil, coal and ash characterisation, ground water quality monitoring, leachate and soil sorption and ground water modelling The existing and proposed ash ponds of VSTPP are located on the weathered and soil-covered Archean schist and gneiss and Gondwana sedimentary rocks. Standard leaching tests show that under rigorous environmental conditions (worst case) leaching of these elements from the ash is very low due to high pH of ash water (7-7.5) as well as huge dilution (ash. water = 1:10) with large quantity natural water. Therefore concentration of these elements in the ash pond effluent is very low when compared to concentration in the ash.

In view of this, irnpact on ground water quality due to ash pond for VSTPP Stage-IV would be insignificant.

Air Quality

Particulate matter and NOx (due to excavations, handling and transport of earth and construction materials, movement of construction equipment and traffic etc.) will be the main pollutants during the construction phase. However, the impact is likely to be for short duration and limited to the construction site only.

The predictions for air quality during operation phase were carried out for increase in ground level concentration (GLC) of suspended particulate matter (SPM), sulphur dioxide (SO_2) and oxides of Nitrogen (NOx) using 'Industrial Source Complex Model" developed by the US Environmental protection Agency (USEPA) and meteorological data recorded at site the maximum predicted incremental GLCs for SPM, SO_2 and NOx are 1.83, 20.75 and 13.88 $\mu g/m^3$, respectively (refer Table-11.30) and these were observed in the NNW direction at a distance of 1.5 km.

The Maximum GLCs for SPM, SO_2 and NOx after implementation of Vindhyachal STPS, Stage-IV are estimated to be 352.83 $\mu g/m^3$, 76 75 $\mu g/m^3$ and 78.88 $\mu g/m^3$, respectively. These values are within the National Ambient Air Quality Standards for industrial areas.

Table 11.30 Resultant Maximum Ground Level Concentration

Pollutant	Maximum AAQ Concentration recorded during study period ($\mu g/m^3$)	Maximurn Incremental Concentration due to the Proposed Power Proiect ($\mu g/m^3$)	Resultant Concentration ($\mu g/m^3$)
SO_2	56	20.75	76.75
NOx	65	13.88	78.88
SPM	351	1.83	352.83

Mitigation Measures for Air Pollution Control

The various proposed air pollution control measures are as follows:

- Provision of high efficiency Electrostatic Precipitators (ESP)

- Provision of dust extraction and dust suppression system in coal stock area and coal handling plant.

- Stack of 275-m height would be provided for wider dispersal of air pollutants.

- All the internal roads have been asphalted during the implernentation of the existing plant. Therefore, vehicular movement may not generate fugitive dust. However, water spraying will be practiced frequently at all dust generating areas during construction period.

- The emissions from the stack will be continuously monitored for particulate matter, sulphur dioxide and oxide of nitrogen.

- The concentration of SPM, SO₂ and NOx in the ambient air quatity is being monitored as per the direction of the State Pollution Control Board for existing project. Monitoring of AAQ will be further continued as per the stipulation of SPCB during the Stage-IV.

- Space provision for retrofitting Flue Gas De-sulphurization (FGD) systern, if required in future.

- Plantation in the available spaces.

Terrestrial Ecology

Deposition of dust generated due to construction activities on vegetation in the surrounding area may reduce photosynthetic activity. However, this will be temporary and confined to the surroundings of the construction area and construction period only.

The particulate rnatter emitted through the stack may also settle on vegetation and may interfere with the gaseous exchange through foliar openings. However, the predicted maximum incremental ground level concentration of 1.83 $\mu g/m^3$ indicates that the irnpact will be insignificant. The maximurn resultant ground level concentration of SO₂ and NOx, after operation of project has been estimated to be 76.75 $\mu g/m^3$ and 78.88 $\mu g/rn^3$, respectively, which are unlikely to cause any injury to the surrounding vegetation.

Aquatic Ecology

Hot Water Discharge channel of VSTPP, has been identified as a source of water for the project, which is an artificial canal and does not support any aquatic resource. Therefore, withdrawal of water from the channel will not have any impact on aquatic ecosystem.

The quantity of treated effluents from VSTPP, Stage-IV (industrial effluents: 235 m^3/hr and domestic effluents: 100 m^3/hr), conforming to regulatory standards shall be discharged into Surya Nala leading to Rihand Reservoir. However, as the quantity is negligibly small as compared to the water available in reservoir and the effluents shall be fully treated, the water quality of the reservoir is not likely to change significantly. Therefore, no tangible impact on the aquatic eco-syslem of Rihand Reservoir is expected.

Further, as the project will have a closed cycle cooling systern with cooling towers and clarified water as make-up to the cooling system, there will be no thermal impact on aquatic ecosystem due to operation of the project.

Noise

The major noise generating sources during the construction phase are vehicular traffic, construction equipment like, dozer, scrapers, concrete mixers, cranes, generators, pumps, compressors, rock drills, pneumatic tools, vibrators etc. The operation of these equipment will generate noise ranging between 75-90 dB (A). The predicted noise level due to operation of such equipment at a distance of 1.8 km frorn the source is 340 dB(A).

The ambient noise level recorded during field studies in the nearby area located at a distance of 1.8 km from VSTPP Township ranges between 43.2 to 52.4 dB(A). As the ambient noise levels are higher than the predicted noise levels, due to masking effect, no increase in the ambient noise levels durlng construction and operation phase is envisaged.

Thus, there would not be any adverse impact due to construction and operation of the plant on the residents in the nearby villages.

Mitigation Measures for Noise:

To minimize the impact of noise, following measures will be adopted.

- The specifications of machines/equipment will include in-built design to meet statutory requirement. Appropriate noise barriers/silencers, enclosures, insulation etc., will be provided in the equipment, wherever required to attenuate noise levels;
- Earplugs witl be provided to workmen working near high sources;
- Provision of separate cabins for workers/operators working in the high noise area.

11.16.5 Environmental Monitoring Programme

An Environmental Monitoring Programme is already in progress at Vindhyachal STPP Stage - I, II & III and all the facilities for monitoring exist at site. The monitoring shall be strengthened further to include requirement of units under Stage-lV. Monitoring will be done as per stipulation of the State Pollution Control Board.

Project Benefits

Commissioning of Vindhyachal Super Thermal Power Project, Stage-IV (2×500 MW) will improve the power supply position in Northern Region, which is vital for economic growth as well as improving the quality of life. The improved power supply will reduce the dependence of general public and cornmercial establishments on DG Sets thereby reducing the noise pollution as well as air pollution at local levels.

Environmental Management Plan

Institutional Set-Up for Environmental Managernent

The environmental groups in NTPC have a three tier organization structure: Environmental Engineering (EEG), Environmental Manager (EMG), Rehabilitation and Resettlement (R&R), Horticulture, Medical and Safety Groups at Corporate Center (CC); EMG at Regional Headquarters (RHQ) as co-coordinator and EMG, Chemistry Group, ESP Maintenance Group, Ash Management, R&R Group, Horticulture, Medical and Safety Groups at site

The responsibility of environmental management of an operating station lies mainly with Environmental Management Groups at site, which acts as co-coordinator with all other groups at site, SRHQ and CC for environmental matters as well as outside agencies like State Pollution Control Board. The environmental management group at Vindhyachal STPP, Stage - I, II & III is headed by Dy. General Manager in order to deal with the additional work load for environment management of Stage-IV, the above group will be suitably strengthened.

Disaster Management Plan

To tackle the consequences of major emergency inside the factory or immediate vicinity of project site, a Comprehensive Disaster Management Plan for the existing Stage-I, II & III projects has been formulated. This will be revised to meet the requirement of Stage-lV also.

Ash Utilization

VSTPP, Stage-IV (2×500 MW) shall produce about 5000 tonne of ash per day

In order to meet the requirement of Gazette Notification for Ash utilization following actions are proposed:

1. The company shall provide system for 100% extraction of dry fly ash along with suitable storage facilities. Provision shall also be kept for segregation of coarse and fine ash, loading this ash it to closed/open trucks and also for loading rail wagons. This will ensure availability of dry fly ash required for manufacture of Fly Ash based Portland Pozzolana Cement (FAPPC), asbestos cement products, ash based building products and other uses of ash.

2. The company shall rnake efforts to motivate and encourage entrepreneurs to set up ash based building products such as fly ash bricks etc.

3. NTPC has set up Fly ash brick manufacturing plant at Vindhyachal and fly ash bricks produced are being utilized in construction activity. For VSTPP, Stage -IV construction works, fly ash bricks shall be manufactured on large scale in the plant to meet the requirement.

4. To promote use of ash in agriculture/wasteland development - show case project shall be taken up in the vicinity of power stations.

5. Mining authorities will be persuaded to identify and allot abandoned mines for back filling with ash.

6. NTPC shall also explore possibilities of transporting fly ash in bulk quantity to urban/metro cities to achieve ash utilizatron in line with the notification.

7. All government/private agencies responsible for construction/design of buildings, development of low lying areas, and construction of road embankments etc, within 100 km of the plant area shall be persuaded to use ash and ash based products in compliance of MoEF's Gazette Notification.

Plantation and Greenbelt

Units under Stage-IV of VSTPP will be located within the premises of existing plant boundary for Stage-I, II & III protects and green belt development activities have already been undertaken under existing projects.

About 1780000 trees have already been planted (survival rate about 90%) in and around the VSTPP plant and township.

Cost Provision For Environmental Protection Measures

A provision of Rs 5262.56 Million has been kept in the Feasibilily Report towards implementation of environmental protection measures.

References

1. Marstrand, P. K. 1976 "Ecological and Social Evaluation of Industrial Development." Environmental Conservation, vol. 3, no. 4, pp. 303-09.

2. United Nations Economic and Social Council (n.d.). 2018 "Subsidiary Bodies of ECOSOC". United Nations Economic and Social Council. United Nations. Retrieved 27.

3. United Nations Economic and Social Council Resolution 37(IV). 2018 Economic Commission for Asia and the Far East **E/RES/37(IV)** 28 March 1947. Retrieved 27 December.

4. United Nations Economic and Social Commission for Asia and the Pacific (n.d.). 2018 "ESCAP Member States and Associate Members". United Nations Economic and Social Commission for Asia and the Pacific. United Nations. Retrieved 27 December.

5. Burchell, R.W. and D. Listokin. 1992 "Fiscal Impact Procedures State of the Art: The Subset Questions of Nonresidential and Open Space Costs," The Center for Urban Policy Research: New Brunswick, NJ, , p 43.

6. Burchell, R.W., D. Listokin, & W.R. Dolphin 1985. "The New Practitioner's Guide to Fiscal Impact Analysis." Center for Urban Policy Research: New Brunswick, NJ.

7. Burchell, R.W. 1978 "The Fiscal Impact Handbook." The Center for Urban Policy Research: New Brunswick, NJ,.

8. Community guide to Development Impact Analysis; 2000 Mary M. Edwards Wisconsin Land Use Research Program on Agricultural Technology Studies University of Wisconsin–Madison Taylor Hall 427 Lorch Street, Room 202 Madison, WI 53706 March.

9. Traffic Impact Analysis Guidelines Methodology Nov 2017

10. Khandker, Shahidur R.; Koolwal, Gayatri B.; Samad, Hussain A.. 2010. Handbook on Impact Evaluation : Quantitative Methods and Practices. World Bank. © World Bank. https://openknowledge.worldbank.org/handle/10986/2693 License: CC BY 3.0 IGO."

11. Wischmeier WH, Smith DD (1978) Predicting rainfall erosion losses: a guide to conservation planning. USDA, Agriculture Handbook No. 537, Washington, DC

12. Wischmeier, W. H. & Smith, D. D. (1958) Rainfall energy and its relationship to soil loss. Trans. AGU 3J> 285-291

13. The World Bank Sector Policy and Research staff Environment Department; 1991 "Environmentally Sustainable Economic Development Building on Brundtland" Compiled and Edited by Robert Goodland, Herman Daly and Salah El Serafy; July.

14. Mock, J.G. and P. Bolton 1993. The ICID Environmental Check-list to Identify Environmental Effects of Irrigation, Drainage and Flood Control Projects. HR Wallingford [Available from HR wallingford at £16.50 incl P & P. Single copies available free to developing country institutions, courtesy of ODA].

15. Combining facts and values in environmental impact assessment: Theories and techniques; Bruce Stiftel D. H. Moreau R. C. Nichols Book-1986

16. Burrough PA. Monographs on Soil and Resources Survey No. 12. Oxford Science Publications; New York: 1986. Principles of Geographic Information Systems for Land Resource Assessment.

17. Bergmann H. et Boussard J. 1976 Guide de evaluation economique des projects d irrigation. OCDE, Paris. 261 p.

www.ingramcontent.com/pod-product-compliance
Lightning Source LLC
LaVergne TN
LVHW081926160726
843514LV00006B/934